AUTOMATA AND ALGEBRAS IN CATEGORIES

Automata and Algebras in Categories

by

JIŘÍ ADÁMEK

Faculty of Electrical Engineering, Technical University, Prague, Czechoslovakia

and

VĚRA TRNKOVÁ

Faculty of Mathematics and Physics, Charles University, Prague, Czechoslovakia

KLUWER ACADEMIC PUBLISHERS

DORDRECHT / BOSTON / LONDON

Library of Congress Cataloging in Publication Data

CIP

Adámek, Jiří, Dr.
Automata and algebras in categories / by Jiří Adámek and Věra Trnková.
p. cm. — — (Mathematics and its applications. East European series; 37)
Bibliography: p.
Includes index.
ISBN 0-7923-0010-6
1. Machine theory. 2. Categories (Mathematics) 3. Functor theory. I. Trnková, Věra. II. Title. III. Series: Mathematics and its applications (D. Reidel Publishing Company). East European series: v. 37.
QA267.A32 1989
511--dc19 88-29780

Published by Kluwer Academic Publishers,
P.O. Box 17, 3300 AA Dordrecht, The Netherlands
in co-edition with SNTL—Publishers of Technical Literature, Prague

Kluwer Academic Publishers incorporates
the publishing programmes of
D. Reidel, Martinus Nijhoff, Dr. W. Junk and MTP Press.

Sold and distributed in the U.S.A. and Canada
by Kluwer Academic Publishers,
101 Philip Drive, Norwell, MA 02061, U.S.A.

Sold and distributed in Albania, Bulgaria, China, Czechoslovakia, German Democratic Republic, Hungary, Mongolia, Northern Korea, Poland, Rumania, U.S.S.R., Vietnam, and Yugoslavia by ARTIA, Foreign Trade Corporation for the Import and Export of Cultural Commodities, Ve Smečkách 30, 111 27 Prague 1, Czechoslovakia

In all other countries, sold and distributed
by Kluwer Academic Publishers Group,
P.O. Box 322, 3300 AH Dordrecht, The Netherlands

Printed in Czechoslovakia

Series Editor's Preface

'Et moi, ..., si j'avait su comment en revenir, je n'y serais point allé.'
Jules Verne

The series is divergent; therefore we may be able to do something with it.
O. Heaviside

One service mathematics has rendered the human race. It has put common sense back where it belongs, on the topmost shelf next to the dusty canister labelled 'discarded non-sense'.
Eric T. Bell

Mathematics is a tool for thought. A highly necessary tool in a world where both feedback and nonlinearities abound. Similarly, all kinds of parts of mathematics serve as tools for other parts and for other sciences.

Applying a simple rewriting rule to the quote on the right above one finds such statements as: 'One service topology has rendered mathematical physics ...'; 'One service logic has rendered computer science ...'; 'One service category theory has rendered mathematics ...'. All arguably true. And all statements obtainable this way form part the raison d'être of this series.

This series, ***Mathematics and Its Applications***, started in 1977. Now that over one hundred volumes have appeared it seems opportune to reexamine its scope. At the time I wrote

> "Growing specialization and diversification have brought a host of monographs and textbooks on increasingly specialized topics. However, the 'tree' of knowledge of mathematics and related fields does not grow only by putting forth new branches. It also happens, quite often in fact, that branches which were thought to be completely disparate are suddenly seen to be related. Further, the kind and level of sophistication of mathematics applied in various sciences has changed drastically in recent years: measure theory is used (non-trivially) in regional and theoretical economics; algebraic geometry interacts with physics; the Minkowsky lemma, coding theory and the structure of water meet one another in packing and covering theory; quantum fields, crystal defects and mathematical programming profit from homotopy theory; Lie algebras are relevant to filtering; and prediction and electrical engineering can use Stein spaces. And in addition to this there are such new emerging subdisciplines as 'experimental mathematics', 'CFD', 'completely integrable systems', 'chaos synergetics and large-scale order', which are almost impossible to fit into the existing classification schemes. They draw upon widely different sections of mathematics."

By and large, all still applies today. It is still true that at first sight mathematics seems rather fragmented and that to find, see, and exploit the deeper underlying interrelations more effort is needed and so are books that can help mathematicians and scientists do so. Accordingly MIA will continue to try to make such books available.

If anything, the description I gave in 1977 is now an understatement. To the examples of interaction areas one should add string where Riemann surfaces, algebraic geometry, modular functions, knots, quantum field theory, Kac-Moody algebras, monstrous moonshine (and more) all come together. And to the examples of things which can be usefully applied let me add the topic 'finite geometry'; a combination of words which sounds like it might not even exist, let alone be applicable. And yet it is being applied: to statistics via designs, to radar/sonar detection arrays (via finite projective planes), and to bus connections of VLSI chips (via difference sets). There seems to be no part of (so-called pure) mathematics that is not in immediate danger of being applied. And, accordingly, the applied mathematician needs to be aware of much more. Besides analysis and numerics, the traditional workhorses, he may need all kinds of combinatorics, algebra, probability, and so on.

In addition, the applied scientist needs to cope increasingly with the nonlinear world and the extra mathematical sophistication that this requiers. For that is where the rewards are. Linear models are honest and a bit sad and depressing: proportional efforts and results. It is in the nonlinear world that infinitesimal inputs may result in macroscopic outputs (or vice versa). To appreciate what I am hinting at: if electronics were linear we would have no fun with transistors and computers; we would have no TV; in fact you would not be reading these lines.

There is also no safety in ignoring such outlandish thing as nonstandard analysis, superspace and anticommuting integration, p-adic and ultrametric space. All three have applications in both electrical engineering and physics. Once, complex numbers were equally outlandish, but they frequently proved the shortest path between 'real' results. Similarly, the first two topics named have already provided a number of 'wormhole' paths. There is no telling where all this is leading—fortunately.

Thus the original scope of the series, which for various (sound) reasons now comprises five subseries: white (Japan), yellow (China), red (USSR), blue (Eastern Europe), and green (everything else), still applies. It has been enlarged a bit to include books treating of the tools from one subdiscipline which are used in others. Thus the series still aims at books dealing with:

— a central concept which plays an important role in several different mathematical and/or scientific specialization areas;
— new applications of the results and ideas from one area of scientific endeavour into another;

— influences which the results, problems and concepts of one field of enquiry have, and have had, on the development of another.

Automata, more precisely, sequential automata, have their importance in logic, computer science and control and signal processing theory. They act on (input) strings (over a given alphabet) producing a 'behaviour map' and important questions concern (minimal) realizations of behaviours and the recognizability of formal languages by deterministic and nondeterministic automata.

In many situations, more general objects than strings need to be processed. A most important class being trees (think, for example, of types). Just as sequential automata act on strings, tree automata act on trees; more precisely Σ-tree automata act on Σ-trees where Σ is a given set of n-ary operations (n not necessarily constant over Σ). And of course there are the natural questions concerning realization, behaviour, and recognizability also in this case.

Both classes of automata are a special case of the much more general idea of F-automata where F is an endofunctor of some category. These objects and the associated questions and old results concerning behaviour, realization and recognizability (Kleene type theorems) from the subject matter of this book, which therefore has something to offer to all mathematicians, engineers and computer scientists with an interest in automata, categories, formal languages, or universal algebra.

Perusing the present volume is not guaranteed to turn you into an instant expert, but it will help, though perhaps only in the sense of the last quote on the right below.

The shortest path between two truths in the real domain passes through the complex domain.

J. Hadamard

La physique ne nous donne pas seulement l'occasion de résoudre des problèmes ... elle nous fait pressentir la solution.

H. Poincaré

Never lend books, for no one ever returns them; the only books I have in my library are books that other folk have lent me.

Anatole France

The function of an expert is not to be more right than other people, but to be wrong for more sophisticated reasons.

David Butler

Bussum, January 1989

Michiel Hazewinkel

Contents

Preface

What is Generalized

The theory of automata has developed rapidly in the last decades: from the first endeavour to describe formally the input-output behavior, to a clear algebraic insight into the basic concepts and their interrelationship. The original notion of a sequential automaton has been generalized in a number of directions. The motivating directions for the present monograph are two:

(i) linear sequential automata, arising from the theory of dynamical systems, and

(ii) tree automata, the basic structure of which is an arbitrary algebra (whereas the structure of a sequential automaton is a unary algebra).

The first example shows that sets with structure and structure-preserving maps play an important role; the latter indicates that "types" more complex than an input set and an output set are needed.

A model of automata based on categories and functors, and encompassing the above examples, has been presented by M. A. Arbib and G. E. Manes in a series of papers since 1974. They study automata in a category which will be the category of sets for the sequential automata and the tree automata, or the category of modules for the linear sequential automata. The fundamental idea is to express the type of automata under study by a suitable functor. This makes the basic notions concise and general. We present the concepts of Arbib and Manes in the third chapter, and thereafter we develop a theory of functorial automata, based on the research of the Prague Seminar on General Mathematical Structures since 1970. The first two chapters present the motivation: the first one is devoted to standard facts concerning sequential automata, and in the latter we study tree automata (with some results appearing for the first time in a book).

What Results are Obtained

Our monograph presents a study of functors motivated by automata-theoretic concepts. An F-automaton in a category $\mathcal{K}$ (where $F\colon \mathcal{K} \to \mathcal{K}$ is a functor expressing the type) is, roughly speaking, an algebra of type F endowed with an output. We discuss

(a) the existence and construction of free F-algebras which play the role that the monoid of words does for sequential automata (Chapter IV),

(b) the existence of minimal realizations for all behaviors (Chapter V), and their construction and universality (Chapter VI), and

(c) the languages recognizable by finite deterministic and nondeterministic automata (Chapter VII).

Each of these problems turns out to be very difficult when investigated in a general category with a general type functor. We try, in each case, to obtain not only sufficient conditions under which the individual construction can be performed, but also necessary conditions on the functor. This serves to shed new light on the boundaries of the automata-theoretic concepts. For example, we prove that if F is a finitarity functor, then all behaviors have minimal realizations. But we are interested also in the converse: does minimal realization imply finitarity? We prove that it does, under additional hypotheses, which shows e.g. the "handicap" of infinitary tree automata.

An analogous situations arises in each of the fields of problems we investigate. We present necessary and sufficient conditions for

(a) the "constructive" and "finitary" existence of free algebras,

(b) the existence and universality of minimal realizations,

(c) the description of the languages recognized by finite automata using rational operations, and the coincidence of these languages with those recognized by nondeterministic finite automata.

The obtained results have diverse degrees of generality: in (a) the categories and functors are quite general, in (b) we often have to assume restrictive additional hypotheses, and (c) is studied only in the category of sets. Nevertheless, in each case the presented condition shows boundaries beyond which automata theory cannot be extended. Particularly sharp results are achieved for the category **Set** of sets and R-**Vect** of vector spaces (over a field R). For example, the main result of Chapter V is that F-automata have minimal realizations iff F is a finitary functor; in case of **Set** and R-**Vect**, this holds for all functors, and otherwise additional hypotheses are needed. Analogously, the main result of Chapter VI is that F-automata have universal minimal realization iff F preserves unions; again, this holds for all functors in case of **Set** and R-**Vect**. And the free-algebra construction of Chapter IV also converges for all functors in **Set** and R-**Vect** for which the free algebras actually exist; these are just the functors with arbitrarily large fixed points.

The special case when the functor F has an adjoint subsumes the theory of automata in closed categories investigated for example by J. A. Goguen, L. Budach and J.-H. Hoehnke, and H. Ehrig *et al.* (see References).

Organization

The interdependence of the chapters of our monograph can be depicted as follows:

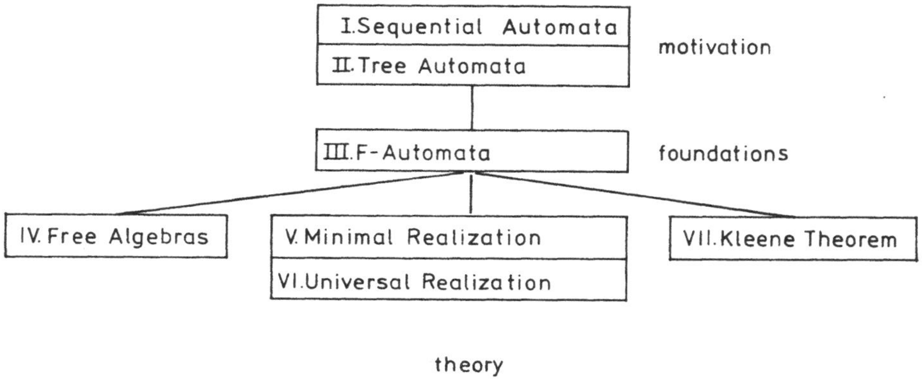

We have endeavoured to make our book self-contained. All concepts of the theory of automata we use can be found in the first two chapters. The reader is expected to be familiar with the fundamentals of category theory. Everything beyond "common knowledge" of categorists is carefully introduced in the text.

Each chapter is numbered by a Roman numeral, and is divided into sections numbered by Arabic numerals. Thus, III.5 denotes the fifth section of Chapter III. Sections, listed in the contents, are subdivided into numbered subsections (for example, III.5.2 is the second subsection of the Section III.5) and they are concluded by exercises, denoted by capital letters (for example, Exercise III.5.B is the Exercise B. in the Section III.5).

All historical comments are concentrated at the end of each chapter. We include a list of references which is very detailed as far as papers of the Prague Seminar are concerned. Other references were chosen so as to cover all the papers we used, or were inspired by, but we have not provided an exhaustive list of the extensive body of literature connected with our monograph.

Acknowledgements

We are deeply indebted to the members of the Prague Seminar of General Mathematical Structures, in particular to V. Koubek and J. Reiterman. Although we have tried to be accurate in our references to their papers, their work has influenced ours throughout the years beyond these bare facts.

Our thanks are due to M. A. Arbib and E. G. Manes for their interest and stimulation of our work. On a suggestion of the latter we wrote a survey paper which became the basis of our monograph. Also G. Grätzer followed our work with a close interest. He has encouraged us on different occasions, particularly during the lectures of the second author on this topic at the Algebra Seminar of the University of Manitoba.

Chapter I: Sequential Automata

I.1. Automata and Behavior

1.1. A sequential automaton is, roughly speaking, a device which is at one of its states and, receiving an input signal (from a specified set Σ, called the input alphabet) changes its state to another state and emits an output signal (from a specified set Γ, called the output alphabet). Each state q and each input signal σ determine the next state $q' = (q, \sigma)\delta$, in which the automaton will be after receiving the input σ in the state q. Formally:

Definition. Let Σ be a non-empty set, called the *input alphabet.* A *sequential Σ-automaton* is a quintuple $A = (Q, \delta, \Gamma, \gamma, q_0)$, where
Q is a set, called the *set of states*;
$\delta: Q \times \Sigma \to Q$ is a map, called the *next-state map*;
Γ is a set, called the *output alphabet*;
$\gamma: Q \to \Gamma$ is a map, called the *output map*;
q_0 is an element of Q, called the *initial state.*

The automaton A is supposed to start its work always in the state q_0. Receiving the first input $\sigma_1 \in \Sigma$, it changes its state to

$$q_1 = (q_0, \sigma_1)\delta$$

and emits the output $(q_1)\gamma$. Receiving the second input $\sigma_2 \in \Sigma$, it changes its state to

$$q_2 = (q_1, \sigma_2)\delta$$

and emits the output $(q_2)\gamma$. Etc.

1.2. Denote by Σ^* the set of all words in the alphabet Σ. That is, the elements of Σ^* are
(0) the empty word $\emptyset$;
(1) all the one-letter words, i.e., elements of Σ;
(2) all the two-letter words $\sigma_1\sigma_2$ (with $\sigma_1, \sigma_2 \in \Sigma$);
(3) all the three-letter words $\sigma_1\sigma_2\sigma_3$ (with $\sigma_1, \sigma_2, \sigma_3 \in \Sigma$), etc.
(We write words without commas or brackets which leads to no confusion provided that Σ does not contain symbols $\sigma_1, \sigma_2, \ldots, \sigma_n$ such that the symbol $\sigma_1\sigma_2 \ldots \sigma_n$ is in Σ too, which we shall always assume.) There is a naturally de-

fined binary operation on the set Σ^* called the *concatenation* which assigns to words $w = \sigma_1\sigma_2 \ldots \sigma_n$ and $w' = \tau_1\tau_2 \ldots \tau_m$ in Σ^* the word $ww' = \sigma_1\sigma_2 \ldots \sigma_n\tau_1\tau_2 \ldots \tau_m$. This operation is associative, i.e. $(ww')w'' = w(w'w'')$ for all w, w', w'' in Σ^*, and the empty word $\emptyset$ is a unit, i.e.

$$\emptyset w = w\emptyset = w$$

for all $w \in \Sigma^*$. Hence, the set Σ^* endowed with this operation is a monoid.

Convention. When writing $\sigma_1\sigma_2 \ldots \sigma_n$ we consider $n = 0, 1, 2, \ldots$, where the case $n = 0$ describes the empty word.

1.3. The reaction of an automation to sequences of input symbols, i. e., to elements of Σ^*, can be described formally as follows:

Definition. The *run map* of a sequential Σ-automaton $A = (Q, \delta, \Gamma, \gamma, q_0)$ is the map

$$\rho: \Sigma^* \to Q$$

defined by the following induction

$$(\emptyset)\rho = q_0;$$
$$(\sigma_1 \ldots \sigma_n)\rho = q_n \quad \text{implies} \quad (\sigma_1 \ldots \sigma_n\sigma_{n+1})\rho = (q_n, \sigma_{n+1})\delta$$

for all $\sigma_1, \ldots, \sigma_n, \sigma_{n+1} \in \Sigma$ and all $q_n \in Q$.

The *behavior* of A is the map

$$\beta = \rho \cdot \gamma: \Sigma^* \to \Gamma.$$

Thus, when receiving inputs $\sigma_1, \sigma_2, \ldots, \sigma_n$, the automaton terminates in the state $q_n = (\sigma_1\sigma_2 \ldots \sigma_n)\rho$, and emits the final output $(\sigma_1\sigma_2 \ldots \sigma_n)\beta = (q_n)\gamma$.

1.4. A "small" automaton can be depicted by a labelled graph, where vertices correspond to the states (and are labelled by pairs (q, y), where $q \in Q$ and $y = (q)\gamma \in \Gamma$) and arrows correspond to δ. That is, an arrow leads from q to q' and carries a label $\sigma \in \Sigma$ iff $(q, \sigma)\delta = q'$. The initial state q_0 is indicated by a small arrow $\mapsto$.

Examples

(i)

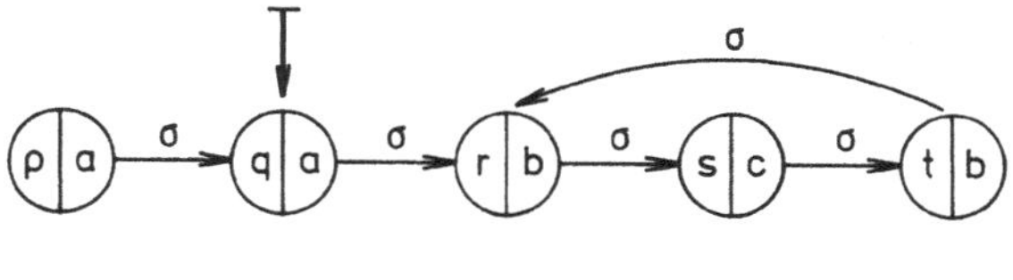

Here $Q = \{p, q, r, s, t\}$, $\Sigma = \{\sigma\}$ and $\Gamma = \{a, b, c\}$. The word monoid $\Sigma^* = \{\sigma\}^*$ consists of the words $\emptyset$, σ, $\sigma\sigma = \sigma^2$, $\sigma\sigma\sigma = \sigma^3$, etc.

The run map $\rho: \{\sigma\}^* \to Q$ is given by the following table:

$\emptyset$	σ	σ^2	σ^3	σ^4	σ^5	σ^6	σ^7	
q	r	s	t	r	s	t	r	...

The behavior $\beta: \{\sigma\}^* \to \{a, b, c\}$ is the following map:

$$(\sigma^n)\beta = \begin{cases} a & \text{if } n = 0 \\ c & \text{if } n = 2 + 3k, \quad k = 0, 1, 2, \ldots \\ b & \text{else.} \end{cases}$$

(ii) 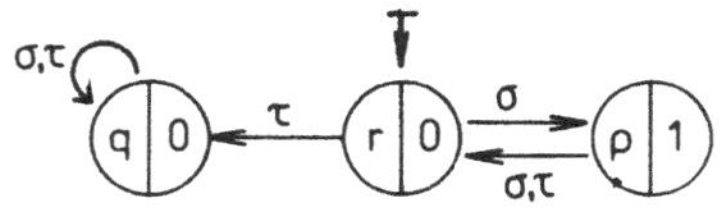

Here $Q = \{p, q, r\}$, $\Sigma = \{\sigma, \tau\}$ and $\Gamma = \{0, 1\}$.
The behavior $\beta: \{\sigma, \tau\}^* \to \{0, 1\}$ is the following map:

$$(s_1 \ldots s_n)\beta = \begin{cases} 1 & \text{if } n \text{ is odd and } s_1 = s_3 = s_5 = \ldots s_n = \sigma \\ 0 & \text{else.} \end{cases}$$

1.5. *Automata are unary algebras.* A unary operation on a set Q is a map of Q into itself. A unary Σ-algebra is a set Q with a collection of unary operations indexed by the elements of Σ, say $\{\delta_\sigma\}_{\sigma \in \Sigma}$. The next-state map $\delta: Q \times \Sigma \to Q$ of a sequential Σ-automaton $A = (Q, \delta, \Gamma, \gamma, q_0)$ defines a unary algebra on Q with the operations

$$(-)\delta_\sigma = (-, \sigma)\delta: Q \to Q.$$

Conversely, each unary Σ-algebra can be obviously described by a map $\delta: Q \times \Sigma \to Q$.

The word monoid Σ^* is a unary Σ-algebra, the σ-operation of which is defined by the concatenation:

$$\sigma_1\sigma_2 \ldots \sigma_n \mapsto \sigma_1\sigma_2 \ldots \sigma_n\sigma.$$

This defines a map $\varphi: \Sigma^* \times \Sigma \to \Sigma$ (which is a restriction of the concatenation of words).

Given unary algebras (Q, δ) and (Q', δ'), a *homomorphism* $f: (Q, \delta) \to (Q', \delta')$ is a map of Q into Q' which commutes with the operations:

$$((q, \sigma)\delta)f = ((q)f, \sigma)\delta' \qquad (q \in Q, \sigma \in \Sigma)$$

or, equivalently, fulfils

$$\delta \cdot f = [f \times \mathrm{id}_\Sigma] \cdot \delta'.$$

This can be expressed by the commutation of the following square

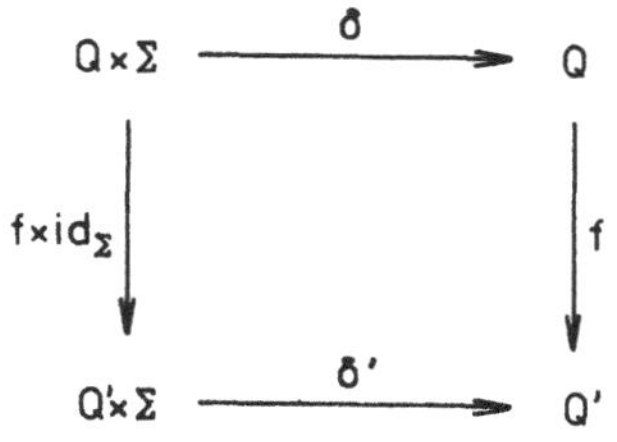

1.6. Proposition. The Σ-algebra (Σ^*, φ) is the free unary algebra on one generator $\emptyset$. That is, for each unary Σ-algebra (Q, δ) and each element $q_0 \in Q$ there exists a unique homomorphism $\rho: (\Sigma^*, \varphi) \to (Q, \delta)$ with $(\emptyset)\rho = q_0$.

Proof. Let us verify that the map ρ of Definition 1.3 is a homomorphism: for each $w = \sigma_1\sigma_2 \ldots \sigma_n$ in Σ^* and each $\sigma \in \Sigma$ we have

$$\begin{aligned} ((w, \sigma)\varphi)\rho &= (\sigma_1\sigma_2 \ldots \sigma_n\sigma)\rho \\ &= (q_n, \sigma)\delta \\ &= ((\sigma_1\sigma_2 \ldots \sigma_n)\rho, \sigma)\delta \\ &= [(w, \sigma)(\rho \times \mathrm{id}_\Sigma)]\delta . \end{aligned}$$

Hence, $\varphi \cdot \rho = (\rho \times \mathrm{id}_\Sigma) \cdot \delta$.

Conversely, any homomorphism assigning q_0 to $\emptyset$ clearly fulfils the inductive condition of Definition 1.3. □

Thus, the run map of a sequential Σ-automaton can be defined as the unique homomorphism which maps $\emptyset$ to the initial state.

1.7. Definition. A *morphism* from a Σ-automaton $A = (Q, \delta, \Gamma, \gamma, q_0)$ to a Σ-automaton $A' = (Q', \delta', \Gamma', \gamma', q'_0)$ is a pair of maps $(f, f_{\mathrm{out}}): A \to A'$ with $f: Q \to Q'$ and $f_{\mathrm{out}}: \Gamma \to \Gamma'$ such that

(i) $f: (Q, \delta) \to (Q', \delta')$ is a homomorphism;

(ii) f and f_{out} commute with the outputs, i.e., $f_{\mathrm{out}} \cdot \gamma = \gamma' \cdot f$;

(iii) f preserves the initial state, i.e., $(q_0)f = q'_0$.

The conditions (i) and (ii) just state that the following diagram

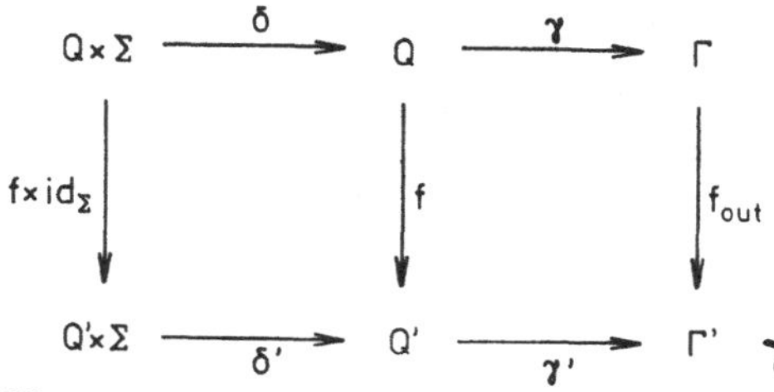

commutes.

Convention. In case $\Gamma = \Gamma'$ and $f_{\text{out}} = \text{id}_\Gamma$, we write $f: A \to A'$ in place of $(f, \text{id}_\Gamma): A \to A'$.

Examples. (i) Consider the following automata A and A' and the following map f (denoted by $\rightsquigarrow$):

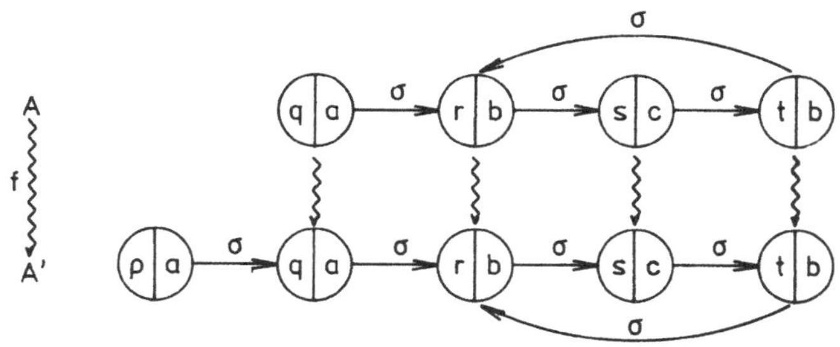

Then $f: A \to A'$ is a morphism.

(ii) Consider the following A, A' and f:

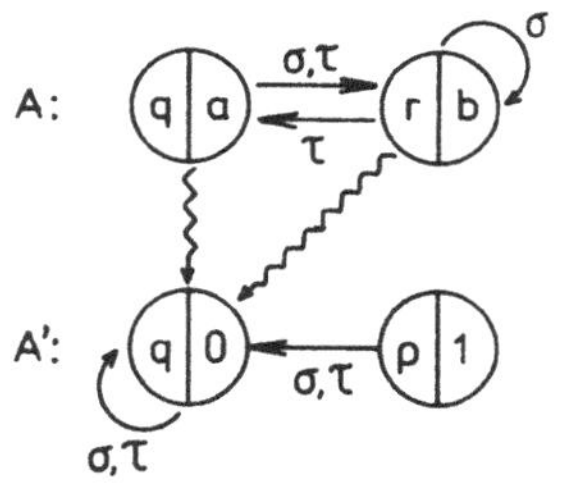

Here $\Gamma = \{a, b\}$ and $\Gamma' = \{0, 1\}$; let $f_{\text{out}}: \Gamma \to \Gamma'$ be the constant map with the value 0, then $(f, f_{\text{out}}): A \to A'$ is a morphism.

1.8. Proposition. For each morphism $(f, f_{\text{out}}): A \to A'$ of Σ-automata, the run map ρ of A is related to the run map ρ' of A' by

$$\rho' = \rho \cdot f,$$

hence, the behavior β of A is related to the behavior β' of A' by

$$\beta' = \beta \cdot f_{\text{out}}.$$

Proof. Put $A = (Q, \delta, \Gamma, \gamma, q_0)$ and $A' = (Q', \delta', \Gamma', \gamma', q_0')$. Since the run map $\rho: (\Sigma^*, \varphi) \to (Q, \delta)$ of A is a homomorphism, the composite map is a homomorphism

$$\rho \cdot f: (\Sigma^*, \varphi) \to (Q', \delta').$$

Moreover,

$$(\emptyset)\rho \cdot f = (q_0)f = q_0'.$$

Hence, $\rho \cdot f$ is the run map of A'—indeed, the run map is the *unique* homomorphism mapping $\emptyset$ to q_0'.

The behavior of A is $\beta = \rho \cdot \gamma$ and the behavior of A' is $\beta' = (\rho \cdot f) \cdot \gamma'$. Since $f \cdot \gamma' = \gamma \cdot f_{out}$, we get $\beta' = \rho \cdot f \cdot \gamma' = \rho \cdot \gamma \cdot f_{out} = \beta \cdot f_{out}$. □

Corollary. Given Σ-automata A and A' with a joint output alphabet, the existence of a morphism $f: A \to A'$ guarantees that A and A' have the same behavior.

1.9. *Isomorphism* of automata is a morphism $(f, f_{out}): A \to A'$ such that f and f_{out} are bijections. It is easy to see that the inverse maps form a morphism $(f^{-1}, f_{out}^{-1}): A' \to A$, again.

If A and A' have the same output alphabet and there is an isomorphism $f: A \to A'$, then A and A' are called *isomorphic*. By 1.8, isomorphic automata have the same behavior.

Exercises I.1

A. Composition of morphisms. (i) Prove that, given morphisms of Σ-automata $(f, f_{out}): A \to A'$ and $(g, g_{out}): A' \to A''$, then $(f \cdot g, f_{out} \cdot g_{out}): A \to A''$ is also a morphism.

(ii) Conversely, if $(f, f_{out}): A \to A'$ is a morphism with both f and f_{out} surjective, and given a pair of maps (g, g_{out}) such that $(f, f_{out}) \cdot (g, g_{out}): A \to A''$ is a morphism, prove that also $(g, g_{out}): A' \to A''$ is a morphism.

B. Automata with a single input. (i) Let a Σ-automaton with $\Sigma = \{\sigma\}$ have finitely many states. Prove that the behavior $\beta: \Sigma^* \to \Gamma$ has a "cycle", i.e. there exists $k = 1, 2, \ldots$ such that $(\sigma^n)\beta = (\sigma^{n+k})\beta$ holds for all sufficiently large n. (Hint: Consider the map $(-, \sigma)\delta: Q \to Q$: since Q is finite, this map has a cycle.)

(ii) Conversely, for each map $\beta: \{\sigma^n\}_{n=0}^{\infty} \to \Gamma$ which has a cycle prove that there is a finite-state Σ-automaton with the behavior β. (Hint: If $(\sigma^n)\beta = (\sigma^{n+k})\beta$ for all $n \geq n_0$, there will be $n_0 + k$ states and the next-state map returns from the $(n_0 + k)$-th state to the $(n_0 + 1)$-st one.)

(iii) Prove that *every* map $\beta: \{\sigma^n\}_{n=0}^{\infty} \to \Gamma$ is the behavior of a (possibly infinite) Σ-automaton. (Hint: Put $Q = \Sigma^*$.)

I.2. Minimal Realization

2.1. Given a behavior, i.e., a map

$$\beta \colon \Sigma^* \to \Gamma,$$

does it have a realization?, that is, can we find a sequential automaton whose behavior is β? And can we find the realization with the minimum number of states? The answers to these questions are affirmative, and we proceed as follows. We first present a "free" realization (which is always infinite), and then we show how to minimize each realization to obtain the (unique) minimal one. The minimization takes two steps: we first discard all superfluous states, and then we merge all pairs of states which behave in the same way.

2.2. Free realization. For each behavior map β, we define the following sequential automaton:

$$A(\beta) = (\Sigma^*, \varphi, \Gamma, \beta, \emptyset).$$

The states of $A(\beta)$ are words in Σ^* and

$$\varphi \colon \Sigma^* \times \Sigma \to \Sigma$$

is the concatenation map, the output is the given behavior, and $\emptyset$ is the initial state.

Remarks. (i) $A(\beta)$ realizes β because the run map of $A(\beta)$ is id_{Σ^*}. In fact, $\mathrm{id}_{\Sigma^*} \colon (\Sigma^*, \varphi) \to (\Sigma^*, \varphi)$ is a homomorphism mapping $\emptyset$ to the initial state of $A(\beta)$ and thus, $\rho = \mathrm{id}_{\Sigma^*}$. Consequently, the behavior of $A(\beta)$ is $\mathrm{id}_{\Sigma^*} \cdot \beta = \beta$. Thus, we see that each behavior has an (infinite) realization.

(ii) For each realization $A = (Q, \delta, \Gamma, \gamma, q_0)$ of β, the run map $\rho \colon \Sigma^* \to Q$ defines a morphism

$$\rho \colon A(\beta) \to A.$$

In fact, ρ is a homomorphism which commutes with the outputs because

$$\beta = \rho \cdot \gamma$$

(for A realizes β), and ρ preserves the initial state:

$$(\emptyset)\rho = q_0.$$

2.3. Reachable part. By a *subautomaton* of a sequential automaton $(Q, \delta, \Gamma, \gamma, q_0)$ is meant an automaton $(Q', \delta', \Gamma, \gamma', q_0)$ where Q' is a subset of Q (containing q_0), δ' is a restriction of δ, and γ' is a restriction of γ. Thus, a subautomaton is given by a set

$$Q' \subset Q$$

such that

$$(Q' \times \Sigma)\delta \subset Q'$$

and

$$q_0 \in Q'.$$

The ***reachable part*** of a sequential automaton $A = (Q, \delta, \Gamma, \gamma, q_0)$ is the subautomaton on the image of ρ, i.e.,

$$Q' = \{q \in Q; q = (\sigma_1 \dots \sigma_n)\rho \quad \text{for some } \sigma_1 \dots \sigma_n \in \Sigma^*\}.$$

Thus, a state is in Q' iff it can be reached from the initial state $q_0 = (\emptyset)\rho$ by some input sequence $\sigma_1 \dots \sigma_n \in \Sigma^*$. We have

$$(Q' \times \Sigma)\delta \subset Q'$$

because for each pair

$$(q, \sigma) \in Q' \times \Sigma$$

we have an input sequence $\sigma_1 \dots \sigma_n \in \Sigma^*$ with $q = (\sigma_1 \dots \sigma_n)\rho$ and then

$$(q, \sigma)\delta = ((\sigma_1 \dots \sigma_n)\rho, \sigma)\delta = (\sigma_1 \dots \sigma_n\sigma)\rho \in Q'.$$

An automaton is ***reachable*** if $Q' = Q$, i.e., each state can be reached from the initial state.

Examples. (i) The reachable part of the automaton in I.1.3, Example (i), is the subautomaton with

$$Q' = \{q, r, s, t\} = Q - \{p\}.$$

(ii) The free realization is always reachable.

Proposition. The reachable part of a sequential automaton A is both the unique reachable subautomaton of A, and the smallest subautomaton of A. Further, it realizes the same behavior as A.

Proof. Let A' be an arbitrary subautomaton of $A = (Q, \delta, \Gamma, \gamma, q_0)$. The inclusion map $v: Q' \to Q$ of the state set of A' is clearly a morphism

$$v: A' \to A.$$

By I.1.8, we have $\rho = \rho' \cdot v$ (where ρ is the run map of A and ρ' that of A'), in other words, ρ' is just the restriction of ρ. It follows, that

(a) A' and A have the same behavior (I.1.8),

(b) Q' contains the state set $(\Sigma^*)\rho = (\Sigma^*)\rho'$ of the reachable part of A, and

(c) if A' is reachable, then $Q' = (\Sigma^*)\rho$ and hence, A' is the reachable part of A.

□

Remark. The preceeding proposition shows that by restricting a given automaton A to its reachable part A', we obtain an automaton having no proper subautomata. Thus, there are no states of A' we can simply discard.

2.4. Congruences. The next step of minimization is achieved by merging pairs of states q and q' as far as it does not ruin the structure of the automaton. Thus, we must expect that by merging q with q', we shall have to merge also

$$(q, \sigma)\delta \quad \text{and} \quad (q', \sigma)\delta \qquad (\sigma \in \Sigma),$$

and further, that two merged states must have the same output. The first condition indicates that we cannot consider the two states q and q' isolated, but we must work with an equivalence relation $\sim$ on Q (where $q \sim q'$ means that q will be merged with q'). For each state q, we have the equivalence class of all states merged with q:

$$[q] = \{q' \in Q;\ q \sim q'\}.$$

We expect to obtain a new automaton $A/\sim$ from $A = (Q, \delta, \Gamma, \gamma, q_0)$ whose state set is the quotient set of Q,

$$Q/\sim\ = \{[q];\ q \in Q\}$$

and whose structure is derived from A. Thus,

$$A/\sim\ = (Q/\sim, \bar{\delta}, \Gamma, \bar{\gamma}, [q_0])$$

where

$$([q], \sigma)\bar{\delta} = [(q, \sigma)\delta]$$

and

$$([q])\bar{\gamma} = (q)\gamma.$$

The following concept describes precisely what equivalences are "admissible" for this approach.

Definition. A *congruence* on an automaton $(Q, \delta, \Gamma, \gamma, q_0)$ is an equivalence relation $\sim$ on Q such that

(a) $q \sim q'$ implies $(q, \sigma)\delta \sim (q', \sigma)\delta$ for all $\sigma \in \Sigma$;
(b) $q \sim q'$ implies $(q)\gamma = (q')\gamma$.

For each congruence $\sim$ we define the *quotient automaton*

$$A/\sim\ = (Q/\sim, \bar{\delta}, \Gamma, \bar{\gamma}, [q_0])$$

as above. An automaton is *reduced* if it has no congruence except the trivial one ($q \sim q'$ iff $q = q'$).

Example. Consider the following automaton

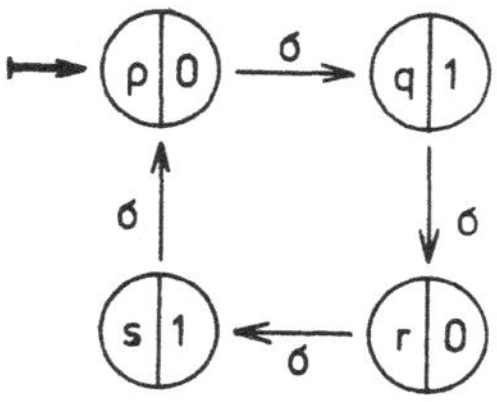

The equivalence with two classes $\{p, r\}$ and $\{q, s\}$ is a congruence. The quotient automaton is the following:

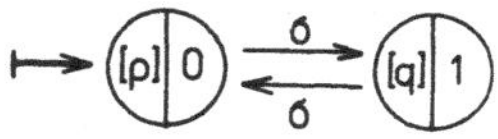

Remark. Congruences are closely related to morphisms of automata:

(i) For each congruence $\sim$ on an automaton A, the canonical map $c: Q \to Q/\sim$, defined by

$$(q)c = [q] \qquad \text{for } q \in Q,$$

is a morphism

$$c: A \to A/\sim.$$

(ii) For each morphism $f: A \to A'$, the kernel equivalence

$$q \sim q' \quad \text{iff} \quad (q)f = (q')f$$

is a congruence.

2.5. Let q be a state of an automaton $A = (Q, \delta, \Gamma, \gamma, q_0)$. By changing the initial state of A to q, we get a new automaton

$$A_q = (Q, \delta, \Gamma, \gamma, q)$$

whose behavior

$$\beta_q: \Sigma^* \to \Gamma$$

is called the *behavior of the state q* in A. Explicitly, $\beta_q = \rho_q \cdot \gamma$ where $(\emptyset)\rho = q$ [and thus, $(\emptyset)\beta_q = (q)\gamma$], and

$$(\sigma_1 \ldots \sigma_n\sigma_{n+1})\rho_q = ((\sigma_1 \ldots \sigma_n)\rho_q, \sigma_{n+1})\delta.$$

The best "merging procedure" is to merge two states iff they have the same behavior:

Theorem. For each sequential automaton A, the equivalence

$$q_1 \approx q_2 \quad \text{iff} \quad \beta_{q_1} = \beta_{q_2}$$

is the largest congruence on A (i.e., given a congruence $\sim$, then $q_1 \sim q_2$ implies $q_1 \approx q_2$).

Proof. I. $\approx$ is a congruence. Let

$$q \approx q',$$

then $(q)\gamma = (q')\gamma$ because $(\emptyset)\beta_q = (\emptyset)\beta_{q'}$. Given $\sigma \in \Sigma$, we put $q_1 = (q, \sigma)\delta$ and $q'_1 = (q', \sigma)\delta$, and we verify that

$$\beta_{q_1} = \beta_{q'_1}.$$

We prove by induction on n that

$$(\sigma_1 \dots \sigma_n)\rho_{q_1} = (\sigma\sigma_1 \dots \sigma_n)\rho_q$$

for each $\sigma_1 \dots \sigma_n \in \Sigma^*$. In fact, for $n = 0$ we have

$$(\emptyset)\rho_{q_1} = q_1 = (q, \sigma)\delta = (\sigma)\rho_q.$$

The induction step follows from the inductive definition (of ρ_{q_1} as well as ρ_q):

$$\begin{aligned}(\sigma_1 \dots \sigma_n\sigma_{n+1})\rho_{q_1} &= ((\sigma_1 \dots \sigma_n)\rho_{q_1}, \sigma_{n+1})\delta \\ &= ((\sigma\sigma_1 \dots \sigma_n)\rho_q, \sigma_{n+1})\delta \\ &= (\sigma\sigma_1 \dots \sigma_{n+1})\rho_q.\end{aligned}$$

Consequently,

$$(\sigma_1 \dots \sigma_n)\beta_{q_1} = (\sigma\sigma_1 \dots \sigma_n)\beta_q \qquad (\text{for all } \sigma_1 \dots \sigma_n \in \Sigma^*).$$

Analogously,

$$(\sigma_1 \dots \sigma_n)\beta_{q'_1} = (\sigma\sigma_1 \dots \sigma_n)\beta_{q'} \qquad (\text{for all } \sigma_1 \dots \sigma_n \in \Sigma^*).$$

Thus, $\beta_q = \beta_{q'}$ implies $\beta_{q_1} = \beta_{q'_1}$.

II. $\approx$ is largest. Let $\sim$ be a congruence on A. For each state q, we have a quotient automaton of A_q under the equivalence $\sim$ (which, of course, is a congruence on A_q, too). Let

$$c_q \colon A_q \to A_q/\sim$$

be the canonical morphism. Then $A_q/\sim$ has the same behavior as A_q, viz., β_q (I.1.8).

Given states $q \sim q'$, the two automata $A_q/\sim$ and $A_{q'}/\sim$ coincide (because they would differ only in the initial state, but $[q] = [q']$). Hence, β_q is the behavior of $A_{q'}/\sim$, in other words, $\beta_q = \beta_{q'}$. □

Example. Consider the following automaton

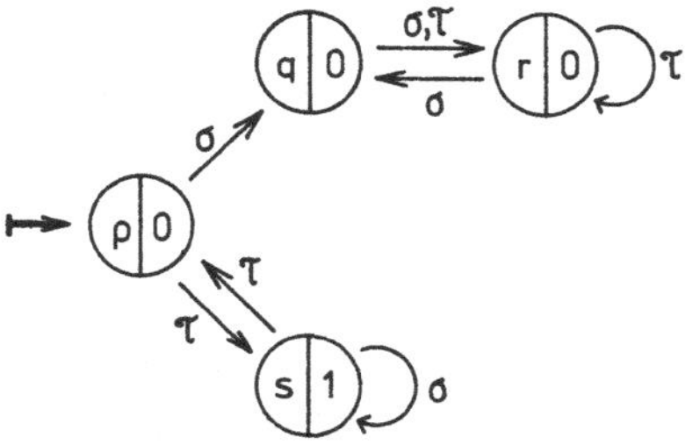

The behavior of p is the following map

$$(w)\beta_p = \begin{cases} 1 & \text{if } w = \tau\sigma^{n_1}\tau^{m_1}\sigma^{n_2}\tau^{m_2}\ldots\sigma^{n_k}\tau^{m_k} \text{ with all } m_1,\ldots, m_k \text{ even,} \\ 0 & \text{else.} \end{cases}$$

The behaviors of q and r are constantly 0, and the behavior of s is

$$(w)\beta_s = (\tau w)\beta_p .$$

Hence, the largest congruence merges q and r, resulting in the following three-state automaton:

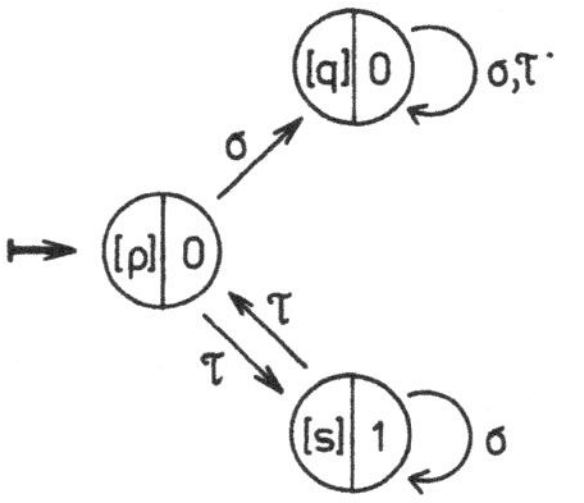

Remarks. (i) We shall prove below that the minimization of each automaton is obtained by factoring its reachable part through the largest congruence.

(ii) Since it is sometimes inconvenient to have equivalence classes of states as new states, we introduce the following relaxation of the concept of quotient automaton. A *reduction* of an automaton A is any surjective morphism

$$e: A \twoheadrightarrow A'.$$

Thus, on the one hand, each congruence $\sim$ defines the (canonical) reduction

$$c: A \twoheadrightarrow A/\!\sim .$$

On the other hand, each reduction $e: A \twoheadrightarrow A'$ defines a congruence

$$q \sim q' \quad \text{iff} \quad (q)e = (q')e$$

such that A' is isomorphic to $A/\sim$. In fact, the map

$$j\colon Q/\sim \to Q'$$

defined by

$$([q])j = (q)e \qquad \text{for } q \in Q,$$

i.e., by

$$c \cdot j = e,$$

is clearly one-to-one, and it is surjective since e is. Thus, j is a bijection and because c and $e = c \cdot j$ are morphisms,

$$j\colon A/\sim \to A'$$

is an isomorphism (see Exercise I.1.A).

2.6. Definition. The *minimal realization* of a behavior β is a reachable realization A_0 of β such that any reachable realization A has a reduction

$$e\colon A \to A_0.$$

The minimal realization is obtained by applying the minimization procedure to the free realization $A(\beta)$. Since $A(\beta)$ is a reachable realization of β (I.2.2), it is sufficient to merge states with the same behavior. Let us have a look at the behavior of an arbitrary state (i.e., word) $w \in \Sigma^*$. The run map

$$\rho_w\colon \Sigma^* \to \Sigma^*$$

of $A(\beta)_w$ is defined by

$$(\sigma_1 \ldots \sigma_n)\rho_w = w\sigma_1 \ldots \sigma_n \qquad \text{for } \sigma_1 \ldots \sigma_n \in \Sigma^*$$

or, more symmetrically,

$$(v)\rho_w = wv \qquad \text{for } v \in \Sigma^*.$$

In fact, we have $(\emptyset)\rho_w = w\emptyset = w$, which is the initial state, and it is easy to see that ρ_w is a homomorphism on (Σ^*, φ). Thus, the behavior map is $v \mapsto (wv)\beta$. Two words w_1 and w_2 have the same behavior in $A(\beta)$ iff

$$(w_1 v)\beta = (w_2 v)\beta \qquad \text{for all } v \in \Sigma^*.$$

We now obtain the minimal realization by factoring $A(\beta)$ through the corresponding congruence:

Construction. For each behavior $\beta\colon \Sigma^* \to \Gamma$ we define the *Nerode equivalence* $\approx$ on the set Σ^* as follows:

$$w_1 \approx w_2 \quad \text{iff} \quad (w_1 v)\beta = (w_2 v)\beta \qquad \text{for all } v \in \Sigma^*.$$

Then the following automaton

$$A(\beta)/\approx \;=(\Sigma^*/\approx, \bar{\varphi}, \Gamma, \bar{\beta}, [\emptyset])$$

where

$$([\sigma_1 \dots \sigma_n], \sigma)\bar{\varphi} = [\sigma_1 \dots \sigma_n \sigma]$$

and

$$([\sigma_1 \dots \sigma_n])\bar{\beta} = (\sigma_1 \dots \sigma_n)\beta$$

is the minimal realization of β.

Proof. Since $w_1 \approx w_2$ iff w_1 and w_2 have the same behavior in $A(\beta)$, it follows that $\approx$ is the largest congruence on $A(\beta)$. In particular, $A(\beta)/\approx$ is well-defined, and it has the same behavior as $A(\beta)$, viz., β.

For each reachable realization $A = (Q, \delta, \Gamma, \gamma, q_0)$ of β, the run map is a morphism

$$\rho: A(\beta) \to A$$

of automata, see I.2.2. Therefore, the kernel equivalence of ρ is a congruence on $A(\beta)$ and hence,

$$(w_1)\rho = (w_2)\rho \quad \text{implies} \quad w_1 \approx w_2.$$

We can define

$$e: Q \to \Sigma^*/\approx$$

by $\rho \cdot e = c$, i.e.,

$$((w)\rho)\, e = [w] \qquad\qquad \text{for } w \in \Sigma^*.$$

Then

$$e: A \to A(\beta)/\approx$$

is a morphism (since ρ is a surjective morphism, see I.1.A) and hence, $A(\beta)/\approx$ is the minimal realization. □

Example. (i) Find the minimal realization of the following behavior $\beta: \{\sigma, \tau\}^* \to \{0, 1\}: (w)\beta = 1$ iff w contains at most two σ's. To do so, we shall go through all the words in $\{\sigma, \tau\}^*$ (from smaller lengths to larger) and we shall try to collect a set of representatives for the Nerode equivalence of β.

First, the empty word $\emptyset$ will correspond to the initial state. The word σ is not equivalent to $\emptyset$ since

$$(\sigma\sigma\sigma)\beta = 0 \text{ while } (\emptyset\sigma\sigma)\beta = 1;$$

thus, $[\sigma]$ will be another state. The word τ is obviously equivalent to $\emptyset$, since a word $\alpha_1 \dots \alpha_n \in \{\sigma, \tau\}^*$ has at most two σ's iff so does $\tau\alpha_1 \dots \alpha_n$. Next, $\sigma\sigma$

is not equivalent to $\emptyset$:

$$(\sigma\sigma\sigma)\beta = 0 \text{ while } (\emptyset\sigma)\beta = 1$$

and neither to σ:

$$(\sigma\sigma\sigma)\beta = 0 \quad \text{while} \quad (\sigma\sigma)\beta = 1.$$

But

$$\sigma\tau \sim \sigma,\ \tau\sigma \sim \sigma \quad \text{and} \quad \tau\tau \sim \emptyset.$$

Finally, the word $\sigma\sigma\sigma$ is not equivalent to $\emptyset$ nor to σ nor to $\sigma\sigma$; but each word is equivalent to one of the words

$$\emptyset,\ \sigma,\ \sigma\sigma \quad \text{and} \quad \sigma\sigma\sigma.$$

Thus, the minimal realization has four states.

The next-state map is given by concatenation, and the output map is a restriction of β. Thus, the following is a minimal realization of β:

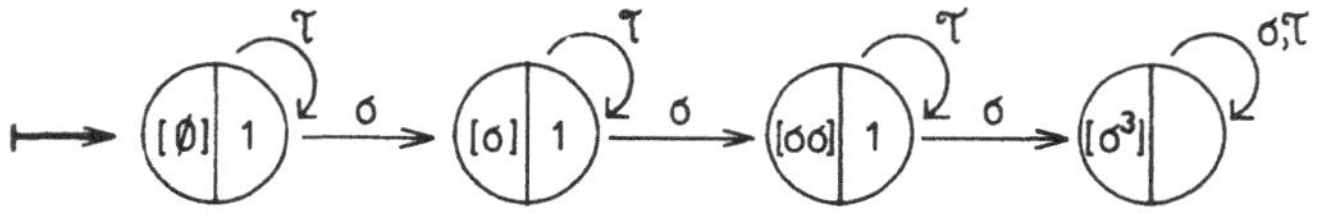

(ii) Find the minimal realization of the behavior

$$\beta: \{\sigma\}^* \rightarrow \{a, b, c\}$$

given by the following table:

$\emptyset$	σ	$\sigma\sigma$	σ^3	σ^4	σ^5	σ^6	σ^7	σ^8	σ^9	σ^{10}	σ^{11}	σ^{12}	σ^{13}	σ^{14}	...
a	b	c	a	a	b	b	c	c	a	a	a	b	b	b	

We try again to find a set of representatives for the Nerode equivalence. The initial state is $[\emptyset]$; the words $\emptyset$, σ, σ^2 are non-equivalent because β has distinct values on them; the word σ^3 is not equivalent to $\emptyset$ since

$$(\sigma^3\sigma)\beta = a \quad \text{while} \quad (\emptyset\sigma)\beta = b;$$

the word σ^4 is not equivalent to $\emptyset$, since

$$(\sigma^4\sigma^2)\beta = b \quad \text{while} \quad (\emptyset\sigma^2)\beta = c$$

nor to σ^3 since

$$(\sigma^4\sigma)\beta = b \quad \text{while} \quad (\sigma^3\sigma)\beta = a,$$

etc. We find out that all the words in $\{\sigma\}^*$ are pairwise non-equivalent. Thus, the free realization is minimal:

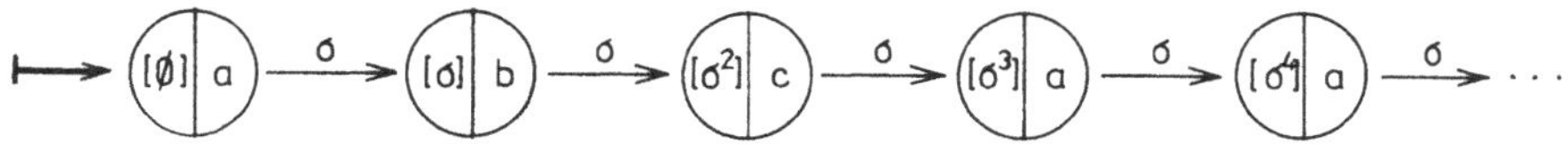

Remark. The above behavior has no finite realization.

2.7. The minimal realization of β can be constructed from any realization. Given an automaton A, let us call a reduction

$$e_0 \colon A \twoheadrightarrow A_0$$

minimal if each reduction $e \colon A \twoheadrightarrow A'$ can be further reduced to A_0. (That is, there is a reduction $e' \colon A' \twoheadrightarrow A_0$ with $e_0 = e \cdot e'$.) The minimal reduction is unique up to isomorphism, and one of the possibilities is

$$c \colon A \twoheadrightarrow A/{\approx}$$

where

$$q \approx q' \quad \text{iff} \quad q \text{ and } q' \text{ have the same behavior.}$$

Proposition. Let A' be the reachable part of an automaton A. The minimal reduction of A' is the minimal realization of the behavior of A.

Proof. Each reduction of a reachable automaton A' is clearly reachable. Let A_0 be the minimal reduction of A', then A_0 is a reachable realization of the behavior β of A' (or, of A). Thus, A_0 can be reduced to the minimal realization of β. But any minimal reduction is clearly reduced, hence, A_0 is the minimal realization. □

Examples. (i) Find the minimal realization of the behavior of the following automaton:

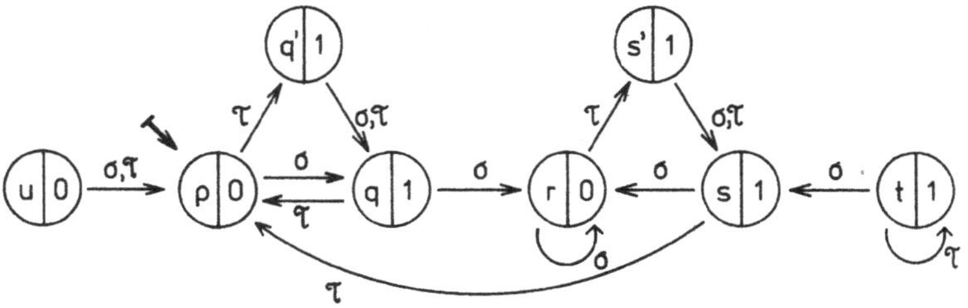

First, the reachable part is the following subautomaton:

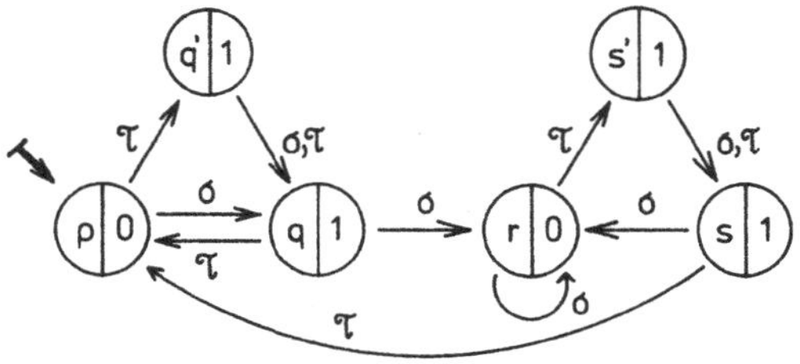

The minimal reduction gives the required minimal realization; it is the following automaton:

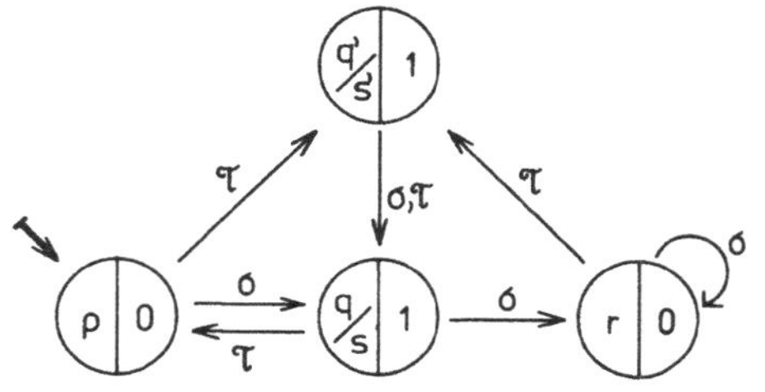

(ii) Given the input alphabet Σ and the output alphabet Γ, we define an automaton realizing all behaviors (with appropriate initial states). Put

$$\Gamma_{\#} = \hom(\Sigma^*, \Gamma) = \{\beta;\ \beta:\Sigma^* \to \Gamma\}.$$

This will be the state set. The next-state map

$$\psi:\Gamma_{\#} \times \Sigma \to \Gamma_{\#}$$

assigns to each behavior $\beta \in \Gamma_{\#}$ and each letter $\sigma \in \Sigma$ the behavior $(\beta, \sigma)\psi = (\sigma\text{—})\beta:\Sigma^* \to \Gamma$ defined by

$$v \mapsto (\sigma v)\beta \quad \text{for } v \in \Sigma^*.$$

The output map

$$\gamma:\Gamma_{\#} \to \Gamma$$

is given by

$$(\beta)\gamma = (\emptyset)\beta.$$

Thus, for each behavior $\beta \in \Gamma_{\#}$ we have a Σ-automaton

$$\bar{A}_\beta = (\Gamma_{\#}, \psi, \Gamma, \gamma, \beta).$$

Its run map $\rho:\Sigma^* \to \Gamma_{\#}$ is defined by

$$(w)\rho = (w\text{—})\beta,$$

where $(w\text{—})\beta:\Sigma^* \to \Gamma$ assigns to each $v \in \Sigma^*$ the value $(wv)\ \beta$. In fact, ρ is a homomorphism with $(\emptyset)\rho = \beta$. It follows that no two distinct states (behaviors) of $\bar{A}_\beta$ have the same behavior.

Consequently, the minimal realization of β is just the reachable part of $\bar{A}_\beta$. This is the subautomaton with the following state set

$$\{(w\text{—})\beta;\ w \in \Sigma^*\}.$$

We obtain another description of the Nerode equivalence: instead of the class $[w]$ we work with the map $(w\text{—})\beta$.

2.8. Remark. Minimal realization is unique up to isomorphism: an automaton A is a minimal realization of β iff A is isomorphic with $A(\beta)/\approx$. In fact, if A is a minimal realization, then there is a reduction $e: A(\beta)/\approx \; \to A$, and e is one-to-one (and hence, a bijection) because the following equivalence

$$w_1 \sim w_2 \quad \text{iff} \quad ([w_1])e = ([w_2])e$$

is a congruence on $A(\beta)$. This implies that $\approx$ is larger than $\sim$ and thus, $([w_1])e = ([w_2])e$ implies $[w_1] = [w_2]$.

Corollary. Let $\beta: \Sigma^* \to \Gamma$ have a realization with a finite state-set. Then its minimal realization is characterized (up to isomorphism) as the realization with the least possible number of states.

In fact, let A be the realization with the least number of states. Then A is reachable (else, the reachable part would have less states) and the reduction $e: A \to A(\beta)/\approx$ is one-to-one [else, $A(\beta)/\approx$ would have less states]. Thus, A is isomorphic to $A(\beta)/\approx$.

Exercises I.2

A. Minimal reduction and realization. (i) Prove that for each reachable automaton A, the minimal realization of the behavior of A is the minimal reduction of A. (This is a converse to I.2.7.)

(ii) Verify that minimal reduction is unique up to isomorphism (commuting with the corresponding morphisms).

(iii) Prove that an automaton is minimal (i.e., is the minimal realization of its behavior) iff it is reachable and reduced.

B. Observability map. For each Σ-automaton $A = (Q, \delta, \Gamma, \gamma, q_0)$ define $b: Q \to \Gamma_\#$ (see I.2.7) by $(q)b = \beta_q$, the behavior of q. A is said to be *observable* if b is one-to-one.

(i) Prove that an automaton is minimal iff it is both reachable and observable.

(ii) Verify that $b: A \to \bar{A}_\beta$ is a morphism of automata (where β is the behavior of A) and prove that b is the unique morphism from A to $\bar{A}_\beta$.

(iii) For each behavior $\beta: \Sigma^* \to \Sigma$ denote by

$$\beta_\# : \Sigma^* \to \Gamma_\#$$

the observability map of the free realization $A(\beta)$ (I.2.2.). Verify that the minimal realization of β is obtained by image factorization of $\beta_\#$: it is (a) the subautomaton of A on the state set $(\Sigma^*)\beta_\#$ and (b) the quotient automaton of $A(\beta)$ under the kernel equivalence of $\beta_\#$.

C. Construct the minimal realizations for each of the following behaviors with $\Sigma = \{\sigma, \tau\}$:

(i) $\Gamma = \{0, 1\}$ and $(w)\beta = 1$ iff $w = \sigma\tau\sigma^n$ for some $n = 1, 2, 3, \ldots$;

(ii) $\Gamma = \{0, 1\}$ and $(w)\beta = 1$ iff $w = \sigma\tau\sigma^n$ for some $n = 0, 1, 2, \ldots$;

(iii) $\Gamma = \{0, 1, 2, \ldots\}$ and $(w)\beta$ is the number of letters in w.

I.3. Finite Automata and Languages

3.1. We know that every behavior has a realization (I.2.2.). In the present section we study the question which behaviors can be realized by a finite automaton, i.e., an automaton with finitely many states. First, we prove that, instead of mappings $\beta: \Sigma^* \to \Gamma$, it suffices to study subsets of Σ^*.

Convention. Subsets of Σ^* are called *languages* in the alphabet Σ.

3.2. Definition. A language $L \subset \Sigma^*$ is *recognizable* if there exists a finite Σ-automaton with the output alphabet $\Gamma = \{0, 1\}$, the behavior of which is the characteristic function of L, i.e.,

$$\beta: \Sigma^* \to \{0, 1\}; \ (w)\beta = 1 \quad \text{iff} \quad w \in L.$$

Remark. Finite Σ-automata with the output alphabet $\Gamma = \{0, 1\}$ are called *acceptors*; their output map

$$\gamma: Q \to \{0, 1\}$$

is the characteristic function of the set

$$T = (1)\gamma^{-1} \subset Q$$

of so called *terminal states.* Thus, acceptors are usually described as quadruples $A = (Q, \delta, T, q)$ with $T \subset Q$ the set of terminal states and $q \in Q$ the initial state.

The *behavior* of an acceptor A is the language L_A of all words in Σ^* which, received in the initial state of A, transfer A to one of its terminal states. We say that A *accepts* such words; globally, A *recognizes* the language L_A. Note that if there is no terminal state ($T = \emptyset$), then the recognized language is empty.

3.3. Proposition. A behavior $\beta: \Sigma^* \to \Gamma$ has a finite realization iff
(1) the language $(y)\beta^{-1}$ is recognizable for each $y \in \Gamma$;
(2) the set $(\Sigma^*)\beta \subset \Gamma$ is finite.

Proof. (a) Let β have a finite realization, say $A = (Q, \delta, \Gamma, \gamma, q)$.
(1) For each $y \in \Gamma$ define an acceptor

$$A_y = (Q, \delta, (y)\gamma^{-1}, q).$$

We prove that $(y)\beta^{-1}$ is the language recognized by A_y. Denote by $\rho: \Sigma^* \to Q$ the run map of A; then

$$\beta = \rho \cdot \gamma.$$

Moreover, ρ is the run map of A_y as well; its output map $\gamma_y: Q \to \{0, 1\}$ is defined by

$$(q)\gamma_y = 1 \text{ iff } (q)\gamma = y.$$

Thus, the behavior of A_y,

$$\beta_y = \rho \cdot \gamma_y,$$

maps $w \in \Sigma^*$ to 1 iff $y = ((w)\rho)\gamma = (w)\beta$. Therefore,

$$L_{A_y} = (y)\beta^{-1}.$$

This proves that $(y)\beta^{-1}$ is recognizable.

(2) The set $(\Sigma^*)\beta = (Q)\gamma$ is finite, since Q is.

(b) Let β have properties (1), (2). Put

$$(\Sigma^*)\beta = \{y_1, \ldots, y_n\}.$$

By hypothesis, for each $i = 1, \ldots, n$ the language

$$L_i = (y_i)\beta^{-1}$$

has a finite realization, say,

$$A_i = (Q_i, \delta_i, T_i, q_i^0),\ i = 1, \ldots, n.$$

Define a finite automaton $A = (Q, \delta, \Gamma, \gamma, q^0)$ as follows:

$$Q = Q_1 \times \ldots \times Q_n;$$

$$((q_1, \ldots, q_n), \sigma)\delta = ((q_1, \sigma)\delta_1, \ldots, (q_n, \sigma)\delta_n)$$

for all $(q_1, \ldots, q_n) \in Q$ and $\sigma \in \Sigma$;

$$\gamma: Q_1 \times \ldots \times Q_n \to \Gamma$$

is an arbitrary map such that, for each $(q_1, \ldots, q_n) \in Q$,

$$(q_1, \ldots, q_n)\gamma = y_i \text{ whenever } q_i \in T_i \text{ but } q_j \in Q_j - T_j \text{ for all } j \neq i;$$

finally,

$$q^0 = (q_1^0, \ldots, q_n^0) \in Q.$$

We shall prove that A realizes β.

Denote by $\rho_i: \Sigma^* \to Q_i$ the run map of A_i, $i = 1, \ldots, n$. The map $\rho: \Sigma^* \to Q_1 \times \ldots \times Q_n$ defined by

$$(w)\rho = ((w)\rho_1, \ldots, (w)\rho_n) \text{ for all } w \in \Sigma^*$$

is a homomorphism, since for each $w \in \Sigma^*$ and $\sigma \in \Sigma$

$$\begin{aligned}(w\sigma)\rho &= ((w\sigma)\rho_1, \ldots, (w\sigma)\rho_n)\\ &= (((w)\rho_1, \sigma)\delta_1, \ldots, ((w)\rho_n, \sigma)\delta_n)\\ &= (((w)\rho_1, \ldots, (w)\rho_n), \sigma)\delta\\ &= ((w)\rho, \sigma)\delta.\end{aligned}$$

Since $(\emptyset)\rho = (q_1^0, \ldots, q_n^0) = q^0$, we see that ρ is the run map of A. Hence, the behavior of A is $\bar{\beta} = \gamma \cdot \rho$. For each $w \in \Sigma^*$ there exists $i = 1, \ldots, n$ with $(w)\beta = y_i$; then w belongs to the language of A_i and does not belong to the language of any A_j, $j \neq i$. Thus

$$(w)\rho_i \in T_i \quad \text{while} \quad (w)\rho_j \in Q_j - T_j \quad \text{for all } j \neq i.$$

Then $((w)\rho)\gamma = ((w)\rho_1, \ldots, (w)\rho_n)\gamma = y_i$. This proves that

$$(w)\beta = y_i \quad \text{implies} \quad (w)\bar{\beta} = ((w)\rho)\gamma = y_i,$$

in other words, $\beta = \bar{\beta}$. □

Example. In I.2.6 we have exhibited a behavior $\beta: \{\sigma\}^* \to \{a, b, c\}$ which has no finite realization. It follows that at least one of the languages $(a)\beta^{-1}$, $(b)\beta^{-1}$ and $(c)\beta^{-1}$ is not recognizable.
In fact, the language

$$(a)\beta^{-1} = \{\emptyset, \sigma^3, \sigma^4, \sigma^9, \sigma^{10}, \sigma^{11}, \ldots\}$$

is not recognizable, see Exercise I.1.B.

3.4. Next, we want to describe the recognizable languages. To do so, we introduce nondeterministic acceptors.

First, recall that a *relation* with a domain X and codomain Y is a triple (X, f, Y) where f is a subset of the cartesian product $X \times Y$. We write $f: X \rightharpoonup Y$ instead of (X, f, Y) and, for each $x \in X$, we put

$$(x)f = \{y \in Y; (x, y) \in f\}.$$

Clearly, any relation $f: X \rightharpoonup Y$ is determined by its domain and codomain and by the collection $\{(x)f; x \in X\}$. We denote by

$$\tilde{f}: X \to \exp Y = \{M \mid M \subset Y\}$$

the map with $(x)\tilde{f} = (x)f$ for all $x \in X$. Note that any map from X to $\exp Y$ uniquely determines a relation $X \rightharpoonup Y$.

The composition of relations is defined in the usual way: given relations $f: X \rightharpoonup Y$ and $g: Y \rightharpoonup Z$, their composition is the relation $f \cdot g: X \rightharpoonup Z$ with

$$(x)f \cdot g = \bigcup_{y \in (x)f} (y)g \qquad \text{for all } x \in X.$$

3.5. Definition. A *nondeterministic Σ-acceptor* is a quadruple $A = (Q, \delta, T, I)$ where

Q is a finite set (of states);
$\delta: Q \times \Sigma \rightharpoonup Q$ is a (next-state) relation;
$T \subset Q$ is the set of terminal states;
$I \subset Q$ is the set of initial states.

While a deterministic acceptor A accepts a word iff this transfers A from the initial state to one of its terminal states, for nondeterministic acceptors we can choose various possibilities and we are happy if at least one works.

3.6. Definition. The *run relation* of a nondeterministic Σ-acceptor (Q, δ, T, I) is the relation $\rho: \Sigma^* \rightharpoonup Q$, defined by the following induction:

$$(\emptyset)\rho = I;$$

$$(\sigma_1, \dots \sigma_n)\rho = D \quad \text{implies} \quad (\sigma_1 \dots \sigma_n\sigma_{n+1})\rho = \bigcup_{q \in D} (q, \sigma_{n+1})\delta$$

for arbitrary $\sigma_1, \dots, \sigma_n, \sigma_{n+1} \in \Sigma$ and $D \subset Q$.

The *language recognized* by A consists of precisely those words $w \in \Sigma^*$ with $(w)\rho \cap T \neq \emptyset$.

Remark. Nondeterministic Σ-acceptors can be depicted by graphs analogous to those we have used for Σ-automata. The nodes carry a single label (of a state) and the terminal states are denoted by small out-comming arrows (replacing the output labels, of course).

Example. Consider the following nondeterministic $\{\sigma, \tau\}$-acceptor:

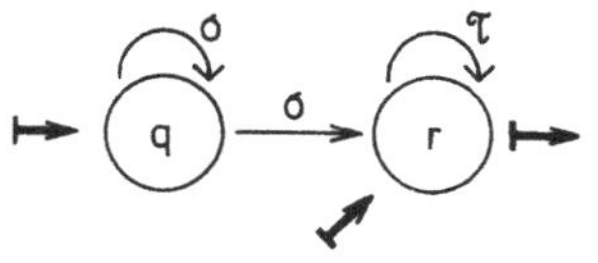

It accepts the word $\emptyset$ (and τ, $\tau\tau$, . . .) since r is initial as well as terminal; it accepts σ since from q we can go to r; it accepts $\sigma\sigma$ since we can remain in q once and then go to r, etc. It does not accept $\tau\sigma$ since this word brings us to no state at all. The language recognized by the given acceptor is

$$L = \{\sigma^n\tau^m \mid n, m = 0, 1, 2, \dots\}.$$

Can L be recognized by a (deterministic) acceptor? We use the Nerode equivalence of the characteristic function of L to construct the minimal realization (see I.2.6). The initial state is the equivalence class of $\emptyset$. The word σ is equivalent to $\emptyset$, since for each $w \in \{\sigma, \tau\}$,

$$w \in L \text{ iff } \sigma w \in L.$$

The word τ is not equivalent to $\emptyset$, since

$$\emptyset\sigma \in L \text{ while } \tau\sigma \notin L.$$

The word $\tau\sigma$ is not equivalent to $\emptyset$ or τ, since

$$\tau\sigma \notin L \text{ and } \emptyset \in L.$$

But each word is easily seen to be equivalent to $\emptyset$, τ or $\tau\sigma$: if $w \in \{\sigma, \tau\}^* - L$ then $w \sim \tau\sigma$; if $w = \sigma^n\tau^m$ with $n \neq 0$ then $w \sim \emptyset$ and if $w = \tau^m$ then $w \sim \tau$. Here is the resulting minimal realization:

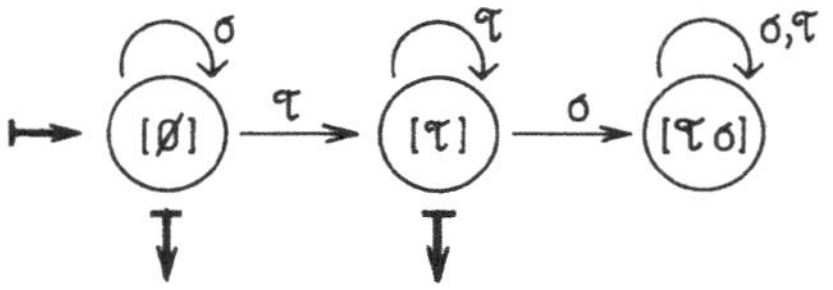

Remark. Observe that nondeterministic acceptors can be smaller than the deterministic ones: no deterministic acceptor recognizing the above language has two states. In the proof of the next proposition we shall see that if an n-state nondeterministic acceptor recognizes a language, then a 2^n-state deterministic acceptor can do the same.

3.7. Proposition. Each language, recognizable by a nondeterministic Σ-acceptor, is recognizable by a deterministic one.

Proof. Let L be a langugae recognized by a nondeterministic Σ-acceptor $A = (Q, \delta, T, I)$. Define a Σ-acceptor $\bar{A} = (\exp Q, \bar{\delta}, \bar{T}, q_0)$, where $\exp Q$ is the set of all subsets of Q and

$$(D, \sigma)\bar{\delta} = \bigcup_{q \in D} (q, \sigma)\delta \qquad \text{for each } D \subset Q \text{ and } \sigma \in \Sigma;$$
$$\bar{T} = \{D \subset Q \mid D \cap T \neq \emptyset\};$$
$$q_0 = I.$$

Let $\rho: \Sigma^* \rightharpoonup Q$ be the run relation of A, let $\tilde{\rho}: \Sigma^* \to \exp Q$ be the corresponding map. We prove that $\tilde{\rho}$ is the run map of $\bar{A}$. Indeed, by the inductive definition I.3.6 we have

$$(\emptyset)\tilde{\rho} = I = q_0;$$
$$(\sigma_1, \ldots \sigma_n)\tilde{\rho} = D \text{ implies}$$
$$(\sigma_1 \ldots \sigma_n\sigma_{n+1})\tilde{\rho} = \bigcup_{q \in D} (q, \sigma_{n+1})\bar{\delta} = (D, \sigma_{n+1})\bar{\delta}.$$

Thus, $\tilde{\rho}: (\Sigma^*, \varphi) \to (\exp Q, \bar{\delta})$ is a homomorphism mapping $\emptyset$ to the initial state q_0.

We conclude that a word $w \in \Sigma^*$ is accepted by $\bar{A}$ iff $(w)\tilde{\rho} \in \exp Q$ is a terminal state of $\bar{A}$, i.e., iff $(w)\bar{\rho} \cap T \neq \emptyset$. This is equivalent to A accepting this word. □

3.8. Operations on acceptors. The reason why nondeterministic acceptors are useful in the study of languages is that some operations on them can be easily described (and used to introduce the corresponding operations on recognizable languages). We mention some examples needed below.

(A) *Union.* Let $A = (Q, \delta, T, I)$ and $A' = (Q', \delta', T', I')$ be nondeterministic Σ-acceptors with $Q \cap Q' = \emptyset$. Their union is the nondeterministic acceptor $A \cup A' = (Q \cup Q', \delta \cup \delta', T \cup T', I \cup I')$. Note that $\delta \cup \delta'$ denotes the set-theoretical union:

$$(q, \sigma)[\delta \cup \delta'] = (q, \sigma)\delta \cup (q, \sigma)\delta' = \begin{cases} (q, \sigma)\delta & \text{if } q \in Q \\ (q, \sigma)\delta' & \text{if } q \in Q'. \end{cases}$$

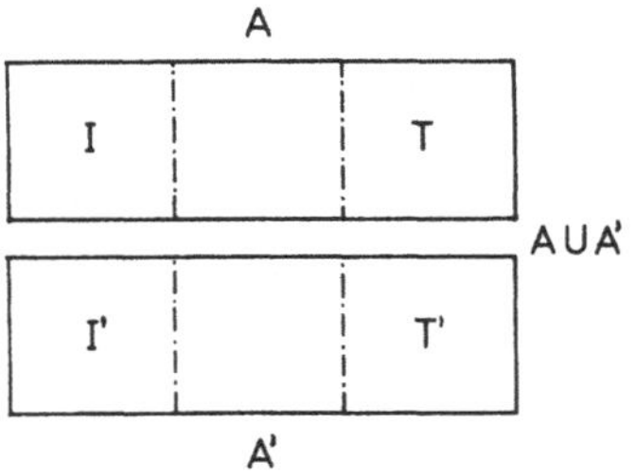

It is easy to verify that the language recognized by $A \cup A'$ is just the union of languages recognized by A and A':

$$L_{A \cup A'} = L_A \cup L_{A'}.$$

(B) *Serial connection.* Let A and A' be as in (A). Their serial connection is the nondeterministic Σ-acceptor $A \cdot A' = (Q \cup Q', \bar{\delta}, \bar{T}, \bar{I})$ where, for each $q \in Q \cup Q'$ and each $\delta \in \Sigma$,

$$(q, \sigma)\bar{\delta} = \begin{cases} (q, \sigma)\delta & \text{if } q \in Q \text{ and } (q, \sigma)\delta \cap T = \emptyset \\ (q, \sigma)\delta \cup I' & \text{if } q \in Q \text{ and } (q, \sigma)\delta \cap T \neq \emptyset \\ (q, \sigma)\delta' & \text{if } q \in Q'. \end{cases}$$

Moreover,

$$\bar{T} = T',$$
$$\bar{I} = I \text{ in case } I \cap T = \emptyset;\ \bar{I} = I \cup I' \text{ in case } I \cap T \neq \emptyset.$$

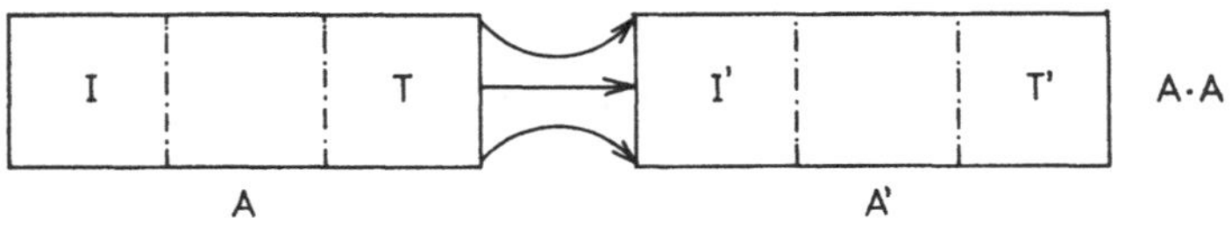

Proposition. The language recognized by $A \cdot A'$ is

$$L_A \cdot L_{A'} = \{vw \in \Sigma^* \mid v \in L_A \text{ and } w \in L_{A'}\}.$$

Proof. Denote by ρ and ρ' the run relations of A and A', respectively. The run relation $\bar{\rho}$ of $A \cdot A'$ assigns to each $\sigma_1 \ldots \sigma_n \in \Sigma^*$ the set

$$(\sigma_1 \ldots \sigma_n)\bar{\rho} = (\sigma_1 \ldots \sigma_n)\rho \cup \bigcup_{\sigma_1 \ldots \sigma_i \in L_A} (\sigma_{i+1} \ldots \sigma_n)\rho'.$$

This is easy to prove by induction on n (just distinguishing the cases $\emptyset \in L$, i.e., $I \cap T \neq \emptyset$, and $\emptyset \neq L$).

Thus, $\sigma_1 \ldots \sigma_n$ is accepted by $A \cdot A'$ iff $(\sigma_1 \ldots \sigma_n)\bar{\rho} \cap \bar{T} \neq \emptyset$, i.e., iff there is $\sigma_1 \ldots \sigma_i \in L_A$ with $\sigma_{i+1} \ldots \sigma_n \in L_{A'}$. Equivalently, $\sigma_1 \ldots \sigma_n \in L_A \cdot L_{A'}$. □

Convention. Given languages $L, L' \subset \Sigma^*$, the language $L \cdot L' = \{vw \in \Sigma^* \mid v \in L \text{ and } w \in L'\}$ is called their *concatenation.*

(C) *Feedback* of a nondeterministic acceptor $A = (Q, \delta, T, I)$ is the nondeterministic Σ-acceptor

$$A^+ = (Q, \delta^+, T, I)$$

where, for each $q \in Q$ and $\sigma \in \Sigma$,

$$(q, \sigma)\delta^+ = \begin{cases} (q, \sigma)\delta & \text{if } (q, \sigma)\delta \cap T = \emptyset; \\ (q, \sigma)\delta \cup I & \text{if } (q, \sigma)\delta \cap T \neq \emptyset. \end{cases}$$

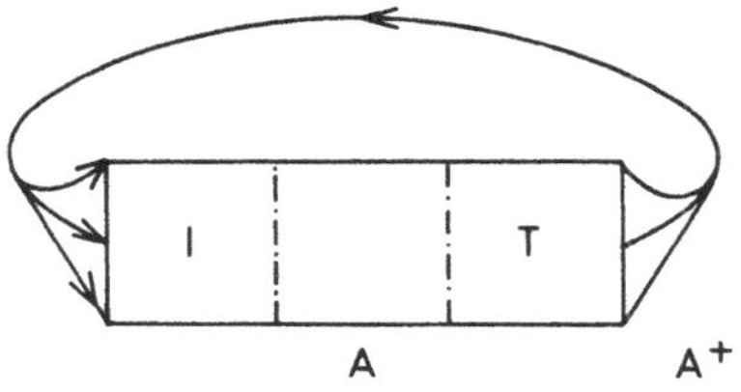

Proposition. The language recognized by A^+ is

$$\begin{aligned} L_A^+ &= L_A \cup (L_A \cdot L_A) \cup (L_A \cdot L_A \cdot L_A) \cup \ldots \\ &= \{w_1 w_2 \ldots w_k \in \Sigma^* \mid w_1, \ldots, w_k \in L_A;\ k = 1, 2, 3, \ldots\}. \end{aligned}$$

Proof. Denote by ρ the run relation of A. The run relation ρ^+ of A^+ can be described as follows:

$$(\emptyset)\rho^+ = I$$

and

$$(\sigma_1 \ldots \sigma_n)\rho^+ = \bigcup (\sigma_i \ldots \sigma_n)\rho \quad \text{for each } \sigma_1 \ldots \sigma_n \in \Sigma^*,\ n \neq 0,$$

where the union ranges over all $i = 1, \ldots, n$ for which there exist

$$j_1 < j_2 < \ldots < j_k < i \quad (k = 0, 1, 2, \ldots)$$

with

$$\sigma_1 \ldots \sigma_{j_1} \in L_A;\ \sigma_{j_1+1} \ldots \sigma_{j_2} \in L_A;\ \ldots;\ \sigma_{j_k+1} \ldots \sigma_i \in L_A.$$

This is easy to prove by induction on n.

Thus, $\sigma_1 \ldots \sigma_n$ is recognized iff there exist $j_1 < \ldots < j_k < i$ with $\sigma_1 \ldots \sigma_{j_1}$; $\sigma_{j_1+1} \ldots \sigma_{j_2}$; ...; $\sigma_{j_k+1} \ldots \sigma_i$; $\sigma_{i+1} \ldots \sigma_n$ in L_A. Equivalently, $\sigma_1 \ldots \sigma_n \in L_A^+$. □

Convention. Given a language $L \subset \Sigma^*$, the language

$$L^* = \{\emptyset\} \cup L^* = \{\emptyset\} \cup L \cup (L \cdot L) \cup (L \cdot L \cdot L) \cup \ldots$$

is called its *iteration.*

3.9. Kleene Theorem. For each finite alphabet Σ, the class of all recognizable languages in Σ^* is the smallest class of languages which

(i) contains the singleton languages $\{\emptyset\}$ and $\{\sigma\}$ for each $\sigma \in \Sigma$ and the empty language $\emptyset$;

(ii) is closed under the formation of union, concatenation and iteration.

Remark. The operations $\cup$, $\cdot$ and $*$ are called *rational operations* on languages. A language $L \subset \Sigma^*$ is said to be *rational* if it can be expressed by a finite expression using the letters of the alphabet Σ, the symbols $\emptyset$, $\{,\}$, and the rational operations. Example:

$$L = \{\sigma\}^* \cup \{\tau\}^* \cdot \{\sigma\} \cup \{\emptyset\}.$$

Kleene's theorem then states:

A language is recognizable iff it is rational.

Proof. (a) Rational languages are recognizable. We use Proposition I.3.7. First, $\{\emptyset\}$ and $\{\sigma\}$ and $\emptyset$ are recognizable languages:

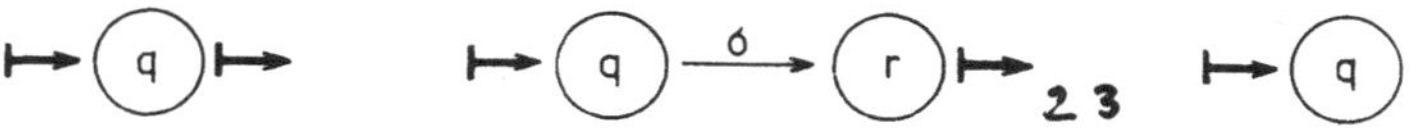

By I.3.8, the class of all recognizable languages is closed under rational operations.

(b) Recognizable languages are rational. First, observe that every subset of $\Sigma \cup \{\emptyset\}$ is rational (since it is a finite union of singleton, one-letter languages).

For each Σ-acceptor

$$A = (Q, \delta, T, q_0)$$

we are going to prove that the language L_A is rational. Put

$$Q = \{q_0, q_1, \ldots, q_m\}.$$

For arbitary $i, j = 0, \ldots, m$ we denote by L_{ij} the set of all words in Σ^* which transfer A from q_i to q_j. More precisely, for each $\sigma_1 \ldots \sigma_n \in \Sigma^*$ we put

$$r_0 = q_i\,;\ r_1 = (r_0, \sigma_1)\delta\,;\ \ldots;\ r_n = (r_{n-1}, \sigma_n)\delta$$

and we define L_{ij} as follows:

$$\sigma_1 \ldots \sigma_n \in L_{ij} \quad \text{iff} \quad r_n = q_j\,.$$

Clearly,

$$L_A = \bigcup_{q_j \in T} L_{0j}.$$

Hence, to prove that L_A is rational, it suffices to prove the rationality of each L_{ij}. To do so, we denote by

$$L_{ij}^k,\ k = 0, \ldots, m + 1,$$

the set of all words in Σ^* which transfer A from q_i to q_j without passing through the states $q_k, \ldots, q_m$. More precisely, $\sigma_1 \ldots \sigma_n \in L_{ij}^k$ iff the above states $r_0, \ldots, r_n$ fulfil

$$r_1, \ldots, r_{n-1} \in \{q_0, \ldots, q_{k-1}\}.$$

Note that $L_{ij} = L_{ij}^{m+1}$. We are going to prove that L_{ij}^k is a rational language by induction on k.

First, L_{ij}^0 is a rational language: here $r_1, \ldots, r_{n-1} \in \emptyset$, thus $n \leq 1$, therefore $L_{ij}^0 \subset \Sigma \cup \{\emptyset\}$.

Next, if L_{ij}^{k-1} is rational then, to prove that L_{ij}^k is rational, we shall verify the following formula:

$$L_{ij}^k = L_{ij}^{k-1} \cup L_{ik}^{k-1} \cdot (L_{kk}^{k-1})^* \cdot L_{kj}^{k-1}.$$

Indeed, let $\sigma_1 \ldots \sigma_n \in L_{ij}^k$ and consider the above states $r_0, \ldots, r_n$. If all the states $r_1, \ldots, r_{n-1}$ are distinct from q_k, then $\sigma_1 \ldots \sigma_n \in L_{ij}^{k-1}$. Else, denote by $t_1 < t_2 < \ldots < t_s$ all of the indices $t = 1, \ldots, n - 1$ with $r_t = q_k$. Then the states $r_1, \ldots, r_{t_1 - 1}$ are distinct from q_k, thus

$$\sigma_1 \ldots \sigma_{t_1} \in L_{ik}^{k-1}.$$

Analogously,

$$\sigma_{t_1+2} \ldots \sigma_{t_2}, \sigma_{t_2+1} \ldots \sigma_{t_3}, \ldots, \sigma_{t_{s-1}+1} \ldots \sigma_{t_s} \in L_{kk}^{k-1}$$

as well as

$$\sigma_{t_s+1} \ldots \sigma_n \in L_{kj}^{k-1}.$$

This shows that

$$\sigma_1 \ldots \sigma_n \in L_{ik}^{k-1} \cdot (L_{kk}^{k-1})^* \cdot L_{kj}^{k-1}.$$

The reverse inclusion $L_{ij}^{k-1} \cup L_{ik}^{k-1} \cdot (L_{kk}^{k-1})^* \cdot L_{kj}^{k-1} \subset L_{ij}^k$ is obvious. This proves the above formula, thus, L_{ij}^k is rational. □

Exercises I.3

A. Infinite alphabets. Let Σ be an infinite input alphabet. Prove that all recognizable languages in Σ^* form the least class containing all languages $L \subset \Sigma \cup \{\emptyset\}$ and closed under the rational operations.

B. Operations preserving the recognizability. If $L_1, L_2 \subset \Sigma^*$ are recognizable languages, prove that the following languages are also recognizable:

(i) $\Sigma^* - L_1$. Hint: Interchange the terminal and the nonterminal states.

(ii) $L_1 \cap L_2$. Hint: Use (i).

(iii) rev $L_1 = \{\sigma_1 \ldots \sigma_k \mid \sigma_k \ldots \sigma_1 \in L_1\}$. Hint: reverse the arrows of δ and interchange the terminal and initial states.

C. Find a rational expression for each of the following languages:

(i) $\Sigma^* - \{\sigma\}$;

(ii) $\{\sigma, \tau\}^* - \{\sigma\}^*$;

(iii) rev $(\{\sigma\} \cdot \{\tau\}^* \cup \{\sigma, \rho\}^* \cdot \{\rho\sigma\rho\})$.

Notes to Chapter I

This chapter presents just a standard introduction to sequential automata. The interested reader can find more information for example in M. A. Arbib [1969] or S. Eilenberg [1974].

Chapter II: Tree Automata

II.1. Finitary Tree Automata

1.1. Tree automata are devices which handle labelled trees analogously as sequential automata handle sequences (words) of input symbols. The internal structure of a sequential automaton is a unary algebra; for a tree automaton, it is an algebra of an arbitrary type.

A *finitary type* is a set Σ of operation symbols together with an arity map assigning to each $\sigma \in \Sigma$ a natural number

$$|\sigma| = 0, 1, 2, \ldots .$$

The set of all *n-ary operation symbols* (such that $|\sigma| = n$) is denoted by Σ_n. A Σ-algebra consits of a set Q and operations of the prescribed arities:

$$\delta_\sigma : Q^n = Q \times Q \times \ldots \times Q \to Q \quad (\sigma \in \Sigma, |\sigma| = n).$$

For $n = 1$ we have a *unary operation* $\delta_\sigma : Q \to Q$, for $n = 2$ a *binary operation* $\delta_\sigma : Q \times Q \to Q$, etc. For $n = 0$, the set Q^0 has just one element, and a *nullary operation*

$$\delta_\sigma : Q^0 \to Q$$

is usually identified with the element of Q which forms $(Q^0)\delta_\sigma$.

Operation-preserving maps are called *homomorphisms*. Thus, a map $f: (Q, \{\delta_\delta\}_{\sigma \in \Sigma}) \to (Q' \{\delta'_\sigma\}_{\sigma \in \Sigma})$ is a homomorphism if for each $\sigma \in \Sigma_n$,

$$(q_0, \ldots, q_{n-1})\delta_\sigma = q \quad \text{implies} \quad ((q_0)f, \ldots, (q_{n-1})f)\delta_\sigma = (q)f.$$

1.2. Definition. A *Σ-tree automaton* is a sixtuple $A = (Q, \{\delta_\sigma\}_{\sigma \in \Sigma}, \Gamma, \gamma, I, \lambda)$ where

Q is a set, called the *set of states*;
$\delta_\sigma : Q^n \to Q \quad (\sigma \in \Sigma, |\sigma| = n)$ are operations on Q;
Γ is a set, called the *output alphabet*;
$\gamma : Q \to \Gamma$ is a map, called the *output map*;
I is a set, called the *set of variables*;
$\lambda : I \to Q$ is map, called the *initialization*.

If all operations are unary (i.e., $\Sigma = \Sigma_1$) and if I contains just one variable,

$I = \{x\}$, then a Σ-tree automaton is precisely a sequential Σ-automaton with the initial state $(x)\lambda$.

1.3. Example. Let $\Sigma = \Sigma_2 = \{+\}$ and let $(Z, +)$ be the additive Σ-algebra of integers. Put $\Gamma = \{0, 1\}$, and let γ be the parity map:

$$(z)\gamma = \begin{cases} 1 & \text{if } z \text{ is odd} \\ 0 & \text{if } z \text{ is even.} \end{cases}$$

Then we have a Σ-tree automaton

$$A = (Z, +, \Gamma, \gamma, \{x, y\}, \lambda)$$

with

$$(x)\lambda = -1 \text{ and } (y)\lambda = 1.$$

The "action" of this automaton (to be made precise below) consists of taking any binary tree with leaves labelled by x and y, computing the tree and giving an output. For example, the following tree

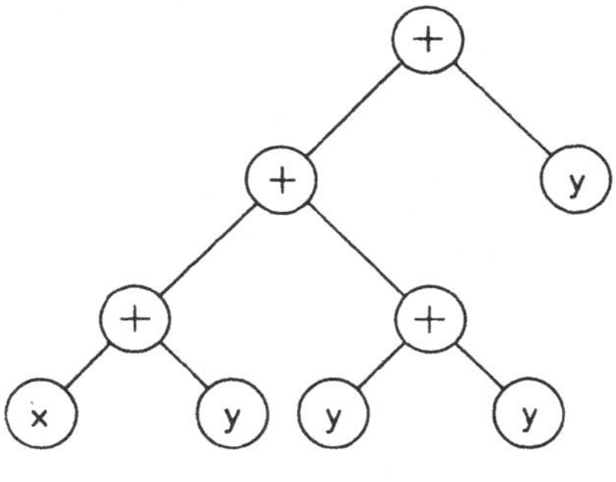

is computed as follows:

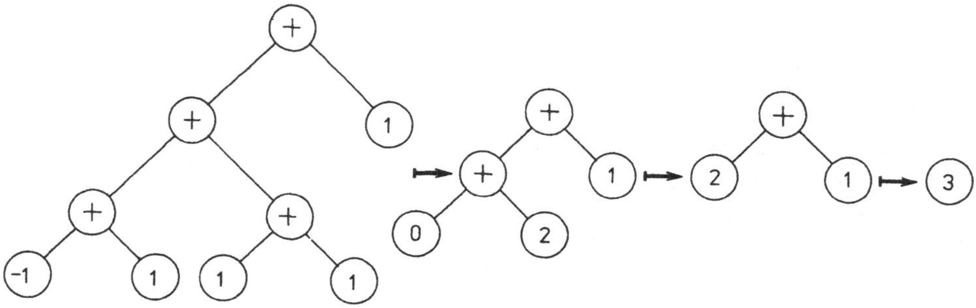

The resulting output is $(3)\gamma = 1$.

The external behavior of the automaton A is expressed by the map β assigning to each of these trees t the value

$$(t)\beta \in \Gamma$$

of the output which results after the computation of t. In our present automaton clearly $(t)\beta = 0$ iff the number of x-labelled leaves is congruent to the number of y-labelled ones modulo 2.

1.4. Σ-tree automata act on finite trees labelled as follows: each node with n successors ($n > 0$) is labelled by an n-ary operation symbol, and each leaf is labelled by a variable or a nullary symbol. We formalize these trees by introducing a non-labelled "base" tree and defining its admissible labellings.

Put

$$m = \bigvee_{\sigma \in \Sigma} |\sigma|,$$

i.e., m is the maximal arity if such exists, and $m = \omega$ if arities are unbounded. The nodes of the base tree are all sequences $p_1 \ldots p_k$ of numbers smaller than m (for $k = 0, 1, 2, \ldots$, where $k = 0$ stands for the empty sequence, the root). More precisely, recall that m is the set of all natural numbers smaller than m (i.e., $m = \{0, 1, \ldots, m - 1\}$ if m is finite, and ω is the set of all natural numbers). The nodes of the base tree form the set

$$m^*$$

of all words in m, with $p_1 \ldots p_k \in m^*$ preceded precisely by

$$\emptyset, p_1, p_1p_2, \ldots, p_1 \ldots p_{k-1}.$$

For example

$$2^* = \{0, 1\}^*$$

is the complete binary tree:

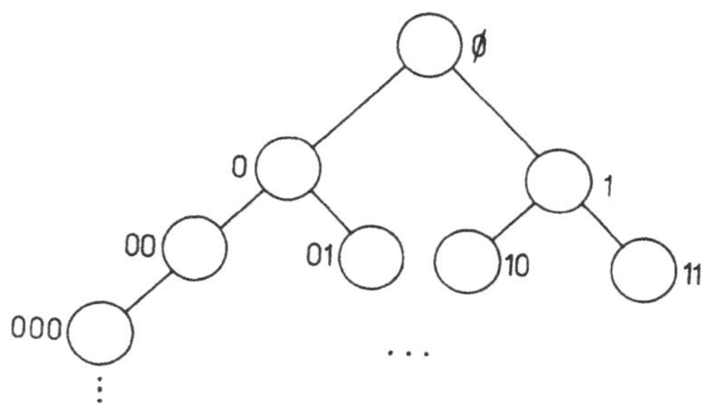

Here $\emptyset$ is the root, and each node $p_1 \ldots p_k \in 2^*$ has two immediate successors: $p_1 \dots p_k 0$ and $p_1 \dots p_k 1$. (In ω^* each element has countably many immediate successors.)

The binary tree of the above example can be considered as a partial labelling of 2^* by the labels $+$, x and y, i.e., a partial map

$$t: 2^* \to \{+, x, y\}.$$

The map is defined as follows

	$\emptyset$	0	1	00	01	000	001	010	011
t	$+$	$+$	y	$+$	$+$	x	y	y	y

The domain of definition D_t of t is $\{\emptyset, 0, 1, 00, 01, 000, 001, 010, 011\}$. In general, Σ-trees will be partial maps from m^* with values either in Σ or I. For each partial map $t: X \to Y$, put

$$D_t = \{x \in X;\ (x)t \text{ is defined}\}.$$

Definition. Let Σ be a finitary type with $m = \bigvee_{\sigma \in \Sigma} |\sigma|$. Let I be a set (of variables) with $I \cap \Sigma = \emptyset$. A *finite Σ-tree* is a partial map

$$t: m^* \to I \cup \Sigma$$

such that

(i) the domain of definition D_t is non-empty and finite;

(ii) given $p_1 \dots p_k p_{k+1} \in m^*$, then

$$p_1 \dots p_{k+1} \in D_t \quad \text{iff} \quad (p_1 \dots p_k)t \in \Sigma_n \quad \text{for some } n > p_{k+1}.$$

The meaning of (ii) is that:

(A) Labels are assigned from left to right: if $p_1 \dots p_k p_{k+1}$ has a label (i.e., is in D_t), then also $p_1 \dots p_k 0, p_1 \dots p_k 1, \dots, p_1 \dots p_k p_{k+1}$ have labels.

(B) Each node $p_1 \dots p_k$ labelled by $\sigma[= (p_1 \dots p_k)t]$ in Σ_n has precisely n successors, viz, $p_1 \dots p_k 0, \dots, p_1 \dots p_k(n-1)$.

(C) Each node $p_1 \dots p_k$ labelled by $x[= (p_1 \dots p_k)t]$ in I is a *leaf* (i.e., $p_1 \dots p_k p_{k+1} \notin D_t$ for any p_{k+1}); the same holds for Σ_0-labels, of course.

Furthermore,

(D) $\emptyset$ always has a label: there exists $p_1 \dots p_k \in D_t$ (because $D_t \neq \emptyset$) and it follows that $p_1 \dots p_{k-1} \in D_t$ and hence $p_1 \dots p_{k-2} \in D_t$, etc.

1.5. Notation. (i) The set of all Σ-trees over I is denoted by

$$I^\#.$$

This set carries a natural structure of a Σ-algebra. For each $\sigma \in \Sigma_n$ we have the operation

$$\varphi_\sigma: (I^\#)^n \to I^\#$$

of tree-tupling: given trees $t_0 \ldots t_{n-1} \in I^\#$, we form the following tree

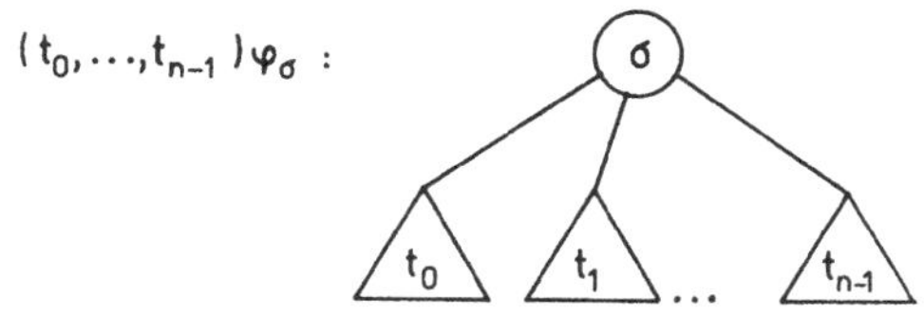

Formally, the tree

$$t = (t_0, \ldots, t_{n-1})\varphi_\sigma$$

has the following domain of definition

$$D_t = \{p_1 p_2 \ldots p_k \in m^*;\ p_1 < n \quad \text{and} \quad p_2 \ldots p_k \in D_{t_{p_1}}\} \cup \{\emptyset\}$$

and is defined by

$$(\emptyset)t = \sigma;$$
$$(p_1 p_2 \ldots p_k)t = (p_2 \ldots p_k)t_{p_1}$$

for all $p_1 p_2 \ldots p_k$ with $p_1 < n$ and $p_2 \ldots p_k \in D_{t_{p_1}}$. The conditions (i) and (ii) of II.1.4 are easily verified for $(t_0, \ldots, t_{n-1})\ \varphi_\sigma$; in particular each $\sigma \in \Sigma_0$ defines the singleton tree

(ii) We consider I as a subset of $I^\#$ by identifying each variable $x \in I$ with the singleton tree

(iii) The *depth* $|t|$ of a tree $t \in I^\#$ is the largest number $k = 0, 1, 2, \ldots$ for which there exists $p_1 p_2 \ldots p_k \in D_t$. Thus, each variable and each operation symbol in Σ_0 have depth 0. The tree in Example II.1.3 has depth 3.

(iv) For each node w of a tree t, i.e., each

$$w = q_1 \ldots q_m \in D_t$$

we define the *branch*

$$\partial_w t$$

of t at w as the following tree:

$$(p_1 \ldots p_k)\partial_w t = (q_1 \ldots q_m p_1 \ldots p_k)t$$

(where the left-hand side is defined iff the right-hand one is). For example, the following tree t:

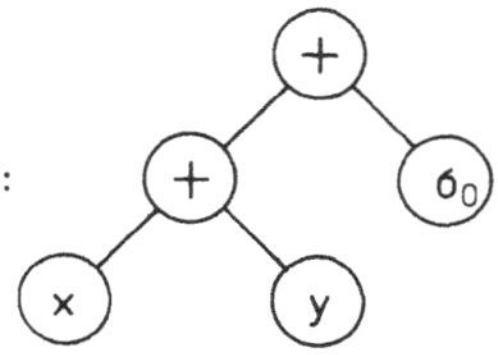

has the following branches:

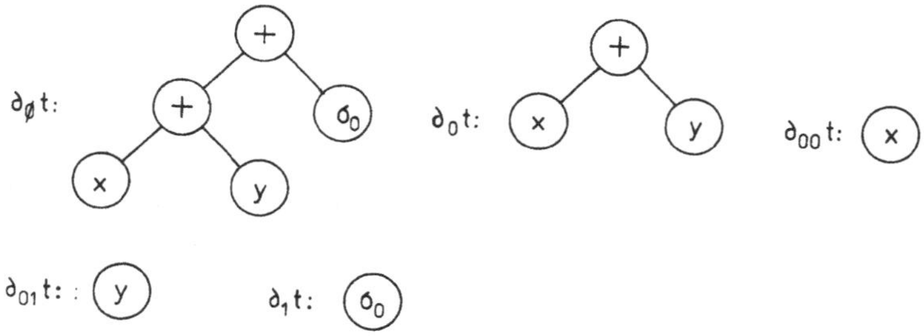

(iv) We define sets of trees

$$W_n \subset I^{\#}\,(n = 0, 1, 2, \ldots)$$

by the following induction:

$$W_0 = I$$

and

$$W_{n+1} = I \cup \{(t_0, \ldots, t_{k-1})\varphi_\sigma;\ \sigma \in \Sigma, |\sigma| = k, \text{and } t_0, \ldots, t_{k-1} \in W_n\}.$$

We have

$$I^{\#} = \bigcup_{n=0}^{\infty} W_n$$

because for each $t \in I^{\#}$,

$$|t| \leq n \quad \text{implies} \quad t \in W_{n+1}.$$

[This is clear if $n = 0$. Each tree t of depth $n + 1$ has the form $t = (t_0, \ldots, t_{k-1})\varphi_\sigma$ for some $\sigma \in \Sigma_k$, $k > 0$, where $|t_i| \leq n$ and hence by induction hypothesis, $t_i \in W_{n+1}$ for $i = 0, \ldots, k - 1$. Thus, $t \in W_{n+2}$.]

Proposition. The algebra of finite trees

$$(I^{\#}, \{\varphi_\sigma\}_{\sigma \in \Sigma})$$

is the free Σ-algebra generated by the set I. That is, for each Σ-algebra

$(Q, \{\delta_\sigma\})$ and each map $f: I \to Q$ there is a unique homomorphism

$$f^{\#} : (I^{\#}, \{\varphi_\sigma\}) \to (Q, \{\delta_\sigma\})$$

extending f.

Proof. We define $f^{\#}$ on each W_n by induction on n. First, $f^{\#} = f$ on $I = W_0$. Given $f^{\#}$ on W_n, for each $t \in W_{n+1} - I$ we have $t = (t_0, \ldots, t_{k-1})\varphi_\sigma$ with $t_0, \ldots, t_{k-1} \in W_n$, and we put

$$(t)f^{\#} = ((t_0)f^{\#}, \ldots, (t_{k-1})f^{\#})\varphi_\sigma.$$

It is obvious that this is how $f^{\#}$ has to be defined (i.e., $f^{\#}$ is unique) and that $f^{\#}$ is a homomorphism, provided only that $f^{\#}$ is well-defined. Thus, the proof is concluded by the following

Observation. The algebra $(I^{\#}, \{\varphi_\sigma\})$ has the following *Peano properties*:

(i) Each element of $I^{\#} - I$ has the form $(t_0, \ldots, t_{k-1})\varphi_\sigma$ for a unique operation symbol $\sigma \in \Sigma$, and a unique $|\sigma|$-tuple $t_0, \ldots, t_{k-1} \in I^{\#}$;

(ii) conversely, each element $(t_0, \ldots, t_{k-1})\varphi_\sigma$ lies in $I^{\#} - I$.

In fact, given $t \in I^{\#} - I$ then $(\emptyset)t = \sigma$ for some $\sigma \in \Sigma$. If $|\sigma| = k$, it follows that $t = (t_0, \ldots, t_{k-1})\varphi_\sigma$ iff $t_i = \partial_i t$ for $i = 0, \ldots, k-1$. Conversely, the root of each $(t_0, \ldots, t_{k-1})\varphi_\sigma$ is labelled by σ and hence, this is not an element of I.

Examples. (i) Let $\Sigma = \Sigma_1$ and $I = \{x\}$ (the case of sequential automata). The free algebra $I^{\#}$ consists of the following trees:

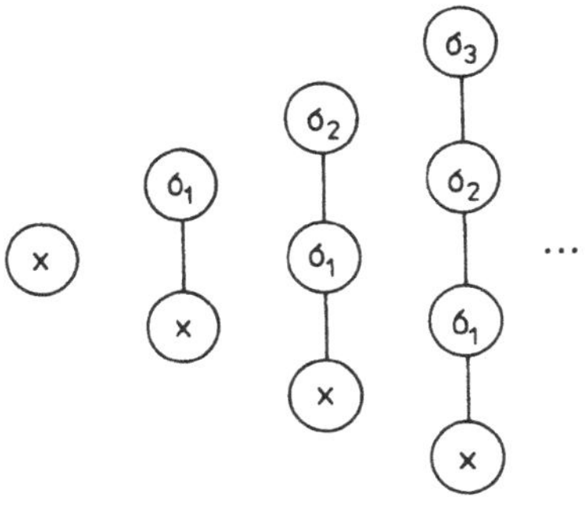

and the operations φ_σ act as follows

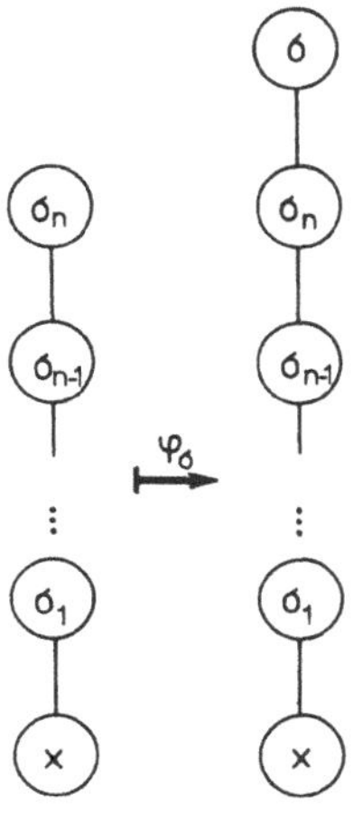

This is just the algebra Σ^*, except for the (superfluous) symbol x.

If we consider an arbitary set I of variables, then words starting with different variables act quite separately. Therefore,

$$I^\# = I \times S^\#$$

with

$$(x, \sigma_1 \ldots \sigma_n)\varphi_\sigma = (x, \sigma_1 \ldots \sigma_n\sigma) \quad (x \in I \text{ and } \sigma_1 \ldots \sigma_n \in \Sigma^\#).$$

(ii) Let Σ be an arbitrary type and $I = \emptyset$. Then the free algebra, known as the *initial* Σ-algebra, consists of finite Σ-labelled trees : all leaves are labelled in Σ_0 and each node with n successors is labelled in Σ_n. The characteristic property of the initial algebra is that each Σ-algebra A has a unique homomorphism from the initial algebra to A.

1.6. Definition. Let $A = (Q, \{\delta_\sigma\}, \Gamma, \gamma, I, \lambda)$ be a Σ-tree automaton. The unique homomorphism

$$\rho: (I^\#, \{\varphi_\sigma\}) \to (Q, \{\delta_\sigma\})$$

extending the initialization map λ is called the *run map* of A. The map

$$\beta = \rho \cdot \gamma : I^\# \to \Gamma$$

is called the *behavior* of A.

For each tree t, the result of the computation of t [after interpreting the variables x as the states $(x)\lambda$] is the state $(t)\rho$. And the resulting output is $(t)\beta$.

Example. Let Σ consist of a binary symbol σ and a nullary symbol τ. Consider the set

$$Z_4 = \{0, 1, 2, 3\}$$

with the addition $\sigma = +$ modulo 4 (which is the usual addition with 4 subtracted if the result would exceed 3) and with $\tau = 1$; let

$$\gamma : Z_4 \to \{0, 1\}$$

be the parity map. Denote by

$$A = (Z_4, \{+, 1\}, \{0, 1\}, \gamma, \{x, y\}, \lambda)$$

the automaton with

$$(x)\lambda = 1 \text{ and } (y)\lambda = 0.$$

For the tree

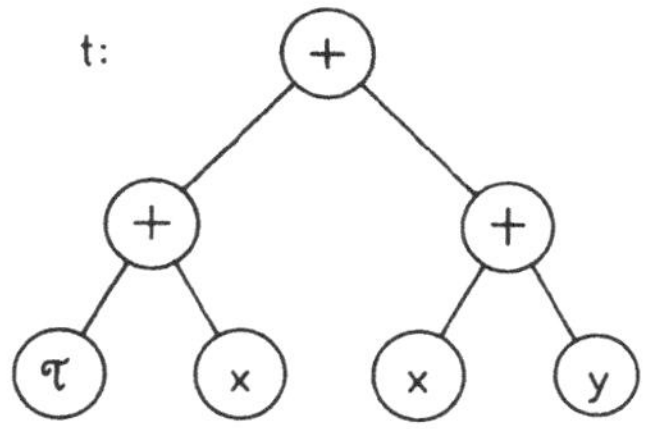

the computation in A yields

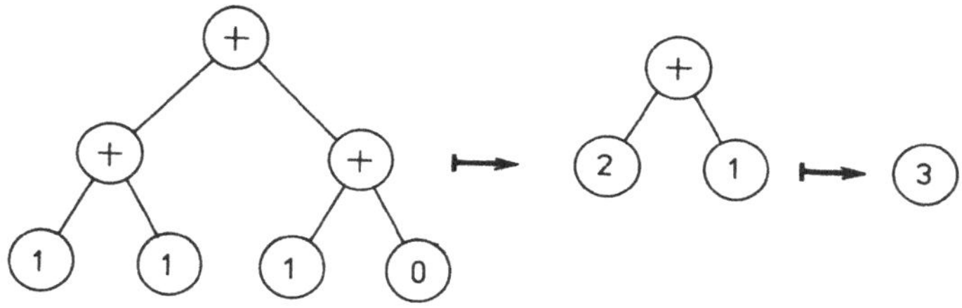

i.e.,

$$(t)\rho = 3 \quad \text{and} \quad (t)\beta = 1.$$

In general, $(t)\rho = i$ iff the number of all leaves labelled by τ or x is congruent to i modulo 4. Thus,

$$\beta: \{x, y\}^{\#} \to \{0, 1\}$$

is given by

$$(t)\beta = 0 \text{ iff the number of leaves labelled by either } x \text{ or } \tau \text{ is even.}$$

1.7. Example. Let $\Sigma = \Sigma_2 = \{\vee, \wedge\}$ and for a non-empty set M, consider the set

$$Q = \exp M$$

of all subsets of M with the operations union $\cup$ and intersection $\cap$. Define an automaton

$$A = (\exp M, \{\cup, \cap\}, \{a_0, a_1, b\}, \gamma, \{x, y\}, \lambda)$$

where

$$(T)\gamma = \begin{cases} a_0 & \text{if } T = \emptyset \\ a_1 & \text{if } T = M \\ b & \text{if } T \in \exp M - \{M, \emptyset\} \end{cases}$$

and

$$(x)\lambda = \emptyset, \ (y)\lambda = M.$$

The tree

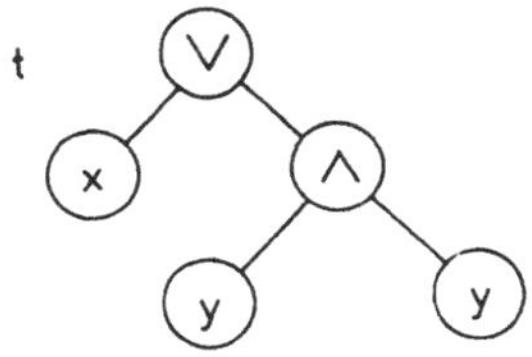

is computed as follows:

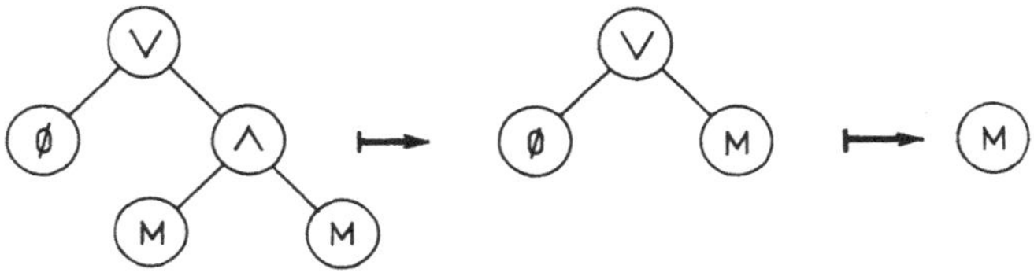

and hence, $(t)\beta = a_1$.
The tree

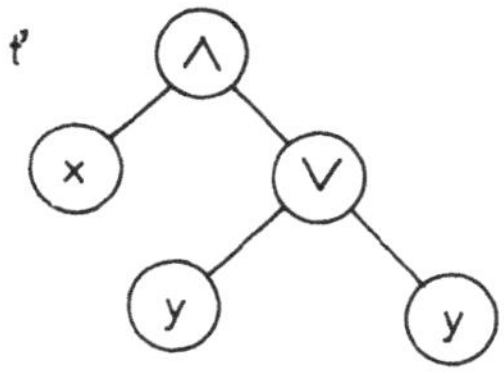

is computed as follows

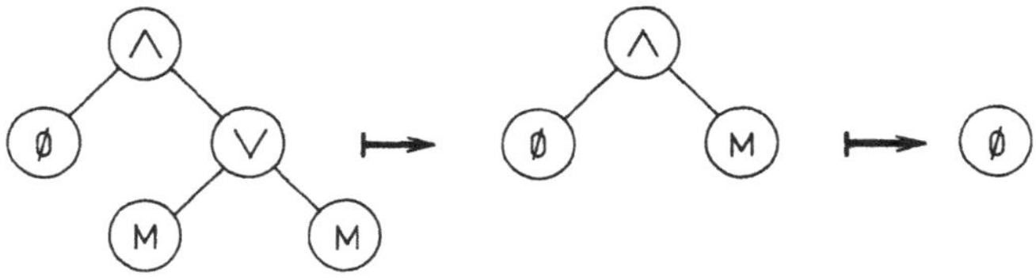

and $(t')\beta = a_0$.

Let us define *majority trees* in $\{x, y\}^{\#}$ by induction: y is a majority tree, and $\vee\ (t_1, t_2)$ is a majority tree iff t_1 or t_2 is, while $\wedge\ (t_1, t_2)$ is a majority tree iff both t_1 and t_2 are. Then

$$\beta\colon \{x, y\}^{\#} \to \{a_0, a_1, b\}$$

is defined by

$$(t)\beta = \begin{cases} a_1 & t \text{ is a majority tree} \\ a_0 & \text{else.} \end{cases}$$

1.8. Let Σ be a finitary type, and let

$$A = (Q, \{\delta_\sigma\}, \Gamma, \gamma, I, \lambda) \text{ and } A' = (Q', \{\delta'_\sigma\}, \Gamma', \delta', I', \lambda')$$

be Σ-tree automata. A *morphism* from A to A' is a triple

$$(f, f_{in}, f_{out}) \colon A \to A'$$

of maps such that

$$f \colon (Q, \{\delta_\sigma\}) \to (Q', \{\delta_\sigma\})$$

is a homomorphism and $f_{\text{in}} \colon I \to I'$, $f_{\text{out}} \colon \Gamma \to \Gamma'$ fulfil

$$f \cdot \gamma' = \gamma \cdot f_{\text{out}},$$

$$\lambda \cdot f = f_{\text{in}} \cdot \lambda'.$$

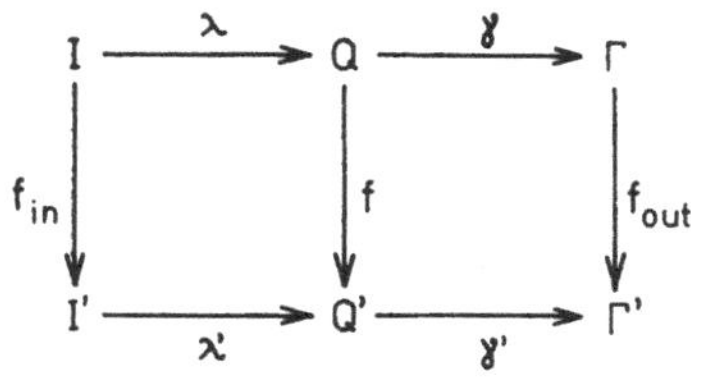

In case $\Gamma = \Gamma'$ and $f_{\text{out}} = \text{id}_\Gamma$ as well as $I = I'$ and $f_{\text{in}} = \text{id}_I$, we write simply

$$f \colon A \to A'.$$

Examples. (i) In II.1.6 we had the automaton $A = (Z_4, \{+, 1\}, \{0, 1\}, \gamma, \{x, y\}, \lambda)$. Define an analogous automaton

$$A' = (Z_2, \{+, 1\}, \{0, 1\}, \gamma', \{x, y\}, \lambda')$$

where $+$ is the addition modulo 2 on $Z_2 = \{0, 1\}$, γ' is the identity map and $(x)\lambda' = 1$, $(y)\lambda' = 0$. Then we have a morphism

$$f \colon A \to A'$$

defined by

$$f \colon \frac{0 \quad 1 \quad 2 \quad 3}{0 \quad 1 \quad 0 \quad 1}$$

Note that A and A' have the same behavior β.

(ii) Define an automaton A' analogously as A in II.1.7 except that the output alphabet is $\Gamma' = \{a, b\}$ and

$$(T)\gamma' = \begin{cases} a & \text{if } T = \emptyset \text{ or } M, \\ b & \text{if } T \in \exp M - \{M, \emptyset\}. \end{cases}$$

Let $f_{\text{out}}\colon \Gamma \to \Gamma'$ be the following map

$$(a_0)f_{\text{out}} = (a_1)f_{\text{out}} = a \quad \text{and} \quad (b)f_{\text{out}} = b.$$

Then

$$(\text{id}_{\exp M}, \text{id}_{\{x, y\}}, f_{\text{out}})\colon A \to A'$$

is a morphism. The behavior of A' is the constant map to a.

1.9. Proposition. For each morphism

$$f\colon A \to A'$$

of Σ-tree automata, A and A' have the same behavior, and if $\rho\colon I^{\#} \to Q$ is the run map of A, then $\rho \cdot f\colon I^{\#} \to Q'$ is the run map of A'.

Proof. Proving the latter statement, the former follows:

$$\beta = \rho \cdot \gamma = \rho \cdot f \cdot \gamma' = \rho' \cdot \gamma' = \beta'.$$

It is sufficient to note that since $\rho\colon (I^{\#}, \{\varphi_\sigma\}) \to (Q, \{\delta_\sigma\})$ and $f\colon (Q, \{\delta_\sigma\}) \to (Q', \{\delta'_\sigma\})$ are both homomorphisms, $\rho \cdot f\colon (I^{\#}, \{\varphi_\sigma\}) \to (Q', \{\delta'_\sigma\})$ is also a homomorphism. For each $x \in I$, we have

$$(x)\rho \cdot f = (x)\lambda \cdot f = (x)\lambda'$$

and hence, $\rho \cdot f$ is an extension of $\lambda'\colon I \to Q'$. But also the run map ρ' of A' is an extension of λ', and since λ' has a unique extension to a homomorphism, we conclude that

$$\rho \cdot f = \rho'. \qquad \square$$

1.10. We have seen in this section that there are close analogies between sequential automata and tree automata. We shall see more of these analogies in the subsequent sections. Let us conclude by having a look on the role of the variables.

Given a type Σ of algebras and a set I of variables ($I \cap \Sigma = \emptyset$), we can extend Σ by "co-opting" the variables. Let Σ' be the following type

$$\Sigma'_0 = \Sigma_0 \cup I \quad \text{and} \quad \Sigma'_n = \Sigma_n \quad \text{for all } n > 0.$$

A Σ'-algebra on a set Q is given by a Σ-algebra on Q plus a map $\lambda\colon I \to Q$. Thus, Σ-automata with the set I of variables are precisely Σ'-automata without variables (i.e. with the empty set of variables). Also behaviors correspond naturally:

Observation. The free Σ-algebra $(I^\#, \{\varphi_\sigma\})$ is precisely the initial Σ'-algebra (II.1.5). In fact, $I^\#$ consists of finite Σ-trees with leaves labelled in $\Sigma_0 \cup I = \Sigma_0'$, and this is precisely $(\emptyset^\#, \{\varphi_\sigma\})$ in the extended type Σ'.

Remark. The concept of variables for tree automata is nevertheless useful, since we want to consider a fixed type Σ. It will turn out that both for minimal realizations (II.2) and for recognizability of languages (II.4) it is of crucial importance that we can enlarge the set of variables. In the approach without variables, this would change the type.

Example. Sequential automata: Here $\Sigma = \Sigma_1$ and $I = \{x\}$. We can consider them as Σ'-automata without variables where $\Sigma' = \Sigma_1 \cup \{x\}$. In fact, the nullary operation x is just the initial state.

Exercises II.1

A. Composition of morphisms. Consider tree automata of a given type and with given Γ and I.

(i) Prove that the composition of morphisms $f: A \to A'$ and $g: A' \to A''$ is a morphism $f \cdot g: A \to A''$.

(ii) For a surjective morphism $f: A \to A'$, conversely, a map g is a morphism $g: A' \to A''$ whenever $f \cdot g: A \to A''$ is a morphism.

(iii) Can (ii) be generalized to the morphisms $(f, f_{\text{in}}, f_{\text{out}})$?

B. Subalgebras of $I^\#$. A subalgebra of a Σ-algebra $(Q, \{\delta_\sigma\})$ is a subset $Q_0 \subset Q$ closed under the operations [i.e., $(q_i)\delta_\sigma \in Q_0$ for all $\sigma \in \Sigma$, $q_i \in Q_0$].

(i) Prove that the set of all trees of depth ≥ 3 is a subalgebra of $I^\#$.

(ii) For each $x \in I$ prove that the set of all trees with x as a label of some leaf is a subalgebra of $I^\#$.

(iii) Prove that each subalgebra of $I^\#$, containing all of I, is $I^\#$.

C. Uniform trees are trees such that all leaves have the same distance from the root. Formulate this precisely. For which types Σ is each tree in $I^\#$ uniform? For which types Σ do all uniform trees form a subalgebra od $I^\#$?

D. Run maps and morphisms. (i) Generalize Proposition II.1.9 to morphisms $(f, f_{\text{in}}, f_{\text{out}},): A \to A'$.

(ii) Conclude that given a Σ-tree automaton $A = (Q, \{\delta_\sigma\}, \Gamma, \gamma, I, \lambda)$ and a subset $I_0 \subset I$, then the run map of the corresponding automaton $A_0 = (Q, \{\delta_\sigma\}, \Gamma, \gamma, I_0, \lambda_0)$ (where λ_0 is the restriction of λ) is the restriction of the run map of A.

E. Sequential automata with resets. A *reset* in a sequential automaton is an input σ such that the map $(-, \sigma)\delta : Q \to Q$ is constant.

(i) Let $\Sigma = \Sigma_0 \cup \Sigma_1$ be a type (with arities 0 or 1). Verify that Σ-tree automata are just sequential Σ-automata with resets in Σ_0.

(ii) Describe the initial algebra for the type in (i). (Hint: Using II.1.10, we get $\emptyset^{\#} = \Sigma_0 \times \Sigma_1^{\#}$.)

II.2. Minimal Realization

2.1. Throughout this section, a fixed finitary type Σ is considered. Analogously to the case of sequential automata, we show that each behavior map, i.e., a map

$$\beta: I^{\#} \to \Gamma$$

has a realization (i.e. there exists a Σ-tree automaton with the behavior β) and we then apply the minimization procedure to get the minimal realization of β.

We start by defining the *free realization*

$$A(\beta) = (I^{\#}, \{\varphi_\sigma\}_{\sigma \in \Sigma}, \Gamma, \beta, I, \eta),$$

where $\eta : I \to I^{\#}$ is the inclusion map (see II.1.5 (ii)). Since the run map of $A(\beta)$ is clearly $\mathrm{id}_{I^{\#}}$, it realizes β. Let

$$A = (Q, \{\delta_\sigma\}, \Gamma, \gamma, I, \lambda)$$

be another realization of β. The run map of A is a morphism

$$\rho : A(\beta) \to A.$$

In fact,

(i) $\rho : (I^{\#}, \{\varphi_\sigma\}) \to (Q, \{\delta_\sigma\})$ is a homomorphism

with

(ii) $\eta \cdot \rho = \lambda$,

and since A realizes β, we also have

(iii) $\rho \cdot \gamma = \beta$.

Since (i) and (ii) actually define the run map, we see that ρ is the *unique* morphism $A(\beta) \to A$.

2.2. We turn to the minimization procedure. As in the case of sequential automata, the minimization of a tree automaton is performed in two steps : the first (easy) one is to discard all superflouous states, and the latter is to merge pairs of states which behave in the same way.

By a *subautomaton* of a Σ-tree automaton $A = (Q, \{\delta_\sigma\}, \Gamma, \gamma, I, \lambda)$ we understand a Σ-tree automaton (with the same Γ and I)

$$A' = (Q', \{\delta'_\sigma\}, \Gamma, \gamma', I, \lambda')$$

such that;

(i) Q' is a subalgebra (Exercise II.1.B);

(ii) Q' contains $(I)\lambda$ and λ' is range-restriction of λ;

(iii) γ' is the restriction of γ.

Shortly, A' is a subautomaton of A if $Q' \subset Q$ and the inclusion map $i : Q' \rightarrow Q$ is a morphism

$$i : A' \rightarrow A.$$

It follows that the behavior of an automaton A is the same as that of each subautomaton (II.1.9).

A Σ-tree automaton is *reachable* if each state is the result of the computation of a tree; in other words, if ρ is surjective. The automaton in II.1.3 is clearly reachable.

Proposition. Each Σ-tree automaton has a unique reachable subautomaton which is also its smallest subautomaton.

Proof. Let $A = (Q, \{\delta_\sigma\}, \Gamma, \gamma, I, \lambda)$ be a Σ-tree automaton. The image of its run map

$$Q_0 = (I^{\#})\rho \subset Q$$

is a subalgebra of $(Q, \{\delta_\sigma\})$. Indeed, given $\sigma \in \Sigma_n$ and $q_i \in Q_0$ $(i < n)$, we have trees $t_i \in I^{\#}$ with

$$(t_i)\rho = q_i \qquad (i < n).$$

Put

$$t = (t_i)_{i<n}\varphi_\sigma \in I^{\#}.$$

Since ρ is a homomorphism, we have

$$\begin{aligned}(t)\rho &= ((t_i)\varphi_\sigma)\rho \\ &= ((t_i)\rho)\delta_\sigma \\ &= (q_i)\delta_\sigma;\end{aligned}$$

hence, $(q_i)_{i<n}\,\delta_\sigma \in Q_0$.

Thus, we obtain a subautomaton A_0 of the automaton A with the state set Q_0. Then A_0 is reachable since its run map is a restriction of ρ (II.1.9).

A_0 is the smallest subautomaton of A because for each subautomaton A'_0 with the state set Q'_0 we have $(I)\lambda \subset Q'_0$, and this implies $Q_0 = (I^{\#})\rho \subset Q'_0$ (because the run map of A'_0 is also a restriction of ρ). □

Remark. The subautomaton above is called the *reachable part* of A. For example, the automaton in II.1.7 is not reachable (if M has more than 1 point). The reachable part of A is the subautomaton with two states, $\emptyset$ and M.

A reachable automaton has no proper subautomata. The free realization $A(\beta)$ is an example of a reachable automaton.

For the next step of minimization, we need the concept of an "admissible" equivalence $\sim$ on a Σ-tree automaton A. Our aim is to construct a new (smaller) automaton $A/{\sim}$ with the state set $Q/{\sim}$ (of all equivalence classes $[q]$ of states $q \in Q$), and with the structure derived from A. That is

$$A/{\sim} = (Q/{\sim}, \{\bar{\delta}_\sigma\}, \Gamma, \bar{\gamma}, I, \bar{\lambda})$$

where

(a) $([q_0], \ldots, [q_{k-1}])\bar{\delta}_\sigma = [(q_0, \ldots, q_{k-1})\delta_\sigma]$;
(b) $([q])\bar{\gamma} = (q)\gamma$;
(c) $(x)\bar{\lambda} = [(x)\lambda]$.

Thus, "admissible" are those equivalences for which (a) and (b) are well-defined:

Definition. A *congruence* on a Σ-tree automaton $A = (Q, \{\delta_\sigma\}, \Gamma, \gamma, I, \lambda)$ is an equivalence $\sim$ on Q such that

(A) given $\sigma \in \Sigma_k$ and $q_i \sim q'_i$ in Q $(i = 0, \ldots, k-1)$, then

$$(q_0, \ldots, q_{k-1})\delta_\sigma \sim (q'_0, \ldots, q'_{k-1})\delta_\sigma;$$

(B) given $q \sim q'$ in Q, then

$$(q)\gamma \sim (q')\gamma.$$

For each congruence $\sim$, we define the *quotient automaton* $A/{\sim}$ by (a), (b) and (c) above.

Example. For the automaton of II.1.6, the equivalence with two "parity" classes

$$\{0, 2\} \text{ and } \{1, 3\}$$

is a congruence (because the operation $+$ respects parity, the nullary operation τ makes no difference, and γ is the parity map). The quotient automaton is

$$A/{\sim} = (\{[0], [1]\}, \{+, [1]\}, \{0, 1\}, \gamma, \{x, y\}, \lambda),$$

where

$$([0])\gamma = 0 \quad \text{and} \quad ([1])\gamma = 1$$

and

$$(x)\lambda = [1] \quad \text{and} \quad (y)\lambda = [0].$$

Note that $A/{\sim}$ has the same behavior as A.

Remark. Congruences are closely related to morphisms:

(i) For each congruence $\sim$ on a tree automaton A, the canonical map $c: Q \to Q/\sim$ (with $(q)c = [q]$) is a morphism

$$c: A \to A/\sim.$$

Hence, A has the same behavior as any quotient automaton (see II.1.9).

(ii) For each morphism

$$f: A \to A'$$

of Σ-tree automata, the kernel equivalence

$$q_1 \sim q_2 \quad \text{iff} \quad (q_1)f = (q_2)f$$

is a congruence on A.

2.3. The concept of two states having the same behavior was crucial in I.2. We consider the corresponding concept of interchangeable states q_1 and q_2: these are states for which a substitution of q_1 for q_2 in the interpretation of variables does not influence the behavior.

Definition. Given states $q_1, q_2 \in Q$ of a Σ-tree automaton $A = (Q, \{\delta_\sigma\}, \Gamma, \gamma, I, \lambda)$, choose a variable $y \notin I$, and put

$$A_i = (Q, \{\delta_\sigma\}, \Gamma, \gamma, I \cup \{y\}, \lambda_i) \qquad \text{for } i = 1, 2$$

where λ_i extends λ by $(y)\lambda_i = q_i$. If A_1 and A_2 have the same behavior, we say that q_1 and q_2 are *interchangeable.*

Remark. Interchangeable states have the same output : if β_i is the behavior of A_i above, then

$$(q_1)\gamma = (y)\beta_1 = (y)\beta_2 = (q_2)\gamma.$$

Example. In the automaton of II.1.3, any two even numbers are interchangeable (and so are any two odd ones). The states 1 and 2 are not interchangeable because $(1)\gamma \neq (2)\gamma$.

Proposition. For each congruence on a Σ-tree automaton, any two congruent states are interchangeable.

Proof. Let $\sim$ be a congruence on A. Then $\sim$ is clearly a congruence on each of the automata A_i above ($i = 1, 2$). Therefore, A_i has the same behavior as $A_i/\sim$. Since $[q_1] = [q_2]$, clearly $A_1/\sim$ is the same automaton as $A_2/\sim$. □

2.4. We are now ready to minimize reachable tree automata: we prove that interchangeability is a congruence and hence, the largest congruence. Consequently, the quotient automaton is minimal.

Definition. A *reduction* of a tree automaton A is a surjective morphism $e : A \twoheadrightarrow A'$. The *minimal reduction* is a reduction $e_0 : A \twoheadrightarrow A_0$ such that for each reduction $e : A \twoheadrightarrow A'$ we can further reduce A' to A_0, i.e., there exists a reduction $f : A' \twoheadrightarrow A_0$ with $e_0 = e \cdot f$.

Theorem. Let Σ be a finitary type. Each reachable Σ-tree automaton A has a minimal reduction $c : A \twoheadrightarrow A/\sim$ obtained from the following congruence

$$q_1 \sim q_2 \quad \text{iff} \quad q_1 \text{ and } q_2 \text{ are interchangeable.}$$

Proof. I. The interchangeability equivalence $\sim$ is a congruence. In fact, by Remark II.2.3 we know that $\sim$ respects the outputs; it is sufficient to prove that it respects the operations. Let $\sigma \in \Sigma_k$ and $q_i \sim q'_i$ in Q be given for $i = 0, \ldots, k-1$. To prove that

$$(q_0, \ldots, q_{k-1})\delta_\sigma \sim (q'_0, \ldots, q'_{k-1})\delta_\sigma,$$

it is clearly sufficient to show that for each $n = 0, \ldots, k-1$,

$$(q_0, \ldots, q_{n-1}, q_n, q'_{n+1}, \ldots, q'_{k-1})\delta_\sigma \sim (q_0, \ldots, q_{n-1}, q'_n, q'_{n+1}, \ldots, q'_{k-1})\delta_\sigma.$$

Denote by $d : Q \to Q$ the map defined by

$$(q)d = (q_0, \ldots, q_{n-1}, q, q'_{n+1}, \ldots, q'_{k-1})\delta_\sigma \qquad (q \in Q).$$

The proof of I. will be concluded if we show that

$$(*) \qquad q \sim q' \quad \text{implies} \quad (q)d \sim (q')d \qquad (q, q' \in Q).$$

For each state q denote by

$$\lambda_q : \mathrm{I} \cup \{y\} \to Q$$

the extension of λ (the initialization map of A) with

$$(y)\lambda_q = q.$$

The "extended" automaton has run map $\lambda_q^{\#}$ and behavior $\lambda_q^{\#} \cdot \gamma : (I \cup \{y\})^{\#} \to \Gamma$. Thus, $(*)$ states that

$$\lambda_q^{\#} \cdot \gamma = \lambda_{q'}^{\#} \cdot \gamma \quad \text{implies} \quad \lambda_{(q)d}^{\#} \cdot \gamma = \lambda_{(q')d}^{\#} \cdot \gamma.$$

Since A is reachable, there exist trees $s_i, s'_i \in I^{\#}$ with

$$q_i = (s_i)\lambda^{\#} \quad \text{and} \quad q'_i = (s'_i)\lambda^{\#} \qquad (i = 0, \ldots, k-1).$$

Since each λ_q extends λ, we also have $q_i = (s_i)\lambda_q^{\#}$ *and* $q'_i = (s'_i)\lambda_q^{\#}$ (see Exercise II.1.D). Put

$$s = (s_0, \ldots, s_{n-1}, y, s'_{n-1}, \ldots, s'_{k-1})\varphi_\sigma \in (I \cup \{y\})^{\#}.$$

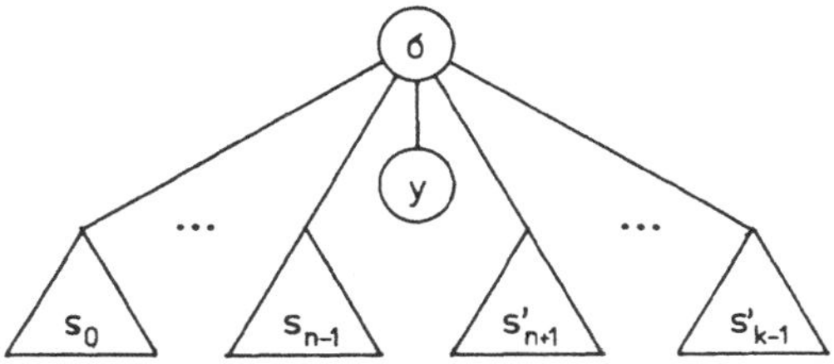

Then for each $q \in Q$,

$$(s)\lambda_q^\# = ((s_0)\lambda_q^\#, \ldots, (s_{n-1})\lambda_q^\#, q, (s'_{n+1})\lambda_q^\#, \ldots, (s'_{k-1})\lambda_q^\#)\delta_\sigma = (q)d.$$

Denote by

$$h : (I \cup \{y\})^\# \rightarrow (I \cup \{y\})^\#$$

the unique homomorphism with

$$(x)h = x \quad (x \in I) \quad \text{and} \quad (y)h = s.$$

Then

$$h \cdot \lambda_q^\# = \lambda_{(q)d}^\# : (I \cup \{y\})^\# \rightarrow Q.$$

In fact, $h \cdot \lambda_q^\#$ is a homomorphism which coincides with the homomorphism $\lambda_{(q)d}^\#$ on the set of all variables:

$$(x)h \cdot \lambda_q^\# = (x)\lambda_q^\# = (x)\lambda = (x)\lambda_{(q)d}^\# \qquad (x \in I)$$

and

$$(y)h \cdot \lambda_q^\# = (s)\lambda_q^\# = (q)d = (y)\lambda_{(q)d}^\#.$$

Thus, if $\lambda_q^\# \cdot \gamma = \lambda_{q'}^\# \cdot \gamma$, then

$$\lambda_{(q)d}^\# \cdot \gamma = h \cdot \lambda_q^\# \cdot \gamma = h \cdot \lambda_{q'}^\# \cdot \gamma = \lambda_{(q')d}^\# \cdot \gamma.$$

II. The reduction $c : A \rightarrow A/\sim$ is minimal. In fact, for each reduction $e : A \rightarrow A'$, the kernel equivalence $\approx$ of e is a congruence on A and hence, by Proposition II.2.3,

$$q \approx q' \quad \text{implies} \quad q \sim q' \qquad (q, q' \in Q).$$

Thus, we can define a map f by the condition

$$c = e \cdot f,$$

i.e., for each $q \in Q$,

$$((q)e)f = [q].$$

Since e is a surjective morphism, and c is a morphism, it follows that f is a (surjective) morphism by Exercise II.1.A. □

Example. The minimal reduction of the automaton II.1.3 has two states, [0] (= the class of all even numbers) and [1] (= the class of all odd numbers). Here

$$[0] + [0] = [0 + 0] = [0] \qquad [1] + [1] = [1 + 1] = [0];$$
$$[0] + [1] = [0 + 1] = [1] \qquad [1] + [0] = [1 + 0] = [1].$$

The output is given by $([0])\gamma = 0$ and $([1])\gamma = 1$, the initialization by $(x)\lambda = (y)\lambda = [1]$.

2.5. We now apply the minimization procedure to the free realization $A(\beta)$ of II.2.1. Given a state (i.e., a tree) $t \in I^{\#}$, we consider the behavior of the automaton

$$A(\beta)_t = ((I \cup \{y\})^{\#}, \{\varphi_\sigma\}, \Gamma, \beta, I \cup \{y\}, \lambda_t)$$

where

$$(x)\lambda_t = \begin{cases} x & \text{for } x \in I \\ t & \text{for } x = y. \end{cases}$$

The morphism

$$\lambda_t^{\#} : (I \cup \{y\})^{\#} \to I^{\#}$$

changes each tree $s \in (I \cup \{y\})^{\#}$ by substituting every y-labelled leaf by the tree t. We use the notation

$$t \cdot_y s = (s)\lambda_t^{\#}.$$

Thus, for each node $w \in D_t$ we have

$$(w)(t \cdot_y s) = \begin{cases} (w)s & \text{if } (w)s \neq y \\ (u)t & \text{if } w = v \cdot u \text{ and } (v)s = y \\ \text{undefined} & \text{else.} \end{cases}$$

Example: For

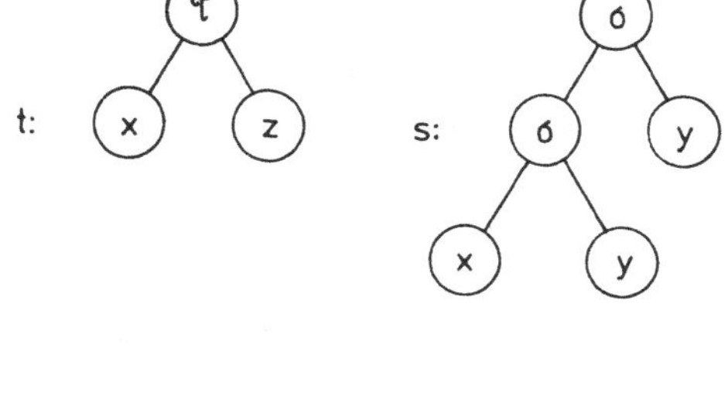

we have

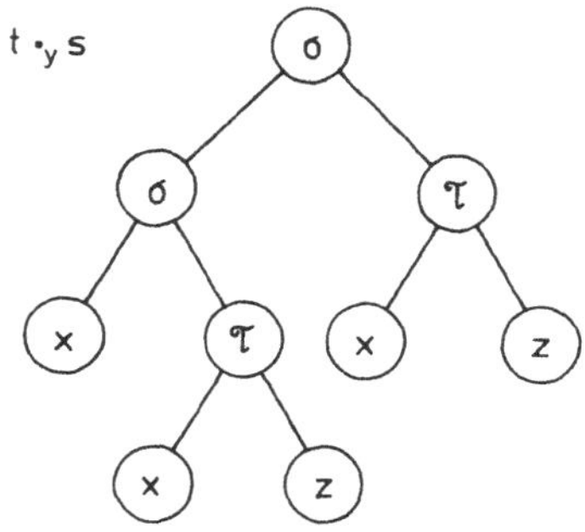

The behavior of $A(\beta)_t$ is the map $s \mapsto (t \cdot_y s)\beta$. Therefore, two trees $t_1, t_2 \in I^\#$ are interchangeable in $A(\beta)$ iff

$$(t_1 \cdot_y s)\beta = (t_2 \cdot_y s)\beta \qquad \text{for each } s \in (I \cup \{y\})^\#.$$

Construction of the minimal realization of a map

$$\beta : I^\# \to \Gamma.$$

We define the *Nerode equivalence* $\approx$ on $I^\#$ by

$$t_1 \approx t_2 \quad \text{iff} \quad (t_1 \cdot_y s)\beta = (t_2 \cdot_y s)\beta \quad \text{for all } s \in (I \cup \{y\})^\#$$

(where $y \notin I$). The minimal realization of β is the Σ-tree automaton

$$A_0 = (I^\# /\approx, \{\bar{\varphi}_\sigma\}, \Gamma, \bar{\beta}, I, \bar{\lambda})$$

with

$$([t_0], \ldots, [t_{|\sigma|-1}])\bar{\varphi}_\sigma = [(t_0, \ldots, t_{|\sigma|-1})\varphi_\sigma],$$
$$([t])\bar{\beta} = (t)\beta$$

and

$$(x)\bar{\lambda} = [x].$$

Proof. Since the Nerode equivalence is just the interchangeability congruence on $A(\beta)$, we know already that A_0 is a minimal reduction of $A(\beta)$. For each reachable realization A of β, the run map is a morphism

$$\rho : A(\beta) \to A$$

(see II.2.1). Since ρ is surjective, A is a reduction of $A(\beta)$, and we can reduce A to A_0. □

Example. Put

$$\Sigma = \Sigma_2 = \{\sigma, \tau\}.$$

We construct the minimal realization of the behavior

$$\beta: \{x, u, v\}^\# \to \{0, 1\}$$

defined as follows:

$$(t)\beta = \begin{cases} 1 & \text{if } t \text{ has the label } \sigma\text{, and the left-most leaf is } x \\ 0 & \text{else.} \end{cases}$$

We inspect the Nerode equivalence classes of the simplest trees.

First, clearly

$$[x] \neq [u] = [v].$$

The class $[u]$ consists of those trees which neither have the label σ, nor have their left-most leaf x. Further, the tree

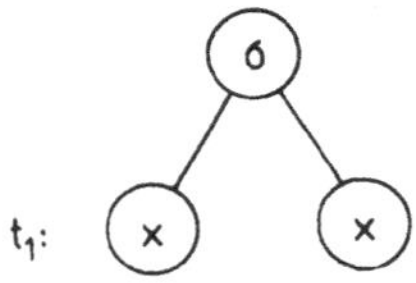

is not equivalent to x or u because $(t_1)\beta = 1$; the class $[t_1]$ is precisely $(1)\beta^{-1}$. The next tree

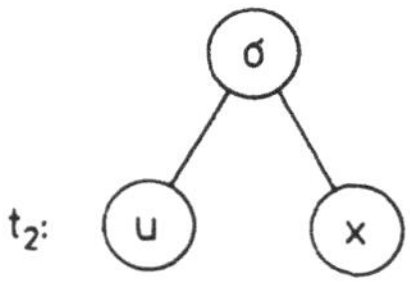

is non-equivalent to x or u: given $y \in \{x, u, v\}$, consider the following trees

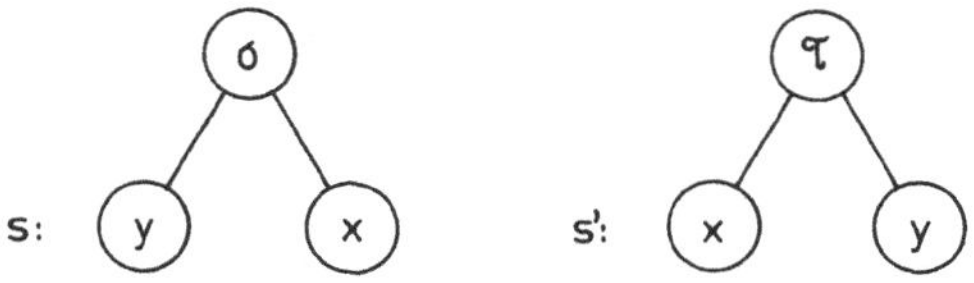

Then $(t_2 \cdot_y s)\beta \neq (x \cdot_y s)\beta$ and $(t_2 \cdot_y s')\beta \neq (x \cdot_y s')\beta$. The class $[t_2]$ consists of all trees having the label σ but not having the left-most leaf x. Finally, $[x]$ is the class of all trees not having the label σ but having the left-most leaf x. We see that

$$\{x, u, v\}^\# = [x] \cup [u] \cup [t_1] \cup [t_2].$$

The minimal realization of β has the state set

$$Q = \{[x], [u], [t_1], [t_2]\}.$$

The operation σ depends on the first variable only:

$$([x], -)\sigma = ([t_1], -)\sigma = [t_1]$$

and

$$([u], -)\sigma = ([t_2], -)\sigma = [t_2].$$

Also τ depends on the first variable only: it is the first projection. Further, γ and λ are given by the following tables

γ	$[x]$	$[u]$	$[t_1]$	$[t_2]$
	0	0	1	0

λ	x	u	v
	$[x]$	$[u]$	$[u]$

Exercises II.2

A. Minimal realization. Find the minimal realization of the following behaviors:

(i) $\Sigma = \Sigma_3 = \{\sigma\}$, $\beta: \{x\}^\# \to \{0, 1\}$, $(t)\beta = 1$ iff the depth of t is at least 2.

(ii) $\Sigma_n = \{\sigma_n\}$ for all $n = 0, 1, 2, \ldots$, $\beta: \{x\}^\# \to \{0, 1\}$, $(t)\beta = 1$ iff the root has more immediate successors than any other node.

(iii) $\Sigma = \Sigma_2 = \{\sigma, \tau\}$, $\beta: \{x, y\}^\# \to \{0, 1, 2\}$, $(t)\beta = 0$ if t has only labels σ and x, $(t)\beta = 1$ if t has the label τ but not y, and else $(t)\beta = 2$.

B. Generation. A subalgebra Q' of a Σ-algebra $(Q, \{\delta_\sigma\})$ is said to be *generated* by a set $M \subset Q$ if Q' is the least subalgebra with $M \subset Q'$.

(i) Verify that the reachable part of a Σ-tree automaton is just the subalgebra generated by $(I)\lambda \subset Q$.

(ii) Describe the subalgebra of $(I^\#, \{\varphi_\sigma\})$ generated by $\{x\}$ for a given $x \in I$, and that generated by $\emptyset$.

C. Cogeneration. A congruence $\approx$ on a Σ-algebra $(Q, \{\delta_\sigma\})$ is said to be *cogenerated* by an equivalence $\sim$ (on Q) if $\approx$ is the largest congruence contained in $\sim$ (i.e., such that $q \approx q'$ implies $q \sim q'$).

(i) Verify that the Nerode equivalence of β is cogenerated by the kernel equivalence of β (on $I^\#$).

(ii) Describe the congruence cogenerated by an arbitrary equivalence on an arbitrary Σ-algebra. (Hint: See II.2.4.)

II.3. Infinitary Tree Automata

3.1. In this section we consider types Σ with infinitary arities of operations. The definition of Σ-tree automata is naturally extended to these types, and the run map is defined on the (free) Σ-algebra $I^{\#}$ of all finite-path Σ-trees. It turns out, however, that minimization cannot be extended: there are behaviors which do not have a minimal realization.

3.2. By a *type* we understand a set Σ of operation symbols, together with an arity map assigning to each $\sigma \in \Sigma$ a cardinal number $|\sigma|$. Again, Σ_n denotes the set of all n-ary symbols, $\Sigma_n = \{\sigma \in S; |\sigma| = n\}$. A Σ-algebra is a set Q equipped with operations

$$\delta_\sigma\colon Q^n \to Q \qquad\qquad (\sigma \in \Sigma, |\sigma| = n),$$

where the elements of the n-fold cartesian product Q^n are all n-tuples

$$(q_i)_{i<n}$$

with $q_i \in Q$ for each $i < n$. A homomorphism

$$f\colon (Q, \{\delta_\sigma\}_{\sigma \in S}) \to (Q', \{\delta'_\sigma\}_{\sigma \in \Sigma})$$

is a map such that for each $\sigma \in \Sigma_n$,

$$(q_i)_{i<n}\delta_\sigma = q \quad \text{implies} \quad ((q_i)f)_{i<n}\delta'_\sigma = (q)f.$$

A *Σ-tree automaton* is defined precisely as in II.1.2: it is a Σ-algebra $(Q, \{\delta_\sigma\}_{\sigma \in \Sigma})$ together with an output map $\gamma\colon Q \to \Gamma$ and an initialization map $\lambda\colon I \to Q$.

Example. Let $\Sigma = \{\sigma\}$ where σ is ω-ary. Put

$$A = (\omega \cup \{\infty\}, \delta_\sigma, \{0, 1\}, \gamma, \{x\}, \lambda)$$

where δ_σ is the following operation on sequences in $\omega \cup \{\infty\}$:

$$(q_0, q_1, q_2, \ldots)\delta_\sigma = \left(\bigvee_{n=0}^{\infty} q_n\right) + 1$$

(i.e., δ_σ is the maximum plus 1 if the sequence is bounded, else $\delta_\sigma = \infty$). Further,

$$(q)\gamma = \begin{cases} 0 & \text{if } q \in \omega \\ 1 & \text{if } q = \infty \end{cases}$$

and $(x)\lambda = 0$.

3.3. The construction of the free algebra $I^{\#}$ (formed, for finitary types, by all *finite* Σ-trees over I) requires a revision.

If $\sigma \in \Sigma$ is an infinitary operation, then the trees in a free Σ-algebra will no longer be finite—consider the following tree:

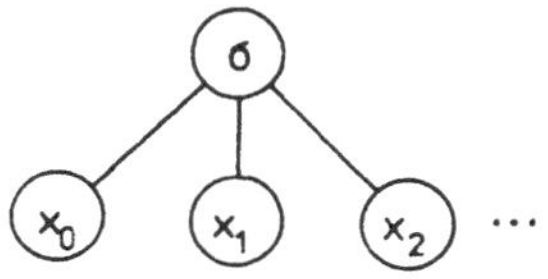

Neither are they going to have finite depths—consider the following tree with λ unary:

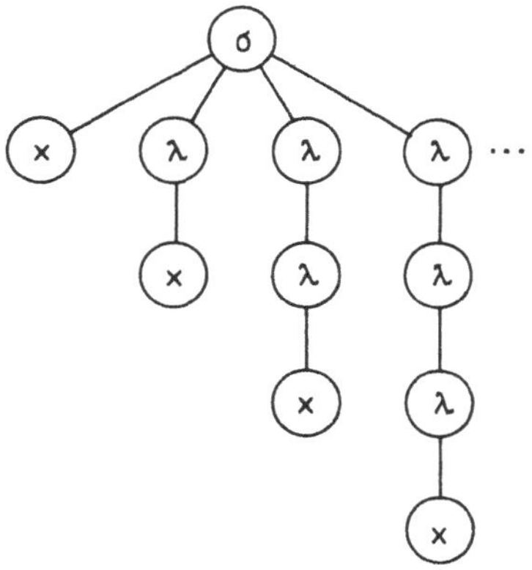

It turns out that the trees in a free Σ-algebra are just those which have finite paths.

As in the finitary case, we shall work with the complete m-ary tree

$$m^*$$

with $m = \bigvee_{\sigma \in \Sigma} |\sigma|$. (**Remark**: The case $m = \omega$ can indicate either that Σ is an infinitary type with all operations at most ω-ary, or that Σ is a finitary type with unbounded arities.)

A *path* in the tree m^* (from the root $\emptyset$ downwards) is a sequence of elements of m^* of the following form

$$\emptyset, p_1, p_1p_2, p_1p_2p_3, \ldots$$

where $p_1, p_2, p_3, \ldots$ are elements of m (i.e., ordinals smaller than m). An example of a path in ω^*:

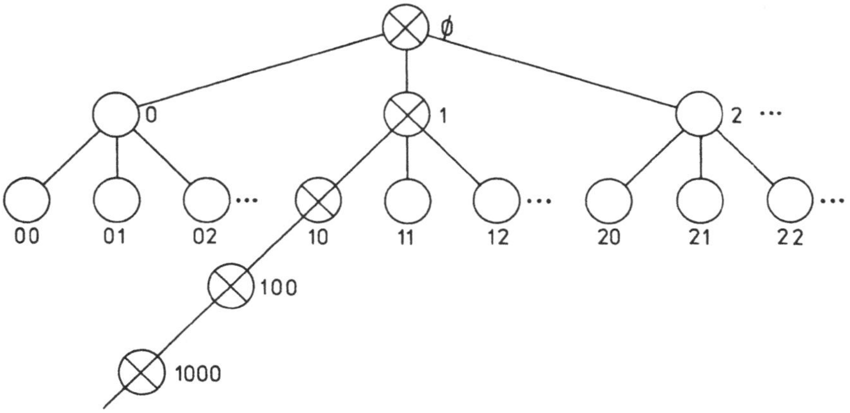

3.4. Definition. Let Σ be a type and let I be a set (of variables) with $\Sigma \cap I = \emptyset$; put $m = \bigvee_{\sigma \in \Sigma} |\sigma|$. A *finite-path Σ-tree over I* (shortly, a Σ-tree) is a partial map $t: m^* \to I \cup \Sigma$ such that

(i) the domain of definition D_t is non-empty and its intersection with any path in m^* is finite;

(ii) given $p_1 \dots p_k p_{k+1} \in m^*$, then

$$p_1 \dots p_{k+1} \in D_t \quad \text{iff} \quad (p_1 \dots p_k)t \in \Sigma_n \quad \text{for some } n > p_{k+1}.$$

Remark. The condition (i) can [in view of (ii)] be reformulated as follows: $(\emptyset)t$ is defined, and for each path $\emptyset, p_1, p_1 p_2, \dots$ there exists $k_0 \in \omega$ such that $(p_1 p_2 \dots p_k)t$ is defined iff $k \leq k_0$. Note that for Σ finitary, the latter is equivalent to the finiteness of D_t.

The set of all finite-path Σ-trees over I is, again, denoted by $I^\#$.

Example. Let $\Sigma = \{\sigma, \lambda\}$ with $|\sigma| = \omega$ and $|\lambda| = 1$. The following trees

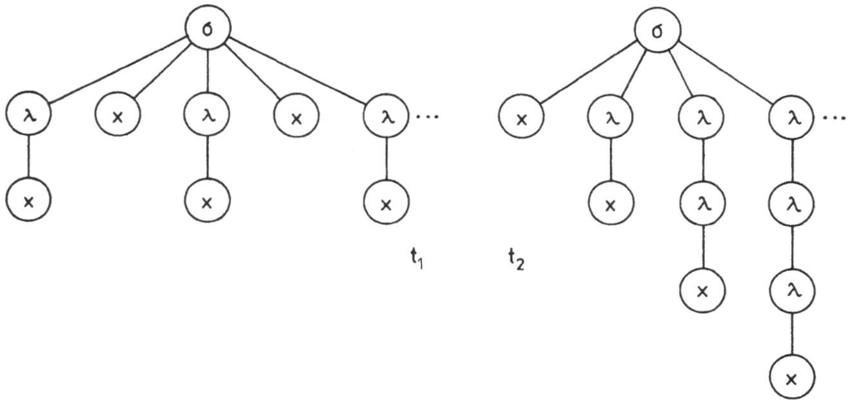

are elements of $\{x\}^\#$. In t_1, for each path we have either $k_0 = 1$ or $k_0 = 2$. In t_2, for each path $\emptyset, p_1, p_1p_2, \ldots$ we have $k_0 = p_1 + 1$. In contrast, the following trees

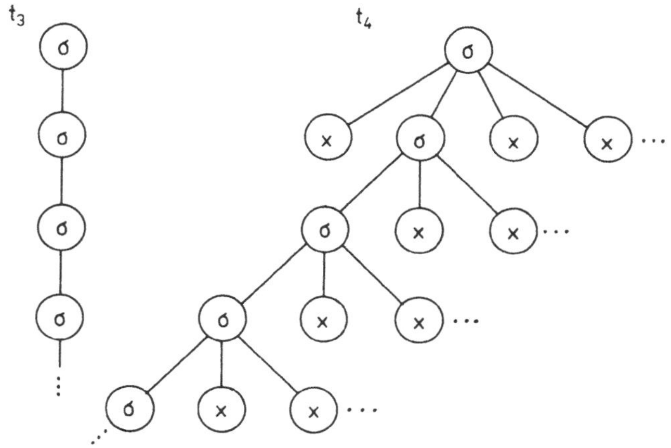

are not elements of $\{x\}^\#$: consider the path $\emptyset$, 0, 00, 000, . . . for t_3 and the path $\emptyset$, 1, 10, 100, 1 000, . . . for t_4.

3.5. The properties of $I^\#$ are analogous in case of infinitary types. The *branch* of a tree $t \in I^\#$ with the root $w \in D_t$ is the tree $\partial_w t$ defined by

$$(v)\partial_w t = (wv)t \quad \text{for all } v \in m^*.$$

Clearly $\partial_w t \in I^\#$. The operations $\varphi_\sigma: (I^\#)^n \to I^\#$, for all $\sigma \in \Sigma_n$, are again defined as follows: given $t_i \in I^\#$, $i < n$, denote by

$$t = (t_i)_{i<n}\varphi_\sigma$$

the tree with

$$D_t = \{p_1p_2 \ldots p_k \in m^*;\ p_1 < n \text{ and } p_2 \ldots p_k \in D_{t_{p_1}}\} \cup \{\emptyset\}$$

where

$$(\emptyset)t = \sigma,$$
$$(p_1p_2 \ldots p_k)t = (p_2 \ldots p_k)t_{p_1}.$$

We must verify the condition (i): For each path $\emptyset, p_1, p_1p_2 \ldots$, either $(p_1)t$ is undefined (then $k_0 = 0$) or $p_1 < n$; since $t_{p_1} \in I^\#$, there is k_0 such that

$$(p_2 \ldots p_k)t_{p_1} \text{ is defined iff } k \leq k_0.$$

Then

$$(p_1p_2 \ldots p_k)t \text{ is defined iff } k \leq k_0.$$

Thus, we get an algebra

$$(I^{\#}, \{\varphi_\sigma\})$$

which clearly has the Peano properties (II.1.5). To prove that this is a free algebra (as in II.1.5), we use *transfinite induction*: in order to define sets W_k for all ordinals k, we must define

(a) W_0;
(b) W_{k+1}, given W_k;
(c) W_k, for each limit ordinal k, given $W_j \quad (j < k)$.

Analogously, to prove a statement about each ordinal k, we must (a) prove the statement for $k = 0$; (b) derive the case $k + 1$ from the case k and (c) for each limit ordinal k, derive the case k from the preceding cases.

We now define sets W_k analogous to those in II.1.5:

$$W_0 = I,$$
$$W_{k+1} = I \cup \{(t_i)_{i<n}\varphi_\sigma;\ \sigma \in \Sigma_n,\ t_i = W_k\}$$

and for each limit ordinal k,

$$W_k = \bigcup_{j<k} W_j.$$

Whereas in the finitary case we had $I^{\#} = W_\omega$, in general we have $I^{\#} = W_k$ for some ordinal k, but the proof is more technical here.

We say that a Σ-tree t has *finite depth* if it is an element of W_k, k finite. Then the depth of t equals to the largest number k for which there exists $p_1 \ldots p_k \in D_t$.

3.6. Theorem. The algebra of finite-path trees

$$(I^{\#}, \{\varphi_\sigma\})$$

is the free Σ-algebra generated by the set I.

Proof. For each Σ-algebra $(Q, \{\delta_\sigma\})$ and each map $f\colon I \to Q$, we are to show that f has a unique extension to a homomorphism $f^{\#}\colon (I^{\#}, \{\varphi_\sigma\}) \to (Q, \{\delta_\sigma\})$. It is sufficient to prove that there exists a cardinal $\bar{m}$ with

$$I^{\#} = W_{\bar{m}}.$$

Then we proceed by extending f to each W_k:

(a) $f^{\#} = f$ on $W_0 = I$;
(b) given $f^{\#}$ on W_k and given $t \in W_{k+1}$, we have $t = (t_i)_{i<n}\varphi_\sigma$ with $t_i \in W_k$, and we put $(t)f^{\#} = ((t_i)f^{\#})_{i<n}\delta_\sigma$;
(c) the extension on the limit ordinals is clear.

The Peano properties guarantee that $f^{\#}$ is well-defined, and (b) implies that $f^{\#}$ is a homomorphism. Moreover, $f^{\#}$ is unique because the rule for the extension above is just a consequence of the expected properties of $f^{\#}$.

We prove

$$I^{\#} = W_{\bar{m}}$$

where $\bar{m}$ is the first regular cardinal larger than

$$m = \bigvee_{\sigma \in \Sigma} |\sigma|.$$

For each tree $t \in I^{\#}$ we define sets

$$(t)S_k \subset m^*$$

by transfinite induction as follows:

$$(t)S_0 = m^* - D_t;$$

$$(t)S_{k+1} = \{p_1 \dots p_s \in m^*; p_1 \dots p_s p_{s+1} \in (t)S_k \text{ for each } p_{s+1} < m\};$$

$$(t)S_k = \bigcup_{j<k} (t)S_j \quad \text{if } k \text{ is a limit ordinal.}$$

We are going to prove the following statements for each tree $t \in I^{\#}$ and each ordinal k:

I. $(t)S_{k'} \subset (t)S_k$ for each ordinal $k' \leq k$;
II. if $(t)S_k = (t)S_{k+1}$, then $\emptyset \in (t)S_k$;
III. if $\emptyset \in (t)S_k$, then $t \in W_k$.

This will prove the proposition: for each $t \in I^{\#}$ we have by I. a monotone chain of sets $(t)S_0 \subset (t)S_1 \subset \dots \subset (t)S_k \subset \dots \subset m^*$; necessarily, there exists an ordinal k of cardinality $\leq m$ (= card m^*) with $(t)S_k = (t)S_{k+1}$. Then II. and III. imply $t \in W_k$. Since m $< \bar{m}$ and $\bar{m}$ is a regular cardinal, the ordinal k is smaller than the ordinal $\bar{m}$ and hence, $W_k \subset W_{\bar{m}}$. This proves $I^{\#} \subset W_{\bar{m}}$.

I. $(t)S_{k'} \subset (t)S_k$. We prove this by transfinite induction on k. This is clear if $k = 0$, since than $k' = k$. Assume that $k' \leq k$ implies $(t)S_{k'} \subset (t)S_k$; to prove that this also holds for $k + 1$, it suffices to prove $(t)S_k \subset (t)S_{k+1}$. Let $p_1 \dots p_s \in (t)S_k$.

(a) If $k \neq 0$, then for each $p_{s+1} \in m$ there exists $k' < k$ with $p_1 \dots p_s p_{s+1} \in (t)S_{k'} \subset (t)S_k$; this proves $p_1 \dots p_s \in (t)S_{k+1}$.

(b) If $k = 0$, then $(p_1 \dots p_s)t$ is undefined and hence, for each $p_{s+1} \in m$ also $(p_1 \dots p_s p_{s+1})t$ is undefined. Thus, $p_1 \dots p_s p_{s+1} \in (t)S_0$, which proves $p_1 \dots p_s \in (t)S_1$.

The proposition is obvious for each limit ordinal k, since $(t)S_k = \bigcup_{k'<k} (t)S_k$.

II. $(t)S_k = (t)S_{k+1}$ implies $\emptyset \in (t)S_k$. Assume the contrary: there exists a finite-path Σ-tree $t \in I^{\#}$ and an ordinal k with

$$\emptyset \notin (t)S_k = (t)S_{k+1}.$$

The statement $\emptyset \notin (t)S_{k+1}$ means that there exists a $p_1 < m$ such that $p_1 \notin (t)S_k$. Hence, $p_1 \notin (t)S_{k+1}$; this means that exists a $p_2 < m$ such that $p_1p_2 \notin (t)S_k$, etc. We obtain an infinite sequence $p_1, p_2, p_3, \ldots$ in m such that $p_1 \ldots p_s \notin (t)S_k$ for all $s \in \omega$. It follows that $(p_1 \ldots p_s)t$ is defined for all $s \in \omega$: by I. if $p_1 \ldots p_s \in (t)S_0$ then $p_1 \ldots p_s \in (t)S_k$. This contradicts to the condition (i) in the definition of finite-path Σ-trees (II.3.4).

III. $\emptyset \in (t)S_k$ implies $t \in W_k$.
We first prove that given $p_1 \ldots p_s \in m^*$ with $p_1 \in D_t$, then

$$p_1 \ldots p_s \in (t)S_k \quad \text{implies} \quad p_2 \ldots p_s \in (\partial_{p_1} t)S_k$$

by transfinite induction on k. (Remark: For $s = 0$, $p_1 \ldots p_s = \emptyset$ as well as $p_2 \ldots p_s = \emptyset$.)

Let $k = 0$. Then $p_1 \ldots p_s \in (t)S_0$ means that $(p_1 \ldots p_s)t$ is undefined; this implies $(p_2 \ldots p_s)\partial_{p_1} t$ is undefined (i.e., $p_2 \ldots p_s \in (\partial_{p_1} t)S_0$) whenever $p_1 \in D_t$. If the statement holds for k (and all $t \in I^{\#}$) then it holds for $k+1$: $p_1 \ldots p_s \in (t)S_{k+1}$ means that for all $p_{s+1} \in m$ we have $p_1 \ldots p_s\, p_{s+1} \in (t)S_k$; this implies $p_2 \ldots p_s\, p_{s+1} \in (\partial_{p_1} t)S_k$ (for all p_{s+1}) or $p_2 \ldots p_s \in (\partial_{p_1} t)S_{k+1}$, whenever $p_1 \in D_t$. The situation with a limit ordinal k is clear.

Now, we prove that

$$\emptyset \in (t)S_k \quad \text{implies} \quad t \in W_k$$

by transfinite induction again. This statement holds trivially for $k = 0$, since $\emptyset \in (t)S_0$ cannot occur. Assuming the statement holds for k, let $\emptyset \in (t)S_{k+1}$. Then for each $p \in m$ we have $p \in (t)S_k$, i.e., $\emptyset \in (\partial_p t)S_k$, whenever $p \in D_t$; for these p we have $\partial_p t \in W_k$ by induction hypothesis. Now, either $t \in I \subset W_{k'+1}$, or $(\emptyset)t = \sigma \in \Sigma$ with $|\sigma| = n$ and

$$t = (\partial_p t)_{p<n}\varphi_\sigma \in W_{k+1}.$$

Finally, the situation with a limit ordinal k is clear. □

3.7. As in the finitary case, Σ-tree automata work with the trees in $I^{\#}$: they compute the trees and give a resulting output. But the computation can now be infinitely long.

We define the run map of a Σ-tree automaton A as the unique homomorphism $\rho : (I^{\#}, \{\varphi_\sigma\}) \to (Q, \{\delta_\sigma\})$ extending the initialization map, and the behavior as

$$\beta = \rho \cdot \gamma : I^{\#} \to \Gamma.$$

Example. For the automaton of II.3.2, the following tree

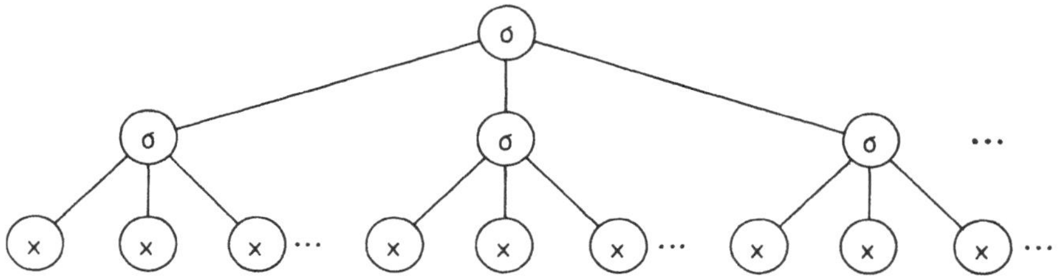

yields the output 0:

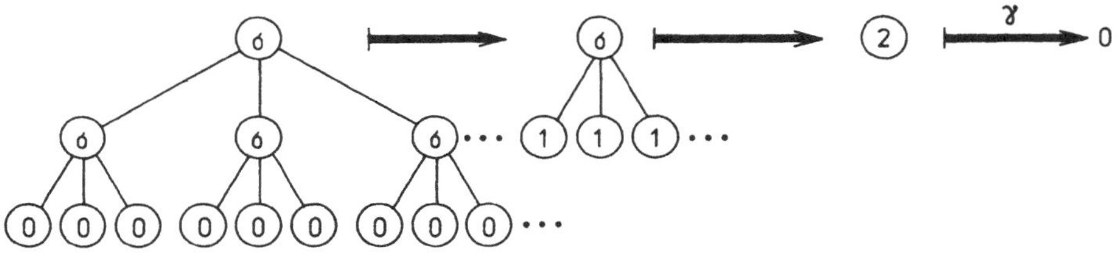

The same holds, evidently, for each Σ-tree of finite depth (II.3.5). On the other hand, the following tree

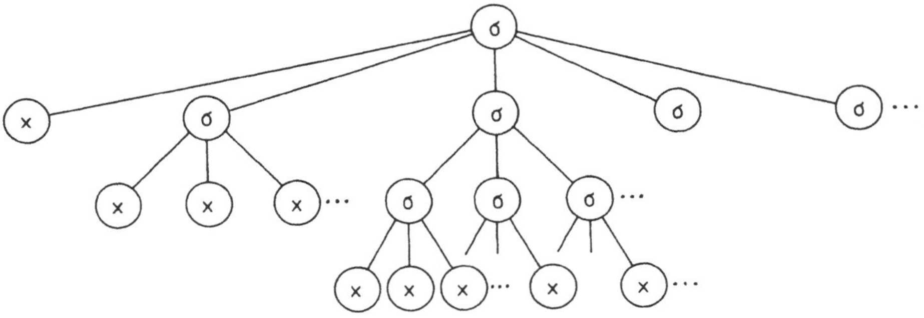

yields the output 1 (after a countable computation):

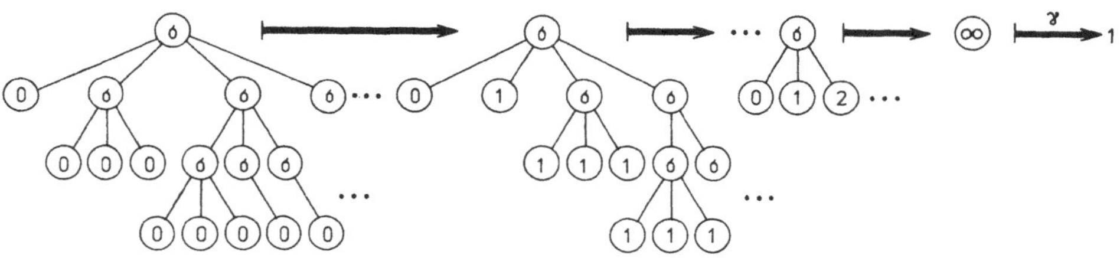

The same holds for each tree of infinite depth. Thus, the behavior of A is defined as follows:

$$(t)\beta = \begin{cases} 0 & \text{if } t \text{ has finite depth;} \\ 1 & \text{else.} \end{cases}$$

3.8. The concepts of congruence, reduction, interchangeability and minimal realization are defined for infinitary automata in the same way as for the finitary ones. But interchangeability is no longer a congruence on each automaton. In fact, any two states $q_1, q_2 \in \omega$ in the automaton of II.3.2 are clearly interchangeable, and ∞ is not interchangeable with any $q \in \omega$. Yet, the equivalence $\approx$ with the classes ω and $\{\infty\}$ is not a congruence: we have

$$0 \approx 0,\ 0 \approx 1,\ 0 \approx 2, \ldots$$

but

$$(0, 0, 0, \ldots)\delta_\sigma = 1 \not\approx \infty = (0,1, 2, \ldots)\delta_\sigma .$$

It turns out that the behavior of this automaton does not have a minimal realization:

Theorem. For each type Σ, the following are equivalent:

(i) Σ is finitary;
(ii) each behavior $\beta : I^{\#} \to \Gamma$ has a minimal realization;
(iii) the behavior $\beta : \{x\}^{\#} \to \{0, 1\}$ defined by $(t)\beta = 0$ iff t has finite depth, has a minimal realization.

Proof. (i) → (ii) See II.2.5.
(ii) → (iii) This is clear.
(iii) → (i) Assuming that Σ is infinitary, we shall prove that the behavior β in (iii) does not have a minimal realization.

(a) Let $\Sigma = \{\sigma\}$ with $|\sigma| = \omega$. The automaton A of II.3.2 realizes β, see II.3.7. And A is reachable : we have

$$\begin{aligned} 0 &= (x)\lambda, \\ 1 &= (0, 0, 0, \ldots)\delta_\sigma, \\ 2 &= (1, 1, 1, \ldots)\delta_\sigma, \end{aligned}$$

etc., and finally,

$$\infty = (0, 1, 2, \ldots)\delta_\sigma.$$

Suppose B is the minimal realization of β. There is a morphism

$$f : A \to B.$$

Since for $q \in \omega$ we have $(q)\gamma = 0 \neq (\infty)\gamma$, it is clear that $f(q) \neq f(\infty)$. We shall prove that

$$f(q) = f(q') \quad \text{for all } q, q' \in \omega.$$

This is a contradiction: the kernel equivalence of f has just two classes, ω and $\{\infty\}$, and we have observed above that this is not a congruence on A.

For each $k = 1, 2, 3, \ldots$ denote by $\sim_k$ the following equivalence on $\omega \cup \{\infty\}$:

$$q \sim_k q' \quad \text{iff} \quad q < k \text{ and } q' < k, \text{ or } q = q'.$$

This is a congruence on A [because for each bounded sequence $(q_0, q_1, q_2, \ldots)$ in $\omega \cup \{\infty\}$, all sequences $(q'_0, q'_1, q'_2, \ldots)$ with $q_n \sim_k q'_n$ for $n < \omega$ are also bounded]. Hence, the quotient automaton $A/\sim_k$ is a reachable realization of β and consequently, there are morphisms

$$f_k\colon A/\sim_k \to B \quad \text{for } k = 1, 2, 3, \ldots.$$

Denote by $c_k\colon A \to A/\sim_k$ the canonical map, then

$$f = c_k \cdot f_k \text{ for } k = 1, 2, 3, \ldots.$$

In fact, if $\rho : \{x\}^\# \to \omega \cup \{\infty\}$ is the run map of A, then $\rho \cdot f$ is the run map of B(II.1.9). Since $\rho \cdot c_k \cdot f_k$ is a homomorphism, $\rho \cdot c_k \cdot f_k$ is also the run map of B. Hence, $\rho \cdot f = \rho \cdot c'_k \cdot f_k$, and since ρ is onto (because A is reachable), we conclude $f = c_k \cdot f_k$. Thus for arbitary $q, q' \in \omega$ we choose k such that $q < k$ and $q' < k$, and we have $(q)c_k = (q')c_k$; consequently,

$$(q)f = (q)c_k \cdot f_k = (q')c_k \cdot f_k = (q')f,$$

which was to be proved.

(b) Let Σ be arbitrary. Then the above automaton A is readily adjusted: choose $\sigma \in \Sigma$ of arity $n \geq \omega$, and put

$$(q_i)_{i<n}\delta_\sigma = \left(\bigvee_{i<n} q_i\right) + 1.$$

All the remaining operations are chosen as the constant map to ∞. The proof then proceeds as above. □

Remarks. (i) Even for Σ infinitary, each Σ-tree automaton A for which the equivalence

$$q_1 \approx q_2 \quad \text{iff } q_1 \text{ and } q_2 \text{ are interchangeable}$$

is a congruence, has the minimal reduction $A/\approx$. (The proof is as in II.2.4.) This is in particular the case if A is finite.

(ii) Each behavior β with a finite realization does have a minimal realization: it is the minimal reduction $A/\approx$ of (any) finite realization A of β.

3.9. Although minimal reductions exist for finitary types Σ, they do not have an important "universal" property encountered in case of sequential automata. This property can be formulated as the possibility to minimize not only automata but also morphisms of automata.

Definition. Let Σ be a type. We say that Σ-tree automata have *universal minimal reduction* if for each morphism

$$(f, f_{\text{in}}, f_{\text{out}}) : A \to A'$$

of Σ-tree automata with minimal reductions $r_A : A \to A_0$ and $r'_A : A' \to A'_0$ there is a morphism $A_0 \to A'_0$ such that the following square

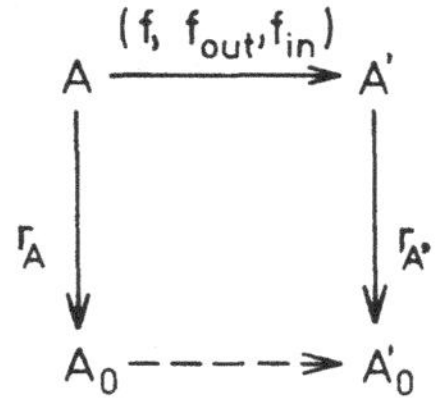

commutes.
Since r_A abbreviates $(r_A, \text{id}_I, \text{id}_\Gamma)$, the new morphism must have the form $(f_0, f_{\text{in}}, f_{\text{out}})$ for the (unique, if any) map f_0 with

$$r_A \cdot f_0 = f \cdot r_{A'}.$$

In the categorical language, let

$$\mathbf{Aut}_i(\Sigma)$$

denote the category of reachable Σ-tree automata and automata morphisms. Composition of morphisms

$$(f, f_{\text{in}}, f_{\text{out}}) : A \to A' \text{ and } (g, g_{\text{in}}, g_{\text{out}}) : A' \to A''$$

is componentwise, i.e.,

$$(f \cdot g, f_{\text{in}} \cdot g_{\text{in}}, f_{\text{out}} \cdot g_{\text{out}}) : A \to A''.$$

(It is easy to check that the last triple is in fact a morphism.) The identity morphisms are $(\text{id}_Q, \text{id}_I, \text{id}_\Gamma)$. Then minimal reduction is universal iff the minimal reduction $r_A : A \to A_0$ of each automaton is a universal arrow [and thus, reduced automata form a full reflective subcategory of $\mathbf{Aut}_i(\Sigma)$]. We shall prove that sequential automata with resets (Exercise II.1.E) present the only case of tree automata with universal realization.

Theorem. If all arities in Σ are 1 or 0 (i.e., $\Sigma = \Sigma_1 \cup \Sigma_0$), then minimal reduction is universal.

In particular, sequential automata have universal minimal reduction.

Proof. Let $\Sigma = \Sigma_1 \cup \Sigma_0$. We can view Σ-tree automata A as sequential Σ-automata with resets in Σ_0 (Exercise II.1.E). Extending the minimization procedure of I.2.6, we introduce the behavior of a state q of A as

$$\beta_q = \rho_q \cdot \gamma: \Sigma_1^* \to \Gamma$$

where $\rho_q : \Sigma_1^* \to Q$ is defined by the following induction:

$$(\emptyset)\rho_q = q;$$
$$(\sigma_1 \ldots \sigma_{n+1})\rho_q = ((\sigma_1 \ldots \sigma_n)\rho_q, \sigma_{n+1})\delta.$$

(We form words only from letters in Σ_1.) It is easy to verify that the results concerning sequential automata generalize to our case: the equivalence

$$q \approx q' \quad \text{iff} \quad \beta_q = \beta_{q'}$$

is a congruence on A, and the minimal reduction is the canonical map

$$r_A: A \to A/\approx, \ (q)r_A = [q].$$

Let

$$(f, f_{\text{in}}, f_{\text{out}}): A \to A'$$

be a morphism of automata. The behavior β_q of a state q in A is related to the behavior $\beta'_{(q)f}$ of $(q)f$ in A' by

$$\beta'_{(q)f} = \beta_q \cdot f_{\text{out}}: \Sigma_1^* \to \Gamma'.$$

This follows from $\rho'_{(q)f} = \rho_q \cdot f$ (which is easily proved by induction):

$$\begin{aligned} \beta'_{(q)f} &= \rho'_{(q)f} \cdot \gamma' \\ &= \rho_q \cdot f \cdot \gamma' \\ &= \rho_q \cdot \gamma \cdot f_{\text{out}} \\ &= \beta_q \cdot f_{\text{out}}. \end{aligned}$$

Consequently, for arbitrary states q and q' of A,

$$\beta_q = \beta_{q'} \qquad \text{implies} \qquad \beta'_{(q)f} = \beta'_{(q')f}.$$

Thus, if $\approx$ is the congruence above for A, and $\approx'$ the corresponding congruence for A', then

$$q \approx q' \qquad \text{implies} \qquad (q)f \approx' (q')f.$$

Define

$$f_0: Q/\approx \to Q'/\approx'$$

by $r_A \cdot f_0 = f \cdot r_{A'}$, i.e., by

$$([q])f_0 = [(q)f].$$

Then

$$(f_0, f_{\text{in}}, f_{\text{out}})\colon A/\approx \;\to A'/\approx'$$

is a morphism such that

$$(*) \qquad r_A \cdot (f_0, f_{\text{in}} \cdot f_{\text{out}}) = (f, f_{\text{out}}, f_{\text{in}}) \cdot r_{A'}.$$

In fact, $(*)$ is clear, and since r_A is a surjective morphism, it follows that $(f_0, f_{\text{in}}, f_{\text{out}})$ is a morphism (Exercise I.1.A).

This proves the universality. □

Theorem. If minimal realization is universal, then all arities are 1 or 0.

Proof. Let Σ be a type with a symbol τ of arity $n > 1$. We define a Σ-tree automaton A and its subautomaton $\bar{A}$ such that the inclusion morphism

$$v\colon \bar{A} \to A$$

cannot be "minimized".

Put

$$A = (\{0, 1, 2, 3\}, \{\delta_\sigma\}, \{0, 1\}, \gamma, \{x\}, \lambda)$$

where δ_σ is constant to 0 for all $\sigma \in \Sigma - \{\tau\}$ and

$$(q_0, q_1, \ldots)\delta_\tau = q_0 + q_1$$

(where $+$ is the addition modulo 4), and

$$(1)\gamma = 1;\ (0)\gamma = (2)\gamma = (3)\gamma = 0$$

$$(x)\lambda = 2.$$

We clearly have a subautomaton $\bar{A}$ of A on the set $\{0, 2\}$; since the output map of $\bar{A}$ is constant to 0, the minimal reduction

$$r_{\bar{A}}\colon \bar{A} \to \bar{A}_0$$

is the constant map. On the other hand, A does not have any non-trivial congruence: since $(1)\gamma^{-1} = 1$, one congruence class must be $\{1\}$, and then it is easy to verify that all congruence classes are singleton sets. Thus,

$$\text{id}\colon A \to A$$

is a minimal reduction. There is no commutative square

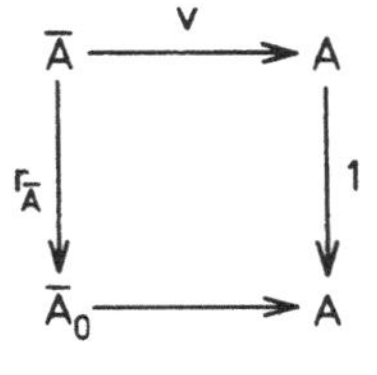

Corollary. Among tree automata, realization is universal just for sequential automata with resets.

Exercises II.3

A. Infinite-path trees. Denote by $I^{\S}$ the set of all (not necessarily finite-path) trees, i.e., maps t defined in $\emptyset$ and satisfying (ii) in Definition II.3.4. Define the operations on $I^{\S}$ analogously as on $I^{\#}$. Why is $I^{\S}$ not the free algebra?

Hint: Try to extend the inclusion map $I \to I^{\#}$.

B. Universality of minimal realization. While behavior has a "functorial" nature (it defines a functor from the category of automata to the category of behaviors), minimal realization is seldom "functorial". We make these ideas precise. Let Σ be a finitary type.

(i) Let $\mathbf{Beh}(\Sigma)$ be the category whose objects are (I, β, Γ), where $\beta: I^{\#} \to \Gamma$ is a behavior, and morphisms $(f_{\text{in}}, f_{\text{out}}): (I, \beta, \Gamma) \to (I', \beta', \Gamma')$ are maps $f_{\text{in}}: I \to I'$, $f_{\text{out}}: \Gamma \to \Gamma'$ with

$$\beta \cdot f_{\text{out}} = f_{\text{in}}^{\#} \cdot \beta'.$$

Verify that the following *behavior functor*

$$\boldsymbol{B}: \mathbf{Aut}_i(\Sigma) \to \mathbf{Beh}(\Sigma)$$

is well-defined: $A\boldsymbol{B}$ is the behavior of A and $(f, f_{\text{in}}, f_{\text{out}})\boldsymbol{B} = (f_{\text{in}}, f_{\text{out}})$.

(ii) In case $\Sigma = \Sigma_1 \cup \Sigma_0$, verify that there is a functor

$$\boldsymbol{M}: \mathbf{Beh}(\Sigma) \to \mathbf{Aut}_i(\Sigma)$$

which assigns to each behavior its minimal realization, and fulfils $\boldsymbol{M} \cdot \boldsymbol{B} = 1$ [i.e., for each morphism $(f_{\text{in}}, f_{\text{out}})$ there is f such that $(f_{\text{in}}, f_{\text{out}})\boldsymbol{M} = (f, f_{\text{in}}, f_{\text{out}})$].

Hint: Proceed as in II.3.10 using the fact that the minimal realization of β is just the minimal reduction of the free realization $A(\beta)$.

(iii) Verify that $\boldsymbol{M}$ and $\boldsymbol{B}$ are adjoint functors.

(iv) Prove that whenever Σ contains a symbol of arity > 1, then there does not exist a functor $\boldsymbol{M}$ with the properties of (ii).

Hint: Try to find $(V, \text{id}_I, \text{id}_\Gamma)\boldsymbol{M}$ in II.3.11.

II.4. Finite Automata and Languages

4.1. Throughout this section, Σ denotes a fixed type of finitely many finitary operational symbols; such types are called *super-finitary*. We denote by m the maximum arity in Σ.

We want to characterize the behaviors of finite Σ-tree automata, i.e., automata with the sets Q (of states) and I (of variables) both finite. We proceed in a close analogy to the case of sequential automata. First, we introduce acceptors and languages, and we prove that instead of behaviors it suffices to study recognizable languages.

Informally, a Σ-tree acceptor is a Σ-tree automaton such that:

(i) variables form a subset of the state set and λ is just the inclusion map;

(ii) the output alphabet is $\Gamma = \{0, 1\}$.

The first restriction means that we need not interpret variables as states, since they are states already. The latter means that instead of a map $\gamma: Q \to \{0, 1\}$ we can just consider the set $T = (1)\gamma^{-1}$.

4.2. Definition. A *Σ-tree acceptor* is a quadruple $A = (Q, \{\delta_\sigma\}_{\sigma \in \Sigma}, T, I)$ where

Q is a finite set (of states);

$\delta_\sigma: Q^n \to Q$ is an operation for each $\sigma \in \Sigma_n$;

T and I are subsets of Q (of terminal and initial states, respectively).

The *run map* of A is the unique homomorphism $\rho: I^\# \to Q$ with $(q)\rho = q$ for each initial state q. In other words, for each Σ-tree $t \in I^\#$ (with leaves labelled by initial states or nullary operations), $(t)\rho$ is the result of the computation of t. The tree t is *accepted* if $(t)\rho$ is a terminal state. The *language recognized* by A is

$$L_A = \{t \in I^\#;\ (t)\rho \in T\}.$$

In general, languages are sets of Σ-trees (i.e., subsets of $I^\#$ where I is a set of variables).

4.3. Definition. Let I be a finite set (of variables). A language $L \subset I^\#$ is said to be *recognizable* if there exists a Σ-tree acceptor A with $L = L_A$.

Remark. In the above definition, it has not been required for the acceptor A that the set I be the set of *all* initial states. Indeed, A can have a larger set of initial states (none of which happens to be a label in any accepted tree, of course).

4.4. Proposition. Let I be a finite set and Γ be an arbitrary set. A behavior

$$\beta: I^\# \to \Gamma$$

has a realization by a finite Σ-tree automaton iff

(1) the language $(y)\beta^{-1}$ is recognizable for each $y \in \Gamma$;

(2) the set $(I^\#)\beta \subset \Gamma$ is finite.

Proof. I. Let β be a behavior with a finite realization, say, $A = (Q, \{\delta_\sigma\}, \Gamma, \gamma, I, \lambda)$.

(a) We prove that β has a finite realization

$$A' = (Q', \{\delta'_\sigma\}, \Gamma, \gamma', I, \lambda')$$

such that $I \subset Q'$ and $(x)\lambda' = x$ for all $x \in I$.

First, assume that $\lambda: I \to Q$ is one-to-one. Then A' is defined by a formal "re-labelling" of the states of A. For each state $q \in Q - (I)\lambda$, choose an element q' such that

$$q_1 \neq q_2 \quad \text{implies} \quad q'_1 \neq q'_2 \notin I;$$

the new state set will be

$$Q' = I \cup \{q';\ q \in Q - (I)\lambda\}.$$

For symmetry, given $q \in (I)\lambda$, put $q' = x$ for the (unique) $x \in I$ with $(x)\lambda = q$. Define

$$\delta'_\sigma: (Q')^n \to Q' \quad (\sigma \in \Sigma_n)$$

by

$$(q'_0, \ldots, q'_{n-1})\delta'_\sigma = [(q_0, \ldots, q_{n-1})\delta_\sigma]'$$

and put

$$\gamma': Q' \to \Gamma;\ (q')\gamma' = (q)\gamma.$$

Finally, $\lambda': I \to Q'$ is the inclusion map. It is easy to see that this automaton A' has the same behavior as A.

Next, let λ be arbitrary. We shall find a realization of β,

$$\bar{A} = (\bar{Q}, \{\bar{\delta}_\sigma\}, \bar{\Gamma}, \bar{\gamma}, I, \bar{\lambda})$$

with $\bar{\lambda}$ one-to-one. Put $m = \text{card } I$; there clearly exists a map

$$\bar{\lambda}: I \to Q \times \{0, 1, \ldots, m-1\}$$

which is one-to-one and such that

$$(x)\bar{\lambda} = (q, j) \quad \text{implies} \quad (x)\lambda = q \quad \text{for all } x \in I.$$

Put

$$\bar{Q} = Q \times \{0, 1, \ldots, m-1\}$$

and define

$$\bar{\delta}_\sigma: \bar{Q}^n \to \bar{Q} \quad (\sigma \in \Sigma_n)$$

by

$$((q_0, j_0), (q_1, j_1), \ldots, (q_{n-1}, j_{n-1}))\bar{\delta}_\sigma = (q, 0)$$

where $q = (q_0, q_1, \ldots, q_{n-1})\delta_\sigma$. Finally define

$$\bar{\gamma}: \bar{Q} \to \Gamma;\ (q, j)\bar{\gamma} = (q)\gamma.$$

Denote by $\bar{\rho}: I^{\#} \to \bar{Q}$ the run map of $\bar{A}$ and define

$$\rho: I^{\#} \to Q$$

by

$$(t)\rho = q \quad \text{iff} \quad (t)\bar{\rho} = (q, j) \quad \text{for some } j.$$

It is easy to check that ρ is a homomorphism; for each $x \in I$, $(x)\rho = (x)\lambda$ since, by the choice of $\bar{\lambda}$,

$$(x)\bar{\rho} = (x)\bar{\lambda} = (q, j) \quad \text{implies} \quad (x)\lambda = q.$$

Hence, ρ is the run map of A. Clearly,

$$\rho \cdot \gamma = \bar{\rho} \cdot \bar{\gamma}: I^{\#} \to \Gamma,$$

hence A and $\bar{A}$ realize the same behavior.

(b) We prove that $(y)\beta^{-1}$ is a recognizable language for each $y \in \Gamma$. We use the realization A', and we define an acceptor

$$A'_y = (Q', \{\delta'_\sigma\}, T, I), \text{ where } T = (y)(\gamma')^{-1}.$$

Let $\rho': I^{\#} \to Q$ be the run map of A', then ρ' is the run map of A'_y, too, and since A' realizes $\beta(= \rho' \cdot \gamma')$, clearly

$$(t)\beta = y \quad \text{iff} \quad (t)\rho' \in T \text{ (for each } t \in I^{\#}).$$

Hence, A'_y recognizes $(y)\beta^{-1}$.

(c) The set $(I^{\#})\beta = ((I^{\#})\rho)\gamma \subset (Q)\gamma$ is finite.

II. Let β have the properties (1) and (2). The proof that β has a finite realization is quite analogous to the sequential case (I.3.3). Put

$$(I^{\#})\beta = \{y_1, \ldots, y_k\};$$

the language

$$L_i = (y_i)\beta^{-1} \; (i = 1, \ldots, k)$$

has a realization by an acceptor, say,

$$A_i = (Q_i, \{\delta^i_\sigma\}, T_i, I_i).$$

Define a finite Σ-tree automaton

$$A = (Q, \{\delta_\sigma\}, \Gamma, \gamma, I, \lambda)$$

as follows:

$$Q = Q_1 \times Q_2 \times \ldots \times Q_k;$$

for each $\sigma \in \Sigma_n$, $\delta_\sigma = \delta^1_\sigma \times \ldots \times \delta^k_\sigma$ — more exactly, given $q_0, \ldots, q_{n-1} \in Q$ [where $q_j = (q_j^1, \ldots, q_j^k)$],

$$(q_0, \ldots, q_{n-1})\delta_\sigma = ((q_0^1, \ldots, q_{n-1}^1)\delta_\sigma^1, \ldots, (q_0^k, \ldots, q_{n-1}^k)\delta_\sigma^k);$$

further,

$$\gamma: Q \to \Gamma$$

is an arbitrary map such that $(q_1, \ldots, q_k)\gamma = y_i$ whenever $q_i \in T_i$ while $q_j \in Q_j - T_j$ for all $j \neq i$; finally

$$\lambda: I \to Q$$

is defined by

$$(q)\lambda = (q, q, \ldots, q) \text{ for each } q \in I.$$

(Note that since A_i realizes a language in variables I, we have $I \subset I_i$ for all i.)

Let $\rho_i: I^\# \to Q_i$ denote the restriction of the run map of A_i, $i = 1, \ldots, k$, then the map

$$\rho: I^\# \to Q;\ (t)\rho = ((t)\rho_1, \ldots, (t)\rho_k) \quad (t \in I^\#)$$

is the run map of A (it suffices to verify that ρ is a homomorphism). The behavior of A is β: given $t \in I^\#$ with $(t)\beta = y_i$, then

$$(t)\rho_i \in T \text{ and } (t)\rho_j \in Q_j - T_j \quad \text{for all } j \neq i$$

(since t is accepted by A_i but not by A_j, $j \neq i$); hence,

$$(t)\gamma = y_i. \qquad \square$$

4.5. We introduce nondeterministic acceptors similarly to the sequential case, and we prove that they accept nothing more than their deterministic relatives. We return to arbitrary variables (to which states are assigned in a nondeterministic manner).

Definiton. A *nondeterministic Σ-tree acceptor* is a quintuple

$$A = (Q, \{\delta_\sigma\}_{\sigma \in \Sigma}, T, I, \lambda)$$

where Q is a finite set (of states), for each $\sigma \in \Sigma_n$

$$\delta_\sigma : Q^n \rightharpoonup Q \text{ is a relation;}$$

$T \subset Q$ is a subset (of terminal states), I is a finite set (of variables) and

$$\lambda : I \rightharpoonup Q$$

is a relation.

The *run relation* of A is the relation

$$\rho : I^\# \rightharpoonup Q$$

defined by induction (on k, where $I^{\#} = \bigcup_{k=0}^{\infty} W_k$, see II.1.5):

$$\text{if } x \in I \text{ then } (x)\rho = (x)\lambda;$$
$$\text{if } t \in W_{k+1} - W_k;\ t = (t_0, \ldots, t_{n-1})\varphi_\sigma \text{ then}$$
$$(t)\rho = \bigcup (s_0, \ldots, s_{n-1})\delta_\sigma$$

where the union ranges over all $s_0 \in (t_0)\rho, \ldots, s_{n-1} \in (t_{n-1})\rho$. The *language recognized by A* is

$$L_A = \{t \in I^{\#};\ (t)\rho \cap T \neq \emptyset\}.$$

4.6. Proposition. Each language, recognizable by a nondeterministic Σ-tree acceptor, is recognizable (by a deterministic one).

Proof. Let $A = (Q, \{\delta_\sigma\}, T, I, \lambda)$ be a nondeterministic Σ-tree acceptor, recognizing a language $L \subset I^{\#}$. We define a Σ-tree automaton with the output alphabet $\Gamma = \{0, 1\}$ such that its behavior is the characteristic map of L; in view of the preceding proposition, L is then recognizable.

Put

$$\bar{A} = (\exp Q, \{\bar{\delta}_\sigma\}, \{0, 1\}, \gamma, I, \bar{\lambda})$$

where, for each $\sigma \in \Sigma_n$ and $D_0, D_1, \ldots, D_{n-1} \in \exp Q$,

$$(D_0, D_1, \ldots, D_{n-1})\ \bar{\delta}_\sigma = \bigcup (q_0, q_1, \ldots, q_{n-1})\delta_\sigma$$

the union ranging over all $q_0 \in D_0, \ldots, q_{n-1} \in D_{n-1}$, and

$$\gamma : \exp Q \to \{0, 1\}$$

is defined by $(D)\gamma = 1$ iff $D \cap T \neq \emptyset$. Moreover, the relation $\lambda : X \rightharpoonup Q$ yields a map $\bar{\lambda} : I \to \exp Q$.

It is easy to prove that the run relation of A,

$$\rho : I^{\#} \rightharpoonup Q$$

yields the run map $\bar{\rho} : I^{\#} \to \exp Q$ of $\bar{A}$. Thus, a tree $t \in I^{\#}$ is accepted by A iff $(t)\rho \cap T \neq \emptyset$, i.e., iff $((t)\bar{\rho})\gamma = 1$. □

4.7. Operations on acceptors and languages. We introduce now operations on acceptors corresponding to the union, concatenation and iteration of languages. While these notions are parallel to those used for sequential automata (I.3.8), the proofs for the last two operations are more technical.

Union. Let

$$A = (Q, \{\delta_\sigma\}, T, I, \lambda) \text{ and } A' = (Q', \{\delta'_\sigma\}, T', I, \lambda')$$

be nondeterministic Σ-tree acceptors with disjoint state sets, $Q \cap Q' = \emptyset$, and with a common variable set I.

Their union is the following acceptor

$$A \cup A' = (Q \cup Q', \{\bar{\delta}_\sigma\}, T \cup T', I, \bar{\lambda})$$

where $\bar{\delta}_\sigma = \delta_\sigma \cup \delta'_\sigma$ and $\bar{\lambda} = \lambda \cup \lambda'$ (as set-theoretical unions), i.e., for each $\sigma \in \Sigma_n$

$$(q_0, \ldots, q_{n-1})\bar{\delta}_\sigma = \begin{cases} (q_0, \ldots, q_{n-1})\delta_\sigma & \text{if } q_0, \ldots, q_{n-1} \in Q \\ (q_0, \ldots, q_{n-1})\delta'_\sigma & \text{if } q_0, \ldots, q_{n-1} \in Q' \\ \emptyset & \text{else} \end{cases}$$

in case $n > 0$, and

$$\bar{\delta}_\sigma = \delta_\sigma \cup \delta'_\sigma$$

in case $n = 0$.

Further,

$$(x)\bar{\lambda} = (x)\lambda \cup (x)\lambda' \qquad (x \in I).$$

It is easy to see that $A \cup A'$ accepts a tree iff either A or A' does; hence,

$$L_{A \cup A'} = L_A \cup L_{A'}.$$

4.8. Serial *x*-connection. Let A and A' be acceptors as above and let $x \in I$. The serial x-connection $A \cdot_x A'$ is the acceptor resulting from connecting the terminal states of A with the $\lambda'(x)$-states of A'. More precisely:

$$A \cdot_x A' = (Q \cup Q', \{\bar{\delta}_\sigma\}, T', I, \bar{\lambda}),$$

where, for each $\delta \in \Sigma_n$,

$$\bar{\delta}_\sigma = \delta_\sigma \cup \delta'_\sigma \cup \psi_\sigma$$

with

$$(q_0, \ldots, q_{n-1})\psi_\sigma = \begin{cases} (x)\lambda' & \text{if } (q_0, \ldots, q_{n-1})\delta_\sigma \cap T \neq \emptyset \\ \emptyset & \text{if } (q_0, \ldots, q_{n-1})\delta_\sigma \cap T = \emptyset \end{cases}$$

and for each $y \in I$,

$$(y)\bar{\lambda} = \begin{cases} (y)\lambda & \text{if } (y)\lambda \cap T = \emptyset \\ (y)\lambda \cup (x)\lambda' & \text{if } (y)\lambda \cap T \neq \emptyset. \end{cases}$$

The operation on languages, corresponding to the serial connection, is the concatenation (generalizing the substitution):

First, given a tree $t \in I^{\#}$ and a language $L \subset I^{\#}$, denote by

$$L \cdot_x t$$

the language of all trees obtained from t by substituting each x-labelled leaf by a tree in L.

Example:

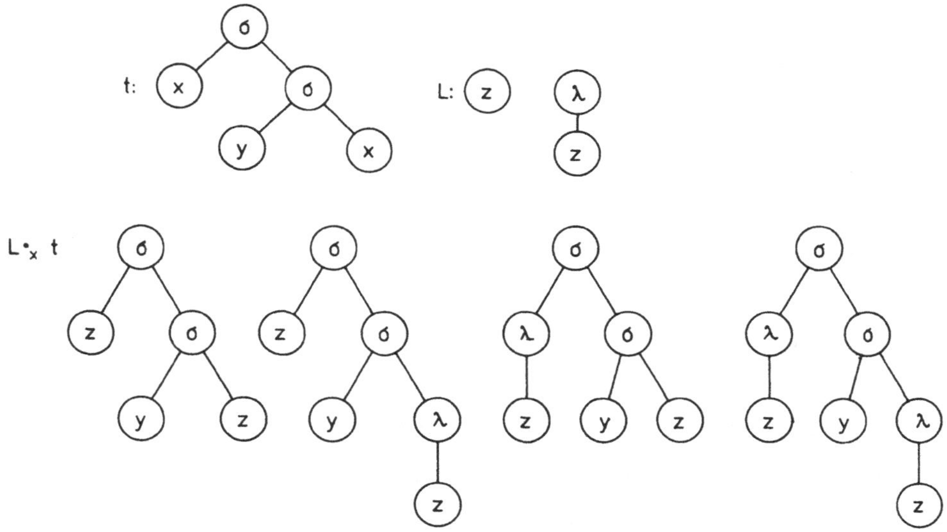

We define this concept formally. Recall that m denotes the join of all arities, $m = \bigvee_{\sigma \in \Sigma} |\sigma|$.

Definition. For each $L \subset I^\#$ and $t \in I^\#$ and for each $x \in I$, denote by $L \cdot_x t \subset I^\#$ the language of all trees $\bar{t}$ such that for all $a \in m^*$,

(i) $(a)t = x$ implies $\partial_a \bar{t} \in L$;

(ii) $(a)t \in \Sigma \cup I - \{x\}$ implies $(a)t = (a)\bar{t}$.

The *x-concatenation* of two languages $L, L' \subset I^\#$ is the language

$$L \cdot_x L' = \bigcup_{t \in L'} L \cdot_x t.$$

Proposition. The language recognized by the serial x-connection $A \cdot_x A'$ is the x-concatenation of the languages L_A and $L_{A'}$:

$$L_{A \cdot_x A'} = L_A \cdot_x L_{A'}.$$

Proof. Denote by ρ the run relation of A, and by ρ' that of A'. For each tree $t \in I^\#$, put

$$L_A \partial_t = \{\bar{t} \in I^\#;\ t \in L_A \cdot_x \bar{t}\}.$$

We are going to prove that the run relation of $A \cdot_x A'$ is defined by

$$(t)\bar{\rho} = (t)\rho \cup \bigcup_{\bar{t} \in L_A \partial_t} (\bar{t})\rho'$$

for each $t \in I^\#$.

Then a tree t is accepted by $A \cdot_x A'$ iff $(t)\rho \cap T' \neq \emptyset$, i.e., iff there exists

$\bar{t} \in L_A\partial_t$ with $(\bar{t})\rho' \cap T' \neq \emptyset$; equivalently, with $\bar{t} \in L_A$. That will prove the proposition.

First, we must prove that $\bar{\rho}$ extends $\bar{\lambda}$. For each $y \in I - \{x\}$, clearly $L_A \cdot_x y = \{y\}$ and $L_A \cdot_x x = L_A$. On the other hand, $y \notin L \cdot_x \bar{t}$ for any $\bar{t} \neq x, y$. Thus,

(i) if $y \in L_A$, then $L_A\partial_y = \{x, y\}$ and $(y)\bar{\rho} = (y)\rho \cup (y)\rho' \cup (x)\rho' = (y)\bar{\lambda}$;
(ii) if $y \notin L_A$, then $L_A\partial_y = \{y\}$ and $(y)\bar{\rho} = (y)\rho \cup (y)\rho' = (y)\bar{\lambda}'$.

Analogously with x: if $x \in L_A$, then $L_A\partial_x = \{x\}$; if $x \notin L_A$, then $L_A\partial_x = \emptyset$. In both cases, $(x)\bar{\rho} = (x)\bar{\lambda}$.

Next, we prove the inductive formula: if $t = (t_0 \ldots, t_{n-1})\varphi_\sigma$, then

$$(t)\bar{\rho} = \bigcup_{s_i \in (t_i)\bar{\rho}} (s_0, \ldots, s_{n-1})\bar{\delta}_\sigma$$

for $n > 0$, and $(\sigma)\bar{\rho} = \bar{\delta}_\sigma$ for $n = 0$. The latter is clear since $L_A\partial_\sigma = \{\sigma\}$. Let us prove the former.

I. $t \notin L_A$. Let us start with the right-hand side. Since $(t_i)\bar{\rho} = (t_i)\rho \cup \bigcup(\bar{t}_i)\rho'$, and since $(s_0, \ldots, s_{n-1})\bar{\delta}_\sigma = \emptyset$ unless all of $s_0, \ldots, s_{n-1}$ are in one of the sets Q and Q', the right-hand side equals to

$$\bigcup_{s_i \in (t_i)\rho} (s_0, \ldots, s_{n-1})\delta_\sigma \cup \bigcup_{\bar{t}_i \in L_A\partial_{t_i}} \bigcup_{s_i \in (\bar{t}_i)\rho'} (s_0, \ldots, s_{n-1})\delta'_\sigma .$$

Now, both ρ and ρ' satisfy the inductive formula. Moreover, $t \notin L_A$ implies $(s_0, \ldots, s_{n-1})\delta_\sigma \cap T = \emptyset$ for any $s_i \in (t_i)\rho$ and hence $(s_0, \ldots, s_{n-1})\bar{\delta}_\sigma = (s_0, \ldots, s_{n-1})\delta'_\sigma$. Thus, the right-hand side can be written as

$$(t)\rho \cup \bigcup_{\bar{t}_i \in L_A\partial_{t_i}} ((\bar{t}_0, \ldots, \bar{t}_{n-1})\varphi_\sigma)\rho'.$$

To analyse the left-hand side, consider an arbitrary tree $\bar{t} \in L_A\partial_t$. Since $t \notin L_A$, clearly $\bar{t} \notin I$ and thus, $\bar{t} = (\bar{t}_0, \ldots, \bar{t}_{n-1})\varphi_\sigma$ where $\bar{t}_i = \partial_i\bar{t}$ for $i = 0, \ldots, n - 1$. We claim that $\bar{t}_i \in L_A\partial_{t_i}$. Indeed, since $t \in L \cdot_x t$, we obtain t from $\bar{t}$ by substituting x-labelled leaves by trees in L_A and hence, the i-th branch t_i is obtained from $\bar{t}_i$ by the same substitutions (in the nodes $p_1 \ldots p_k$ with $p_1 = i$). Conversely, given trees $\bar{t}_i \in L_A\partial_{t_i}$, $i = 0, \ldots, n - 1$, then the tree $\bar{t} = (\bar{t}_0, \ldots, \bar{t}_{n-1})\varphi_\sigma$ is in $L_A\partial_t$. Hence,

$$\begin{aligned}(t)\bar{\rho} &= (t)\rho \cup \bigcup_{\bar{t} \in L_A\partial_t} (\bar{t})\rho' \\ &= (t)\rho \cup \bigcup_{\bar{t}_i \in L_A\partial t_i} ((\bar{t}_0, \ldots, \bar{t}_{n-1})\varphi_\sigma)\rho'.\end{aligned}$$

We see that both sides are equal.

II. $t \in L_A$. The proof is quite analogous, only both sides are "enlarged" by $(x)\lambda'$: the left-hand side since $t \in L_A \cdot_x x$, i.e., $x \in L_A\partial_t$; the right-hand side, since there exist $s_i \in (t)\rho_i$ with $(s_0, \ldots, s_{n-1})\delta_\sigma \cap T \neq \emptyset$. □

4.9. x-Feedback. Let $A = (Q, \{\delta_\sigma\}, T, I, \lambda)$ be a nondeterministic acceptor with $(x)\lambda \cap T \neq \emptyset$. The x-feedback of A is the acceptor

$$A^{*x} = (Q, \{\delta_\sigma^*\}, T, I, \lambda^*)$$

where for each $\sigma \in \Sigma_n$,

$$(q_0, \ldots, q_{n-1})\delta_\sigma^* = \begin{cases} (q_0, \ldots, q_{n-1})\delta_\sigma & \text{if } (q_0, \ldots, q_{n-1})\delta_\sigma \cap T = \emptyset \\ (q_0, \ldots, q_{n-1})\delta_\sigma \cup (x)\lambda & \text{if } (q_0, \ldots, q_{n-1})\delta_\sigma \cap T \neq \emptyset, \end{cases}$$

and for each $y \in I$,

$$(y)\lambda^* = \begin{cases} (y)\lambda & \text{if } (y)\lambda \cap T = \emptyset; \\ (y)\lambda \cup (x)\lambda & \text{if } (y)\lambda \cap T \neq \emptyset. \end{cases}$$

The corresponding operation on languages is the following:

Definition. For each language $L \subset I^\#$ and each $x \in I$ with $x \in L$, the *x-iteration* is the language

$$L^{*x} = L \cup (L \cdot_x L) \cup ((L \cdot_x L) \cdot_x L) \cup \ldots .$$

Example.

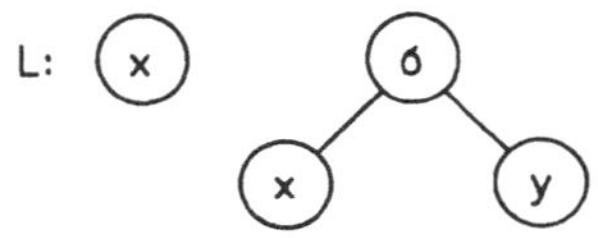

then

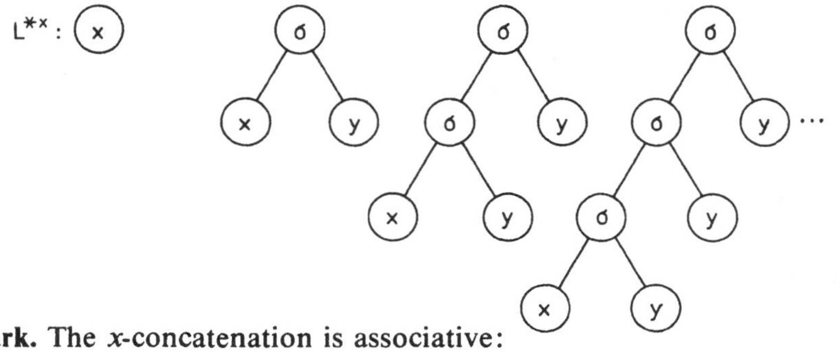

Remark. The x-concatenation is associative:

$$H \cdot_x (K \cdot_x L) = (H \cdot_x K) \cdot_x L$$

(therefore, the brackets in the preceding definition can be omitted). Indeed, if $t \in H \cdot_x (K \cdot_x L)$ then there exists $s \in K \cdot_x L$ with $t \in H \cdot_x s$, and there exists $r \in L$ with $s \in K \cdot_x r$. This implies

$$t \in (H \cdot_x K) \cdot_x r$$

as follows:

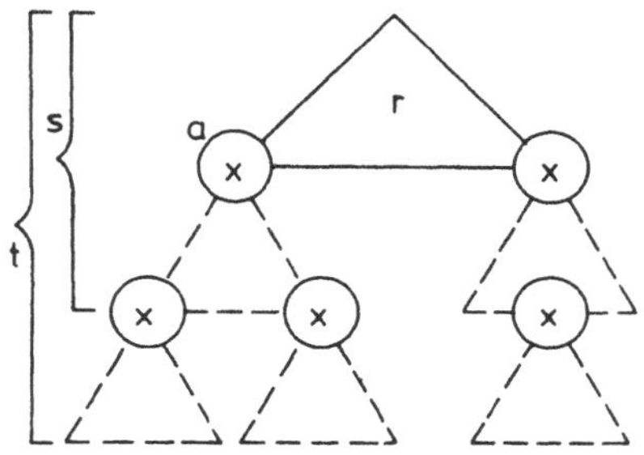

If $(a)r = x$, then $\partial_a s \in K$ and since $t \in H \cdot_x s$, we have

$$\partial_a t \in H \cdot_x \partial_a s \subset H \cdot_x K;$$

if $(a)r \in (I - \{x\}) \cup \Sigma$, then $(a)r = (a)s$ implies $(a)s \in (I - \{x\}) \cup \Sigma$, hence, $(a)r = (a)s = (a)t$.

Analogously, from $t \in (H \cdot_x K) \cdot_x L$ we derive $t \in H \cdot_x (K \cdot_x L)$.

Proposition. The language recognized by the x-feedback A^{*x} is the x-iteration of the language L_A:

$$L_{A^{*x}} = (L_A)^{*x}.$$

Proof. If ρ is the run relation of A, then the run relation of A^{*x} is defined by

$$(t)\bar{\rho} = \bigcup_{\bar{t} \in L^{*x}{}_A \partial_t} (\bar{t})\rho.$$

The proof is quite analogous to the preceding one. Consequently, a tree t is accepted by A^{*x} iff there exists a tree $\bar{t} \in L_A^{*x}\partial_t$ with $(\bar{t})\rho \cap T \neq \emptyset$, i.e., with $\bar{t} \in L_A$. Thus,

$$t \in L_A^{*x} \text{ iff } t \in L_A^{*x} \cdot_x L_A.$$

Since $x \in L_A$ (by the hypothesis on A), clearly $L_A^{*x} \cdot_x L_A = L_A^{*x}$ and hence,

$$L_{A^{*x}} = L_A^{*x}. \qquad \square$$

4.10. Definition. The class of *rational languages* is defined as the least class of languages in arbitrary (finitely many) variables, which contains all singleton languages $\{\emptyset\}$ and $\{t\}$, $t \in I^{\#}$, and the language $\emptyset$, and is closed under the formation of union, x-concatenation and x-iteration for all variables x.

Remarks. (i) Every finite language is rational.

(ii) Trees in W_1 (see II.1.5) are called *basic*; thus, basic trees are of the following kind:

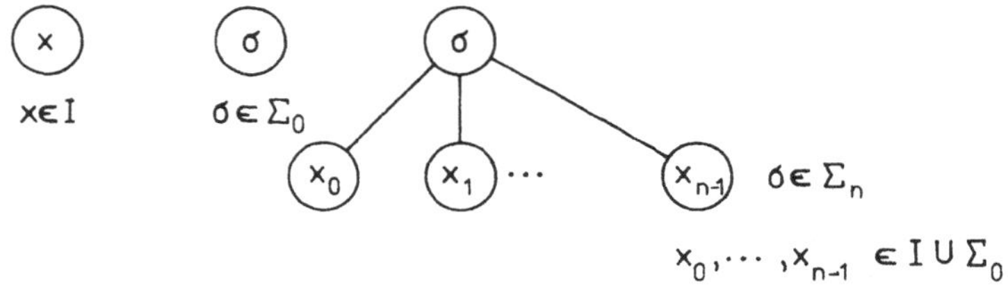

Every tree can be concatenated from basic trees. Example:

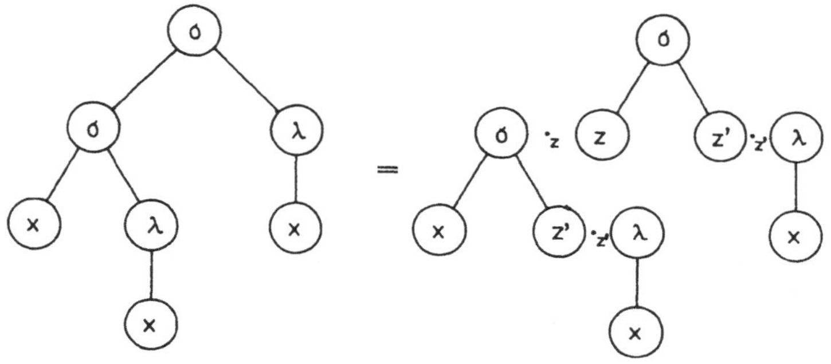

Hence, in the above definition we could start with singleton languages $\{t\}$ where $t \in I^{\#}$ is a basic tree. This corresponds to the sequential case: basic "sequential trees" are $\emptyset$ and σ, for $\sigma \in \Sigma$.

4.11. Kleene Theorem. For each super-finitary type, a language is recognizable iff it is rational.

Proof. I. Rational implies recognizable. We know already that recognizable languages are closed under rational operations, see II.4.7—9. Both $\{\emptyset\}$ and $\emptyset$ are clearly recognizable (compare I.3.9). Thus, it suffices to prove that each singleton language $\{t\}$ is recognizable. Since t can be concatenated from basic trees, we can restrict ourselves to these.

The language $\{x\}$ is recognized by the acceptor

$$A = (\{q\}, \{\delta_\sigma\}, \{q\}, \{x\}, \lambda),$$

where δ_σ is nowhere defined (for all $\sigma \in \Sigma$) and $(x)\lambda = q$.

The language $\{(x_0, \ldots, x_{n-1})\varphi_\sigma\}$ where $\sigma \in \Sigma_n$ (possibly with $n = 0$) is recognized by the acceptor

$$A = (Q, \{\delta_\tau\}_{\tau \in \Sigma}, T, I, \lambda),$$

where

$Q = \{q_0, \ldots, q_n\}$ with $q_i = q_j$ iff $x_i = x_j$;
δ_τ is nowhere defined for all $\tau \in \Sigma - \{\sigma\}$;
$(q_0, \ldots, q_{n-1})\delta_\sigma = q_n$, else δ_σ is not defined;

$I = \{x_0, \ldots, x_{n-1}\}$ and $(x_i)\lambda = q_i$;
$T = \{q_n\}$.

II. Recognizable implies rational. Let $L \subset I^{\#}$ be a language recognized by a (deterministic) Σ-tree acceptor $A = (Q, \{\delta_\sigma\}, T, I)$. We shall prove that L has a rational expression in variables Q.

Let

$$\rho: Q^{\#} \to Q$$

be the run map of the acceptor $\bar{A} = (Q, \{\delta_\sigma\}, T, Q)$. Put

$$Q = \{q_0, \ldots, q_m\},$$

and for each set $M \subset Q$ and each $j = 0, \ldots, m$, put

$$L_{M,j} = \{t \in Q^{\#}; (t)\rho = q_j \text{ and } t \in M^{\#}\}.$$

It suffices to prove that each $L_{M,j}$ is rational language: clearly

$$L = \bigcup_{q_j \in T} L_{M,j}.$$

For each tree $t \in Q^{\#}$ we denote by $B_t \subset Q$ the set of all "intermediate" states in the computation of t, i.e., states computed from the branches $\partial_a t$ of t except $a = \emptyset$ (where $\partial_a t = t$) and the leaves [where $(a)t \in I \cup \Sigma_0$]. That is, $B_t = \{(s)\rho; s = \partial_a t$ for some $a \in n^*$ with $a \neq \emptyset$ and $(a)t \in \Sigma - \Sigma_0\}$. For example, for the following tree

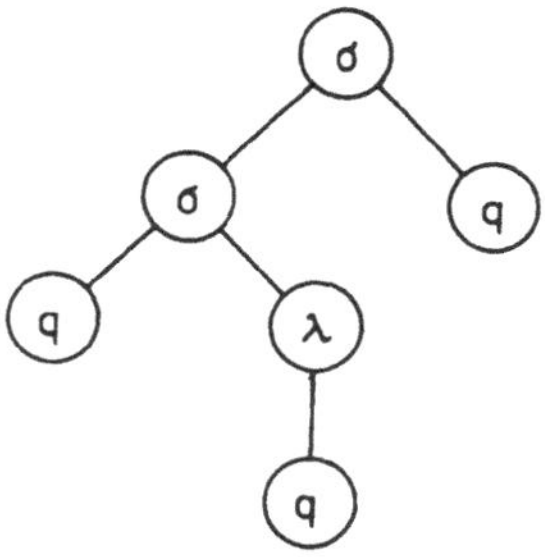

the intermediate states are $(q)\lambda$ and $(q, (q)\lambda)\sigma$. Given $M \subset Q$, $j = 0, \ldots, m$ and $k = 0, \ldots, m$, put

$$L^k_{M,j} = \{t \in L_{M,j}; B_t \text{ does not contain } q_{k+1}, \ldots, q_m\}.$$

We are going to prove that $L^k_{M,j}$ are rational languages (by induction on k); since $L_{M,j} = L^m_{M,j}$, this will prove the statement.

(i) $k = 0$. Since $t \in L^0_{M,j}$ implies $B_t = \emptyset$, all trees in $L^0_{M,j}$ are basic. Hence, $L^0_{M,j}$ is a finite language.

(ii) For the inductive step, we shall verify that

$$L^k_{M,j} = L^{k-1}_{M,k} \cdot_{q_k} (L^{k-1}_{M',k})^{*q_k} \cdot_{q_k} L^{k-1}_{M',j}$$

with

$$M' = M \cup \{q_k\}.$$

It is obvious that $q_k \in L^{k-1}_{M',k}$ (because $q_k \in M'$ and $B_{q_k} = \emptyset$), hence, we can form $(L^{k-1}_{M',k})^{*q_k}$. Note that

$$L^{k-1}_{M,j} \subset L^{k-1}_{M,k} \cdot_{q_k} (L^{k-1}_{M',k})^{*q_k} \cdot_{q_k} L^{k-1}_{M',j}$$

for all M, k, j. Indeed, if $q_k \in M$, then $q_k \in L^0_{M,k} \subset L^{k-1}_{M,k}$ and also $q_k \in (M^{k-1}_{M,k})^{*q_k}$; thus, for each tree $t \in Q^{\#}$ we have

$$t \in \{q_k\} \cdot_{q_k} \{q_k\} \cdot_{q_k} t \subset L^{k-1}_{M,k} \cdot_{q_k} (L^{k-1}_{M',k})^{*q_k} \cdot_{q_k} t.$$

If $q_k \notin M$, then no tree in $L^{k-1}_{M,j}$ has any leaf labelled by q_k, thus, $t \in L^{k-1}_{M,j}$ implies $t \in L_0 \cdot_{q_k} t$ for any language L_0.

Now, we are going to prove the above formula for $L^k_{M,j}$. First, let

$$t \in L^{k-1}_{M,k} \cdot_{q_k} (L^{k-1}_{M',k})^{*q_k} \cdot_{q_k} \bar{t}$$

for some $\bar{t} \in L^{k-1}_{M',j}$. The langugae $L_0 = L^{k-1}_{M',k} \cdot_{q_k} (L^{k-1}_{M',k})$ does not use the variable q_k, unless $q_k \in M$; thus, $L_0 \subset M^{\#}$. This implies $t \in M^{\#}$. Further, for each tree $\tilde{t} \in L_0$, clearly $(\tilde{t})\rho = q_k$; since t is obtained from $\bar{t}$ by substituting such trees $\tilde{t}$ for leaves labelled by q_k, it follows that

$$(t)\rho = (\bar{t})\rho = q_j.$$

Thus, $t \in L_{M,j}$. Finally, the intermediate states of t are, except possibly q_k, all the intermediate states of $\bar{t}$ or the substituting trees (from L_0)—the latter do not include $q_k, \ldots, q_m$. Hence, B_t does not include $q_{k+1}, \ldots, q_m$, i.e., $t \in L^k_{M,j}$.

Conversely, let

$$t \in L^k_{M,j}.$$

Denote the "size" of t by

$$\|t\| = \text{card}\{a \in n^*;\ (a)t \text{ is defined}\};$$

we proceed by induction on $\|t\|$. If $\|t\| = 1$, then t is a variable, thus

$$t \in L^0_{M,j} \subset L^{k-1}_{M,j} \subset L^{k-1}_{M,k} \cdot_{q_k} (L^{k-1}_{M',k})^{*q_k} \cdot_{q_k} L^{k-1}_{M',j}.$$

If $\|t\| = d > 1$ and the proposition holds for trees of a smaller size, choose a node $a \in n^* - \{\emptyset\}$ with $(a)t \in \Sigma - \Sigma_0$ and $\partial_a t = q_k$. (If no such node exists, then $t \in L^{k-1}_{M,j}$ and there is nothing to prove.) Substituting q_k for the branch $\partial_a t$ in t, we obtain a new tree $\bar{t}$:

$(a)\bar{t} = q_k$;

$(ac)\bar{t}$ is undefined for all $c \in n^* - \{\emptyset\}$;

$(b)\bar{t} = (b)t$ for all $b \in n^*$ with $(b)t$ defined and $b \neq ac$ for any $c \neq \emptyset$.

Then $a \neq \emptyset$ implies $\|\partial_a t\| < d$, and $(a)t \in \Sigma - \Sigma_0$ implies $\|\bar{t}\| < d$; hence, we can use the inductive hypothesis on both

$$\bar{t} \in L^k_{M',j} \quad \text{and} \quad \partial_a t \in L^k_{M,k}.$$

Thus,

$$\bar{t} \in L^{k-1}_{M,k} \cdot_{q_k} (L^{k-1}_{M',k})^{*q_k} \cdot_{q_k} L^{k-1}_{M',j} \subset (L^{k-1}_{M',k})^{*q_k} \cdot_{q_k} L^{k-1}_{M',j}$$

and

$$\partial_a t \in L^{k-1}_{M,k} \cdot_{q_k} (L^{k-1}_{M',k})^{*q_k} \cdot_{q_k} L^{k-1}_{M',k} \subset L^{k-1}_{M,k} \cdot_{q_k} (L^{k-1}_{M',k})^{*q_k}.$$

Since clearly

$$t \in \{\partial_a t, q_k\} \cdot_{q_k} \bar{t},$$

it follows that

$$\begin{aligned} t &\in L^{k-1}_{M,k} \cdot_{q_k} [(L^{k-1}_{M',k})^{*q_k} \cdot_{q_k} (L^{k-1}_{M',k})^{*q_k}] \cdot_{q_k} L^{k-1}_{M',j} \\ &\subset L^{k-1}_{M,k} \cdot_{q_k} (L^{k-1}_{M',k})^{*q_k} \cdot_{q_k} L^{k-1}_{M',j}. \end{aligned}$$

□

Exercises II.4

A. Recognizable languages. Describe the language of the following acceptors.

(i) $\Sigma = \Sigma_2 = \{\sigma\}$; $A = (\{q, r\}, \delta_\sigma, \{r\}, \{q\})$ where δ_σ is given by the following table

δ_σ	q	r
q	r	q
r	q	r

(Hint: Count the leaves.)

(ii) $\Sigma = \{\sigma, \tau\}$ with $|\sigma| = 2$ and $|\tau| = 0$;
$A = (\{0, 1, 2, 3\}, \{+, 1\}, 0, \emptyset)$ where $+$ is the addition modulo 4 (i.e., the usual addition with 4 subtracted if the result exceeds 3).

(iii) $\Sigma = \Sigma_3 = \{\sigma\}$;
$A = (\{p, q, r, s\}, \delta_\sigma, \{p, s\}, \{q, s\})$ where the operation is defined as follows:

$$(a, b, c)\delta_\sigma = \begin{cases} p & \text{if } a = p; \\ q & \text{if } a \neq p \text{ and } b = q; \\ r & \text{if } a \neq p,\ b \neq q \text{ and } c = r; \\ s & \text{else.} \end{cases}$$

(Hint: Consider the left-most leaf.)

B. Nondeterministic acceptors. Describe the languages of the following nondeterministic acceptors.

(i) $\Sigma = \Sigma_3 = \{\sigma\}$;

$A = (\{q_1, q_2, q_3\}, \delta_\sigma, \{q_1\}, \{x\}, \lambda)$

where $(x)\lambda = \{q_1, q_2, q_3\}$ and $(a, b, c)\delta_\sigma = q_1$ if a, b, c are pairwise distinct, else δ_σ is undefined.

(Hint: For each $n = 1, 2, 3, \ldots$ count the leaves with distance n from the root.)

(ii) $\Sigma = \Sigma_2 = \{\sigma, \tau\}$;

$A = (\{0, 1, 2, 3\}, \{\delta_\sigma, \delta_\tau\}, \{x, y\}, \lambda)$ where $(x)\lambda = 0$; $(y)\lambda = \{0, 1\}$ and the operations are defined for each $a, b = 0, 1, 2, 3$ as follows: $(a, b)\delta_\sigma = \{0, 1, 2, 3\}$; $(a, b)\delta_\tau = \{a + b + 1\}$ if $a + b < 3$, else undefined. (Hint: Consider the root only.)

C. Rational expressions. Find the languages described by the following rational expressions, where $\Sigma = \Sigma_2 = \{\sigma\}$.

(i) $\{x, t\}^{*x}$ where $t = (x, x)\sigma$.

(ii) $\{y, t\}^{*y}$ for the same tree t;

(iii) $\{x, t\}^{*x} \cdot_y \{x, t\}^{*x}$ where $t = (x, y)\sigma$;

(iv) $\{x, t\}^{*x} \cdot_y t'$ where t and t' are the following trees

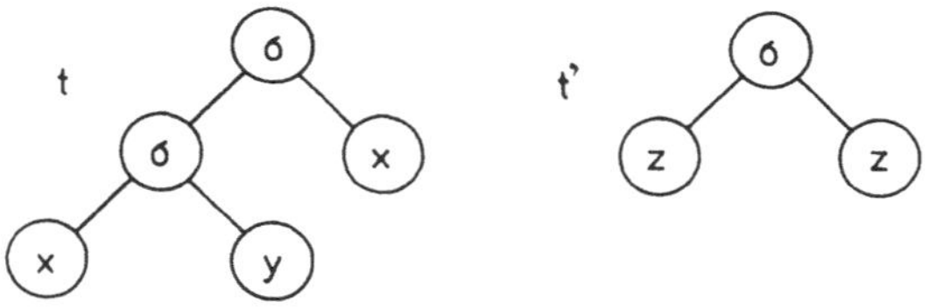

D. Kleene Theorem. The above proof that each recognizable language is rational gives an algorithm for finding a rational expression which describes the language of a given acceptor. Use this procedure to find a rational expression for the languages of Exercise A above.

E. The extension of the set of variables in necessary. (i) Recall Remark II.4.3.

(ii) Put $\Sigma = \Sigma_2 = \{\sigma\}$. Consider the language $L \subset \{x\}^{\#}$ formed by the following sequence of Σ-trees:

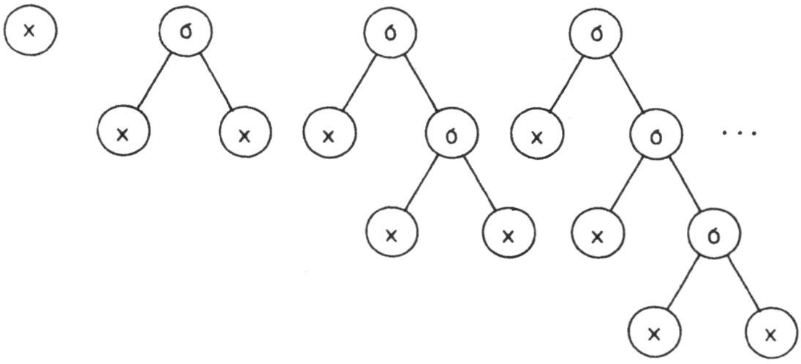

(a) Construct a finite Σ-tree acceptor which recognizes the language L.

(b) Express L as a rational expression.

(c) Prove that L cannot be expressed as a rational language using only $\cdot_x$ and $*^x$.

F. A non-recognizable language. (i) Put $\Sigma = \Sigma_2 = \{\sigma\}$. Prove that the language $L_u \subset I^\#$ of all uniform trees (Ex. II.1.C) is not recognizable (for any set $I \neq \emptyset$). Hint: Let $A = (Q, \delta, T, I)$ be an acceptor with $L_u \subset L_A$. Since Q is finite, there exist distinct trees $t_1, t_2 \in L_u$ with $(t_1)\rho = (t_2)\rho \in T$. Since $(t_1, t_1)\sigma \in L_A$, it follows that $(t_1, t_2)\sigma \in L_A$—but $(t_1, t_2)\sigma \notin L_u$.

(ii) Verify that

$$L_u = \{x\} \cup L \cup (L \cdot_x L) \cup ((L \cdot_x L) \cdot_x L) \cup \ldots$$

for the (finite) language $L = \{(x, x)\sigma\}$. This gives a reason not to define L^{*x} for languages L with $x \notin L$.

Notes to Chapter II

The concept of finitary tree automaton has appeared in the late 1960's in the papers of Arbib and Give'on [1968], Brainerd [1968], Doner [1970], Eilenberg and Wright [1967], Magidor and Moran [1969], Mezei and Wright [1967], and Tchatcher and Wright [1968]. The material presented in Chapter II is, in one form or another, to be found in those papers. A detailed, systematic study of finitary tree automata is presented in the monograph of Gécseg and Steinby [1984]. Tree automata in a variety of algebras are discussed by Eilenberg and Wright [1967] who, however, abandon this generality when dealing with Kleene Theorem. The reason will be seen in Chapter VII below: the validity of Kleene Theorem in a variety will be shown to be a deep problem. The universality theorem II.3.10 is due to J. A. Goguen [1973].

Infinitary tree automata have received but little attention, see Anderson, Arbib and Manes [1976]. The results we have presented on infinitary tree automata are new.

Chapter III: F-Automata

III.1. Introduction

We shall study a general theory of automata in a category $\mathscr{K}$. The main three motivating examples are

(i) sequential Σ-automata in the category of sets;

(ii) tree Σ-automata in the category of sets;

(iii) linear Σ-automata (i.e., linear discrete-time dynamical systems) in the category of modules over a ring.

The idea of the investigated generalization, due to M. A. Arbib and G. E. Manes, is to express the type Σ of automata under study by a functor $F: \mathscr{K} \to \mathscr{K}$. The resulting concepts of F-automaton, behavior, minimal realization, etc. are simply formulated but they often lead to non-trivial categorical problems. For example, we know that Σ-tree automata have minimal realizations iff Σ is a finitary type (II.3.8). It is interesting to know which functors have the property that each behavior has a minimal realization.

The theory we present goes far beyond the three motivating examples above, though not "too far", as we feel and as we hope to persuade the reader. For example, F-automata in the category of sets are much more general than the tree automata. Yet, we prove that the minimal realization problem is solvable essentially only for Σ-tree automata with Σ finitary and for their "varieties".

The aim of the present chapter is to explain all of the basic concepts and problems studied in the subsequent chapters, and to lay category-theoretical foundations for the rest of the book. In the second section we introduce F-automata, free F-algebras (playing the role Σ^* does for sequential automata), behavior, and minimal realization. These concepts are illustrated on the particularly simple case of a coadjoint functor F. The remaining sections present some basic facts about algebras, factorization systems and set functors. After reading the second section, the reader will know what to expect of the subsequent sections (each of which can be skipped without breaking the continuity of the text) and also of the following chapters.

We assume that the reader is familiar with the most basic concepts of category theory, but we discuss in detail all more advanced notions. *Categories* are denoted by script letters $\mathscr{K}$, $\mathscr{L}$, . . ., or by the name of their objects. For example, **Set** denotes the category of sets (as objects) and mappings (as morphisms), **Pos** denotes the category of posets and order-preserving maps, etc. *Objects* are denoted by capital letters and *morphisms*

$$f: A \to B$$

by case letters; here A is the *domain* of f, and B is the *codomain*. For a category $\mathcal{K}$, the class of all objects is denoted by $\mathcal{K}^{o}$, and the class of all morphisms by $\mathcal{K}^{m}$. The composition of morphisms is written from left to right, i.e., for two morphisms $f: A \to B$ and $g: B \to C$, we have $f \cdot g: A \to C$. The identity morphism of an object A is denoted by 1_A or id. Recall that an *isomorphism* is a morphism $f: A \to B$ for which an inverse morphism exists, i.e., a morphism $f^{-1}: B \to A$ with $f \cdot f^{-1} = 1_A$ and $f^{-1} \cdot f = 1_B$. A more general concept is that of a *monomorphism* (shortly: mono) which is a morphism $f: A \to B$ such that in each diagram

if $g_1 \neq g_2$ then $g_1 \cdot f \neq g_2 \cdot f$. And an *epimorphism* (shortly: epi) is a morphism $f: A \to B$ such that in each diagram

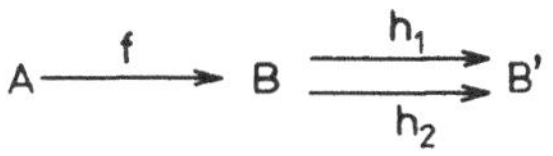

if $h_1 \neq h_2$ then $f \cdot h_1 \neq f \cdot h_2$. For example, let $f: A \to B$ and $g: B \to A$ fulfil $f \cdot g = 1_A$. Then f is a mono (called a *split mono*) and g is an epi (called a *split epi*). (Each isomorphism is a split mono as well as a split epi.)

Functors are denoted by letters $F, G, H, \ldots$; they are also written from left to right, i.e., a functor $F: \mathcal{K} \to \mathcal{L}$ assigns to each object A of $\mathcal{K}$ an object AF of $\mathcal{L}$, and to each morphism $f: A \to B$ of $\mathcal{K}$ a morphism $fF: AF \to BF$ of $\mathcal{L}$ (preserving the composition of morphisms, and the identity morphisms). For example, H_2: **Set** → **Set** is the functor with $XH_2 = X \times X$ and $fH_2 = f \times f$ [sending (x, y) to $((x)f, (y)f)$]. The composition of functors $F: \mathcal{K} \to \mathcal{L}$ and $G: \mathcal{L} \to \mathcal{H}$ is denoted by $F \cdot G = \mathcal{K} \to \mathcal{H}$. For example, $X(H_2 \cdot H_2) = (X \times X) \times (X \times X)$.

Natural transformations are denoted by Greek letters. Given functors $F, G: \mathcal{K} \to \mathcal{L}$, a natural transformation $\tau: F \to G$ is a collection of morphisms $\tau_A: AF \to AG$ in $\mathcal{L}^{m}$ (for A in $\mathcal{K}^{o}$) such that for each morphism $f: A \to B$ in $\mathcal{K}$ the following square

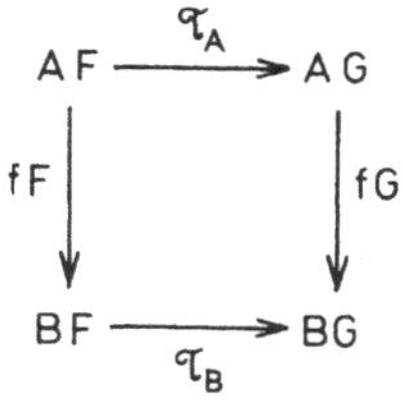

commutes. (A square is said to *commute* if the two passages from the upper left-hand corner to the lower right-hand one compose as the same morphism.) A *natural isomorphism* is a natural transformation τ with τ_A an isomorphism for each A.

For arbitrary objects A and B of a category, the collection of all morphisms from A to B, denoted by hom(A, B), is a set. A category is *small* if its collection of all objects is a set (not a proper class). We list the set-theoretical conventions we use at the end of this book.

Some of the fundamental concepts of category theory are recalled in the following exercises. The interested reader can consult one of the following monographs for further information:

S. Mac Lane: Categories for the Working Mathematician, Springer-Verlag, Berlin—Heidelberg—New York 1971.

H. Herrlich and G. E. Strecker: Category Theory, 2nd Ed., Heldermann Verlag, Berlin 1979.

J. Adámek: Theory of Mathematical Structures, Reidel Publ. Comp., Dordrecht—Boston—Lancaster 1983.

Exercises III.1

A. Classification of functors. Recall that a functor $F: \mathscr{K} \to \mathscr{L}$ is said to be (a) *faithful* if for any two objects A, B of $\mathscr{K}$, the map $(f: A \to B) \mapsto (fF: AF \to BF)$ is one-to-one; (b) an *embedding* if F is one-to-one on the class of all morphisms of $\mathscr{K}$; (c) *full* if each morphism $g: AF \to BF$ in $\mathscr{L}$ has the form $g = fF$ for some $f: A \to B$ in $\mathscr{K}$; and (d) an *isomorphism of categories* if it is bijective.

(a) Verify that the "forgetful functor" U: **Pos** $\to$ **Set** [which forgets the ordering, $(X, \leq)F = XF$] is faithful, but neither an embedding, nor full.

(b) Let **Lat** denote the category of lattices and lattice homomorphisms. Verify that the "inclusion functor" I: **Lat** $\to$ **Pos** (with $AI = A$ and $fI = f$) is an embedding which is not full.

(c) Verify that the functor F: **Set** $\to$ **Pos** assigning to each set X the discretely ordered poset $(X, =)$ is a full embedding.

(d) Prove that $F: \mathscr{K} \to \mathscr{L}$ is an isomorphism of categories iff there is a functor $F^{-1}: \mathscr{L} \to \mathscr{K}$ with $F \cdot F^{-1} = 1_{\mathscr{K}}$ and $F^{-1} \cdot F = 1_{\mathscr{L}}$.

B. Composing natural transformations. (a) For natural transformations $\tau: F \to G$ and $\sigma: G \to H$ (where $F, G, H: \mathscr{K} \to \mathscr{L}$ are functors), the composition $\tau \cdot \sigma: F \to H$ is the natural transformation with $(\tau \cdot \sigma)_A = \tau_A \cdot \sigma_A$. For example, let $\rho: H_2 \to H_2$ (where H_2 is the cartesian-square functor above) be defined by $\rho_A(x, y) = (y, x)$. Verify that $\rho \cdot \rho = 1_{H_2}$, where $1_F: F \to F$ denotes the *identity transformation*.

(b) Verify that a natural transformation $\tau: F \to G$ is a natural isomorphism iff there is a natural transformation τ^{-1} with $\tau \cdot \tau^{-1} = 1_F$ and $\tau^{-1} \cdot \tau = 1_G$.

(c) Let $\tau: F \to G$ be a natural transformation ($F, G: \mathcal{K} \to \mathcal{L}$), and let $H: \mathcal{L} \to \mathcal{H}$ be a functor. Then $\tau H: FH \to GH$ denotes the natural transformation with $(\tau H)_A = \tau_{HA}$. Verify that $\rho H_2: H_2 \cdot H_2 \to H_2 \cdot H_2$ is given by the maps $((x_1, x_2), (y_1, y_2)) \mapsto ((y_1, y_2), (x_1, x_2))$.

(d) Let $\tau: F \to G$ be a natural transformation ($F, G: \mathcal{K} \to \mathcal{L}$), and let $H: \mathcal{H} \to \mathcal{K}$ be a functor. Then $H\tau: HF \to HG$ denotes the natural transformation with $(H\tau)_A = \tau_A H$. Verify that $H_2\rho: H_2 \cdot H_2 \to H_2 \cdot H_2$ is given by the maps $((x_1, x_2), (y_1, y_2)) \mapsto ((x_2, x_1)(y_2, y_1))$.

(e) A natural transformation $\tau: F \to G$ is called an *epitransformation* if all τ_A are epis; prove that then $\tau \cdot \sigma = \tau \cdot \sigma'$ implies $\sigma = \sigma'$. Analogously, for a *monotransformation* τ (all τ_A are monos), $\sigma \cdot \tau = \sigma' \cdot \tau$ implies $\sigma = \sigma'$.

C. Products and coproducts. Recall that a *product* of objects $A_i (i \in I)$ of a category is an object $\prod_{i \in I} A_i$ together with morphisms (called projections) $\pi_j: \prod_{i \in I} A_i \to A_j (j \in I)$ having the following universal property: given an object B and morphisms $f_j: B \to A_j (j \in J)$, there exists a unique morphism $f: B \to \prod_{i \in I} A_i$ with $f_j = f \cdot \pi_j$ for all $j \in J$. Then f_j are the *components* of f. A product of two objects is also denoted by $A_1 \times A_2$.

(a) Verify that in the category **Set**, products are the usual cartesian products, $\prod_{i \in I} A_i = \{(a_i)_{i \in I}; a_j \in A_j \text{ for each } j \in I\}$, and $[(a_i)_{i \in I}]\pi_j = a_j$.

(b) Verify that in the category **Pos**, products are the usual cartesian products ordered coordinate-wise.

(c) The concept of *coproduct* is the *dual* concept of product (as the prefix "co-" indicates), i.e., the direction of all arrows in the above definition is reversed: a coproduct of objects $A_i (i \in I)$ is an object $\coprod_{i \in I} A_i$ together with morphisms (called injections) $\varepsilon_j: A_j \to \coprod_{i \in I} A_i (j \in I)$ with the universal property. A coproduct of two objects is also denoted by $A_1 + A_2$.

Verify that in **Set**, the coproduct of sets $A_i (i \in I)$ is their disjoint union $\coprod_{i \in I} A_i = \bigcup_{i \in I} A_i \times \{i\}$ with $\varepsilon_j: A_j \to \bigcup_{i \in I} A_i \times \{i\}$ given by $(x)\varepsilon_j = (x, j)$ for $x \in A_j$.

(d) Verify that in **Pos**, the coproduct of posets $(A_i, \leq)$, $i \in I$, is their disjoint union ordered in such a way that $(x, j) \leq (x', j')$ iff $j = j'$, and $x \leq x'$ in A_j.

(e) Let R-**Mod** denote the category of modules over a given commutative ring R, and module homomorphisms. Verify that for two modules A_1 and A_2,

$$A_1 \times A_2 = A_1 + A_2$$

is the cartesian product with coordinate-wise operations. Here the projections are $(x, y)\pi_1 = x$ and $(x, y)\pi_2 = y$. The injections are $(x)\varepsilon_1 = (x, 0)$ and $(y)\varepsilon_2 = (0, y)$.

(f) Let $f_j: A_j \to B_j$ be morphisms $(j \in J)$. Verify that there exists a unique morphism, denoted by $\prod_{j \in I} f_j$, from $\prod_{j \in J} A_j$ to $\prod_{j \in J} B_j$ the components of which are $\pi_i \cdot f_i$ for the projections $\pi_i: \prod_{j \in J} A_j \to A_i$ $(i \in J)$.

(g) Dually, $\coprod f_j: \coprod A_j \to \coprod B_j$ denotes the morphism the components of which are determined by f_j—formulate properly!

(h) A coproduct of functors $F_i: \mathscr{K} \to \mathscr{L}$ $(i \in I)$ is the functor $\coprod_{i \in I} F_i: \mathscr{K} \to \mathscr{L}$ given by $A(\coprod_{i \in I} F_i) = \coprod_{i \in I} AF_i$ and $f(\coprod_{i \in I} F_i) = \coprod_{i \in I} fF_i$, provided that $\mathscr{L}$ has coproducts. Verify that coproduct injections $(\varepsilon_j)_A: AF_j \to \coprod_{i \in I} AF_i$ form a natural transformation $\varepsilon_j: F_j \to \coprod_{i \in I} F_i (j \in I)$ with the expected universal property: For each functor $G: \mathscr{K} \to \mathscr{L}$ and arbitary natural transformations $\tau_j: F_j \to G (j \in I)$ there is a unique natural transformation $\tau: \coprod_{i \in I} F_i \to G$ with $\tau_j = \varepsilon_j \cdot \tau (j \in I)$.

Analogously, the product $\prod_{i \in I} F_i$ of functors is defined "coordinatewise".

D. Equalizers and coequalizers. Recall that an *equalizer* of two morphisms f_1, $f_2: A \to B$ is a morphism $e: E \to A$, universal with respect to $e \cdot f_1 = e \cdot f_2$, i.e., any morphism $h: H \to A$ with $h \cdot f_1 = h \cdot f_2$ has the form $h = h' \cdot e$ for a unique $h': H \to E$.

(a) Verify that in **Set**, the equalizer of $f_1, f_2: A \to B$ is the inclusion map of the subset $E = \{a \in A; (a)f_1 = (a)f_2\}$.

(b) The dual concept, the *coequalizer* of f_1, $f_2: A \to B$ is a morphism $c: B \to C$ universal with respect to $f_1 \cdot c = f_2 \cdot c$. Verify that in **Set**, the coequalizer is the canonical map $c: B \to B/\sim$ of the least equivalence $\sim$ on the set B with $(a) f_1 \sim (a)f_2$ for each $a \in A$.

(c) Verify that if a morphism $e: E \to A$ is an equalizer of a pair of morphisms, then e is a mono. Such morphisms are called *regular monos*. Prove the following hierarchy of morphisms:

isomorphism $\Rightarrow$ split mono $\Rightarrow$ regular mono $\Rightarrow$ mono.

Verify that in **Pos**, none of these implications can be reversed. Hint: If $e \cdot g = 1$ then e is the equalizer of $g \cdot e$ and 1.

(d) Verify that in **Set**, monos are exactly the one-to-one maps, and they are all regular. Split monos are exactly the monos with non-empty domain.

(e) Dually, *regular epis* are the coequalizers. Formulate and prove the appropriate hierarchy of epis. Verify that in **Set**, epis are exactly the onto maps, and they are all split epis. In **Pos**, find a non-regular epi, and a regular epi that does not split; find a bijective morphism which is not an isomorphism, and observe that it is both a mono, and an epi.

E. Limits and colimits. By a *diagram* D in a category $\mathcal{K}$ is meant a *scheme*, i.e., a small category $\mathcal{D}$, together with a functor $D: \mathcal{D} \to \mathcal{K}$. A collection of morphisms $f_d: A \to dD$ ($d \in \mathcal{D}^o$) in $\mathcal{K}$ is *compatible* if each $\mathcal{D}$-morphism $\delta: d \to d'$ fulfils $f_{d'} = f_d \cdot \delta D$. A *limit* of D is an object $L = \lim D$ together with compatible morphisms $\pi_d: L \to dD$ ($d \in \mathcal{D}^o$) universal with respect to compatibility. That is, for each compatible collection $f_d: A \to dD$ there exists a unique morphism $f: A \to L$ with $f_d = f \cdot \pi_d (d \in \mathcal{D}^o)$. The morphisms f_d are called *components* of f.

(a) Verify that equalizer is a special case of limit: here $\mathcal{D}$ is the scheme consisting of two objects, two parallel morphisms and (necessarily) two identity morphisms of the given objects.

(b) For *discrete categories* $\mathcal{D}$, i.e., categories which do not have any morphism except the identity morphisms, verify that limits are just products.

(c) Verify that the limit of the empty diagram ($\mathcal{D}$ has no object) is the *terminal* object T, i.e., an object such that each object A has exactly one morphism from A to T. Verify that $\{0\}$ is terminal in **Set**, **Pos**, and R-**Mod**.

(d) A *pullback* of morphisms $f: A \to B$ and $g: C \to B$ is the name of the limit of the corresponding diagram (the scheme of which is just the co-span). Verify that a pullback is a commuting square

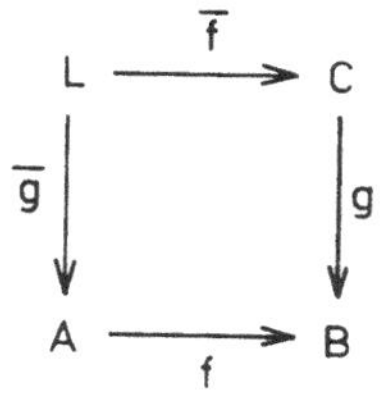

such that for any commuting square $g_1 \cdot f = f_1 \cdot g$ there is a unique morphism t with $g_1 = t \cdot \bar{g}$ and $f_1 = t \cdot \bar{f}$. Verify that a pullback of f and g in **Set** is $L = \{(a, c) \in A \times C; (a)f = (c)g\}$, and $\bar{f}$ and $\bar{g}$ are the projections.

(e) The concept of *colimit* of D is dual to that of limit: it is an object $C = \operatorname{colim} D$ together with a universal compatible collection $\varepsilon_d: dD \to C$ ($d \in \mathcal{D}^o$). Example: Coequalizer is the colimit of a parallel pair of morphisms; coproduct is the colimit of a discrete diagram.

A *pushout* of morphisms $f: A \to B$ and $g: A \to D$ is the colimit of the obvious diagram. Verify that it is a universal commuting square

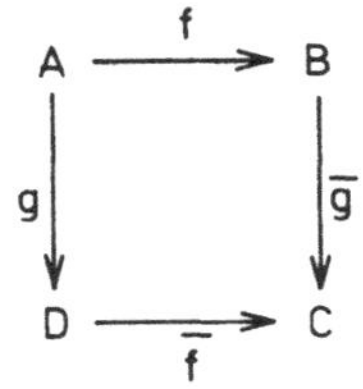

(f) The colimit of the empty diagram is the *initial* object $\perp$, i.e., an object such that each object A has exactly one morphism from $\perp$ to A. Verify that $\emptyset$ is the initial object of **Set** and **Pos**. Find the initial object of R-**Mod**.

(g) A category $\mathcal{K}$ is said to be *complete* if each diagram has a limit. A well-known criterion of completeness: $\mathcal{K}$ is complete iff $\mathcal{K}$ has products and equalizers. The dual concept is *cocomplete*. Verify that **Set** is a complete and cocomplete category.

F. Chain colimits. Each poset $(X, \leq)$ is considered as a small category: objects are the elements of X, and for $x, y \in X$ there is either a unique morphism $x \to y$, if $x \leq y$, or no morphism, if $x \not\leq y$. In particular, each ordinal α is a small category (the well-ordered poset of all ordinals smaller than α). An *α-chain* in a category $\mathcal{K}$ is a diagram D with the scheme α, i.e., a collection of objects $D_i (i < \alpha)$ and morphisms $D_{ij}: D_i \to D_j$ $(i \leq j < \alpha)$ such that $D_{ii} = 1$ and $D_{ij} \cdot D_{jk} = D_{ik}$. A category is said to be *chain-cocomplete* if each chain has a colimit; this includes the initial object (= colimit of the 0-chain).

Verify that each α-chain $D: \alpha \to \mathcal{K}$ has the same colimit as the diagram $D_0: (X, \leq) \to \mathcal{K}$ which is the restriction of D to a cofinal set $X \subset \alpha$ (i.e., for each $i \in \alpha$ there is $j \in X$ with $i \leq j$). In particular, the colimit of an α-chain is not changed by leaving out the "start", i.e., all objects D_i with $i \leq i_0$ (for some $i_0 < \alpha$).

G. Subcategory of a category $\mathcal{K}$ is a category $\mathcal{L}$ with $\mathcal{L}^{\circ} \subseteq \mathcal{K}^{\circ}$ and $\mathcal{L}^{\mathrm{m}} \subseteq \mathcal{K}^{\mathrm{m}}$, and with the same composition and unit morphisms as $\mathcal{K}$. A subcategory $\mathcal{L}$ of $\mathcal{K}$ is *full* iff each $\mathcal{K}$-morphism $f: A \to B$ with $A, B \in \mathcal{L}$ is an $\mathcal{L}$-morphism.

Verify that **Lat** is a non-full subcategory of the category **Pos**. On the other hand, **Pos** is a full subcategory of the category of graphs (= binary relations) and compatible maps.

III.2. Automata in a Category

2.1. A sequential Σ-automaton (I.1.1) can be depicted by the following diagram in the category

Set

of sets and maps:

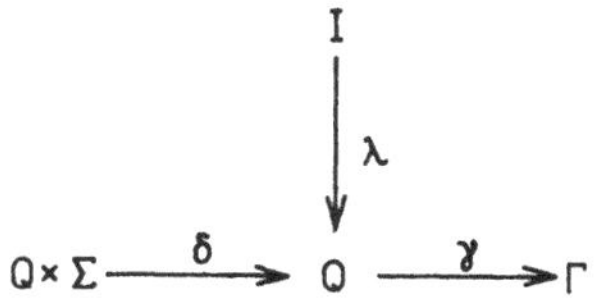

Here Q is the set of states, Σ is the input alphabet, Γ is the output alphabet and $I = \{0\}$ is a singleton set with $(0)\lambda \in Q$ the initial state.

Let us define a functor

$$F\colon \mathbf{Set} \to \mathbf{Set}$$

as follows:

$XF = X \times \Sigma$	for each set X;
$fF = f \times 1_\Sigma$	for each map f.

Then a sequential automaton is a diagram of the following kind:

(*)

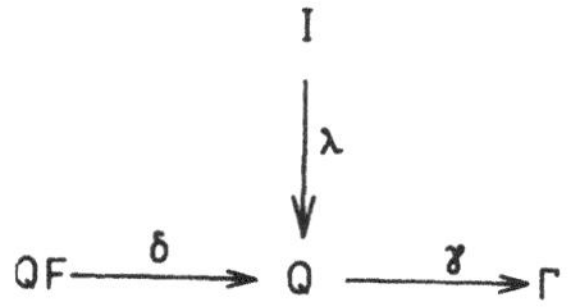

Analogously we can represent Σ-tree automata. Assume, for simplicity, that Σ has just one binary operation. Then a Σ-tree automaton (II.1.2) is a diagram as follows:

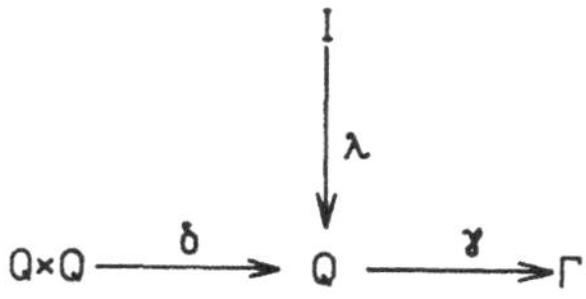

Here, I will be an arbitrary set (of variables). Let us define a functor

$$F\colon \mathbf{Set} \to \mathbf{Set}$$

by

$XF = X \times X$	for each set X;
$fF = f \times f$	for each map f.

Then a Σ-tree automaton is a diagram (*) again.

It is sometimes important to study sequential (or tree) automata with an additional structure on the sets Q, Σ and Γ and with the maps δ and γ preserving the structure. For example, Q, Σ and Γ are modules and the maps δ and γ are linear. Then we get a diagram (∗) again, but not in the category **Set**. These considerations have led to the following concept.

Definition. Let $\mathcal{K}$ be a category and let $F\colon \mathcal{K} \to \mathcal{K}$ be a functor. An *F-automaton* in the category $\mathcal{K}$ is a sixtuple

$$A = (Q, \delta, \Gamma, \gamma, I, \lambda)$$

consisting of objects of $\mathcal{K}$

Q (state object);
Γ (output object);
I (initialization object)

and morphisms of $\mathcal{K}$

$\delta\colon QF \to Q$ (next-state morphism);
$\gamma\colon Q \to \Gamma$ (output morphism);
$\lambda\colon I \to Q$ (initialization morphism).

Remark. We shall often work with less complex concepts: pairs (Q, δ) with $\delta\colon QF \to Q$ are called *F-algebras*; quadruples $(Q, \delta, \Gamma, \gamma)$ with (Q, δ) an F-algebra and $\gamma\colon Q \to \Gamma$ are called *non-initial F-automata.*

2.2. Let us introduce the concept of a morphism. Given two F-algebras (Q, δ) and (Q', δ'), an *F-homomorphism*

$$f\colon (Q, \delta) \to (Q', \delta')$$

is a morphism $f\colon Q \to Q'$ such that the following square

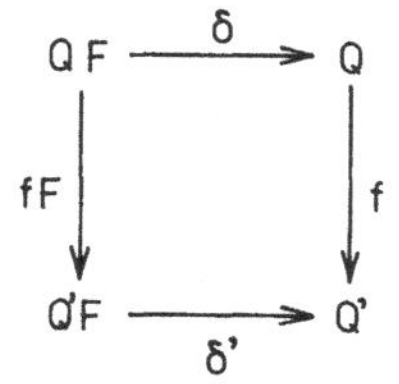

commutes.

Given two F-automata with joint output and initialization objects:

$$A = (Q, \delta, \Gamma, \gamma, I, \lambda) \quad \text{and} \quad A' = (Q', \delta', \Gamma, \gamma', I, \lambda')$$

a *morphism of F-automata*

$$f\colon A \to A'$$

is a homomorphism $f: (Q, \delta) \to (Q', \delta')$ with $\lambda \cdot f = \lambda'$ and $f \cdot \gamma' = \gamma$, i.e., a morphism for which the following diagram

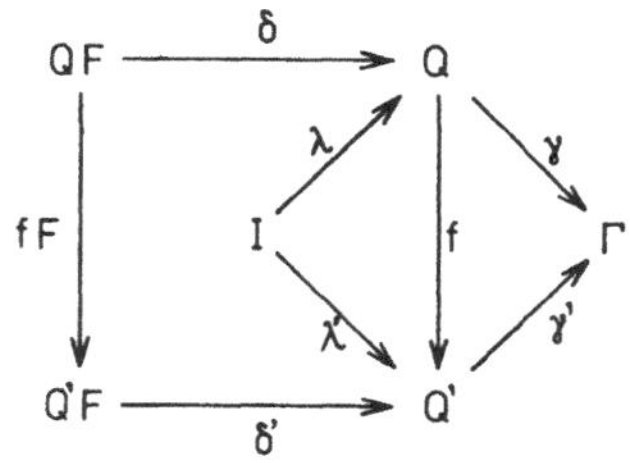

commutes. If f is an isomorphism (in $\mathscr{K}$), we say that A and A' are *isomorphic automata*. It is easy to check that

(a) given a homomorphism $f: (Q, \delta) \to (Q', \delta')$, such that f is an isomorphism in $\mathscr{K}$, then $f^{-1}: (Q', \delta') \to (Q, \delta)$ is a homomorphism;

(b) the composition of two homomorphisms is a homomorphism, too.

Analogous statements hold for morphisms of F-automata. A more general concept of morphism (corresponding to I.1.7) is used in Chapter VI.

2.3. Example: Sequential Σ-automata. Assuming that $\mathscr{K}$ has finite products, we define a functor

$$S_\Sigma: \mathscr{K} \to \mathscr{K}$$

(where Σ is an object of $\mathscr{K}$) as follows:

$$XS_\Sigma = X \times \Sigma \qquad \text{for each object } X;$$
$$fS_\Sigma = f \times 1_\Sigma \qquad \text{for each morphism } f.$$

Then S_Σ-automata are called the *sequential Σ-automata* in $\mathscr{K}$.

(i) $\mathscr{K}$ = **Set**. Here, sequential Σ-automata are precisely the concept studied in Chapter I, except that the initial state is generalized. We have (as in the case of tree automata, II.1.2) a set I of variables and an initialization $\lambda: I \to Q$.

If $I = \{0\}$ is a singleton set, then the morphisms introduced here coincide with those of Chapter I (for the case of a fixed Γ, see I.1.7). Indeed, the condition

$$\lambda' = \lambda \cdot f$$

states that f preserves the initial state $(0)\lambda$.

(ii) $\mathscr{K}$ = R-**Mod**, the category of R-modules and linear maps (for a fixed commutative ring R). In case

$$I = \{0\}$$

is the trivial module, a sequential Σ-automaton consists of

a module Q (of states);
a module Γ (of outputs);
a linear (next-step) map $\delta: Q \times \Sigma \to Q$;
a linear (output) map $\gamma: Q \to \Gamma$.

These are precisely the *linear sequential Σ-automata*, also called linear, discrete-time dynamical systems.

Since in R-**Mod** products and coproducts of pairs of objects coincide (see Exercise III.1.C), the map $\delta: Q + \Sigma \to Q$ is expressed by a pair of linear maps

$$\delta_0: \Sigma \to Q \text{ and } \delta_1: Q \to Q$$

with $(q, \sigma)\delta = (q)\delta_1 + (\sigma)\delta_0$ for each $q \in Q$, $\sigma \in \Sigma$. Here δ_1 represents the "reaction to time": in each time unit the state is changed from q to $(q)\delta_1$ if no input (i.e., input 0) arrives; and δ_0 is the "reaction to input". The addition of both reactions is the resulting change of state in one time unit.

Given two linear sequential Σ-automata

$$A = (Q, \delta_1, \delta_0, \Gamma, \gamma) \quad \text{and} \quad A' = (Q', \delta_1', \delta_0', \Gamma, \gamma'),$$

a morphism

$$f: A \to A'$$

is a linear map $f: Q \to Q'$ such that for each $q \in Q$ and $\sigma \in \Sigma$,

(a) $(\sigma)\delta_0 \cdot f = (\sigma)\delta'_0$;
(b) $(q)\delta_1 \cdot f = (q)f \cdot \delta_1'$;
(c) $(q)\gamma = (q)f \cdot \gamma'$.

Thus, f preserves the reaction to inputs and time, as well as the outputs.

(iii) $\mathscr{K}$ = **Pos**, the category of posets (i.e., partially ordered sets) and order-preserving maps. The product of two posets $(X, \leq)$ and $(Y, \leq)$ is the set $X \times Y$ ordered coordinate-wise: $(x, y) \leq (x', y')$ iff both $x \leq x'$ and $y \leq y'$. Thus, a sequential Σ-automaton in **Pos** is just a sequential Σ-automaton in **Set**, with an additional order on each of the sets Q, Σ, Γ and I such that δ, γ and λ are order-preserving. [For δ this means that $q \leq q'$ and $\sigma \leq \sigma'$ imply $(q, \sigma)\delta \leq (q', \sigma')\delta$.] If we think of $\leq$ as a "preference" relation, then these conditions mean that (i) if a state q' is preferred to a state q, then after one step the same relation will hold: $(q, \sigma)\delta \leq (q', \sigma)\delta$ for each $\sigma \in \Sigma$; (ii) if an input σ' is preferred to an input σ, then the corresponding changes of state are in the same relation: $(q, \sigma)\delta \leq (q, \sigma')\delta$ for each $q \in Q$; (iii) the preference is shown by the outputs: $q \leq q'$ implies $(q)\gamma \leq (q')\gamma$; (iv) preference of variables is preserved by the initialization: $x \leq x'$ implies $(x)\lambda \leq (x')\lambda$.

2.4. The role played by the free monoid of words Σ^* for sequential automata (and by the free Σ-algebra $I^\#$ of Σ-trees for tree automata) is played by the free F-algebras for F-automata:

Definition. Let $F: \mathcal{K} \to \mathcal{K}$ be a functor. An object I of $\mathcal{K}$ is said to *generate a free F-algebra* $(I^{\#}, \varphi)$ if there exists a morphism $\eta: I \to I^{\#}$ (called the *injection of generators*) with the following universal property:

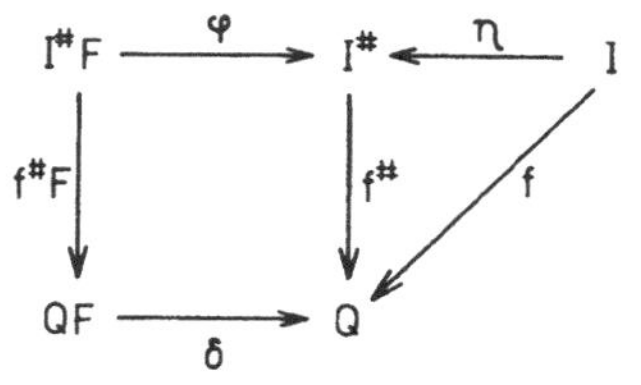

for each F-algebra (Q, δ) and each morphism $f: I \to Q$ there exists a unique homomorphism $f^{\#}: (I^{\#}, \varphi) \to (Q, \delta)$ with $f = \eta \cdot f^{\#}$.

If each object of $\mathcal{K}$ generates a free algebra, then F is called a *varietor* (or input process in the terminology of M. A. Arbib and E. G. Manes). Equivalently, F is a varietor iff the forgetful functor of the category of F-algebras is an adjoint (see Exercice. III.2.B below).

Example. The functor

$$S_\Sigma: \mathbf{Set} \to \mathbf{Set}$$

is a varietor. Here

$$I^{\#} = I \times \Sigma^*$$

and

$$\varphi: (I \times \Sigma^*) \times \Sigma \to I \times \Sigma^*$$

is defined by

$$(i, \sigma_1 \ldots \sigma_n, \sigma)\varphi = (i, \sigma_1 \ldots \sigma_n \sigma)$$

for each $i \in I$, $\sigma_1 \ldots \sigma_n \in \Sigma^*$ and $\sigma \in \Sigma$. Further,

$$(i)\eta = (i, \emptyset) \text{ for each } i \in I.$$

Indeed, for each S_Σ-algebra (Q, δ) and each map $f: I \to Q$, the unique homomorphism $f^{\#}: (I^{\#}, \varphi) \to (Q, \delta)$ with $f = \eta \cdot f^{\#}$ is defined as follows:

$$(i, \emptyset)f^{\#} = (i)\eta \cdot f^{\#} = (i)f;$$
$$(i, \sigma)f^{\#} = (i, \emptyset, \sigma)\varphi \cdot f^{\#} = ((i)f, \sigma)\delta;$$
$$(i, \sigma_1\sigma_2)f^{\#} = (i, \sigma_1, \sigma_2)\varphi \cdot f^{\#} = ((i, \sigma_1)f^{\#}, \sigma_2)\delta;$$
etc.

See I.1.3.

The following Chapter IV is devoted to an explicit construction of the free algebra $I^{\#}$, to a characterization of varietors and to related problems.

Remark. Free algebras are determined uniquely up to an isomorphism:

(i) For each isomorphism of F-algebras

$$h: (I^{\#}, \varphi) \to (Q_0, \delta_0),$$

the algebra (Q_0, δ_0) is a free F-algebra generated by I with the injection $\eta \cdot h: I \to Q_0$.

Proof. Given an F-algebra (Q, δ) and a morphism $f: I \to Q$, there exists a unique homomorphism $f^{\#}: (I^{\#}, \varphi) \to (Q, \delta)$ with $f = \eta \cdot f^{\#}$. Then $h^{-1} \cdot f^{\#}: (Q_0, \delta_0) \to (Q, \delta)$ is the unique homomorphism with $f = (\eta \cdot h) \cdot (h^{-1} \cdot f^{\#})$. □

(ii) If $(I^{\#}, \varphi)$ and $(I_1^{\#}, \varphi_1)$ are both free F-algebras generated by I, then there exists a unique isomorphism $h: (I^{\#}, \varphi) \to (I_1^{\#}, \varphi_1)$ with $\eta \cdot h = \eta_1$.

Proof. Since $(I^{\#}, \varphi)$ is free, the morphism $\eta_1: I \to I_1^{\#}$ can be extended to a homomorphism h: $(I^{\#}, \varphi) \to (I_1^{\#}, \varphi_1)$ with $\eta \cdot h = \eta_1$. To prove that h is an isomorphism, we use the fact that $(I_1^{\#}, \varphi_1)$ is free and hence, $\eta: I \to I^{\#}$ can be extended to a homomorphism k: $(I_1^{\#}, \varphi_1) \to (I^{\#}, \varphi)$ with $\eta_1 \cdot k = \eta$. Then $h \cdot k = 1_{I^{\#}}$ because $h \cdot k: (I^{\#}, \varphi) \to (I^{\#}, \varphi)$ is a homomorphism extending η (since $\eta \cdot h \cdot k = \eta_1 \cdot k = \eta$) and also $1_{I^{\#}}$ is such a homomorphism—but the extension of any morphism is unique. Analogously, $k \cdot h = 1_{I_1^{\#}}$.

We should actually speak about *a* free F-algebra generated by I. We shall, nevertheless, disregard the non-uniqueness of free algebras because this leads to no confusion. For example, let I be a singleton set, then free S_Σ-algebra $I^{\#}$ is either (Σ^*, φ) as in I.1.6, or $(I \times \Sigma^*, \varphi)$ as above.

If $I = \perp$ is the initial object (Exercise III.1.E), then the free algebra $(\perp^{\#}, \varphi)$ is the *initial F-algebra*, i.e., for each F-algebra (Q, δ) there exists a unique homomorphism from $(\perp^{\#}, \varphi)$ to (Q, δ) (extending the unique morphism $\perp \to Q$).

Example. The functor

$$S_\Sigma: R\text{-}\mathbf{Mod} \to R\text{-}\mathbf{Mod}$$

is a varietor for each module Σ.

Let us first describe the initial algebra (i.e., the free algebra generated by the trivial module 0). This is the module

$$0^{\#} = \Sigma[z]$$

of all polynomials $(z)a = \sigma_0 + \sigma_1 z + \ldots + \sigma_n z^n$ in indeterminate z and with coefficients $\sigma_0, \ldots, \sigma_n$ in Σ (the addition and scalar multiplication are defined coordinate-wise). The injection $\eta: 0 \to \Sigma[z]$ is defined by $(0)\eta = 0$ and the operation by

$$((z)a, \sigma)\varphi = \sigma + (z)a \cdot z = \sigma + \sigma_0 z + \sigma_1 z^2 + \ldots + \sigma_n z^{n+1}.$$

To verify this, we are to show that for each S_Σ-algebra (Q, δ) there is a unique homomorphism

$$h: (\Sigma[z], \varphi) \to (Q, \delta).$$

Put

$$(\sigma_0 + \sigma_1 z + \ldots + \sigma_n z^n)h = (\sigma_0)\delta_0 + (\sigma_1)\delta_0 \cdot \delta_1 + \ldots + (\sigma_n)\delta_0 \cdot \delta_1^n.$$

Then h is a homomorphism, i.e., $\varphi \cdot h = (h \times 1_\Sigma) \cdot \delta$, because for each $(z)a \in \Sigma[z]$ and each $\sigma \in \Sigma$ we have

$$\begin{aligned}((z)a, \sigma)\varphi \cdot h &= (\sigma + (z)a \cdot z)h \\ &= (\sigma)\delta_0 + (\sigma_0)\delta_0 \cdot \delta_1 + \ldots + (\sigma_n)\delta_0 \cdot \delta_1^{n+1} \\ &= (\sigma)\delta_0 + [(\sigma_0)\delta_0 + \ldots (\sigma_n)\delta_0 \cdot \delta_1^n]\delta_1 \\ &= (\sigma)\delta_0 + [((z)a)h]\delta_1 \\ &= ((z)a, \sigma)(h \times 1_\Sigma) \cdot \delta.\end{aligned}$$

And h is unique because $\varphi \cdot h = (h \times 1_\Sigma) \cdot \delta$ implies that

$$(\sigma_0)h = (0, \sigma_0)\varphi \cdot h = (0, \sigma_0)\delta = (\sigma_0)\delta_0,$$
$$(\sigma_0 + \sigma_1 z)h = (\sigma_1, \sigma_0)\varphi \cdot h = ((\sigma_1)h, \sigma_0)\delta = (\sigma_1)\delta_0 \cdot \delta_1 + (\sigma_0)\delta_0,$$

etc. Therefore, h is necessarily defined by the rule above.

More in general, for each module I we have

$$I^\# = I[z] \times \Sigma[z]$$

with $\eta: I \to I[z] \times \Sigma[z]$ defined by $(i)\eta = (i, 0)$ and

$$\varphi: I[z] \times \Sigma[z] \times \Sigma \to I[z] \times \Sigma[z]$$

defined by $((z)b, (z)a, \sigma)\varphi = ((z)b, \sigma + (z)a \cdot z)$. For each S_Σ-algebra (Q, δ) and each linear map $f: I \to Q$, the unique extension to a homomorphism $f^\#: I[z] \times \Sigma[z] \to Q$ is defined by

$$(i_0 + i_1 z + \ldots + i_n z^n, \sigma_0 + \sigma_1 z + \ldots + \sigma_n z^n)f^\# = \sum_{k=0}^{n} ((i_k)f + (\sigma_k)\delta_0)\delta_1^k.$$

2.5. The concept of behavior studied in Chapters I and II is generalized as follows.

Definition. Let $F: \mathcal{K} \to \mathcal{K}$ be a varietor. For each F-automaton $A = (Q, \delta, \Gamma, \gamma, I, \lambda)$ we denote by

$$\rho: (I^\#, \varphi) \to (Q, \delta)$$

the unique homomorphism extending λ (i.e., with $\lambda = \eta \cdot \rho$). We call ρ the *run map* of A and

$$\beta = \rho \cdot \gamma: I^\# \to \Gamma$$

the *behavior* of A.

Example: Σ-tree automata. Let first Σ be a type consisting of a single operation of arity n (where n is a cardinal). Define a functor

$$H_n: \mathbf{Set} \to \mathbf{Set}$$

by forming the n-fold cartesian products: for each set X,

$$XH_n = X^n;$$

for each map $f: X \to Y$,

$$fH_n = f^{(n)},$$

where $(x_i)f^{(n)} = ((x_i)f)$. Then H_n-algebras are precisely algebras of type Σ, i.e., pairs (Q, δ), where $\delta: Q^n \to Q$ is an n-ary operation. H_n-homomorphisms are also precisely Σ-homomorphisms.

Now, let Σ be an arbitrary type. We denote by

$$H_\Sigma: \mathbf{Set} \to \mathbf{Set}$$

the coproduct of functors $H_{|\sigma|}$, $\sigma \in \Sigma$ (where $|\sigma|$ denotes the arity of σ):

$$H_\Sigma = \coprod_{\sigma \in \Sigma} H_{|\sigma|}.$$

Thus, for each set X we have

$$XH_\Sigma = \coprod_{\sigma \in \Sigma_n} X^n$$

and analogously with the maps fH_Σ. An H_Σ-algebra consists of a set Q and a map

$$\delta: \coprod_{\sigma \in \Sigma_n} Q^n \to Q$$

or, equivalently, a collection of maps (the components of δ),

$$\delta_\sigma: Q^n \to Q \; (\sigma \in \Sigma, |\sigma| = n).$$

This is precisely an algebra of type Σ. H_Σ-homomorphisms $f: (Q, \delta) \to (Q', \delta')$ are also just the usual homomorphisms because

$$\delta \cdot f = fH_\Sigma \cdot \delta' \quad \text{iff} \quad \delta_\sigma \cdot f = f^{(n)} \cdot \delta'_\sigma \qquad \text{for each } \sigma \in \Sigma.$$

The functors H_Σ are varietors: for each set I the free H_Σ-algebra generated by I is the algebra $(I^\#, \varphi)$ of finite-path Σ-trees over I, see II.3.6.

Σ-tree automata are precisely the H_Σ-automata. The run map $\rho: I^\# \to Q$ and the behavior $\beta: I^\# \to \Gamma$ have the same meaning as in II.3.7.

Remark. More generally, for each category $\mathcal{K}$ with products we define a functor

$$H_n : \mathcal{K} \to \mathcal{K}$$

on objects X by

$$XH_n = X^n$$

(the n-fold product of X) and on morphisms $f: X \to Y$ by

$$fH_n = f^{(n)} : X^n \to Y^n,$$

where $f^{(n)}$ denotes the morphism all components of which are equal to f.

If $\mathcal{K}$ has also coproducts, then for each type Σ we define

$$H_\Sigma : \mathcal{K} \to \mathcal{K}$$

by

$$H_\Sigma = \coprod_{\sigma \in \Sigma} H_{|\sigma|}$$

2.6. Proposition. Let F be a varietor and let

$$f: A \to A'$$

be a morphism of automata. The run maps ρ(of A) and ρ' (of A') fulfil

$$\rho \cdot f = \rho'.$$

Proof. The run map ρ' is the unique homomorphism $\rho' : (I^{\#}, \varphi) \to (Q', \delta')$ with $\eta \cdot \rho' = \lambda'$. Since both $\rho : (I^{\#}, \varphi) \to (Q, \delta)$ and $f: (Q, \delta) \to (Q', \delta')$ are homomorphisms, $\rho \cdot f: (I^{\#}, \varphi) \to (Q', \delta')$ is also a homomorphism. We have

$$\eta \cdot (\rho \cdot f) = \lambda \cdot f = \lambda'.$$

Therefore, $\rho \cdot f = \rho'$. □

Corollary. Any two F-automata connected by a morphism have the same behavior.

In fact, for each morphism $f: A \to A'$ we have $\rho \cdot f = \rho'$ and hence, $\beta = \rho \cdot \gamma = \rho \cdot f \cdot \gamma' = \rho' \cdot \gamma' = \beta'$.

2.7. Another concept we want to generalize is reachability. We shall use image factorizations of morphisms. We begin with two examples.

Examples. (i) $\mathcal{K}$ = **Set**. Each map $f: X \to Y$ can be factorized as $f = e \cdot m$ with

$$e: X \to T \quad \text{epi } (= \text{onto})$$

and

$$m: T \to Y \quad \text{mono } (= \text{one-to-one}).$$

For example, let $T = X/\sim$ be the quotient set of X under the kernel equivalence $\sim$ of f [$x \sim x'$ iff $(x)f = (x')f$]. Let $e: X \to T$ be the quotient map, assigning to each $x \in X$ its equivalence class $[x] \in T$; e is clearly onto. Let $m: T \to Y$ be the map defined by $([x])m = (x)f$ for each $[x] \in T$; m is clearly well-defined and one-to-one. We have $(x)f = ([x])m = (x)e \cdot m$, and thus, $f = e \cdot m$.

This factorization is essentially unique: if $f = e' \cdot m'$ with $e': X \to T'$ onto and $m': T' \to Y$ one-to-one, then the kernel equivalences of f and of e' coincide (because m' is one-to-one) and we have a unique isomorphism $i: T \to T'$ such that the following diagram

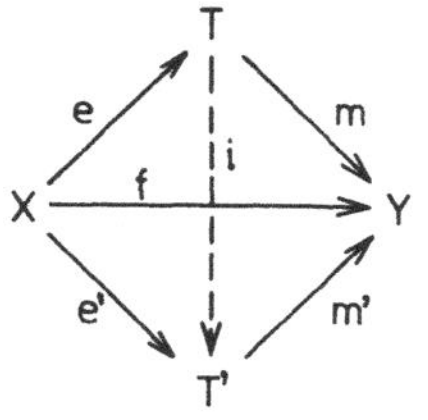

commutes: Put $([x])\, i = (x)e'$ for each $[x] \in T$.

(ii) $\mathcal{K}$ = **Pos**. There are two important ways of factorization of a morphism $f: (X, \leq) \to (Y, \preceq)$:

(a) Epis and embeddings. Let $T = X/\sim$ be as above. Define an ordering $\preceq^*$ on T by

$$[x_1] \preceq^* [x_2] \quad \text{iff} \quad ([x_1])m \preceq ([x_2])\, m.$$

Then $m: (T, \preceq^*) \to (Y, \preceq)$ is an embedding [i.e., an isomorphism of $(T, \preceq^*)$ onto a subposet of $(Y, \preceq)$]. And $e: (X, \leq) \to (T, \preceq^*)$ is an order-preserving map onto, in other words, an epi.

(b) Quotient maps and monos. Define another ordering of T by

$$[x_1] \leq^* [x_2] \quad \text{iff} \quad \bar{x}_1 \leq \bar{x}_2$$

for some $\bar{x}_1 \sim x_1$ and $\bar{x}_2 \sim x_2$. Then $e: (X, \leq) \to (T, \leq^*)$ is a quotient map (i.e., an order-preserving onto map such that the order of the range is induced by the order of the domain). And $m: (T, \leq^*) \to (Y, \preceq)$ is a one-to-one order-preserving map, i.e., a mono.

The example of **Pos** suggests that in order to investigate image factorizations in a general category, we must specify *axiomatically* what factorizations are considered.

Definition. A *factorization system* in a category $\mathcal{K}$ consists of a class $\mathcal{E}$ of epis and a class $\mathcal{M}$ of monos such that

(i) each morphism $f: X \to Y$ can be factorized as $f = e \cdot m$ with $e: X \to T$ in $\mathcal{E}$ and $m: T \to Y$ in $\mathcal{M}$;

(ii) this factorization is essentially unique: if $f = e' \cdot m'$ with $e' \in \mathscr{E}$ and $m' \in \mathscr{M}$, then there exists an isomorphism i such that the following diagram

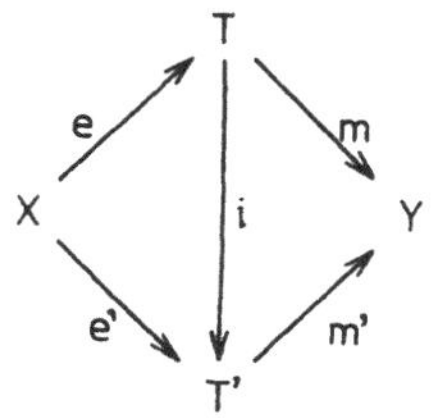

commutes;

(iii) both $\mathscr{E}$ and $\mathscr{M}$ are closed under composition and contain all isomorphisms.

Remarks. (i) Since factorization systems are fundamental in a number of considerations throughout the present book, we devote section III.5 below to their properties.

(ii) We call $\mathscr{K}$ an *$(\mathscr{E}, \mathscr{M})$-category* if a factorization system $(\mathscr{E}, \mathscr{M})$ is specified in $\mathscr{K}$. Thus, **Pos** is an

(epi, embedding)-category

or a

(quotient, mono)-category,

according to which factorizations are considered.

(iii) We often work with functors $F: \mathscr{K} \to \mathscr{K}$ *preserving $\mathscr{E}$-epis*, i.e. such that for each $e: X \to Y$ in $\mathscr{E}$, the morphism $eF: XF \to YF$ is in $\mathscr{E}$, too.

Lemma. Each factorization system $(\mathscr{E}, \mathscr{M})$ has the following *diagonal fill-in property:* for each commuting square

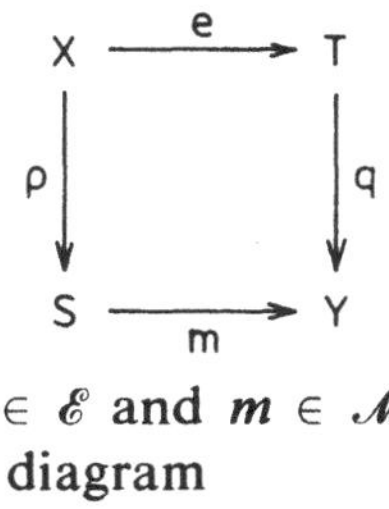

with $e \in \mathscr{E}$ and $m \in \mathscr{M}$ there exists a morphism $d: T \to S$ such that the following diagram

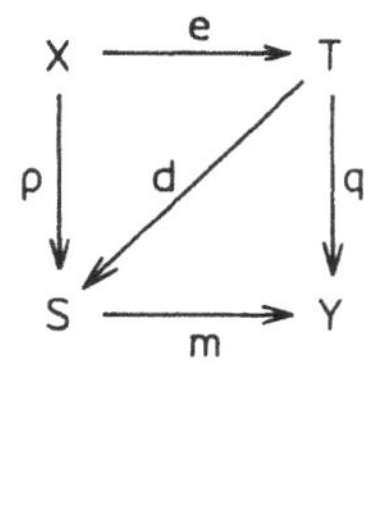

commutes.

Proof. Let $q = q_e \cdot q_m$ be the image factorization of q, and $p = p_e \cdot p_m$ that of p. The following diagram

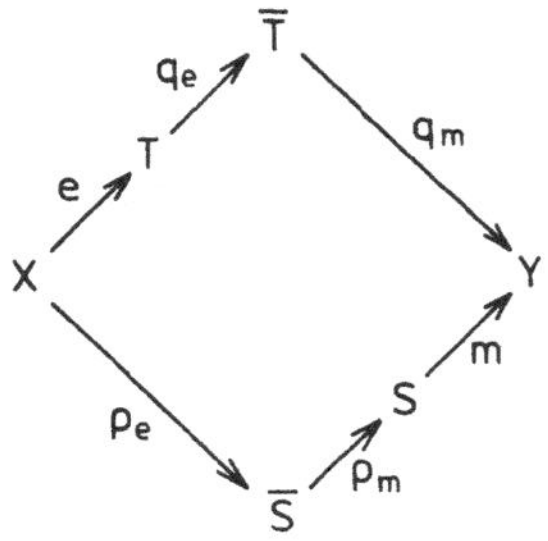

commutes. Since $e \cdot q_e \in \mathscr{E}$ and $q_m \in \mathscr{M}$, as well as $p_e \in \mathscr{E}$ and $p_m \cdot m \in \mathscr{M}$, there is an isomorphism $i\colon \bar{T} \to \bar{S}$ with

$$p_e = e \cdot q_e \cdot i \quad \text{and} \quad q_m = i \cdot p_m \cdot m.$$

Put

$$d = q_e \cdot i \cdot p_m\colon T \to S.$$

Then

$$e \cdot d = e \cdot q_e \cdot i \cdot p_m = p_e \cdot p_m = p$$

and

$$d \cdot m = q_e \cdot i \cdot p_m \cdot m = q_e \cdot q_m = q.$$

This concludes the proof. □

2.8. The reachability of sequential automata (I.2.3) can be generalized as follows.

Definition. Let F be a varietor on an $(\mathscr{E}, \mathscr{M})$-category $\mathscr{K}$. An F-automaton is said to be *reachable* if its run morphism is an $\mathscr{E}$-epi.

Examples. (i) Linear Σ-automata. Let $\mathscr{K} = R$-**Mod** with $\mathscr{E}$ = all onto morphisms and $\mathscr{M}$ = all one-to-one morphisms. This is a (unique, as we prove below) factorization system in R-**Mod**, since for each linear map $f\colon X \to Y$ we have a well-known image factorization

$$X \to X/\ker f \to Y.$$

Let $A = (Q, \delta_0, \delta_1, \Gamma, \gamma)$ be a linear automaton. The run map

$$\rho\colon \Sigma[z] \to Q$$

assigns to each input sequence $\sigma_0\sigma_1 \ldots \sigma_n$, represented by the polynomial $(z)a = \sigma_0 + \sigma_1 z + \ldots + \sigma_n z^n$, the state

$$(a)\rho \in Q$$

at which the automaton stops when receiving the given inputs in the initial state 0.

A linear Σ-automaton is reachable iff each of its states can be reached from the state 0 by a sequence of input symbols.

(ii) Ordered Σ-automata. For each poset Σ the functor

$$S_\Sigma : \mathbf{Pos} \to \mathbf{Pos}$$

is a varietor: given a poset I, then

$$I^\# = I \times \Sigma^*$$

where $I \times \Sigma^*$ is ordered coordinate-wise, i.e., $(i, \sigma_1 \ldots \sigma_n) \leq (j, \tau_1 \ldots \tau_m)$ iff $i \leq j$ in I, $n = m$, and $\sigma_t \leq \tau_t$ in Σ for $t = 1, \ldots, n$. Let I be a singleton poset. Then S_Σ-automata are just sequential Σ-automata with an order on Q and Γ such that both δ and γ are order-preserving maps.

Considering **Pos** as the (epi, embedding)-category, an ordered Σ-automaton is reachable iff for each state $q \in Q$ there exists a word $\sigma_1 \ldots \sigma_n \in \Sigma^*$ with $q = (\sigma_1 \ldots \sigma_n)\rho$. Thus, reachability is precisely the original concept, independent of the order.

For the (quotient, mono)-category **Pos**, an ordered Σ-automaton is reachable iff for arbitrary two states q, $q' \in Q$ with $q \leq q'$ there exist words $\sigma_1 \ldots \sigma_n$ and $\tau_1 \ldots \tau_n$ in Σ^* such that $\sigma_1 \ldots \sigma_n \leq \tau_1 \ldots \tau_n$ and $q = (\sigma_1 \ldots \sigma_n)\rho$, $q' = (\tau_1 \ldots \tau_n)\rho$. For example, if Σ is discretely ordered (i.e., $\sigma \leq \tau$ iff $\sigma = \tau$), then each reachable Σ-automaton has discretely ordered state set.

Remark. By a *subautomaton* of an F-automaton A we understand an F-automaton A_0 together with a morphism $m: A_0 \to A$ in $\mathcal{M}$. Two such morphisms

$$m: A_0 \to A \quad \text{and} \quad m': A_0' \to A$$

are said to represent the same subautomaton if there exists an isomorphism $j: A_0 \to A_0'$ such that $j \cdot m' = m$. (Compare with III.5.2.)

Proposition. Let $\mathcal{K}$ be an $(\mathcal{E}, \mathcal{M})$-category and let F be a varietor preserving $\mathcal{E}$-epis, i.e., $e \in \mathcal{E}$ implies $eF \in \mathcal{E}$. Then each F-automaton has a unique reachable subautomaton.

Proof. Let

$$A = (Q, \delta, \Gamma, \gamma, I, \lambda)$$

be an F-automaton. Its run morphism $\rho: I^\# \to Q$ has an image factorization $\rho = e \cdot m$ with

$$e: I^\# \to \bar{Q} \text{ in } \mathcal{E};$$
$$m: \bar{Q} \to Q \text{ in } \mathcal{M}.$$

Since $\varphi \cdot \rho = \rho F \cdot \delta$ and since $eF \in \mathscr{E}$, we can use the diagonal fill-in property:

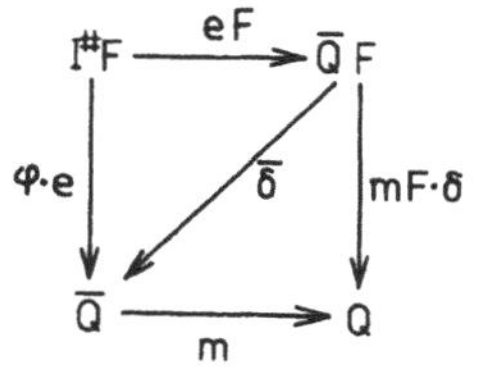

Then $\bar{\delta}$ is a morphism such that

$$e: (I^{\#}, \varphi) \to (\bar{Q}, \bar{\delta}) \text{ and } m: (\bar{Q}, \bar{\delta}) \to (Q, \delta)$$

are homomorphisms.

Put

$$A_0 = (\bar{Q}, \bar{\delta}, \Gamma, m \cdot \gamma, I, \eta \cdot e).$$

A_0 is an F-automaton. Since e is a homomorphism, it is clearly the run morphism of A_0. Thus, A_0 is reachable. Since m is a homomorphism, it is a morphism of automata

$$m: A_0 \to A.$$

Finally, let $m': A_0' \to A$ be another reachable subautomaton of A. The run map $\rho': I^{\#} \to Q'$ of A_0' is an $\mathscr{E}$-epi with

$$\rho' \cdot m' = \rho,$$

by Proposition III.2.6. Since $\rho' \in \mathscr{E}$ and $m' \in \mathscr{M}$ this is an image factorization of ρ and therefore, there is an isomorphism $i: Q \to Q'$ with

$$e \cdot i = \rho' \quad \text{and} \quad m = i \cdot m'.$$

The following diagram

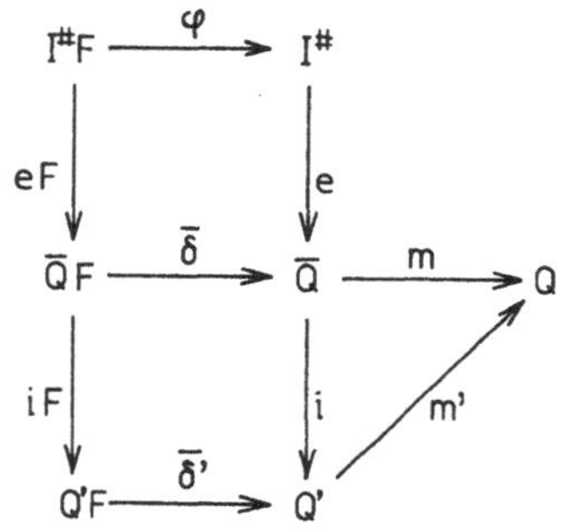

commutes: Since eF is epi, we conclude $\bar{\delta} \cdot i = iF \cdot \bar{\delta}'$ from the following

$$eF \cdot (\bar{\delta} \cdot i) = \varphi \cdot e \cdot i = \varphi \cdot \rho' = \rho' F \cdot \bar{\delta}' = eF \cdot (iF \cdot \bar{\delta}').$$

Thus, $i: A_0 \to A_0'$ is a morphism. This proves the unicity. □

2.9. Definition. Let $\mathcal{K}$ be an $(\mathcal{E}, \mathcal{M})$-category and let $F: \mathcal{K} \to \mathcal{K}$ be a varietor. For each *behavior*, i.e., a morphism

$$\beta: I^{\#} \to \Gamma \qquad (I, \Gamma \in \mathcal{K}^{\circ})$$

the *minimal realization* is a reachable F-automaton A_0 which realizes β (i.e., its behavior is β) and has the following universal property:

> for each reachable realization A of β there exists a morphism $e: A \to A_0$ with $e \in \mathcal{E}$.

Remark. For Σ-tree automata, minimal realizations exist iff Σ is finitary (II.3.8). We devote Chapter V to the general problem which varietors have the property that each behavior has a minimal realization.

Proposition. Minimal realization is unique up to an isomorphism:

(i) If A_0 is a minimal realization of β, then each F-automaton isomorphic to A_0 is also a minimal realization.

(ii) Any two minimal realizations of β are isomorphic.

Proof. Let A_0 and A_1 be minimal realizations of β. There exist $\mathcal{E}$-morphisms $e: A_1 \to A_0$ and $\bar{e}: A_0 \to A_1$. Let us check that $\bar{e} = e^{-1}$. Denote by $\rho_0: I^{\#} \to Q_0$ and $\rho_1: I^{\#} \to Q_1$ the run morphisms of A_0 and A_1, respectively. Then

$$\rho_0 = \rho_1 \cdot e \quad \text{and} \quad \rho_1 = \rho_0 \cdot \bar{e}$$

by Proposition III.2.6. Since ρ_0 is an epi and $\rho_0 \cdot (\bar{e} \cdot e) = \rho_1 \cdot e = \rho_0$, we conclude that $\bar{e} \cdot e = 1$; analogously $e \cdot \bar{e} = 1$.

Conversely, let $i: A_0 \to A_1$ be an isomorphism of automata. Then A_1 is a realization of β (III.2.6) and its run map is $\rho_0 \cdot i \in \mathcal{E}$. Given another reachable realization A of β, there exists a morphism $e: A \to A_0$ with $e \in \mathcal{E}$. Then $e \cdot i: A \to A_1$ is a morphism with $e \cdot i \in \mathcal{E}$. This proves the minimality of A_1. □

2.10. Adjoint functors. The above concepts are much simplified if the functor F has an adjoint. We recall briefly some of the basic facts concerning adjoint functors, but a reader not familiar with this theory can skip the rest of section III.2, not breaking the continuity of the text.

A functor $F: \mathcal{K} \to \mathcal{L}$ is said to *have an adjoint functor* $G: \mathcal{L} \to \mathcal{K}$ if there is a natural transformation $\eta: 1_{\mathcal{K}} \to F \cdot G$ such that each $\mathcal{K}$-morhpism $f: K \to LG$ (with $K \in \mathcal{K}^{\circ}$ and $L \in \mathcal{L}^{\circ}$) has the form $f = \eta_K \cdot f^*G$ for a unique $f^*: KF \to L$ in $\mathcal{L}^{m}$. Equivalent condition: There is a natural transformation $\varepsilon: G \cdot F \to 1_{\mathcal{L}}$ such that each $\mathcal{L}$-morphism $g: KF \to L$ has the form $g = g_*F \cdot \varepsilon_L$ for a unique $g_*: K \to LG$. The functor F is then called a *coad-*

joint of G (or left adjoint) since this is the dual concept of adjoint (also called right adjoint).

Examples. (i) For each set Σ, the functor S_Σ: **Set** $\to$ **Set** has an adjoint hom(Σ, —): **Set** $\to$ **Set** assigning to each set X the set hom(Σ, X) of all maps from Σ to X, and to each map $f: X \to Y$ the map sending $p: \Sigma \to X$ to $p \cdot f: \Sigma \to Y$. Here $\eta_X: X \to \hom(\Sigma, \Sigma \times X)$ assigns to each $x \in X$ the map $\sigma \mapsto (\sigma, x)$ in $\hom(\Sigma, \Sigma \times X)$.

(ii) Let $\mathscr{L}$ be a subcategory of $\mathscr{K}$. The inclusion functor $G: \mathscr{L} \to \mathscr{K}$ is adjoint iff $\mathscr{L}$ is a *reflective subcategory*, i.e., for each $\mathscr{K}$-object K there is an $\mathscr{L}$-object K^* (the reflection of K in $\mathscr{L}$) and a $\mathscr{K}$-morphism $r_K: K \to K^*$ universal in the following sense: for each $\mathscr{K}$-morphism $f: K \to L \in \mathscr{L}$ there is a unique $\mathscr{L}$-morphism $f^*: K^* \to L$ with $f = r_K \cdot f^*$. The functor $F: \mathscr{K} \to \mathscr{L}$ given by $KF = K^*$, and $fF = (r_K \cdot f)^*$ for all $f: K \to K'$, is the corresponding coadjoint.

Coadjoints and colimits. Each coadjoint $F: \mathscr{K} \to \mathscr{L}$ preserves colimits. That is, given a diagram $D: \mathscr{D} \to \mathscr{K}$ with a colimit $dD \xrightarrow{\varepsilon_d} C$ (see Exercise III.1.E), then the diagram $D \cdot F: \mathscr{D} \to \mathscr{L}$ has a colimit $(dD)F \xrightarrow{\varepsilon_d F}$ CF. By *Freyd's Adjoint Functor Theorem*, conversely, a functor $F: \mathscr{K} \to \mathscr{L}$ preserving colimits is a coadjoint whenever $\mathscr{K}$ is cocomplete, and each $\mathscr{L}$-object L has (small) *solution set*, i.e., a set of morphisms $\varepsilon_i: K_i F \to L$ $(i \in I)$ such that any morphism $g: KF \to L$ factors as $g = g_* F \cdot \varepsilon_i$ for some $g_*: K_i \to L$, $i \in I$.

Dually, an adjoint preserves limits, and the converse holds if solution sets and limits exist.

2.11. Adjoint automata. Let $F: \mathscr{K} \to \mathscr{L}$ have an adjoint. We then speak about adjoint automata in $\mathscr{K}$. Their properties strongly resemble those of sequential automata in **Set**. (We prove in III.4 that the only coadjoints F: **Set** $\to$ **Set** are $F = S_\Sigma$.) We first turn to the construction of free F-algebras—here we only need the fact that F preserves countable colimits, i.e., colimits of diagrams with countably many morphisms.

Consider the adjoint pair S_Σ and hom(Σ, —) in **Set** (2.10). The free S_Σ-algebra

$$I^\# = I \times \Sigma^* = (I \times \{\emptyset\}) \cup (I \times \Sigma) \cup (I \times \Sigma \times \Sigma) \cup \dots$$

can be re-written as

$$I^\# = I + IS_\Sigma + IS_\Sigma^2 + \dots = \coprod_{n < \omega} IS_\Sigma^n,$$

where for each $F: \mathscr{K} \to \mathscr{K}$ we put $F^0 = 1_\mathscr{K}$, $F^1 = F$, $F^2 = F \cdot F$, etc.

Analogously, for each coadjoint $F: \mathscr{K} \to \mathscr{K}$ we shall prove that

$$I^\# = I + IF + IF^2 + \dots = \coprod_{n < \omega} IF^n$$

provided that $\mathcal{K}$ has countable coproducts. Since F preserves colimits we have then

$$I^{\#}F = \coprod_{n<\omega} IF^{n+1} = \coprod_{0<n<\omega} IF^{n}$$

and hence,

$$I^{\#} = I + I^{\#}F.$$

The coproduct injections of the last coproduct are

$$\eta: I \to I^{\#} \quad \text{and} \quad \varphi: I^{\#}F \to I^{\#}.$$

We prove this slightly more generally:

Proposition. Let $\mathcal{K}$ be a category with countable coproducts and let $F: \mathcal{K} \to \mathcal{K}$ be a functor preserving countable coproducts. Then each object I generates a free F-algebra, viz.,

$$I^{\#} = \coprod_{n<\omega} IF^{n}$$

with $\eta: I \to I^{\#}$ and $\varphi: I^{\#}F \to I^{\#}$ the coproduct injections of $I^{\#} = I + I^{\#}F$.

Proof. For each F-algebra (Q, δ) and each morphism $f: I \to Q$, define morphisms $f_n = IF^n \to Q$ by the following induction:

$$f_0 = f: I \to Q,$$
$$f_{n+1} = f_nF \cdot \delta: (IF^n)F \to Q.$$

Denote by

$$f^{\#}: I^{\#} \to Q$$

the morphism with components f_n $(n < \omega)$ (Exercise III.1.C.). Since F preserves countable coproducts, the morphism

$$f^{\#}F: \coprod_{0<n<\omega} IF^{n} \to QF$$

has components f_nF and thus, $f^{\#}F \cdot \delta: I^{\#}F \to Q$ has components f_{n+1} $(n < \omega)$. So does $f^{\#} \cdot \varphi: I^{\#}F \to Q$ and hence,

$$f^{\#}F \cdot \delta = f^{\#} \cdot \varphi.$$

We conclude that

$$f^{\#}: (I^{\#}, \varphi) \to (Q, \delta)$$

is a homomorphism; moreover,

$$\eta \cdot f^{\#} = f_0 = f.$$

Conversely, let

$$g:(I^{\#}, \varphi) \to (Q, \delta)$$

be a homomorphism with $\eta \cdot g = f$. Denote by $g_n: IF^n \to Q$ the n-th component of $g (n < \omega)$. Then $g_n F$ is the n-th component of gF. Since g is a homomorphism, we have

$$\varphi \cdot g = gF \cdot \delta: \coprod_{0 < n < \omega} IF^n \to Q,$$

or, by components,

$$g_{n+1} = g_n F \cdot \delta \qquad (n < \omega).$$

Since $g_0 = f = f_0$, we conclude by induction that $g_n = f_n$ $(n < \omega)$, and hence, $g = f$. □

2.12. Duality principle. Before continuing with adjoint automata we recall the duality principle used in category theory.

The *dual (or opposite) category* of a category $\mathscr{K}$ is obtained from $\mathscr{K}$ by reversing the direction of all arrows. That is, we define the dual category $\mathscr{K}^{\text{op}}$ to have the same objects as $\mathscr{K}$, and the morphisms from A to B in $\mathscr{K}^{\text{op}}$ to be precisely the $\mathscr{K}$-morphisms $f: B \to A$. The identity morphisms in $\mathscr{K}$ and $\mathscr{K}^{\text{op}}$ are the same, and the composition of $\mathscr{K}^{\text{op}}$ is inferred from $\mathscr{K}$ as follows

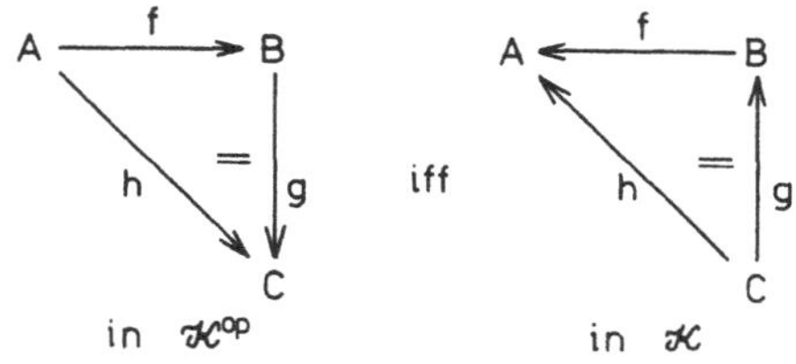

For each concept C concerning categories, the *dual concept* (often called co-C) is obtained by applying C to the dual categories. For example, the dual concept of equalizer is coequalizer (see III.1). For each functor $F: \mathscr{K} \to \mathscr{L}$ we define the *dual functor* $F^{\text{op}}: \mathscr{K}^{\text{op}} \to \mathscr{L}^{\text{op}}$ by $XF = XF^{\text{op}}$ and $fF = fF^{\text{op}}$.

The *duality principle* states that if a proposititon concerning categories and functors is valid, then so is the dual proposition, i.e., the proposition with all concepts substituted by their duals.

2.13. Observability. The observability of sequential automata, studied in I.2.7 and Exercise I.2.B, is captured by duality as follows. Let $F: \mathscr{K} \to \mathscr{K}$ be a coadjoint, and let $H: \mathscr{K} \to \mathscr{K}$ be the corresponding adjoint. For each object Γ put

$$\Gamma_{\#} = \prod_{n < \omega} \Gamma H^n.$$

Then there is an F-algebra $(\Gamma_\#, \psi)$ and a morphism $\pi: \Gamma_\# \to \Gamma$ with the following universal property:

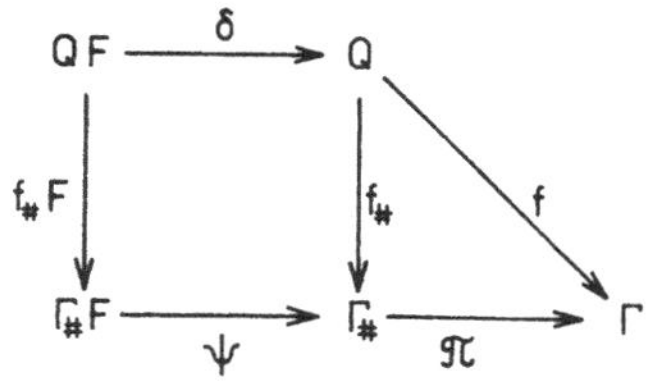

for each F-algebra (Q, δ) and each morphism $f: Q \to \Gamma$ there exists a unique homomorphism $f_\#: (Q, \delta) \to (\Gamma_\#, \psi)$ with $f = f_\# \cdot \pi$.

In fact, the functor

$$H^{\mathrm{op}}: \mathscr{K}^{\mathrm{op}} \to \mathscr{K}^{\mathrm{op}}$$

is a coadjoint (III.2.10) and hence, for each object Γ we have a free H^{op}-algebra

$$\Gamma_\# = \coprod_{n<\omega} \Gamma H^n \qquad \text{in } \mathscr{K}^{\mathrm{op}},$$

in other words,

$$\Gamma_\# = \prod \Gamma H^n \qquad \text{in } \mathscr{K}.$$

Since H preserves the last product, we have $\Gamma_\# = \Gamma \times \Gamma_\# H$ (in $\mathscr{K}$), and the projections

$$\pi: \Gamma_\# \to \Gamma \text{ and } \psi^*: \Gamma_\# \to \Gamma_\# H$$

form the injection ($\pi: \Gamma \to \Gamma_\#$ in $\mathscr{K}^{\mathrm{op}}$) and the operation morphism ($\psi^*: \Gamma_\# H \to I_\#$ in $\mathscr{K}^{\mathrm{op}}$) of the free H^{op}-algebra. The universal property follows from the bijective correspondence between F-algebras in $\mathscr{K}$ and H^{op}-algebras in $\mathscr{K}^{\mathrm{op}}$:

$$\frac{QF \xrightarrow{\delta} Q}{Q \xrightarrow[\delta^*]{} QH}$$

Definition. For each F-automaton $A = (Q, \delta, \Gamma, \gamma, I, \lambda)$ the unique F-homomorphism $\gamma_\#: (Q, \delta) \to (\Gamma_\#, \psi)$ with

$$\gamma = g_\# \cdot \pi$$

is called the *observability map* of A.

Remark. The components of $\gamma_\#: Q \to \prod_{n<\omega} \Gamma H^n$ are the following morphisms $\gamma_n: Q \to \Gamma H^n$:

$$\gamma_0 = \gamma;$$
$$\gamma_{n+1} = \delta^* \cdot \gamma_n H.$$

Example. Let $\mathscr{K}$ = **Set** and

$$F = S_\Sigma \text{ and } H = \hom(\Sigma, -).$$

Using the natural bijection between hom $(A, \hom(B, C))$ and hom $(A \times B, C)$ and between A and hom $(\{\emptyset\}, A)$, we can write

$$\Gamma_\# = \prod_{n<\omega} \Gamma H^n = \Gamma \times \hom(\Sigma, \Gamma) \times \hom(\Sigma, \hom(\Sigma, \Gamma)) \times \dots$$
$$= \hom(\{\emptyset\}, \Gamma) \times \hom(\Sigma, \Gamma) \times \hom(\Sigma \times \Sigma, \Gamma) \times \dots.$$

Further, using the natural bijection between hom $(A_0 + A_1 + A_2 + \dots, B)$ and $\prod_{n<\omega} \hom(A_n, B)$, we get

$$\Gamma_\# = \hom(\{\emptyset\} + \Sigma + \Sigma \times \Sigma + \dots, \Gamma) = \hom(\Sigma^*, \Gamma).$$

The observability map

$$\gamma_\# : Q \to \hom(\Sigma^*, \Gamma)$$

is defined as in Exercise I.2.B: it assigns to each state $q \in Q$ its behavior $(q)\gamma_\# = \beta_q : \Sigma^* \to \Gamma$ in A.

2.14. Theorem (Minimal realization as factorization.) Let $\mathscr{K}$ be an $(\mathscr{E}, \mathscr{M})$-category with countable products and coproducts, and let $F: \mathscr{K} \to \mathscr{K}$ be a coadjoint preserving $\mathscr{E}$-epis.

For each behavior $\beta: I^\# \to \Gamma$ we can obtain the minimal realization by forming the image factorization of $\beta_\# : (I^\#, \varphi) \to (\Gamma_\#, \psi)$:

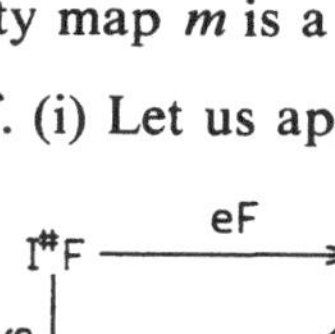

More in detail, the unique automaton with state object Q, run map e and observability map m is a minimal realization of β.

Proof. (i) Let us apply the diagonal fill-in:

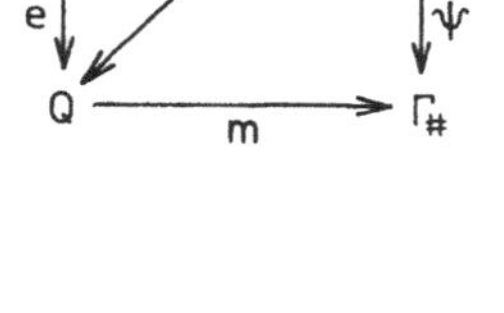

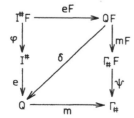

We obtain an *F*-algebra (Q, δ) such that both $e: (I^\#, \varphi) \to (Q, \delta)$ and $m: (Q, \delta) \to (\Gamma_\#, \psi)$ are homomorphisms. It follows that the automaton

$$A_0 = (Q, \delta, \Gamma, m \cdot \pi, I, \eta \cdot e)$$

has run map e and observability map m. Moreover, such an automaton is unique, in other words, δ is uniquely determined by the fact that m and e are homomorphisms. (If $\delta': QF \to Q$ also has this property, then $\delta' \cdot m = mF \cdot \psi = \delta \cdot m$, and m is mono, hence, $\delta = \delta'$.)

The behavior of A_0 is

$$e \cdot m \cdot \pi = \beta_\# \cdot \pi = \beta,$$

and since $e \in \mathscr{E}$, we see that A_0 is a reachable realization of β.

(ii) Let $\bar{A} = (\bar{Q}, \bar{\delta}, \Gamma, \bar{\gamma}, I, \bar{\lambda})$ be another reachable realization of β. The run morphism $\rho: I^\# \to Q$ of A is an $\mathscr{E}$-epi such that

$$\beta = \bar{\rho} \cdot \bar{\gamma}.$$

We have a unique morphism $\bar{\gamma}_\#$ for which the following diagram

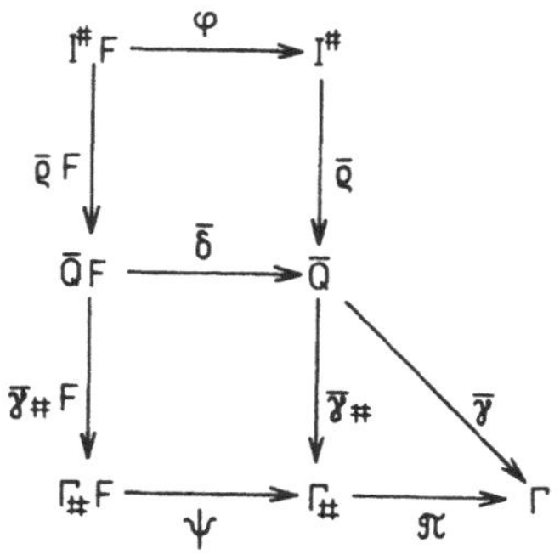

commutes. Further,

$$\beta_\# = \bar{\rho} \cdot \bar{\gamma}_\#$$

because $\beta_\#$ is the unique *F*-homomorphism with $\beta_\# \cdot \pi = \beta$, and we have $\bar{\rho} \cdot \bar{\gamma}_\# \cdot \pi = \bar{\rho} \cdot \bar{\gamma} = \beta$.

Hence, we can use the diagonal fill-in:

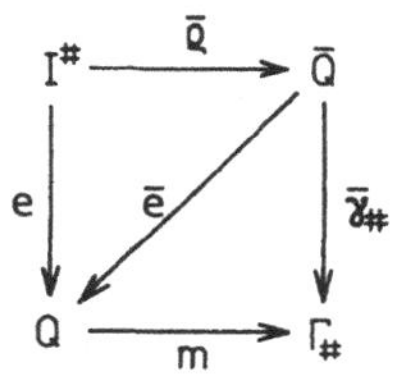

We obtain an $\mathscr{E}$-epi $\bar{e}\colon \bar{Q} \to Q$ [in fact, let $\bar{e} = e_1 \cdot m_1$ be a factorization, then $e = (\bar{\rho} \cdot e_1) \cdot m_1$ is a factorization of $e \in \mathscr{E}$ and therefore, m_1 is an isomorphism, which proves $\bar{e} \in \mathscr{E}$]. Moreover,

$$\bar{e}\colon A \to A_0$$

is a morphism of automata:

(a) $\bar{e}\colon (\bar{Q}, \bar{\delta}) \to (Q, \delta)$ is a homomorphism because $\bar{\rho}F$ is an epi ($\bar{\rho} \in \mathscr{E}$ implies $\bar{\rho}F \in \mathscr{E}$), and we have

$$\bar{\rho}F \cdot (\bar{e}F \cdot \delta) = eF \cdot \delta = \varphi \cdot e = \varphi \cdot \bar{\rho} \cdot \bar{e} = \bar{\rho}F \cdot (\bar{\delta} \cdot e);$$

(b) $\bar{\lambda} \cdot \bar{e} = \eta \cdot \bar{\rho} \cdot \bar{e} = \eta \cdot e;$

(c) $\bar{e} \cdot (m \cdot \pi) = \bar{\gamma}_{\#} \cdot \pi = \bar{\gamma}.$

This proves that A_0 is a minimal realization of β. □

Corollary. For $\mathscr{K}$ and F as above, an automaton is minimal (i.e., a minimal realization of its behavior) iff it is

(i) reachable (the run map is an $\mathscr{E}$-epi)

and

(ii) observable (the observability map is an $\mathscr{M}$-mono).

Conversely, if A has both of these properties, then by $\rho \cdot \gamma = \beta$ we obtain an image factorization of $\beta_{\#}$:

$$\beta_{\#} = \rho \cdot \gamma_{\#} \quad (\rho \in \mathscr{E};\ \gamma_{\#} \in \mathscr{M}).$$

Therefore, there exists an isomorphism i with

$$\rho = e \cdot i \quad \text{and} \quad i \cdot \gamma_{\#} = m.$$

It is easy to check that $i\colon A_0 \to A$ is a morphism of automata. Hence, A is a minimal realization of β.

Remark. In case of sequential automata we have $\Gamma_{\#} = \hom(\Sigma^*, \Gamma)$ (see Exercise I.2.8). Given $\beta\colon \Sigma^* \to \Gamma$, then

$$\beta_{\#}\colon \Sigma^* \to \hom(\Sigma^*, \Gamma)$$

assigns to each word $w \in \Sigma^*$ the behavior $(-w)\beta\colon \Sigma^* \to \Gamma$ defined by $v \mapsto (vw)\beta\ (v \in \Sigma^*)$. The kernel equivalence $\sim$ of this map $\beta_{\#}$ is precisely the Nerode equivalence of β (I.2.6). Thus, our construction of minimal realization is precisely that considered in Chapter I.

2.15. Example: Bilinear sequential Σ-automata. Let R be a commutative ring with 1, and let Σ be an R-module. A bilinear sequential Σ-automaton is defined precisely as the linear one (III.2.3) except that (i) the next-step map

$$\delta\colon Q \times \Sigma \to Q$$

is not supposed to be linear but *bilinear*, which means that for each state $q \in Q$ and each input $\sigma \in \Sigma$

$$(q, -)\delta : \Sigma \to Q \quad \text{and} \quad (-, \sigma)\delta : Q \to Q$$

are linear maps, and (ii) the initial state is arbitrary, not necessarily 0.

To show how bilinear automata fit into our theory, we recall the concept of *tensor product* $\otimes$. Given R-modules X and Y, their tensor product $X \otimes Y$ is defined as an R-module with a bilinear map $\tau: X \times Y \to X \otimes Y$ universal in the following sense: for any bilinear map $f: X \times Y \to Z$ there exists a unique linear map $\hat{f}: X \otimes Y \to Z$ with $f = \tau \cdot \hat{f}$. The tensor product exists for each X and Y, and it is generated by the elements $x \otimes y = (x, y)\tau$.

Define the *tensor-product functor*

$$V_\Sigma: R\text{-}\mathbf{Mod} \to R\text{-}\mathbf{Mod}$$

on objects X by

$$XV_\Sigma = X \otimes \Sigma$$

and on morphisms $f: X \to Y$ by

$$fV_\Sigma = f \otimes 1_\Sigma : x \otimes \sigma \mapsto (x)f \otimes \sigma.$$

Then bilinear Σ-automata are precisely V_Σ-automata

$$A = (Q, \delta, \Gamma, \gamma, R, \lambda)$$

where R is considered as a module over itself. In fact,

$$\delta: Q \otimes \Sigma \to Q$$

describes precisely the bilinear next-state map,

$$\gamma: Q \to \Gamma$$

is the output map, and

$$\lambda: R \to Q$$

is fully determined by $(1)\lambda$, which is the initial state.

The functor V_Σ is a coadjoint. The corresponding adjoint is the *hom-functor*

$$\hom(\Sigma, -): R\text{-}\mathbf{Mod} \to R\text{-}\mathbf{Mod}$$

defined as in **Set** (III.2.10) except that for each R-module X the set $\hom(\Sigma, X)$ of all linear maps receives the usual structure of an R-module. Thus, V_Σ is a varietor with

$$I^\# = \coprod_{n \in \omega} IV_\Sigma^n = I + (I \otimes \Sigma) + (I \otimes \Sigma \otimes \Sigma) + \dots \ .$$

Using the fact that $I \simeq I \otimes R$ and that the tensor product $I \otimes -$ preserves coproducts, we can write

$$\begin{aligned} I^{\#} &= (I \otimes R) + (I \otimes \Sigma) + (I \otimes \Sigma \otimes \Sigma) + \ldots \\ &= I \otimes (R + \Sigma + \Sigma \otimes \Sigma + \ldots). \end{aligned}$$

Denote

$$\Sigma^{\otimes} = R + \Sigma + \Sigma \otimes \Sigma + \ldots .$$

We have, quite analogously as for S_Σ in **Set**:

$$I^{\#} = I \otimes \Sigma^{\otimes}.$$

The analogy goes further: using $\Gamma \simeq \hom(R, \Gamma)$ and the adjunction of $\hom(\Sigma, -)$ and V_Σ, we get

$$\begin{aligned} \Gamma_{\#} &= \prod_{n<\omega} (\Gamma)(\hom(\Sigma, -))^n \\ &= \hom(R, \Gamma) \times \hom(\Sigma, \Gamma) \times \hom(\Sigma, \hom(\Sigma, \Gamma)) \ldots \\ &= \hom(R + \Sigma + \Sigma \otimes \Sigma + \ldots, \Gamma) \\ &= \hom(\Sigma^{\otimes}, \Gamma). \end{aligned}$$

For $I = R$ we have $I^{\#} = \Sigma^{\otimes}$, and the run map of a bilinear automaton is the (unique) linear map

$$\rho : \Sigma^{\otimes} \to Q$$

assigning to each

$$\sigma_1 \otimes \sigma_2 \otimes \ldots \otimes \sigma_n \in \Sigma \otimes \Sigma \otimes \ldots \otimes \Sigma$$

the state obtained from the initial state $(1)\lambda$ by an application of inputs σ_1, $\sigma_2, \ldots, \sigma_n$. The observability map

$$\gamma_{\#} : Q \to \hom(\Sigma^{\otimes}, \Gamma)$$

assigns to each state q the behavior of the automaton initialized in q.

For each behavior $\beta : \Sigma^{\otimes} \to \Gamma$, the homomorphism $\beta_{\#} : (\Sigma^{\otimes}, \varphi) \to ((\hom(\Sigma^{\otimes}, \Gamma), \psi)$ is the unique linear map assigning to each $\sigma_1 \otimes \ldots \otimes \sigma_n$ the map $(\sigma_1 \otimes \sigma_2 \otimes \ldots \otimes \sigma_n \otimes -)\beta : \Sigma^{\otimes} \to \Gamma$. The kernel equivalence $\sim$ of $\beta_{\#}$ is the analogy of the Nerode equivalence: the state algebra of the minimal realization of β is $(\Sigma^{\otimes}, \varphi)/\sim$.

Concluding remark. The concepts of minimal realization, reachability, and observability, introduced for sequential automata, are naturally extended to F-automata with a coadjoint functor F. We find out that observability is the dual concept to reachability.

This special case does not cover tree automata or linear automata. The problem of minimal realizations is much more intricate for general functors F. We shall devote Chapter V to a study of minimal realization.

Exercises III.2

A. Single-input automata are automata of type $F = 1_{\mathscr{K}}$.
(i) If $\mathscr{K}$ has countable coproducts, describe the free algebras for $1_{\mathscr{K}}$. Do this in particular for $\mathscr{K}$ = **Set**.
(ii) Verify that the minimal realization of any morphism $\beta: I^{\#} \to \Gamma$ is obtained from the image factorization of the morphism $\beta_{\#}: I^{\#} \to \Gamma \times \Gamma \times \Gamma \times \ldots$, the components of which are $\beta, \varphi \cdot \beta, \varphi^2 \cdot \beta, \ldots$.
(iii) Let $\mathscr{K}$ = **Pos** as the (epi, embedding)-category. Put $I = \{0\}$ and $\Gamma = \{1, 2\}$, ordered by $1 \leq 2$. Describe the minimal realization of the behavior

$$\beta: I^{\#} = I + I + I + I \ldots \to \Gamma$$

sending the first copy of 0 to 1 and all other copies to 2.
(iv) Describe the minimal realization of the behavior above in the (quotient, mono)-category **Pos**.

B. Forgetful functor. For each functor $F: \mathscr{K} \to \mathscr{K}$ we denote by *F*-**Alg** the category of *F*-algebras and homomorphisms. Let

$$U: F\text{-}\mathbf{Alg} \to \mathscr{K}$$

be the forgetful functor, defined by $(Q, \delta)U = Q$ and $fU = f$. Prove that F is a varietor iff U is an adjoint.

Hint. If F is a varietor, define $\Phi: \mathscr{K} \to F$-**Alg** on objects by $I\Phi = (I^{\#}, \varphi_I)$ and on morphisms $f: I \to J$ by $f\Phi = (f \cdot \eta_J)^{\#}: (I^{\#}, \varphi_I) \to (J^{\#}, \varphi_J)$. Then Φ is coadjoint to U. Conversely, if Φ is coadjoint to U, then $I\Phi$ is a free *F*-algebra generated by I.

III.3. *F*-Algebras

In the present section we discuss various types of algebras which can be expressed as *F*-algebras for suitable functors $F: \mathscr{K} \to \mathscr{K}$. We denote by

F-**Alg**

the category of *F*-algebras and homomorphisms.

3.1. Epitransformations. For arbitary two naturally isomorphic (II. 1) functors $F, G: \mathscr{K} \to \mathscr{K}$, the categories *F*-**Alg** and *G*-**Alg** are also isomorphic: if $\mu: F \to G$ is a natural isomorphism, then each *F*-algebra (Q, δ) defines a *G*-algebra $(Q, \mu_Q^{-1} \cdot \delta)$ and vice versa. Also, a morphism $f: Q \to Q'$ is a homomorphism of *F*-algebras $f: (Q, \delta) \to (Q', \delta')$ iff it is a homomorphism of the corresponding *G*-algebras $f: (Q, \mu_Q^{-1} \cdot \delta) \to (Q', \mu_{Q'}^{-1} \cdot \delta)$. We are going to identify naturally isomorphic functors as well as the corresponding categories of algebras, whenever convenient.

More in general, let $F, G : \mathcal{K} \to \mathcal{K}$ be functors, and let $\varepsilon: F \to G$ be a natural transformation. Then we get a functor

$$\Phi: G\text{-}\mathbf{Alg} \to F\text{-}\mathbf{Alg}$$

assigning to each G-algebra (Q, δ) the F-algebra $(Q, \varepsilon_G \cdot \delta)$, and to each homomorphism $f: (Q, \delta) \to (Q', \delta')$ in G-**Alg** the (same) homomorphism $f: (Q, \varepsilon_Q \cdot \delta) \to (Q', \varepsilon_{Q'} \cdot \delta')$:

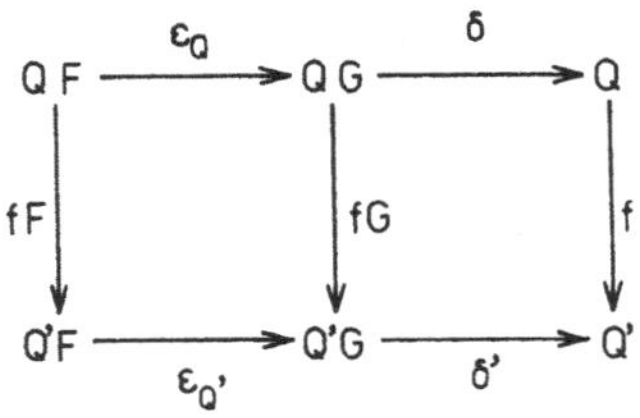

Example. Denote by P_2 : **Set** $\to$ **Set** the functor assigning to each set X the set XP_2 of all non-ordered pairs in X, i.e., sets $\{x_1, x_2\} \subseteq X$, and to each map $f: X \to Y$ the map $fP_2 : \{x_1, x_2\} \mapsto \{(x_1)f, (x_2)f\}$. We have a natural transformation

$$\varepsilon: \mathrm{H}_2 \to P_2$$

defined by

$$(x_1, x_2)\varepsilon_X = \{x_1, x_2\}.$$

Here, H_2-**Alg** is the category of groupoids, and P_2-**Alg** is the category of commutative groupoids, with Φ: P_2-**Alg** $\to$ H_2-**Alg** the full embedding.

As in the example above, if $\varepsilon: F \to G$ is an epitransformation (i.e., each $\varepsilon_X: XF \to XG$ is an epimorphism) then G can be considered as a quotient of F, in the sense made precise in III.5 below, and the functor Φ: G-**Alg** $\to$ F-**Alg** is a full embedding. In fact, Φ is one-to-one on objects since $(Q, \varepsilon_Q \cdot \delta) = (Q', \varepsilon_Q' \cdot \delta')$ implies $Q = Q'$ and $\varepsilon_Q \cdot \delta = \varepsilon_Q \cdot \delta'$, and the latter implies $\delta = \delta'$. It is obvious that Φ is one-to-one on morphisms too.

Convention. For each epitransformation $\varepsilon: F \to G$ we consider G-**Alg** as a full subcategory of F-**Alg** by identifying each G-algebra (Q, δ) with the corresponding F-algebra $(Q, \varepsilon_Q \cdot \delta)$.

3.2. Varieties. We recall the concept of a variety, i.e., a subcategory of the category H_Σ-**Alg** of Σ-algebras (in **Set**) given by equations. Let I be a set (of variables), the cardinality of which is infinite, regular and larger than all arities $|\sigma|$ for $\sigma \in \Sigma$. A pair of Σ-trees (II.3.4)

$$(t_1, t_2) \in I^\# \times I^\#$$

is called an *equation*, and is usually denoted by

$$t_1 = t_2 .$$

An algebra (Q, δ) is said to *satisfy* the equation $t_1 = t_2$ if for each map $f\colon I \to Q$ the extended homomorphism $f^\# : (I^\#, \varphi) \to (Q, \sigma)$ fulfils

$$(t_1)f^\# = (t_2)f^\# .$$

For example, let Σ consist of one binary operation, i.e., let us consider the category of groupoids. A groupoid is commutative iff it satisfies the following equation

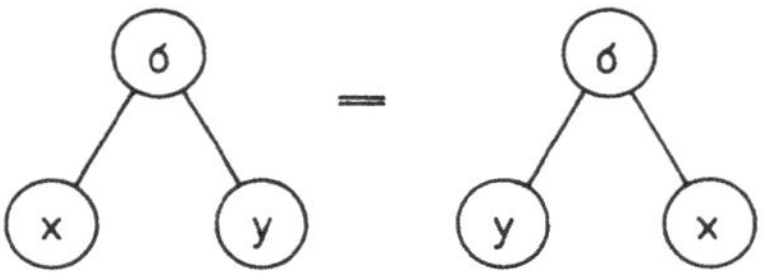

$(x, y \in I$ and $x \neq y)$.

A *variety* of Σ-algebras is a full subcategory $\mathscr{V}$ of H_Σ-**Alg** for which there exists a set of equations $E \subset I^\# \times I^\#$ such that $\mathscr{V}$-algebras are precisely those Σ-algebras which satisfy each of the equations in E. The triple (Σ, I, E) is called an *equational presentation* of $\mathscr{V}$. For example, let $E = \{(t_1, t_2)\}$ for the trees $t_1 = (x, y)\sigma$ and $t_2 = (y, x)\sigma$ above. Then (Σ, I, E) is a presentation of the category of commutative groupoids. The category of commutative semigroups is presented by two equations: the commutativity law

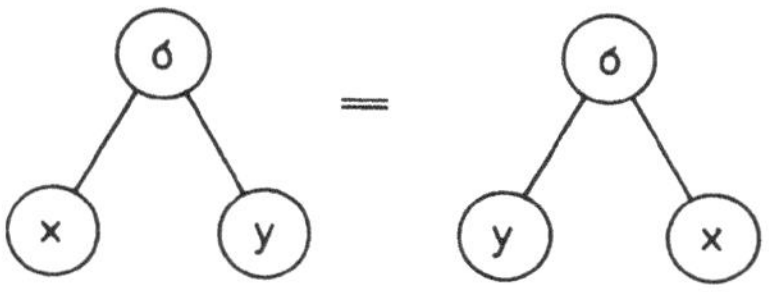

and the associativity law

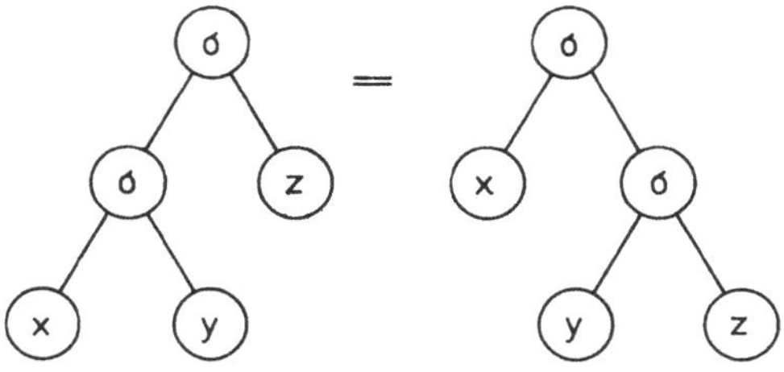

The former example is the category P_2-**Alg**. What about the latter? We are going to characterize varieties of Σ-algebras which have the form *F*-**Alg** for some set functor *F*. The variety of commutative semigroups is not of this form.

Proposition. For each epitransformation $\varepsilon: H_\Sigma \to F$, the category F-**Alg** is a variety of Σ-algebras, presented by the collection of all equations

$$t = t'$$

where $t, t' \in IH_\Sigma (\subset I^\#)$ are arbitary trees with

$$(t)\varepsilon_I = (t')\varepsilon_I ,$$

and I is an infinite set with card $I \geq |\sigma|$ for $\sigma \in \Sigma$.

Proof. (i) Let (Q, δ) be an F-algebra. For each equation

$$t = t'$$

as above and each map $f: I \to Q$, we are to prove that $(t)f^\# = (t')f^\#$.

Put

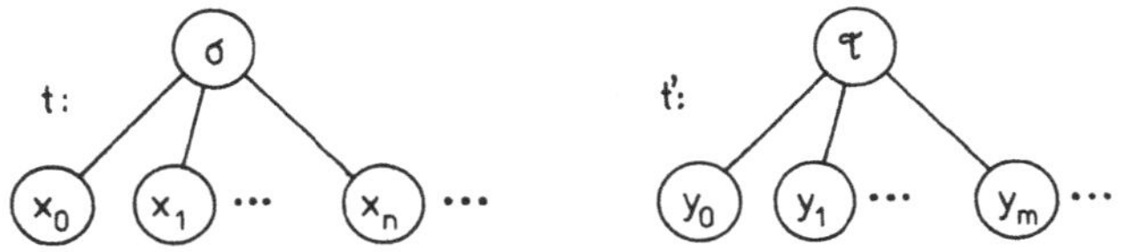

where $\sigma, \tau \in \Sigma$ and $x_n \in I$ $(n < |\sigma|)$, $y_m \in I (m < |\tau|)$. Then $f^\#$ satisfies the following:

$$(t)f^\# = (t)f\,H_\Sigma \cdot \varepsilon_Q \cdot \delta$$

and

$$(t')f^\# = (t')f\,H_\Sigma \cdot \varepsilon_Q \cdot \delta .$$

The following square

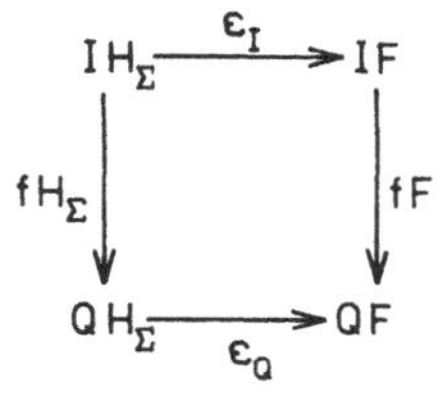

commutes. Thus, if

$$(t)\varepsilon_I = (t')\varepsilon_I ,$$

then

$$(t)f^\# = (t)\varepsilon_I \cdot fF \cdot \delta = (t')\varepsilon_I \cdot fF \cdot \delta = (t')f^\# .$$

(ii) Conversely, let $(Q, \bar{\delta})$ be an algebra which satisfies all of the equations above. We are to prove that given trees $t, t' \in QH_\Sigma$, then

$$(t)\varepsilon_Q = (t')\varepsilon_Q \quad \text{implies} \quad (t)\bar{\delta} = (t')\bar{\delta}.$$

Then we can define $\delta\colon QF \to Q$ by $((t)\varepsilon_Q)\delta = (t)\bar{\delta}$ for each tree $t \in QH_\Sigma$ and this proves that $(Q, \bar{\delta})$ is an F-algebra (since $\bar{\delta} = \varepsilon_Q \cdot \delta$).

Put

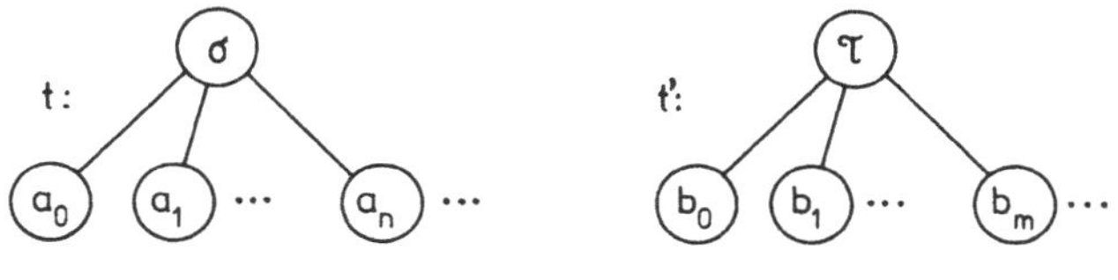

where $\sigma, \tau \in \Sigma$ and $a_n \in Q\,(n < |\sigma|)$, $b_m \in Q(m < |\tau|)$. Since card I is larger than or equal to $|\sigma| + |\tau|$, there exists a map

$$g\colon Q \to I$$

which is one-to-one on the set $\{a_n\}_{n<|\sigma|} \cup \{b_m\}_{m<|\tau|}$. Choose a map

$$f\colon I \to Q$$

with

$$(*) \qquad a_n = (a_n)g \cdot f \quad \text{and} \quad b_m = (b_m)g \cdot f$$

for each $n < |\sigma|$ and $m < |\tau|$. The following square

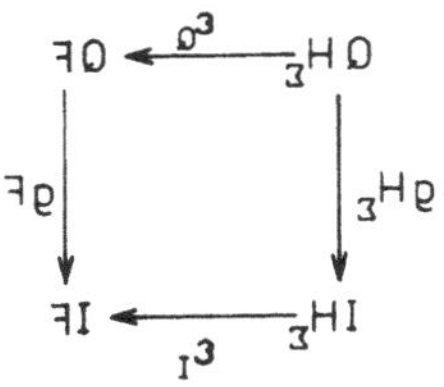

commutes. For $s = (t)gH_\Sigma$ and $s' = (t')gH_\Sigma$, the equation

$$s = s'$$

belongs to the presentation above. Indeed,

$$(s)\varepsilon_I = (t)gH_\Sigma \cdot \varepsilon_I = (t)\varepsilon_Q \cdot gF$$

and, analogously,

$$(s')\varepsilon_I = (t')\varepsilon_Q \cdot gF,$$

thus, $(t)\varepsilon_Q = (t')\varepsilon_Q$ implies $(s)\varepsilon_I = (s')\varepsilon_I$. Since (Q, δ) fulfils this equation, we have

$$(s)f^{\#} = (s')f^{\#},$$

i.e.

$$(s)fH_\Sigma \cdot \delta = (s')fH_\Sigma \cdot \delta.$$

By (*) above clearly

$$t = (t)(g \cdot f)H_\Sigma \quad \text{and} \quad t' = (t')(g \cdot f)H_\Sigma,$$

thus,

$$\begin{aligned}(t)\bar{\delta} &= (t)gH_\Sigma \cdot fH_\Sigma \cdot \bar{\delta} \\ &= (s)fH_\Sigma \cdot \bar{\delta} \\ &= (s')fH_\Sigma \cdot \bar{\delta} \\ &= (t')gH_\Sigma \cdot fH_\Sigma \cdot \bar{\delta} \\ &= (t')\bar{\delta}.\end{aligned}$$

This concludes the proof. □

3.3. Definition. A variety of Σ-algebras is called *basic* if it has an equational presentation (Σ, I, E) such that

$$E \subset IH_\Sigma \times IH_\Sigma,$$

i.e., each of its equations has the following form:

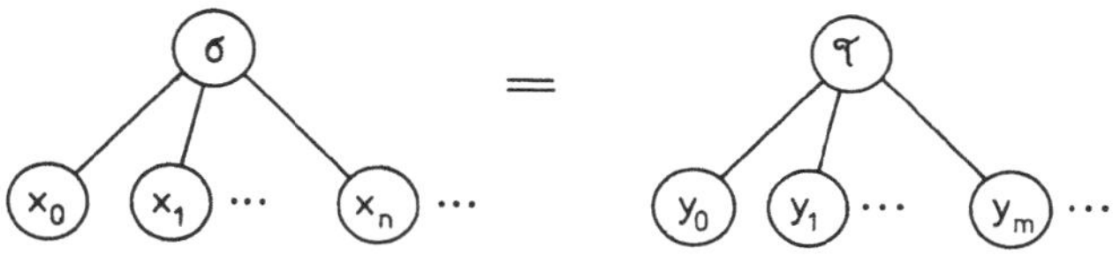

By the proposition above, each variety F-**Alg** given by a quotient F of H_Σ is basic. We shall prove the converse.

Proposition. For each basic variety $\mathcal{V}$ of Σ-algebras there exists an epitransformation $\varepsilon: H_\Sigma \to F$ with $\mathcal{V} = F$-**Alg**.

Proof. Let (Σ, I, E) be an equational presentation of $\mathcal{V}$ with $E \subset IH_\Sigma \times IH_\Sigma$. For each set X let $\sim_X$ be the least equivalence relation on XH_Σ such that

$$(t_1)kH_\Sigma \sim_X (t_2)kH_\Sigma$$

holds for each equation $t_1 = t_2$ in E and each $k: I \to X$. For any map $f: X \to Y$ and arbitrary $a_1, a_2 \in XH_\Sigma$,

(*) $\qquad a_1 \sim_X a_2 \quad \text{implies} \quad (a_1)fH_\Sigma \sim_Y (a_2)fH_\Sigma.$

To prove (*), it clearly suffices to verify this for $a_i = (t_i)kH_\Sigma$, where $k: I \to X$ is a map and $t_1 = t_2$ is in E. Put $\bar{k} = k \cdot f: I \to Y$, then $(t_1)kH \sim_Y (t_2)kH$ and we have

$$(a_i)fH_\Sigma = (t_i)k \cdot fH_\Sigma = (t_i)\bar{k}H_\Sigma \quad (i = 1, 2).$$

We can define a quotient functor F of H_Σ as follows: for each set X,

$$XF = XH_\Sigma/\sim_X$$

is the quotient set, i.e., the set of all equivalence classes $[t]$ with $t \in XH_\Sigma$; for each map $f: X \to Y$ define

$$fF: XF \to YF; \quad [t] \mapsto [(t)fH_\Sigma].$$

This is well-defined due to (*). The corresponding natural transformation $\varepsilon: H_\Sigma \to F$ assigns to each $t \in XH_\Sigma$ the equivalence class $(t)\varepsilon_X = [t]$.

(a) Each $\mathscr{V}$-algebra (Q, δ) is an F-algebra. More precisely, each $\mathscr{V}$-algebra is of the type $(Q, \varepsilon_Q \cdot \bar{\delta})$ for some F-algebra $(Q, \bar{\delta})$. To prove this, we have to show that for $q_1, q_2 \in QH_\Sigma$,

$$q_1 \sim_Q q_2 \quad \text{implies} \quad (q_1)\delta = (q_2)\delta.$$

[Then $\bar{\delta}: QH_\Sigma/\sim_Q \to Q$ is defined by $([q])\bar{\delta} = (q)\delta$.] We can clearly assume that $q_i = (t_i)kH_\Sigma$, where $k: I \to Q$ is a map and $t_1 = t_2$ is in E.

Since (Q, δ) satisfies the equation $t_1 = t_2$, we have $(t_1)k^\# = (t_2)k^\#$. Also, $t_i \in IH_\Sigma$ implies $(t_i)k^\# = (t_i)kH_\Sigma \cdot \delta$. Thus,

$$(q_1)\delta = (t_1)kH_\Sigma \cdot \delta = (t_2)kH_\Sigma \cdot \delta = (q_2)\delta.$$

(ii) Each F-algebra is a $\mathscr{V}$-algebra. More precisely, for each F-algebra $(Q, \bar{\delta})$, the H_Σ-algebra $(Q, \varepsilon_Q \cdot \bar{\delta})$ satisfies each equation in E. Let $t_1 = t_2$ be in E and let $k: I \to Q$ be a map. Then $(t_i)k^\# = (t_i)kH_\Sigma \cdot \delta$ for $\delta = \varepsilon_Q \cdot \bar{\delta}$, and we have $(t_1)kH_\Sigma \sim_Q (t_2)kH_\Sigma$, therefore,

$$\begin{aligned}(t_1)k^\# &= (t_1)kH_\Sigma \cdot \varepsilon_Q \cdot \bar{\delta} \\ &= [(t_1)kH_\Sigma]\bar{\delta} \\ &= [(t_2)kH_\Sigma]\bar{\delta} \\ &= (t_2)k^\#.\end{aligned}$$

□

Remark. In the next section we shall prove that each set functor F such that F-**Alg** is a variety must be a quotient of H_Σ for some type Σ. This will make the picture complete: varieties of the form F-**Alg** are precisely the basic ones. This shows that for example commutative semigroups form a variety distinct from any F-**Alg**.

Example: P_f-algebras. Define a functor

$$P_f: \mathbf{Set} \to \mathbf{Set}$$

on each set X by

$$XP_f = \{M;\ M \subset X,\ M \neq \emptyset \text{ finite}\}$$

and on each map $f: X \to Y$ by

$$(M)fP_f = (M)f \quad \text{for all } M \in XP_f.$$

A P_f-algebra is a set Q together with an operation δ, assigning to each finite, non-empty set $M \subset Q$ an element $(M)\delta \in Q$. A homomorphism $f: (Q, \delta) \to (Q', \delta')$ is a map $f: Q \to Q'$ such that $(M)\delta = q$ implies $((M)f)\delta = (q)f$. An example of a P_f-algebra is any join semilattice, where $(M)\delta = \vee M$; for two semilattices, a P_f-homomorphism is precisely a semilattice homomorphism. Thus, semilattices form a full subcategory of P_f-**Alg**.

Let $\Sigma = \{\sigma_n\}_{1 \le n < \omega}$ be a type with $|\sigma_n| = n$ for each $n = 1, 2, 3, \ldots$. Then P_f is quotient of H_Σ: consider the natural transformation

$$\varepsilon: H_\Sigma \to P_f$$

assigning to each $(x_0, \ldots, x_{n-1}) \in XH_{\sigma_n}$ the set $\{x_0, \ldots, x_{n-1}\} \in XP_f$. Thus, P_f-**Alg** is a basic variety of Σ-algebras. It is presented by the equations

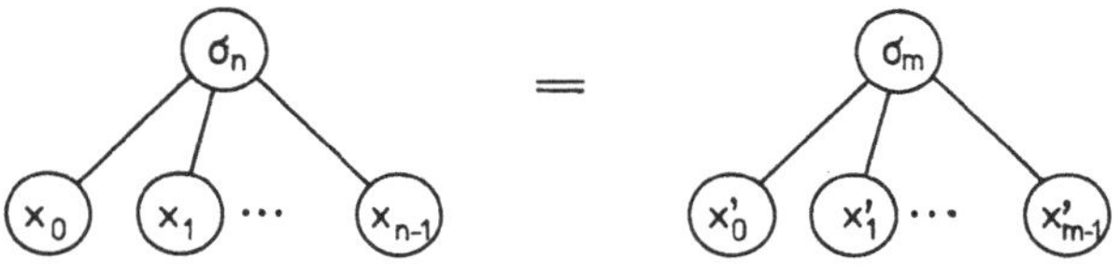

for all $n, m = 1, 2, 3, \ldots$ and $x_0, \ldots, x_{n-1}, x_0', \ldots, x'_{m-1} \in I$ such that $\{x_0, \ldots, x_{n-1}\} = \{x_0', \ldots, x_{m-1}'\}$.

3.4. Example: *P*-algebras. We have seen that quotient functors of the functors H_Σ define precisely the basic varieties of universal algebras. But there are other set functors which are beyond the scope of universal algebra. For example, the *power-set functor*

$$P: \mathbf{Set} \to \mathbf{Set},$$

assigning to each set X the set

$$XP = \exp X = \{M;\ M \subset X\}$$

and to each map $f: X \to Y$ the map fP with

$$(M)fP = (M)f \quad \text{for each } M \subset X.$$

A *P*-algebra (Q, δ) is given by a map $\delta: \exp Q \to Q$. For example, complete join semilattices and complete homomorphisms form a full subcategory of *P*-**Alg**: here $(M)\delta = \vee M$ for each $M \subset Q$.

Forming the coproduct functor

$$P + P: \mathbf{Set} \to \mathbf{Set},$$

we obtain the category $(P + P)$-**Alg** of algebras given by two operations $\exp Q \to Q$. Thus, for example the category of complete lattices and complete homomorphisms is a full subcategory of $(P + P)$-**Alg**.

3.5. Ordered algebras. An ordered algebra is an algebra defined on a poset in such a way that all of its operations are order-preserving. Here, for each poset Q the cartesian power Q^n (n a cardinal) is ordered *component-wise*:

$$(x_i)_{i<n} \leq (y_i)_{i<n} \quad \text{iff} \quad x_i \leq y_i \quad \text{for all } i < n.$$

(This is precisely the product of n copies of the object Q in the category **Pos**.)

Explicitly, for each type Σ, an *ordered Σ-algebra* is a pair $(Q, \{\delta_\sigma\}_{\sigma \in \Sigma})$ which consists of a poset Q and a collection of operations

$$\delta_\sigma : Q^n \to Q \; (\sigma \in \Sigma_n)$$

such that

$$x_i \leq y_i (i < n) \quad \text{implies} \quad (x_1)_{i<n}\delta \leq (y_i)_{i<n}\delta.$$

These are precisely H_Σ-algebras for the functor $H_\Sigma = \coprod_{\sigma \in \Sigma_n} H_n$ of Remark III.2.5; a coproduct in the category **Pos** is just the disjoint union (with the given order on each summand, and with elements of distinct summands pairwise incompatible). An order-preserving map

$$\delta : QH_\Sigma = \coprod_{\sigma \in \Sigma_n} Q^n \to Q$$

is precisely a collection of order-preserving maps $\delta_\sigma : Q^n \to Q$ $(\sigma \in \Sigma_n)$. H_Σ-homomorphisms are also precisely the order-preserving homomorphisms.

3.6. ω-continuous algebras. A poset is said to be *ω-complete* if it has a least element 0 and each increasing ω-sequence has a join. An order-preserving map which preserves the least element and joins of increasing ω-sequences is said to be *ω-continuous.* We denote by

$$\mathbf{Pos}_\omega$$

the category of ω-complete posets and ω-continuous maps. This is the category in which ω-continuous algebras "live".

An ordered Σ-algebra $(Q, \{\delta_\sigma\})$ is *strict ω-continuous* if Q is a ω-complete poset and each $\delta : Q^n \to Q$ is ω-continuous (with respect to the component-wise order of Q^n). Explicitly, an operation $\delta : Q^n \to Q$ is ω-continuous iff $(0, 0, 0, \ldots)\delta = 0$ and

$$\bigvee_{k<\omega} (x_0^k, x_1^k, x_2^k, \ldots)\delta = (y_0, y_1, y_2, \ldots)\delta$$

for any ω-sequence $(x_0^0, x_1^0, x_2^0, \ldots) \leq (x_0^1, x_1^1, x_2^1, \ldots) \leq (x_0^2, x_1^2, x_2^2, \ldots) \ldots$ in Q^n with joins $y_n = \bigvee_{k<\omega} x_n^k$.

Now, for each ω-complete poset Q the powers Q^n are also ω-complete;

hence, we have a functor

$$H_n: \mathbf{Pos}_\omega \to \mathbf{Pos}_\omega$$

which is the restriction of the corresponding functor on **Pos**. And

$$H_n\text{-}\mathbf{Alg}$$

is the category of ω-continuous $\{\sigma\}$-algebras and ω-continuous homomorphisms, where $|\sigma| = n$.

For more than one operation, we use the coproduct. In the category $\mathbf{Pos}_\omega$, coproducts are not disjoint unions (because of the least element): we must first form the coproduct in **Pos** and then we merge all the 0' s. Thus, a coproduct of objects Q_i ($i \in I$) of $\mathbf{Pos}_\omega$ is the set

$$\coprod_{i \in I} Q_i / \sim$$

where $\sim$ is the equivalence, one class of which, denoted by [0], is formed by all the least elements of all Q_i's, and all other classes are singletons (denoted by x rather than $[x]$, for each $x \in Q_i - \{0\}$). The ordering is given as follows:

$$x \leq y \quad \text{iff} \quad x = [0] \text{ or } x, y \in Q_{i_0} \quad \text{and} \quad x \leq y \text{ in } Q_{i_0} \text{ for some } i_0 \in I.$$

The injections $Q_j \to \coprod_{i \in I} Q_i$ assign [0] to 0 and x to any $x \in Q_j - \{0\}$.

Given a type Σ of algebras, we define $H_\Sigma: \mathbf{Pos}_\omega \to \mathbf{Pos}_\omega$ as the coproduct in $\mathbf{Pos}_\omega$:

$$H_\Sigma = \coprod_{\sigma \in \Sigma} H_{|\sigma|}.$$

For each ω-continuous poset $Q = (Q_0, \leq)$ the poset

$$QH_\Sigma = \bigcup_{\sigma \in \Sigma_n} Q_0^n \times \{\sigma\}$$

consists, besides the least element [0], of all trees

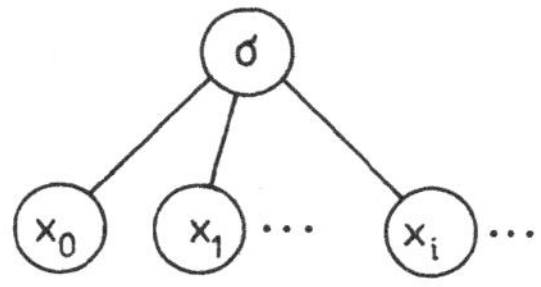

(i<n)

where $\sigma \in \Sigma_n$ and if $n > 0$, there exists $i_0 < n$ with $x_{i_0} \neq 0$. The ordering is such that [0] is the least element, while

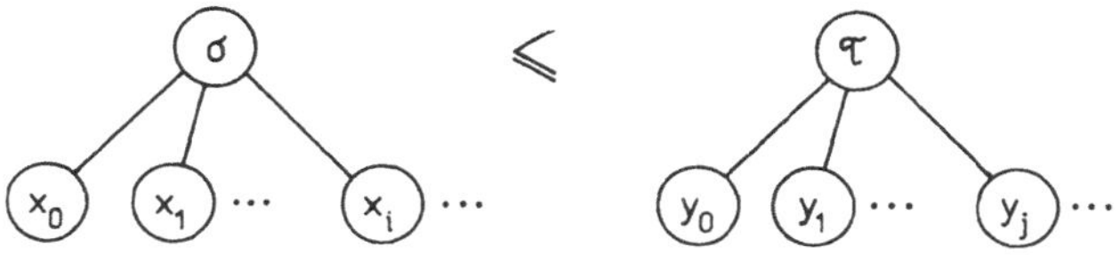

iff $\sigma = \tau$ and $x_i \leq y_i$ for each $i < |\sigma|$. Clearly,

$$H_\Sigma\text{-}\mathbf{Alg}$$

is the category of strict ω-continuous Σ-algebras.

Remark. Algebras over ω-complete posets are investigated in computer science where, however, the requirement on the operations is usually weaker: The least element 0 is not supposed to be idempotent. We call the resulting algebras the *(non-strict) ω-continuous Σ-algebras.* These consist of an ω-complete poset Q and of operations

$$\delta_\sigma : Q^n \to Q \quad (\sigma \in \Sigma_n)$$

which preserve joins of increasing ω-sequences.

Also these algebras are F-algebras. Denote by

$$H^0_\Sigma : \mathbf{Pos}_\omega \to \mathbf{Pos}_\omega$$

the functor given by the disjoint union of all $H_{|\sigma|}$, $\sigma \in \Sigma$, with a *new* element 0^* as the least element. Thus, for each ω-continuous poset Q,

$$QH^0_\Sigma = \bigcup_{\sigma \in \Sigma_n} Q^n \times \{\sigma\} \cup \{0^*\}$$

where the ordering of the basic trees is as above and 0^* is the least element (assumed not to belong to any $Q^n \times \{\sigma\}$). For each ω-continuous map $f: Q \to Q'$ the map

$$fH^0 : QH^0_\Sigma \to Q'H^0_\Sigma$$

equals to $f^{(n)}$ on each $Q^n \times \{\sigma\}$, $\sigma \in \Sigma_n$, and $(0^*)\, fH^0_\Sigma = 0^*$.

An H^0_Σ-algebra (Q, δ) is given by an ω-complete poset Q and an ω-continuous map

$$\delta : QH^0_\Sigma \to Q.$$

This means that 0^* brings no information: $(0^*)\delta = 0$; and for each $\sigma \in \Sigma_n$ we have an n-ary operation

$$\delta_\sigma : Q^n \times \{\sigma\} \to Q$$

which clearly preserves joins of ω-sequences but need not preserve the least element. We see that

$$H^0_\Sigma\text{-}\mathbf{Alg}$$

is the category of non-strict ω-continuous Σ-algebras and ω-continuous homomorphisms.

3.7. Many-sorted algebras. Algebras acting on several sets are called many-sorted or heterogeneous. We discuss 2-sorted algebras, for simplicity; the generalization to more sorts is easy. A 2-sorted algebra consists of two sets X_1 and X_2 and of operations of different arities, e.g.,

$$\sigma: X_1 \times X_2 \to X_1; \ \tau: X_1 \times X_1 \times X_1 \to X_2,$$

etc. The notion of arity of an operation is more difficult than in the one-sorted case. We must (a) give two exponents: of X_1 and of X_2 and (b) state whether the result is in X_1 or X_2. Therefore, a *type* of 2-sorted algebras is a pair $\Sigma = \langle \Sigma_1, \Sigma_2 \rangle$ of disjoint sets; the elements of Σ_i are called operation symbols resulting in X_i ($i = 1, 2$). For each $\sigma \in \Sigma_1 \cup \Sigma_2$, an *arity* is given which is a pair

$$|\sigma| = (n_1, n_2)$$

of cardinals.

A 2-sorted Σ-algebra consists of a pair $\langle X_1, X_2 \rangle$ of sets and for each $\sigma \in \Sigma_i$ of arity (n_1, n_2), a map

$$\delta_\sigma: X_1^{n_1} \times X_2^{n_2} \to X_i.$$

For example, modules over arbitrary commutative rings can be viewed as 2-sorted algebras with X_1 the ring and X_2 the X_1-module. Let $\oplus$, $\otimes$ and 0 be the ring operations, and $+ : X_2 \times X_2 \to X_2$ and $\times : X_1 \times X_2 \to X_2$ the module operation. Then modules are 2-sorted algebras of type $\langle \Sigma_1, \Sigma_2 \rangle$ with

$$\Sigma_1 = \{\oplus, \otimes, 0\} \text{ and } \Sigma_2 = \{+, \times\}$$

where the arity of $\oplus$ is (2, 0), of $+$ is (0, 2), etc.

Homomorphisms of 2-sorted algebras are pairs of maps (one for each sort) preserving the given operations. Thus, let $A = (\langle X_1, X_2 \rangle, \{\delta_\sigma\})$ and $B = (\langle Y_1, Y_2 \rangle, \{\bar{\delta}_\sigma\})$ be two algebras of type $\langle \Sigma_1, \Sigma_2 \rangle$. A homomorphism from A to B is a pair

$$\langle f_1, f_2 \rangle : A \to B$$

of maps $f_1: X_1 \to Y_1$ and $f_2: X_2 \to Y_2$ such that each $\sigma \in \Sigma_i$ of arity (n_1, n_2) and for each $(a_j)_{j<n_1} \in X^{n_1}$ and $(b_j)_{j<n_2} \in X^{n_2}$ we have
$(a_0, a_1, \ldots, b_0, b_1, \ldots)\delta_\sigma \cdot f_i = ((a_0)f_1, (a_1)f_1, \ldots, (b_0)f_2, (b_1)f_2 \ldots)\bar{\delta}_\sigma.$
In other words, the following square

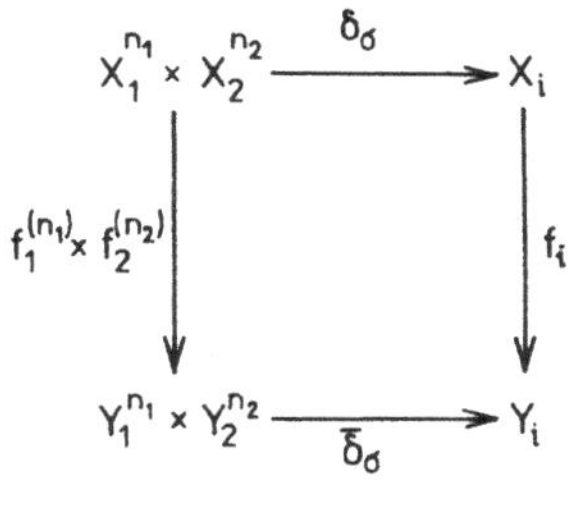

commutes for each $\sigma \in \Sigma_1 \cup \Sigma_2$.

The 2-sorted algebras are *F*-algebras in the category **Set**^{2} of pairs of sets and pairs of maps. (The composition of morphisms $\langle f_1, f_2\rangle : \langle X_1, X_2\rangle \to \langle Y_1, Y_2\rangle$ and $\langle g_1, g_2\rangle : \langle Y_1, Y_2\rangle \to \langle Z_1, Z_2\rangle$ is coordinate-wise, i.e., the resulting morphism is $\langle f_1 \cdot g_1, f_2 \cdot g_2\rangle$.) If Σ has just one operation of arity (n_1, n_2) in Σ, the corresponding functor is

$$H^1_{(n_1, n_2)} : \mathbf{Set}^2 \to \mathbf{Set}^2,$$

defined by

$$\langle X_1, X_2\rangle H^1_{(n_1, n_2)} = \langle X_1^{n_1} \times X_2^{n_1}, \emptyset\rangle;$$
$$\langle f_1, f_2\rangle H^1_{(n_1, n_2)} = \langle f_1^{(n_1)} \times f_2^{(n_2)}, \emptyset\rangle.$$

Analogously, if Σ consists of one operation in Σ_2, we define $H^2_{(n_1, n_2)}$, sending $\langle X_1, X_2\rangle$ to $\langle \emptyset, X_1^{n_1} \times X_2^{n_2}\rangle$.

Finally, for each type $\Sigma = \langle \Sigma_1, \Sigma_2\rangle$ we use the coproduct (which in **Set**^{2} is formed component-wise):

$$H_\Sigma = \coprod_{\sigma \in \Sigma_1} H^1_{|\sigma|} + \coprod_{\sigma \in \Sigma_2} H^2_{|\sigma|}$$

Thus, on objects $\langle X_1, X_2\rangle$,

$$\langle X_1, X_2\rangle H_\Sigma = \left\langle \coprod_{\sigma \in \Sigma_1, |\sigma| = (n_1, n_2)} X_1^{n_1} \times X_2^{n_2}, \coprod_{\tau \in \Sigma_1, |\tau| = (m_1, m_2)} X_1^{m_1} \times X_2^{m_2} \right\rangle,$$

analogously on morphisms. Then

$$H_\Sigma\text{-}\mathbf{Alg}$$

is the category of 2-sorted Σ-algebras.

3.8. Algebras in concrete categories. The algebras in **Set**, **Pos**, **Pos**$_\omega$ and **Set**^{2} have common features which we discuss in general presently.

By a *concrete category* is meant a category $\mathcal{K}$ together with a functor

$$U: \mathcal{K} \to \mathbf{Set}$$

which is faithful, i.e., given distinct morphisms $f_1, f_2: A \to B$ then $f_1 U \neq f_2 U$

(for each pair of objects A, B). We can view the objects of $\mathcal{K}$ as sets endowed with a structure. (Quite formally: given an object A with $AU = X$ then X is a set with structure A.) Since U is faithful, we can use the same symbol for a morphism $f: A \to B$ in $\mathcal{K}$ and the corresponding map in **Set**. Thus, if $AU = X$ and $BU = Y$, then morphisms from A to B are some maps $f: X \to Y$—those maps which "preserve" the given structure. The functor U is called *forgetful* because it forgets the structure.

For example, **Pos** is a concrete category with $(X, \leq)U = X$. Morphisms are precisely the structure (= order) preserving maps. Also $\mathbf{Set}^2$ can be considered as a concrete category. The forgetful functor is given by the disjoint union:

$$\langle X_1, X_2 \rangle U = X_1 + X_2,$$

analogously for morphism.

Given a concrete category $\mathcal{K}$, a set M is said to generate a *free object* M^* of $\mathcal{K}$ if $M \subset M^*U$ and for each object A and each map $f: M \to AU$ there exists a unique morphism $f^*: M^* \to A$ extending f. For example, **Pos** *has free objects* (which means that each set generates a free object): M^* is the set M endowed with the discrete order. Also $\mathbf{Pos}_\omega$ has free objects: $M^* = M \cup \{0\}$ (where $0 \notin M$) with 0 the least element and M discretely ordered.

Assume that $\mathcal{K}$ is a concrete category with *concrete products* (i.e., $\mathcal{K}$ has products and U preserves them). Then we get a natural concept of an n-ary algebraic operation on an object Q of $\mathcal{K}$ as a morphism $\delta: Q^n \to Q$. The corresponding functor is the functor

$$H_n: \mathcal{K} \to \mathcal{K}$$

of the n-th power. The H_n-homomorphisms are the structure-preserving maps which also preserve the operation. This is the case of R-**Mod**, **Pos** and $\mathbf{Pos}_\omega$ above, and a lot of other current concrete categories. For example:

Top,

the category of topological spaces and continuous maps. Here Q^n is the usual topological product (of n copies of Q). An H_n-algebra is a topological space Q together with a continuous n-ary operation on Q. An H_n-homomorphism is a continuous homomorphism.

On the other hand, $\mathbf{Set}^2$ fails to have concrete products: the product of $\langle X_1, X_2 \rangle$ and $\langle Y_1, Y_2 \rangle$ in $\mathbf{Set}^2$ is $\langle X_1 \times Y_1, X_2 \times Y_2 \rangle$. And, of course, the equation

$$(X_1 \times Y_1) + (X_2 \times Y_2) = (X_1 + X_2) \times (Y_1 + Y_2)$$

does not hold in general. This explains why the functors H_n are not "convenient" in $\mathbf{Set}^2$.

Assume that $\mathcal{K}$ has not only concrete products but also (possibly non-concrete) coproducts. Then for each type Σ of algebras we use the coproduct functor

$$H_\Sigma = \coprod_{\sigma \in \Sigma} H_{|\sigma|}$$

introduced in III.2.5. The category H_Σ-**Alg** is then the category of $\mathcal{K}$-structured Σ-algebras (with structure-preserving operations) and structure-preserving Σ-homomorphisms.

For each set functor F: **Set** $\to$ **Set**, the category F-**Alg** is concrete. It has free objects iff F is a varietor (see Exercise III.2.B).

Given a functor $H: \mathcal{K} \to \mathcal{L}$ between concrete categories, we say that H is *concrete* if $U_{\mathcal{K}} = H \cdot U_{\mathcal{L}}$, i.e., if for each object (X, α) of $\mathcal{K}$ we have $(X, \alpha)H = (X, \alpha^*)$ for some structure α^* in $\mathcal{L}$, and for each morphism f we have $fH = f$. Two concrete categories are *concretely isomorphic* if there exists a concrete isomorphism (i.e., a concrete, bijective functor) from one to another. For example, if F and G are two naturally isomorphic set-functors, then F-**Alg** and G-**Alg** are concretely isomorphic categories (see III.3.1).

Remark. Let $(\mathcal{K}, U)$ be a concrete category which has free objects. For each set M denote by $M\Phi = M^*$ a free object generated by M, and for each map $f: M \to M'$ let $f\Phi: M\Phi \to M'\Phi$ be the unique morphism extending f. This defines a functor

$$\Phi: \mathbf{Set} \to \mathcal{K}$$

which is coadjoint to the forgetful functor (III.2.10): the inclusion maps $\eta_M: M \to (M\Phi)U$ define the universal natural transformation

$$\eta: 1_{\mathbf{Set}} \to \Phi \cdot U.$$

(In fact, for each morphism $f: M \to AU$ in **Set** there exists a unique morphism $f^*: M\Phi \to A$ in $\mathcal{K}$ with $f = \eta_M \cdot f^*$.)

For each object A of $\mathcal{K}$ we can extend $1_{AU}: AU \to AU$ to a morphism

$$\varepsilon_A: (AU)\Phi \to A,$$

and this gives a natural transformation

$$\varepsilon: U \cdot \Phi \to 1_{\mathcal{K}}.$$

This is an epitransformation: if $\varepsilon_A \cdot f = \varepsilon_A \cdot g$ for two morphisms $f, g: A \to B$, then $\eta_A \cdot \varepsilon_A U = 1$ implies

$$fU = \eta_A \cdot (\varepsilon_A \cdot f)U = \eta_A \cdot (\varepsilon_A \cdot g)U = gU$$

and, since U is faithful, we conclude that $f = g$.

III.4. Set Functors

4.1. Since set functors, i.e., functors

$$F\colon \mathbf{Set} \to \mathbf{Set}$$

play an important role in our book, we devote the present section to their pro-

We have introduced the set functors H_n and their coproducts H_Σ in III.2.5, further the functors S_Σ (= H_Σ if each $\sigma \in \Sigma$ is unary) in III.2.3 and hom-functors hom(Σ, —) in III.2.10. Note that hom(Σ, —) is naturally isomorphic to H_n for $n = \operatorname{card} \Sigma$. If Σ is a singleton set, then S_Σ and hom(Σ, —) are naturally isomorphic to the identity set functor. Further, recall the functors P, P_2 and P_f of III.3.3 and III.3.4. We define a subfunctor P_n of P for each cardinal n by

$$XP_n = \{M \subset X;\ 0 < \operatorname{card} M \leq n\}.$$

On morphisms: fP_n sends M to $(M)f$.

For each set M we denote by

$$C_M\colon \mathbf{Set} \to \mathbf{Set}$$

the constant functor ($XC_M = M$ and $fC_M = 1_M$). We extend the concept of constant set functor slightly, by disregarding the empty set and empty map. Thus, for each map $h\colon M_0 \to M$ we define

$$C_M^h\colon \mathbf{Set} \to \mathbf{Set}$$

on objects by

$$\begin{aligned} XC_M^h &= M \\ \emptyset C_M^h &= M_0 \end{aligned} \qquad \text{if } X \neq \emptyset;$$

and on morphisms $f\colon X \to Y$ by

$$fC_M^h = \begin{cases} 1_M & \text{if } X \neq \emptyset \\ h & \text{if } X = \emptyset \neq Y \\ 1_{M_0} & \text{if } X = \emptyset = Y. \end{cases}$$

A set functor is said to be *constant* if it is naturally isomorphic to some C_M^h. If $M_0 = \emptyset$ we write C_{0M} instead of C_M^h.

4.2. Proposition. Each set functor $F \neq C_\emptyset$ preserves

(i) non-empty sets ($X \neq \emptyset$ implies $XF \neq \emptyset$);

(ii) epis (if $e\colon X \to Y$ is onto then eF is onto);

(iii) non-empty monos (if $m\colon X \to Y$ is one-to-one, $X \neq \emptyset$, then mF is one-to-one).

Proof. (i) Since $F \neq C_\emptyset$, there exists a set X with $XF \neq \emptyset$. Let $Y \neq \emptyset$ be arbitrary. There exists a map $f: X \to Y$; then $fF: XF \to YF$ is a map and hence, $XF \neq \emptyset$ implies $YF \neq \emptyset$.

(ii) Each epi $e: X \to Y$ splits: for each $y \in Y$ we can choose $(y)f \in X$ with $((y)f)e = y$, and $f: Y \to X$ is a map such that $f \cdot e = 1_Y$. Then $fF \cdot eF = 1_{YF}$, hence, eF is a split epi.

(iii) Each non-empty mono $m: X \to Y$ splits: we choose $x_0 \in X$, and we define $g: Y \to X$ by $(y)g = x_0$ if $y \in Y - (X)m$; $(y)g = x$ if $y = (x)m$. Then $m \cdot g = 1_X$. Hence, $mF \cdot gF = 1_{XF}$ and mF is a split mono. □

Remark. The empty mono $m: \emptyset \to X$ need not be mapped to a mono: consider the functor C_M^h above with h a constant map. Then $mC_M^h = h$.

4.3. Definition. A set functor F is said to be *small* if there exists a cardinal γ such that for each set X we have

$$XF = \bigcup (MF)fF,$$

where the union ranges over all maps $f: M \to X$ with $\operatorname{card} M < \gamma$. If $\gamma = \aleph_0$, then F is said to be *finitary*.

Examples. (i) H_n is small, and it is finitary iff n is finite. In fact, let γ be the least cardinal larger than n. For each set X and each element $(x_i)_{i<n}$ of $XH_n = X^n$ put $M = \{x_i;\ i < n\}$. Then card $M < \gamma$ and $(x_i)_{i<n} \in M^n = (MH_n)fH_n$ for the inclusion map $f: M \to X$.

If n is finite, then $\gamma = n + 1$, hence, we can also choose $\gamma = \aleph_0$, and we see that H_n is finitary. If n is infinite, choose a set X and an n-tuple $(x_i)_{i<n} \in XH_n$ such that the elements x_i are pairwise distinct. Then for each map $f: M \to X$ with M finite clearly $(x_i)_{i<n} \notin (MH_n)fH_n$. Hence, H_n is not finitary.

(ii) The functor P_f is finitary: for each set X and each element of XP_f, i.e., a finite set $M \subset X$, we have $M \in (MP_f)j\,P_f$ for the inclusion map $j: M \to X$.

(iii) The functor P is not small. For each cardinal γ we can choose a set X of power γ. Then $X \in \exp X = XP$ and clearly $X \neq (MP)fP$ for any map $f: M \to X$ with $\operatorname{card} M < \gamma$.

Proposition. A set functor F is small iff it is a quotient functor of H_Σ (i.e., there is an epitransformation $H_\Sigma \to F$) for some type Σ. F is finitary iff it is a quotient of H_Σ for some finitary type Σ.

Proof. I. Let F be a quotient of H_Σ and let γ be the least infinite cardinal larger than all arities in Σ. (Thus, $\gamma = \aleph_0$ iff Σ is finitary.) The sufficiency of both of the statements we are proving will be verified when we show that for each $x \in XF$ there exists a map $f: M \to X$ with $x \in (MF)fM$ and card $M < \gamma$.

Let $\varepsilon: H_\Sigma \to F$ be an epitransformation. There exists $t \in XH_\Sigma$ with $x = (t)\varepsilon_X$. Let $t = (x_i)\sigma$ for $\sigma \in \Sigma$ of arity $n (< \gamma)$ and for $x_i \in X$, $i < n$. Put $M = \{x_i;\ i < n\}$ and let $f: M \to X$ denote the inclusion map. Then card $M \leq n < g$ and we have

$$x = (t)\varepsilon_X = (((x_i)f)\sigma)\varepsilon_X = (x_i)fH_\Sigma \cdot \varepsilon_X.$$

Since the following square

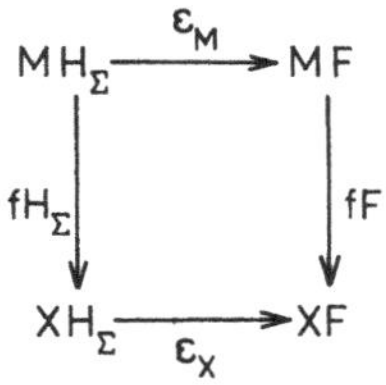

commutes, we conclude that $x \in (MF)fF$.

II. (a) Let F be a small functor, and let γ be the corresponding cardinal. Choose a set M of cardinality γ. Put

$$\Sigma = MF$$

and define the arity of any element of Σ to be γ. We prove that F is a quotient of H_Σ.

For each set X we have $XH_\Sigma = \coprod_{MF} X^\gamma$, and this can be identified with $X^M \times MF$. Then for each map $f\colon X \to Y$ the map $fH_\Sigma\colon X^M \times MF \to Y^M \times MF$ is defined by $(h, m)fH_\Sigma = (h \cdot f, m)$ for all $h\colon M \to X$ and all $m \in MF$. Define a natural transformation

$$\varepsilon\colon H_\Sigma \to F$$

by

$$(h, m)\varepsilon_X = (m)hF$$

for each $h\colon M \to X$ and $m \in MF$. It is easy to verify that this is a well-defined natural transformation, i.e., that for each map $f\colon X \to Y$ the following square

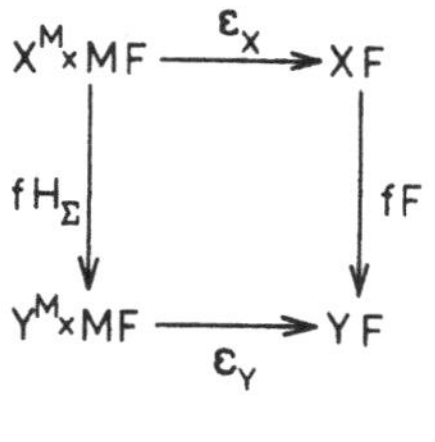

commutes. And the choice of γ guarantees that ε is an epitransformation: for each $x \in XF$ there exists a map $h\colon M \to X$ with $x \in (MF)hF$, i.e., $x = (m)hF = (h, m)\varepsilon_X$ for some $m \in MF$.

II. (b) Let F be a finitary functor. For each $n < \omega$ put $[n] = \{1, 2, \ldots, n\}$ and define a finitary type Σ by

$$\Sigma_n = [n]F \quad \text{for each } n < \omega.$$

For each set X we can identify XH_Σ with $\coprod_{n<\omega} X^{[n]} \times [n]F$. We define a natural transformation

$$\varepsilon: H_\Sigma \to F$$

by

$$(h, m)\varepsilon_X = (m)hF$$

for each $n < \omega$, $h: [n] \to X$ and $m \in [n]F$. Since F is finitary, ε is an epitransformation: for each $x \in XF$ there exists a map from a finite set into X, say, $h: [n] \to X$, and $m \in [n]F$ with $x = (m)hF = (h, m)\varepsilon_X$. □

Remark. A set functor is finitary iff it preserves directed unions (see Exercise III.4.E below) or directed colimits (this will be proved in Chapter V).

4.4. Definition. Let F be a set functor. A point

$$a \in AF,$$

where A is an arbitrary set, is said to be *distinguished* if for arbitrary two maps f, $g: A \to X$ we have

$$(a)fF = (a)gF.$$

Examples. (i) The power-set functor P has $\emptyset (\in AP$ for any set $A)$ as a distinguished point.

(ii) No point of H_n is distinguished if $n > 0$.

(iii) Each point of a constant functor is distinguished (and conversely, if each point of a set functor F is distinguished, then F is constant).

Remarks. (i) Each distinguished point $a \in AF$ defines „related" distinguished points $a_x \in XF$ in all sets $X \neq \emptyset$: choose any map $f: A \to X$ and put

$$a_X = (a)fF \qquad \text{(independent of } f\text{)}.$$

Then

$$(a_X)hF = a_Y$$

for each map $h: X \to Y$. Thus, we obtain a natural transformation

$$a: C_{01} \to F$$

(where $\emptyset C_{01} = \emptyset$ and $XC_{01} = 1 = \{0\}$ for each $X \neq \emptyset$) taking the value a_X for each set $X \neq \emptyset$.

(ii) What about $a_\emptyset$? In other words, can a be extended to a natural transformation $C_1 \to F$? Not in general: the point 0 is distinguished in C_{01} but there is no natural transformation $C_1 \to C_{01}$ at all.
(iii) For each set functor F, each point $a \in \emptyset F$ is distinguished simply because any two maps $f, g: \emptyset \to X$ are equal.

Definition. A distinguished point $a \in AF$ is called *standard* if there is a point $a_0 \in \emptyset F$ with $a = (a_0)fF$ for the empty map $f: \emptyset \to A$ (in other words, if the natural transformation a can be extended to C_1).

4.5. Convention. Let X and Y be sets with $X \subset Y$. Denote by $j_X^Y: X \to Y$ the *inclusion map* defined by $(x)j_X^Y = x$ for each $x \in X$.

Definition. A set functor is said to be *standard* if it preserves inclusion, i.e.,

$$X \subset Y \quad \text{implies} \quad XF \subset YF \quad \text{and} \quad j_X^Y F = j_{XF}^{YF},$$

and each of its distinguished points is standard.

Examples. (i) The power-set functor P is standard.
(ii) The hom-functor hom(M, —) is non-standard: if $X \neq Y$ then actually hom(M, X) $\cap$ hom(M, Y) $= \emptyset$ (because a map f carries, by definition, the information about its codomain). Nevertheless, hom(M, —) is naturally isomorphic to H_n for n = card M, and H_n is standard.
(iii) The functor C_{01} is non-standard because the distinguished point 0 is non-standard. But C_{01} differs from the standard functor C_1 only in the empty set and empty maps.
(iv) H_Σ is standard for each type Σ.

Remark. The reason for introducing the concept of standard functors is to obtain a class of functors which is „reasonably“ representative (this is proved in the following theorem) and which avoids complicated and not really interesting examination of the empty set which behaves somewhat irregularly.

For example, standard set functors preserve monos (also the empty ones) because the empty maps are actually inclusion maps $j_\emptyset^Y$. Further,

$$\emptyset F$$

is just the set of all distinguished points of F: if $a \in AF$ is distinguished, then $a \in \emptyset F \subset AF$ because there exists $a_0 \in \emptyset F$ with

$$a = (a_0) j_\emptyset^X F,$$

and $j_\emptyset^X F$ is the inclusion map.

Theorem. (Each set functor is almost standard). For each set functor F there exists a standard set functor F' such that the restriction of F and F' to all non-empty sets and non-empty maps are naturally isomorphic.

Proof. (a) First, assume that F preserves (empty) monos, and that each distinguished point of F is standard.

On the class of all pairs

$$(X, x)$$

where X is a set and $x \in XF$, we define the following relation:

$$(X, x) \sim (Y, y) \quad \text{iff} \quad (x)j_X^{X \cup Y}F = (y)j_Y^{X \cup Y}F.$$

This is an equivalence relation. The reflexivity and symmetry are obvious. For the transitivity, let

$$(X, x) \sim (Y, y) \sim (Z, z)$$

and consider the following commutative diagram of inclusion maps:

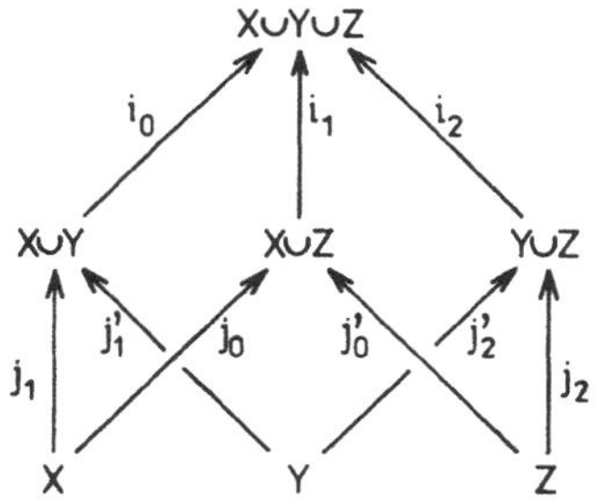

Since $(X, x) \sim (Y, y)$ means that $(x)j_1F = (y)j_1'F$, we get

$$(x)j_0F \cdot i_1F = (x)j_1F \cdot i_1F = (y)j_1'F \cdot i_0F.$$

Also, since $(Y, y) \sim (Z, z)$ means that $(y)j_2'F = (z)j_2F$, we get

$$(z)j_2'F \cdot i_1F = (z)j_2F \cdot i_2F = (y)j_2'F \cdot i_2F = (y)j_1'F \cdot i_0F.$$

Thus, $(x)j_0F \cdot i_1F = (z)j_0' \cdot \mathrm{i}_1F$ and since F preserves monos, it follows that $(x)j_0F = (z)j_0'F$, i.e., that $(X, x) \sim (Z, z)$. For each $x \in XF$ denote by

$$[X, x]$$

the equivalence class of (X, x). Note that

$$(+) \qquad (X, x) \sim (X, \bar{x}) \quad \text{implies} \quad x = \bar{x} \qquad\qquad (\text{for each } x, \bar{x} \in XF)$$

simply because $j_X^{X \cup X} = 1_X$.

Let us define a set functor F' on objects X by

$$XF' = \{[X, x];\ x \in XF\},$$

and on morphisms $f\colon X \to Y$ by

$$fF'\colon [X, x] \mapsto [Y, (x)fF] \qquad\qquad \text{for each } x \in XF.$$

The last map is well-defined [see (+) above], and F' is a functor since F is. Moreover, F and F' are naturally isomorphic: consider the natural transformation

$$\varepsilon_X : XF \to XF'$$

defined by

$$(x)\varepsilon_X = [X, x] \qquad (x \in XF).$$

Each ε_X is a bijection by (+). Let us verify that the functor F' is standard. It preserves inclusion because given $X \subset Y$, then

$$[X, x] = [Y, (x)j_X^Y F] \qquad \text{for each } x \in XF.$$

(This follows from $j_X^{X\cup Y} = j_X^Y$ and $j_Y^{X\cup Y} = 1_Y$.) Therefore $XF' \subset YF'$ and $j_X^Y F'$ is the inclusion map:

$$([X, x])j_X^Y F' = [Y, (x)j_X^Y F] = [X, x].$$

Each distinguished point of F' is standard because the naturally isomorphic functor F has this property.

(b) Let F be an arbitrary set functor. Using (a), it is sufficient to exhibit a functor $\bar{F}$ such that $F = \bar{F}$ on the full subcategory of all non-empty sets, and $\bar{F}$ preserves empty monos and has only standard distinguished points. Define $\emptyset\bar{F}$ to be the set of all natural transformations $\tau: C_{01} \to F$. This is a set (not a proper class) because each such transformation is fully determined by $(0)\tau_1 \in 1F$: given $\tau', \tau'': C_{01} \to F$, then

$$(*) \qquad \text{if } \tau'_X = \tau''_X \text{ for any } X \neq \emptyset, \text{ then } \tau' = \tau''.$$

[To prove (*), consider any set Y. If $Y = \emptyset$, then $\tau'_Y = \tau''_Y$ because $\emptyset C_{01} = \emptyset$. If $Y \neq \emptyset$, choose a map $g: X \to Y$ and then $\tau'_Y = \tau''_Y(: \{0\} \to YF)$ because

$$(0)\tau'_Y = (0)gC_{01} \cdot \tau'_Y = (0)\tau'_X \cdot gF = (0)\tau''_X \cdot gF = (0)\tau''_Y.]$$

For each empty map $f: \emptyset \to X$ define $f\bar{F}: \emptyset\bar{F} \to X\bar{F}$ by $f\bar{F} = 1_\emptyset$ if $X = \emptyset$ and

$$(\tau)f\bar{F} = (0)\tau_X \qquad \text{for each } \tau: C_{01} \to F$$

if $X \neq \emptyset$. By (*), fF is one-to-one. Thus, we obtain a functor $\bar{F}$ with $X\bar{F} = XF$ if $X \neq \emptyset$ and $fF = fF$ if the domain of f is non-empty, and this functor preserves monos. Each distinguished point of $\bar{F}$ is standard because any transformation $\tau: C_{01} \to \bar{F}$ is also a transformation $\tau: C_{01} \to F$, thus, $\tau \in \emptyset\bar{F}$. We can extend τ to a transformation $\bar{\tau}: C_1 \to \bar{F}$ by $(0)\bar{\tau}_\emptyset = \tau$. □

4.6. Proposition. Each standard set functor F preserves finite intersections.

Remark. In the category **Set**, finite intersections are just pullbacks of monos:

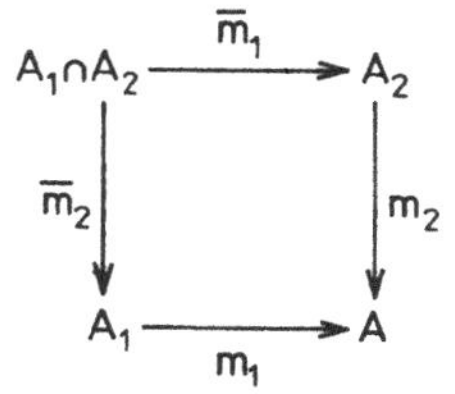

(where m_i are the given monos and $\bar{m}_i$ are the corresponding monos, definding $A_1 \cap A_2$ as a subobject of A_i). Thus, the proposition above states that each standard set functor preserves pullbacks of monos.

On the other hand, the statement can be understood purely set-theoretically:

$$A_1F \cap A_2F = (A_1 \cap A_2)F$$

for arbitrary two sets A_1, A_2. Fortunately, this is equivalent to the formulation above, since F preserves inclusion and we can use $A = A_1 \cup A_2$ and the inclusion maps m_1 and m_2.

Proof. Denote the inclusion maps as follows:

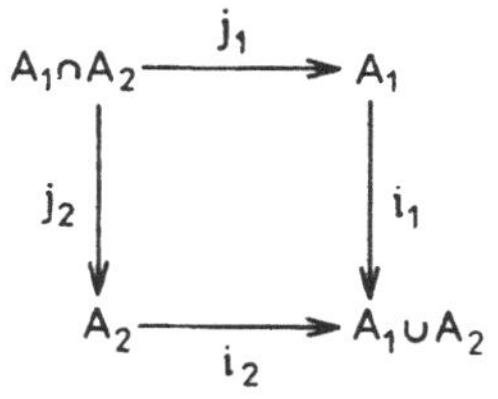

I. Let $A_1 \cap A_2 \neq \emptyset$.
We prove that there exist maps

$$r: A_1 \to A_1 \cap A_2,\ s: A_1 \cup A_2 \to A_2$$

such that

(1) $\quad j_1 \cdot r = 1;\ i_1 \cdot s = r \cdot j_2$ and $i_2 \cdot s = 1$.

In fact, choose $x_0 \in A_1 \cap A_2$ and define r by

$$(x)r = \begin{cases} x & \text{if } x \in A_1 \cap A_2 \\ x_0 & \text{if } x \in A_1 - A_2. \end{cases}$$

Further, define s by

$$(y)s = \begin{cases} y & \text{if } y \in A_2 \\ x_0 & \text{if } y \in A_1 - A_2. \end{cases}$$

Then (1) is obvious.

To prove that F preserves the pullback above, consider arbitrary maps p_1 and p_2 with $p_1 \cdot i_1F = p_1 \cdot i_2F$.

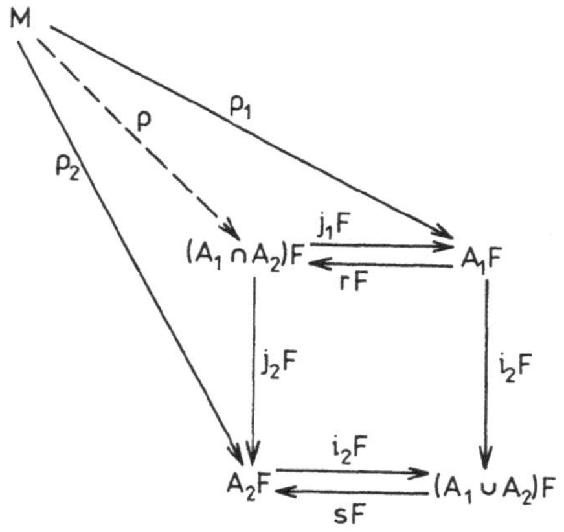

The map

$$p = p_1 \cdot rF$$

fulfils $p_1 = p \cdot j_1F$ and $p_2 = p \cdot j_2F$. The latter follows from (1) above:

$$\begin{aligned} p \cdot j_2F &= p_1 \cdot (r \cdot j_2)F \\ &= p_1 \cdot (i_1 \cdot s)F \\ &= p_2 \cdot i_2F \cdot sF \\ &= p_2. \end{aligned}$$

The former follows from the fact that i_1F is a mono:

$$\begin{aligned} (p \cdot j_1F) \cdot i_1F &= p \cdot j_2F \cdot i_2F \\ &= p_2 \cdot i_2F \\ &= p_1 \cdot i_2F. \end{aligned}$$

Finally, if p' also fulfils $p_1 = p' \cdot j_1F$ and $p_2 = p' \cdot j_2F$, then

$$p = p_1 \cdot rF = p' \cdot j_1F \cdot rF = p'.$$

II. Let $A_1 \cap A_2 = \emptyset$.

Since F is standard, it is sufficient to prove that each point

$$a \in A_1F \cap A_2F \subset (A_1 \cup A_2)F$$

is distinguished, i.e., an element of $\emptyset F$. Indeed, then $A_1F \cap A_2F \subset \emptyset F$, and the reverse inclusion is clear, since $\emptyset \subset A_i$ implies $\emptyset F \subset A_iF\,(i = 1, 2)$.

Let

$$f, g: A_1 \cup A_2 \to X$$

be an arbitrary pair of maps. To prove that

$$(a)fF = (a)gF,$$

let $\quad h: A_1 \cup A_2 \to Y$

be the map defined by

$$(x)h = \begin{cases} (x)f & \text{if } x \in A_1 \\ (x)g & \text{if } x \in A_2. \end{cases}$$

Then $j_1 \cdot h = j_1 \cdot f$ and $j_2 \cdot h = j_2 \cdot g$. Since $a \in A_1F$ implies $a = (a)j_1F$, we have

$$(a)hF = (a)j_1F \cdot hF = (a)j_1F \cdot fF = (a)fF.$$

Analogously,

$$(a)hF = (a)gF.$$

The proof is concluded. □

4.7. Proposition. Let F be a non-constant set functor. For arbitrary sets $X \neq \emptyset$ and Y,

$$\text{card } X \leq \text{card } Y \quad \text{implies} \quad \text{card } XF \leq \text{card } YF.$$

Morevoer, there is a cardinal γ such that for each set X,

$$\text{card } X \geq \gamma \quad \text{implies} \quad \text{card } XF \geq \text{card } X.$$

Proof. By Theorem III.4.5 it is clearly sufficient to work with standard functors F. The first statement is obvious: card $X \leq$ card Y means that there exists a mono $X \to Y$ and hence, a mono $XF \to YF$.

For the latter statement, we use the obvious fact that each non-constant functor F has a point $a \in AF$ which is not distinguished. Put

$$\gamma = \max(\text{card } A, \aleph_0).$$

Given a set X of power $\geq \gamma$, we can assume $A \subset X$. Since, moreover, X is infinite, it has a decomposition

$$X = \bigcup_{i \in I} X_i$$

into pairwise disjoint sets with

$$\text{card } I = \text{card } X_i = \text{card } X \qquad (i \in I).$$

For each $i \in I$ we choose a bijection

$$f_i: X \to X_i$$

and we prove that

$$(a)f_iF \neq (a)f_jF \qquad \text{for } i, j \in I;\ i \neq j.$$

Since $(a)f_iF \in X_iF \subset XF$, this will prove that

$$\text{card } XF \geq \text{card } I = \text{card } X.$$

Since f_i is a bijection, it is clear that $(a)f_iF$ is not a distinguished point. If $i \neq j$, then $X_i \cap X_j = \emptyset$, thus (by the preceding proposition), $X_iF \cap X_jF = \emptyset F$ and each point of $\emptyset F$ is distinguished. Therefore,

$$(a)f_iF \notin X_jF$$

and hence, $(a)f_iF \neq (a)f_jF$. The proof is concluded. □

Remark. Each standard set functor F has further pleasant properties.

(i) F preserves images. That is, given a map $f\colon X \to Y$ with $(X)f = M$, then $(XF)fF = MF$. Let $f'\colon X \to M$ denote the restriction of f and $j\colon M \to X$ the inclusion map, then $f = f' \cdot j$ and hence, $fF = f'F \cdot jF$. Since jF is the inclusion map and $f'F$ is onto (because f' is onto), we conclude that MF is the image of fF.

(ii) F preserves preimages for one-to-one maps. That is, given a one-to-one map $f\colon X \to Y$ then for each $A \subset Y$ we have $[(A)^{-1}f]F = (AF)(fF)^{-1}$. This follows from the preservation of finite intersections: for $f = f' \cdot j$ as above, f' is a bijection and $A \cap M$ is the image of $(A)f^{-1}$ under j.

4.8. We are now going to characterize set functors preserving colimits and unions. The latter means, for a standard set functor F, that

$$\bigcup_{i \in I} M_i = M \quad \text{implies} \quad \bigcup_{i \in I} M_iF = MF$$

for arbitrary sets M_i, $i \in I$.

Example. Given sets Σ_1 and Σ_0, define

$$S_{\Sigma_1\Sigma_0}\colon \mathbf{Set} \to \mathbf{Set}$$

on objects X by

$$XS_{\Sigma_1\Sigma_0} = (X \times \Sigma_1) + \Sigma_0$$

and on morphisms $f\colon X \to Y$ by

$$fS_{\Sigma_1\Sigma_0} = (f \times 1_{\Sigma_1}) + 1_{\Sigma_0}.$$

In other words,

$$S_{\Sigma_1\Sigma_0} = S_{\Sigma_1} + C_{\Sigma_0}.$$

The functor $S_{\Sigma_1\Sigma_0}$ preserves unions: if $\bigcup_{i\in I} M_i = M$, then

$$\left(\bigcup_{i\in I} M_i\right) \times \Sigma_1 = \bigcup_{i\in I} M_i \times \Sigma_1,$$

and hence,

$$\left(\left(\bigcup_{i\in I} M_i\right) \times \Sigma_1\right) + \Sigma_0 = \bigcup_{i\in I}(M_i \times \Sigma_1 + \Sigma_0).$$

The functor

$$S_{\Sigma_1\emptyset} = S_{\Sigma_1}$$

preserves colimits because it is a coadjoint (III.2.10).

Proposition. Let F be a standard set functor. F preserves colimits iff it is naturally isomorphic to S_Σ with $\Sigma = 1F$, and F preserves unions iff it is naturally isomorphic to $S_{\Sigma_1\Sigma_0}$ with $\Sigma_1 = 1F$ and $\Sigma_0 = \emptyset F$.

Proof. (i) Let F preserve unions. For each set X we have

$$X = \bigcup_{x\in X}\{x\}$$

and hence,

$$XF = \bigcup_{x\in X}\{x\}F.$$

By III.4.6, for two distinct points $x, y \in X$,

$$\{x\} \cap \{y\} = \emptyset \quad \text{implies} \quad \{x\}F \cap \{y\}F = \Sigma_0.$$

We define a natural transformation $\tau\colon S_{\Sigma_1\Sigma_0} \to F$ as follows. For each set X and each $x \in X$ let $j_x\colon \{0\} \to X$ denote the map with $(0)j_x = x$; the map

$$\tau_X\colon (X \times \Sigma_1) + \Sigma_0 \to XF$$

is defined by

$$(x, \sigma)\tau_X = (\sigma)j_xF \qquad \text{for each } (x, \sigma) \in X \times \Sigma_1;$$
$$(\bar\sigma)\tau_X = \bar\sigma \qquad \text{for each } \bar\sigma \in \Sigma_0.$$

It is easy to check that τ is well-defined, and it remains to verify that each τ_X is a bijection. The inverse map

$$\bigcup_{x\in X}\{x\}F \to (X \times \Sigma_1) + \Sigma_0$$

sends each $\bar\sigma \in \Sigma_0$ ($= \emptyset F \subset XF$) to $\bar\sigma$ (in the second sumand); to each

$$\sigma \in \{x\}F - \Sigma_0 \qquad\qquad (x \in X)$$

it assigns the pair (x, σ)—note that $\sigma \notin \{y\}F$ for any $y \neq x$.

(ii) If F preserves colimits, it also preserves unions. In fact, if $X = \bigcup_{i \in I} M_i$ then the canonical map

$$f: \coprod_{i \in I} M_i \to X$$

the components of which are the inclusion maps $M_i \to X$, is onto. Hence, the map

$$fF: \coprod_{i \in I} M_iF \to XF$$

is onto, and its components are inclusion maps, again. This proves that

$$XF = \bigcup_{i \in I} M_iF.$$

Therefore, F is naturally isomorphic to $S_{\Sigma_1\Sigma_0}$.

It remains to prove that if $S_{\Sigma_1\Sigma_0}$ preserves colimits, then $\Sigma_0 = \emptyset$. Indeed, the initial object $\emptyset$ is the colimit of the empty diagram and hence, $\emptyset S_{\Sigma_1\Sigma_0} = \emptyset$. We have

$$\emptyset = \emptyset S_{\Sigma_1\Sigma_0} = (\emptyset \times \Sigma_1) + \Sigma_0 = \Sigma_0.$$

This concludes the proof. □

Corollary. A standard set functor is an adjoint iff it is naturally isomorphic to H_n for some cardinal n, and a coadjoint iff it is naturally isomorphic to S_Σ for some set Σ.

The second statement follows from the preceding proposition, the former from the unicity of adjoints (for a given coadjoint): hom(Σ, —) and S_Σ form an adjoint pair, and hom(Σ, —) is naturally isomorphic to H_n with $n = \text{card } \Sigma$.

4.9. We conclude this section by returning to set functors defining varieties of universal algebras (see III.3.2). Recall the concept of concrete isomorphism (III.3.8).

Theorem. Let F be a set functor such that F-**Alg** is concretely isomorphic to a variety of Σ-algebras. Then F is small, and if Σ is finitary, then F is finitary.

Proof. We can assume that F is standard. Let γ be the least infinite cardinal larger than all arities in Σ (thus, $\gamma = \aleph_0$ iff Σ is finitary). For each set X we shall prove that XF is the union of all MF with $M \subset X$ of cardinality smaller than γ.

If X is finite, there is nothing to prove. Assume X infinite, and choose distinct points $a, b \in X$. Define an operation

$$\delta : XF \to X$$

by

$$(x)\delta = \begin{cases} a & \text{if } x \in MF \text{ for some } M \subset X \text{ with card } M < \gamma \\ b & \text{else.} \end{cases}$$

Let us prove that the set

$$Y = X - \{b\}$$

is a subalgebra of (X, δ), more precisely, that $(YF)\delta \subset Y$ and therefore (Y, δ') is a subalgebra of (X, δ), where δ' is the restriction of δ. Let

$$V: F\text{-}\mathbf{Alg} \to \mathscr{V}$$

be a concrete isomorphism with a variety $\mathscr{V}$ of Σ-algebras. Each variety is obviously closed under subalgebras and thus, it is sufficient to show that Y is a subalgebra of the Σ-algebra

$$(X, \bar{\delta}) = (X, \delta)V.$$

In fact, we then have a Σ-algebra $(Y, \bar{\delta}')$ which (as a subalgebra of a $\mathscr{V}$-algebra) is a $\mathscr{V}$-algebra, and $(Y, \bar{\delta}')V^{-1}$ is a subalgebra of (X, δ). For each $\sigma \in \Sigma$ of arity n and for arbitrary $y_i \in Y$, $i < n$, we are to show that $(y_i)\bar{\delta}_\sigma \in Y$. Put

$$M = \{y_i\}_{i<n} \cup \{a\} \subset X.$$

Since card $M \leq n + 1 < \gamma$, we have $(M)\delta = a$ and thus, M is a subalgebra of (X, δ). Consequently, M is a subalgebra of $(X, \bar{\delta})$; thus, $(M)\bar{\delta} \subset M \subset Y$ and this proves that $(y_i)\bar{\delta}_\sigma \in Y$.

Since Y is a subalgebra of (X, δ) and $b \notin Y$, we have $(x)\delta = a$ for any $x \in YF$. In other words, for every $x \in YF$ there exists $M \subset X$ with $x \in MF$ and card $M < \gamma$ [hence $x \in (Y \cap M)F$, by III.4.6]. Since Y and X are isomorphic sets, X also has the same property. □

Corollary. Varieties concretely isomorphic to F-**Alg** for set functors F are precisely those concretely isomorphic to basic varieties.

Each basic variety is concretely isomorphic to F-**Alg** by Proposition III.3.3. Conversely, if F-**Alg** is concretely isomorphic to a variety $\mathscr{V}$, then F is a quotient of some H_Σ, and then $\mathscr{V}$ is a basic variety of Σ-algebras by Proposition III.3.2.

Exercises III.4

A. Vector functors. Let R be a commutative field.

(i) Prove that each epi and each (!) mono in R-**Vect** split. [Hint: If

$e: X \to Y$ is onto, choose a base $B \subset Y$ and for each $b \in B$ choose $(b)m \in X$ with $((b)m)e = b$; extend m to a linear map $m: Y \to X$.]

(ii) Prove that each functor

$$F: R\text{-}\mathbf{Vect} \to R\text{-}\mathbf{Vect}$$

is naturally isomorphic to a functor preserving inclusion—such functors will be called *standard.* (Hint: Proceed as in III.4.5, using bases.)

(iii) Prove that each standard vector functor preserves finite intersections. (Hint: Proceed as in III.4.6, using bases.)

(iv) Prove the analogue of III.4.7 for vector functors, using dimension instead of cardinality.

B. Coproducts and components. (i) A set functor F is said to be *connected* if there exist no functors F_1, F_2 distinct from $C_\emptyset$ such that F is naturally isomorphic to $F_1 + F_2$. Prove that this is the case iff F preserves singletons, i.e., card $X = 1$ implies card $XF = 1$. [Hint: Given $a \in XF$ with card $X = 1$, denote for each set Y by $t_Y: Y \to X$ the unique map and define F_1 on objects Y by $YF_1 = \{x \in YF; (a)t_YF = a\}$; on maps f, fF_1 is a restriction of fF. Then put $F_2 = F - F_1$.]

(ii) Prove that each set functor is a coproduct of its maximal connected subfunctors. [Hint: Choose X of power 1 and for each $a \in XF$ define F_1 as in the hint to (i) above.]

C. Linear functors. Let R be a commutative ring. A functor $F: R\text{-}\mathbf{Mod} \to R\text{-}\mathbf{Mod}$ is *linear* if for arbitrary linear maps $f, g: X \to Y$ and each $r \in R$,

$$(f + g)F = fF + gF \quad \text{and} \quad (r \cdot f)F = r \cdot (fF).$$

(i) Prove that any product and coproduct of linear functors is linear. Conclude that for each type Σ, the functor H_Σ (III.2.5) is linear.

(ii) When is the constant functor C_M linear? When is the functor S_Σ (III.2.4) linear?

(iii) Prove that a functor $F: R\text{-}\mathbf{Mod} \to R\text{-}\mathbf{Mod}$ is a quotient functor of some H_Σ iff F is linear and *small.* The latter means that there is a cardinal γ such that for each module X and each $x \in XF$ there is a linear map $f: M \to X$ with $x \in (MF)fF$ and with M having less than γ generators. Hint: The proof is analogous to that in III.4.3.

D. Finitary functors on R-Mod. A functor $F: R\text{-}\mathbf{Mod} \to R\text{-}\mathbf{Mod}$ is *finitary* if for each $x \in XF$ there exists a morphism $f: M \to X$ with $x \in (MF)fF$ and such that M is finitely generated.

(i) For each module M prove that S_M and V_M (see III.2.15) are finitary functors. For any cardinal n put

$$V_n = V_M,$$

where M is the free module on n generators; verify that V_n is the coproduct of n copies of the identity functor.

(ii) Prove that H_Σ is finitary iff Σ is a finitary type, in which case H_Σ is naturally isomorphic to V_n for n equal to the sum of all arities. (Hint: $H_n = V_n$ for $n < \omega$ because finite products coincide with finite coproducts.)

(iii) Let R be a field. Prove that each linear, finitary functor F: R-**Vect** $\to$ R-**Vect** is naturally isomorphic to V_n for some cardinal n. (Hint: As in III.4.3, prove that F is a quotient of some H_Σ, Σ finitary. Prove that each quotient of $H_\Sigma = V_n$ is naturally isomorphic to some V_k.)

E. Preservation of directed unions. A collection of subobjects $m_i: M_i \to X$ $(i \in I)$ is *directed* if for arbitrary $i, j \in I$ there exists a $k \in I$ with $m_i \subset m_k$ and $m_j \subset m_k$.

(i) Prove that a standard set functor F is finitary iff it preserves directed unions, i.e., $\left(\bigcup_{i \in I} M_i\right) F = \bigcup_{i \in I} M_i F$ for each directed collection $M_i \subset X$.

(ii) Prove that a functor F: R-**Mod** $\to$ R-**Mod** is finitary iff it preserves directed unions, i.e., if $\bigcup_{i \in I} m_i = m$ is a directed union, then $\bigcup_{i \in I} \operatorname{im}(m_i F) = \operatorname{im}(mF)$ (where im means the image). [Hint: Given $x \in \operatorname{im}(mF)$, there exists $f: M \to X$ with $x \in (MF)fF$ and M finitely generated, say, with generators $y_1, \ldots, y_k$. Each $(y_n)f$ belongs to some im m_{i_n} $(i_n \in I)$ and there exists $i \in I$ with $m_{i_n} \subset m_i$ for all n. Then $x \in \operatorname{im}(m_i F)$. The converse inclusion $\bigcup \operatorname{im}(m_i F) \subset \operatorname{im}(mF)$ is clear.]

F. Preservation of countable colimits. Prove the following statements.

(i) Let F be a standard set functor with $\emptyset F = \emptyset$. If F preserves coequalizers, then F preserves the countable coproduct

$$Y = X + X + X + \ldots \text{ for each set } X.$$

[Hint: Since $\emptyset F = \emptyset$, no point of F is distinguished and hence, the copies of XF in YF are pairwise disjoint. It remains to prove that each $y \in YF$ lies in some of these copies. Coding Y as $Y = X \times \mathbf{Z}$ (where $\mathbf{Z}$ is the set of all integers), let $g: Y \to Y$ be the isomorphism with $g(x, z) = (x, z + 1)$. The coequalizer of g and 1_Y is the projection $\pi: Y = X \times \mathbf{Z} \to X$ and hence, $\pi F =$ coeq$(gF, 1_{YF})$. Thus, for each $y \in YF$ there is $n < \omega$ and $y' \in (X \times \{z\})F$ with $(y)g^nF = (y')g^nF$ and hence, $y \in (X \times \{z\})F$.]

(ii) Any set functor preserving finite colimits preserves countable colimits. [Hint: If F preserves finite coproducts, then $\emptyset F = \emptyset$ and F has no distinguished point—we can assume that F is standard. To prove that F preserves countable coproducts $Y = X_1 + X_2 + X_3 + \ldots$, find a set X with $X_n \subset X$ $(i < \omega)$ and use (i): Since F preserves the coproduct $X + X + X + \ldots$ and also F preserves finite intersections, any point $y \in YF$ lies in some $X_i F$.]

(iii) Any functor F: R-**Mod** $\to$ R-**Mod** preserving finite colimits preserves countable colimits.

G. Preservation of colimits. Vector functors, i.e., functors F: R-**Vect** $\to$ R-**Vect**, have properties analogous to set functors if they are linear:

(i) Prove that each linear vector functor preserving coproducts is naturally isomorphic to some V_Σ. [Hint: Put $\Sigma = RF$. For each space X with base B we have $X = \coprod_{x \in B} R_x$ with injections $j_x: R \to X$ defined by $(r)j_x = rx$. Then $XF = \coprod_B \Sigma \simeq \Sigma \otimes X$ with a natural isomorphism $\tau: \Sigma \otimes X \to XF$ defined by $(\sigma \otimes x)\tau = (\sigma)j_x F$.]

(ii) Conclude that the only linear co-adjoints in R-**Vect** are V_Σ (up to a natural isomorphism).

III.5. Factorization Systems

5.1. We have introduced factorization systems in III.2.7. In the present section, we study some of the basic facts about them needed below.

Recall the hierarchy of morphisms from Exercise III.1.D. In particular, each regular (or split) mono which is an epi is an isomorphism.

Proposition. Let

$$(\mathscr{E}, \mathscr{M})$$

be an arbitrary factorization system.

(i) $\mathscr{M}$ is right cancellative, i.e., if $m_1 \cdot m_2 \in \mathscr{M}$ then $m_1 \in \mathscr{M}$;

(ii) $\mathscr{M}$ contains all regular monos, hence, all split monos;

(iii) pullbacks carry $\mathscr{M}$-monos, i.e., in each pullback

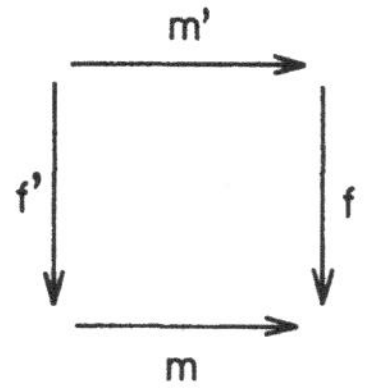

$m \in \mathscr{M}$ implies $m' \in \mathscr{M}$.

Proof. All this is a simple application of the the diagonal fill-in property (III.2.7).

(i) Let $m_1 = e \cdot m$ be an image factorization. We use the diagonal fill-in:

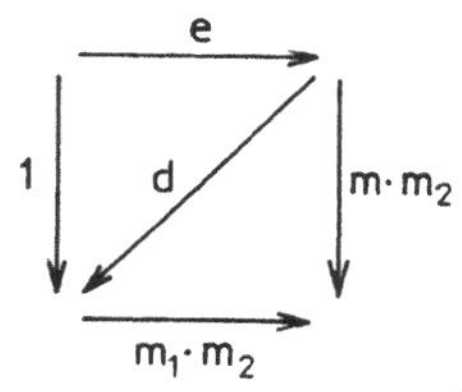

Since e is an epi as well as a split mono ($e \cdot d = 1$), we conclude that e is an isomorphism. Hence,

$$m_1 = e \cdot m \in \mathcal{M}.$$

(ii) Let $m: X \to Y$ be the equalizer of $f, g: Y \to Z$, and let $m = e_0 \cdot m_0$ be its image factorization:

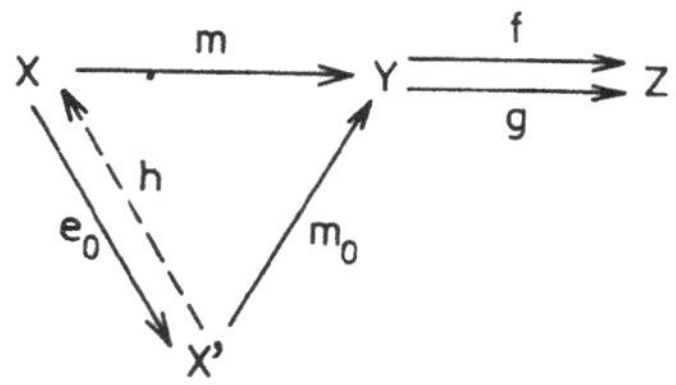

Then $m_0 \cdot f = m_0 \cdot g$ (because e_0 is an epi and $e_0 \cdot m_0 \cdot f = e_0 \cdot m_0 \cdot g$) and thus, there exists h with

$$m_0 = h \cdot m.$$

Since m_0 is a mono and $m_0 = h \cdot e_0 \cdot m_0$, we have $h \cdot e_0 = 1$; since m is a mono and $m = e_0 \cdot m_0 = e_0 \cdot h \cdot m$, we have $e_0 \cdot h = 1$. Thus, $e_0 = h^{-1}$ is an isomorphism which proves that $m \in \mathcal{M}$.

(iii) Let $m' = e_0 \cdot m_0$ be the image factorization in the pullback above:

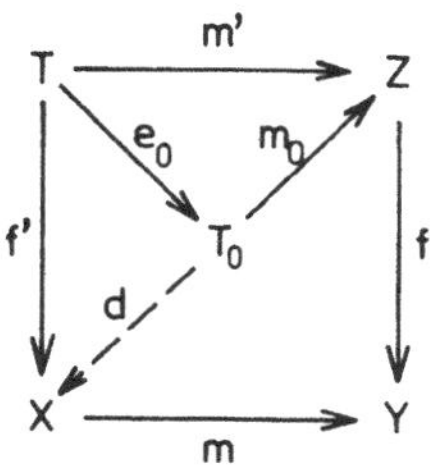

Using the diagonal fill-in we find a morphism d with

$$f' = e_0 \cdot d \quad \text{and} \quad m_0 \cdot f = d \cdot m.$$

The latter equation leads to a unique morphism d_0 such that the following diagram

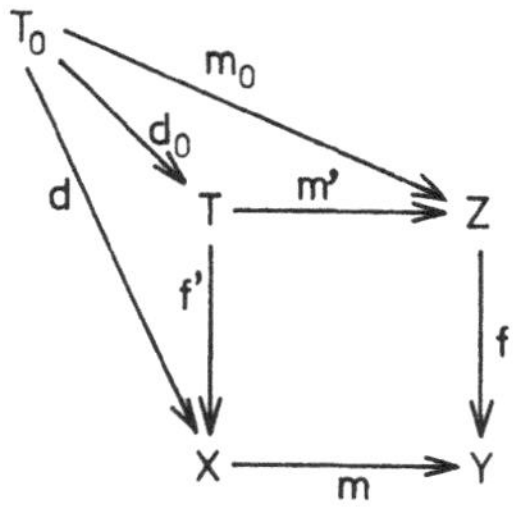

commutes. Then $e_0 \cdot d_0 = 1$ (because the fact that m is mono implies that m' is mono and we have $m' = e_0 \cdot m_0 = e_0 \cdot d_0 \cdot m'$). Thus, e_0 is an isomorphism and we conclude that $m' \in \mathcal{M}$. □

Remark. We shall actually use the dual properties more often:
(i) $\mathcal{E}$ is left cancellative;
(ii) $\mathcal{E}$ contains all regular epis;
(iii) pushouts carry $\mathcal{E}$-epis.
These are proved by the duality principle: if $\mathcal{K}$ is an $(\mathcal{E}, \mathcal{M})$-category, then $\mathcal{K}^{\mathrm{op}}$ (III.2.12) is an $(\mathcal{M}, \mathcal{E})$-category.

Corollary. Each of the following categories:

Set;
R-**Mod** (R a commutative ring);
H_Σ-**Alg** (Σ any type)

has a unique factorizaltion system with $\mathcal{E}$ the class of all epis (= regular epis = onto morphisms) and $\mathcal{M}$ the class of all monos (= regular monos = one-to-one morphisms).

This follows from (ii) above.

5.2. The aim of introducing factorization systems is to obtain a variety of concepts of subobject and quotient object. Let $\mathcal{K}$ be an $(\mathcal{E}, \mathcal{M})$-category. A subobject (more precisely, $\mathcal{M}$-subobject) of an object A is represented by an $\mathcal{M}$-mono

$$m: B \to A.$$

Two $\mathcal{M}$-monos $m: B \to A$ and $m': B' \to A$ represent the same subobject of A iff they are *equivalent,*

$$m \sim m'$$

which means that there is an isomorphism $i: B \to B'$ for which the following triangle

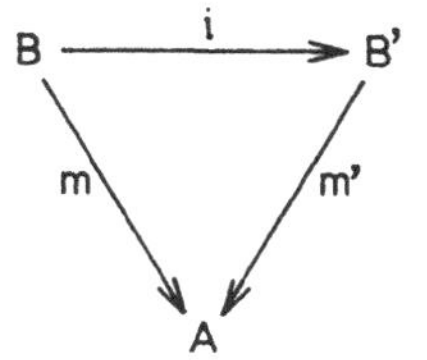

commutes.

Thus, the exact definition of a subobject of A is: an equivalence class of some $\mathcal{M}$-mono $m: B \to A$. (We usually identify the equivalence class with m itself, as an abuse of language.)

Given two subobjects, represented by $\mathcal{M}$-monos $m_1 : B_1 \to A$ and $m_2 : B_2 \to A$, we write

$$m_1 \subset m_2$$

if there is a morphism $f: B_1 \to B_2$ with

$$m_1 = f \cdot m_2.$$

Note that, by the preceding proposition, this implies $f \in \mathcal{M}$ and hence, f represents a subobject of B_2. The fact that we work with equivalence classes implies that whenever

$$m_1 \subset m_2 \quad \text{and} \quad m_2 \subset m_1,$$

then m_1 is the same subobject as m_2. (Proof. We have $m_1 = f \cdot m_2$ as well as $m_2 = g \cdot m_1$. Then f is an isomorphism because $m_1 = f \cdot g \cdot m_1$ implies $f \cdot g = 1$ and $m_2 = g \cdot f \cdot m_2$ implies $g \cdot f = 1$.)

For each object A we obtain an ordered class of all subobjects of A. Its meets are called *intersections* and are denoted by $\bigcap_{i \in I} m_i$ (where I can possibly be a large class); its joins are called *unions* and are denoted by $\bigcup_{i \in I} m_i$.

Given a collection of subobjects

$$m_i : B_i \to A \quad (i \in I)$$

we can consider it as a diagram and we can (possibly) form its limit C with projections

$$u_i : C \to B_i \quad (i \in I)$$

and

$$u = u_i \cdot m_i : C \to A \text{ (independent of } i\text{):}$$

(*) 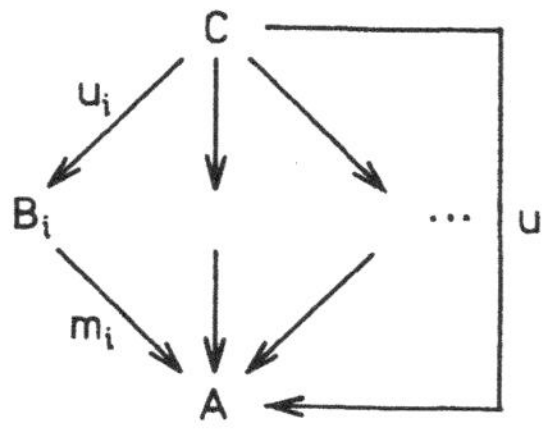

If card $I = 2$, this is a pullback; generally, this is called a *multiple pullback*. The following extends (iii) of Proposition III.5.1:

Proposition. (Intersections are multiple pullbacks.) Let $m_i: B_i \to A$ $(i \in I)$ be a collection of $\mathcal{M}$-subobjects. If (*) is a multiple pullback, then $u_i \in \mathcal{M}$ for each $i \in I$, and

$$u = \bigcap_{i \in I} m_i.$$

Conversely, if $u = \bigcap_{i \in I} m_i$, then for each $i \in I$ there is a unique $u_i \in \mathcal{M}$ with $u = u_i \cdot m_i$, and then (*) is a multiple pullback.

Proof. (i) Let (*) be a multiple pullback. We prove first that

$$u_{i_0} \in \mathcal{M}$$

for any $i_0 \in I$. To do this, let $u_{i_0} = \bar{e} \cdot \bar{m}$ be an image factorization. We use the diagonal fill-in for each $i \in I$:

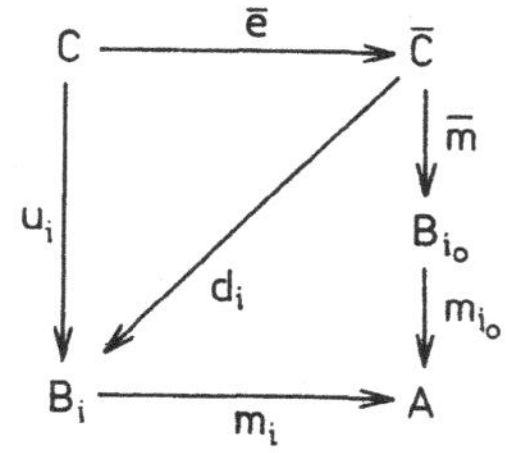

Since $d_i \cdot m_i$ is independent of i, by the definition of multiple pullbacks there exists a unique morphism

$$d: \bar{C} \to C$$

with

$$d_i = d \cdot u_i \ (i \in I).$$

Then $\bar{e} \cdot d = 1_C$ because

$$(\bar{e} \cdot d) \cdot u_i = \bar{e} \cdot d_i = u_i \quad (\mathrm{i} \in I).$$

Hence, $\bar{e}$ is a split mono as well as an epi, i.e., an isomorphism. This proves that $u_{i_0} = \bar{e} \cdot \bar{m} \in \mathcal{M}$.

Next, we prove that $u = \bigcap m_i$. First, $u \subset m_i$ because $u = u_i \cdot m_i$ $(i \in I)$. Furthermore, let $u'\colon C' \to A$ be an $\mathcal{M}$-mono with $u' \subset m_i$ $(i \in I)$ and let $u'_i\colon C' \to B_i$ be the (unique) morphism with

$$u' = u'_i \cdot m_i \qquad (i \in I).$$

Again, there is a unique morphism

$$v\colon C' \to C$$

with

$$u'_i = v \cdot u_i \qquad (i \in I).$$

We have $u' = v \cdot u$ because, for any $i \in I$,

$$u' = u'_i \cdot m_i = v \cdot u_i \cdot m_i = v \cdot u.$$

Hence, $u' \subset u$, which concludes the proof that $u = \bigcap m_i$.

(ii) Let $u = \bigcap_{i \in I} m_i$. For each $i \in I$ we have $u \subset m_i$, thus,

$$u = u_i \cdot m_i$$

for some $u_i\colon C \to B_i$ (which is unique because m_i is mono, and is in $\mathcal{M}$ because $\mathcal{M}$ is right cancellative by III.5.1).

To prove that (*) is a multiple pullback, let $f\colon D \to A$ and $f_i\colon D \to B_i$ $(i \in I)$ be arbitrary morphisms with

$$f = f_i \cdot m_i \qquad (i \in I).$$

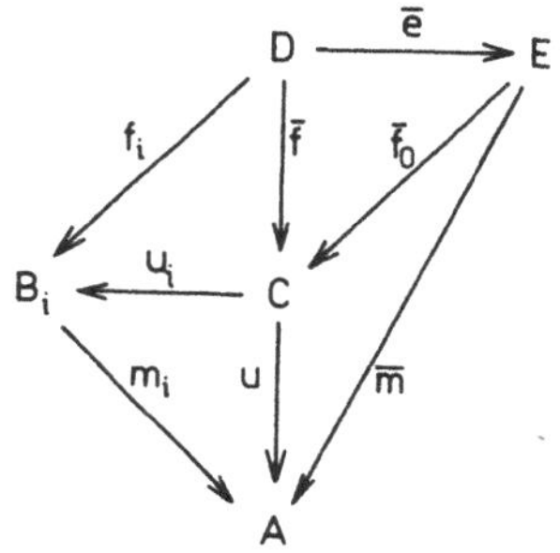

We are to exhibit a morphism $\bar{f}\colon D \to C$ with $f = \bar{f} \cdot u$ and

$$f_i = \bar{f} \cdot u_i \qquad (i \in I).$$

(Then $\bar{f}$ is unique because u is a mono.)

Let $f = \bar{e} \cdot \bar{m}$ be an image factorization. Using the diagonal fill-in:

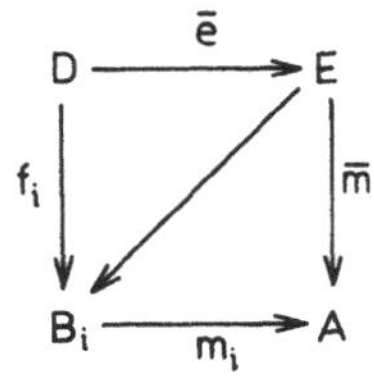

we conclude that $\bar{m} \subset m_i$ $(i \in I)$, hence

$$\bar{m} \subset u.$$

Thus, $\bar{m} = \bar{f}_0 \cdot u$ for some morphism $\bar{f}_0$. Put

$$\bar{f} = \bar{e} \cdot \bar{f}_0.$$

Then

$$f = \bar{e} \cdot \bar{m} = \bar{e} \cdot \bar{f}_0 \cdot u = \bar{f} \cdot u.$$

Moreover, for each $i \in I$ we have $f_i = \bar{f} \cdot u_i$ because m_i is a mono and

$$f_i \cdot m_i = f = \bar{f} \cdot u = \bar{f} \cdot u_i \cdot m_i.$$

This conclude the proof that (∗) is a multiple pullback. □

5.3. Definition. An $(\mathscr{E}, \mathscr{M})$-category $\mathscr{K}$ is said to *have intersections* if each (possibly large) collection of $\mathscr{M}$-subobjects of any object A has an intersection. In other words, if all $\mathscr{M}$-subobjects of A form a large-complete lattice.

$\mathscr{K}$ is said to be *$\mathscr{M}$-well-powered* if each object has only a (small) set of $\mathscr{M}$-subobjects.

Proposition. Let $\mathscr{K}$ be an $\mathscr{M}$-well-powered $(\mathscr{E}, \mathscr{M})$-category. If $\mathscr{K}$ is either complete or cocomplete, then $\mathscr{K}$ has intersections.

Proof. (i) Let $\mathscr{K}$ be complete. Then it has small intersections, since these are multiple pullbacks by the preceding theorem. And each large intersection can be "reduced" to a small one by finding a small set of representatives.

(ii) Let $\mathscr{K}$ be cocomplete. We prove that for each object A the (small) poset of subobjects is a complete lattice. It is sufficient to prove that any (small) collection of subobjects $m_i \colon B_i \to A$ $(i \in I)$ has a union. We form the coproduct

$$B = \coprod_{i \in I} B_i$$

with injections $v_i \colon B_i \to B$ $(i \in I)$, and we define

$$f \colon B \to A$$

by

$$v_i \cdot f = m_i \qquad (i \in I).$$

Let $f = \bar{e} \cdot \bar{m}$ be an image factorization. We prove that

$$\bar{m} = \bigcup_{i \in I} m_i.$$

First, $m_i \subset \bar{m}$ for each $i \in I$ by the diagonal fill-in:

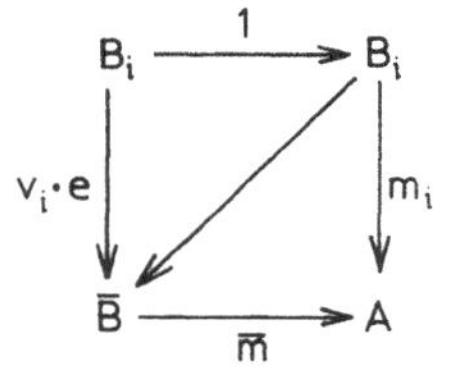

Next, let $\tilde{m}: \tilde{B} \to A$ be a subobject with $m_i \subset \tilde{m}$ for each $i \in I$, i.e.,

$$m_i = u_i \cdot \tilde{m}$$

for some $u_i: B_i \to \tilde{B}$. Then the morphism $u: B \to \tilde{B}$ with components $u_i\,(i \in I)$ fulfils

$$v_i \cdot f = m_i = u_i \cdot \tilde{m} = v_i \cdot u \cdot \tilde{m} \quad (i \in I),$$

hence,

$$f = u \cdot \tilde{m}.$$

Using the diagonal fill-in once more,

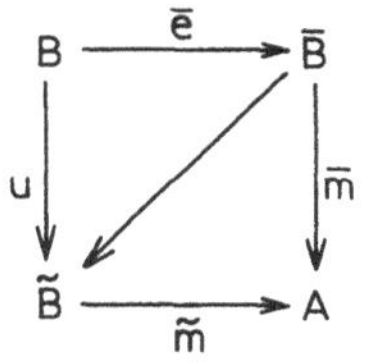

we conclude that $\bar{m} \subset \tilde{m}$. This proves that $\bar{m} = \bigcup m_i$. □

Remark. Let us spell out the dual concepts which will be needed extensively in the subsequent chapters. Epis $e: A \to B$ in $\mathscr{E}$ represent *quotients* of A; we write

$$e \leq e'$$

if the following triangle

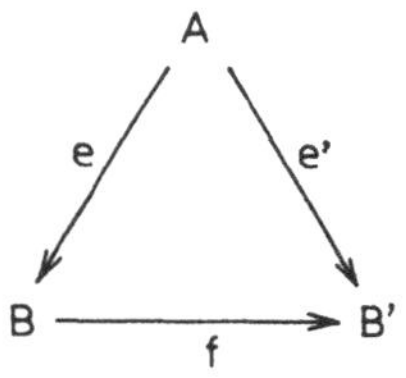

commutes for some f (which is in $\mathscr{E}$ since $\mathscr{E}$ is left cancellative). If $e \leq e'$ and $e' \leq e$, then e and e' represent the same quotient of A.

Joins in the ordered class of all quotients are called *cointersections*. The cointersection of a collection $e_i: A \twoheadrightarrow B_i$ $(i \in I)$ of quotients of A is given by their multiple pushout

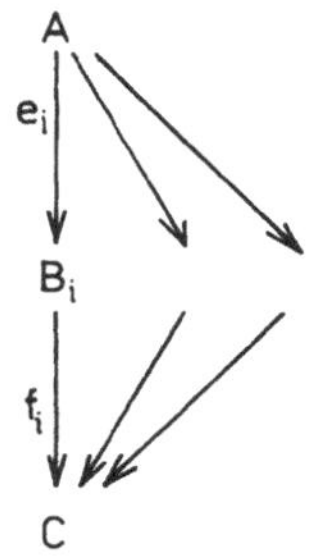

as the $\mathscr{E}$-epi $e = e_i \cdot f_i$ (independent of i). If $\mathscr{K}$ is *$\mathscr{E}$-cowell-powered*, i.e., if each object has only a (small) set of quotients, then it has cointersections whenever it is either complete or cocomplete.

Examples. (i) Quotients in **Set**. We can represent each quotient of a set A by the canonical map

$$e: A \twoheadrightarrow A/E = \{[x];\ x \in A\},$$

where E is an equivalence relation on A (and e assigns to each $x \in A$ its equivalence class $[x]$ under E). Then

$$e \leq e' \text{ iff } E \subset E';$$

thus, the least quotient is 1_A and the largest one has just one equivalence class. Given equivalences E_i on A $(i \in I)$, their cointersection is the equivalence E defined as follows: given $x, y \in A$ then $x\ E\ y$ iff there exist elements $z_0, z_1, \ldots, z_n \in A$ and indices $i_1, \ldots, i_n \in I$ with $x = z_0$, $y = z_n$ and

$$z_{k-1} E_{i_k} z_k \qquad \text{for } k = 1, \ldots, n.$$

(ii) Quotients in R-**Mod**. We can represent each quotient of a module A as the canonical map

$$e: A \twoheadrightarrow A/B = \{x + B;\ x \in A\}$$

where B is a submodule (and e assigns to each $x \in A$ the class $x + B$). Then

$$e \leq e' \text{ iff } B \subset B';$$

thus, the least quotient corresponds to $B = 0$ and the largest to $B = A$. The cointersection of quotients corresponds to union: given quotients

$$e_i: A \twoheadrightarrow A/B_i \qquad (i \in I)$$

let B be the submodule generated by $\bigcup_{i \in I} B_i$; then

$$e: A \to A/B$$

is the cointersection of e_i $(i \in I)$.

5.4. Quotient functors and subfunctors. Let $\mathcal{K}$ be a category with a given factorization system $(\mathcal{E}, \mathcal{M})$. For each functor $F: \mathcal{K} \to \mathcal{K}$, *quotient functors* of F are introduced analogously to quotients of an objects. Every $\mathcal{E}$-epitransformation $\varepsilon: F \to G$ (i.e., natural transformation with $\varepsilon_K \in \mathcal{E}$ for all $K \in \mathcal{K}^0$) represents a quotient functor of F. Two $\mathcal{E}$-epitransformations $\varepsilon: F \to G$ and $\varepsilon': F \to G'$ represent the same quotient iff there is a natural isomorphism $\tau: G \to G'$ with $\varepsilon' = \varepsilon \cdot \tau$. For example, the functor P_2: **Set** $\to$ **Set** (III.4.1) is a quotient of H_2 because we have an epitransformation $\varepsilon: H_2 \to P_2$ given by $(x, y) \mapsto \{x, y\}$.

Dually, *subfunctors* of F are represented by $\mathcal{M}$-monotransformations $\mu: G \to F$.

5.5. We know that $\mathcal{M}$-monos are well-behaved with respect to equalizers and (multiple) pullbacks. We continue with other types of limits. Let D, $D': \mathcal{D} \to \mathcal{K}$ be two diagrams with limits $L = \lim D$ and $\pi_d: L \to dD (d \in \mathcal{D}^o)$; $L' = \lim D'$ and $\pi'_d: L' \to dD' (d \in \mathcal{D}^o)$, respectively. Given a natural transformation

$$\mu: D' \to D,$$

there exists a unique morphism

$$\lim \mu: L' \to L$$

with

$$\lim \mu \cdot \pi_d = \pi'_d \text{ for each } d \in D'^0$$

(because the morphisms $\pi_d \cdot \mu_d: L \to dD'$ form a compatible collection of D').

Proposition. Let $\mathcal{K}$ be an $(\mathcal{E}, \mathcal{M})$-category. Let D' and D be two diagrams which have a limit in $\mathcal{K}$, and let $\mu: D' \to D$ be a natural transformation formed by $\mathcal{M}$-monos. Then $\lim \mu$ is also an $\mathcal{M}$-mono.

Proof. Let $\lim \mu = e \cdot m$ be an image factorization. Using the notation above, we apply the diagonal fill-in

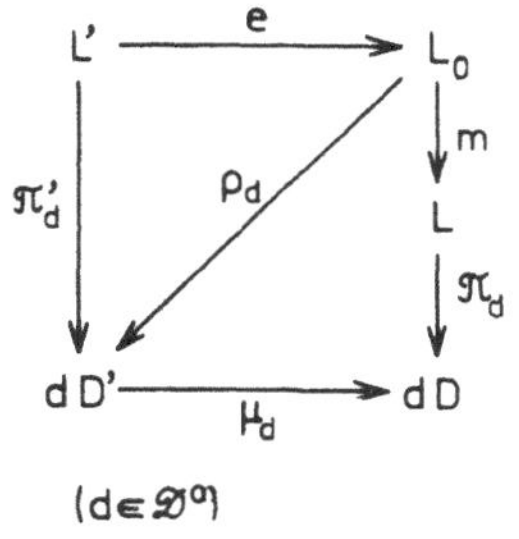

The resulting collection $p_d\colon L'_0 \to dD'$ $(d \in D'^0)$ is compatible with D' because for each $\delta\colon d_1 \to d_2$ in D'^m we have

$$e \cdot p_{d_2} = \pi'_{d_2} = \pi'_{d_1} \cdot \delta D' = e \cdot (p_{d_1} \cdot \delta D'),$$

and e is epi. Therefore, there exists a unique morphism

$$p\colon L'_0 \to L'$$

with

$$p_d = p \cdot \pi'_d \qquad (d \in D'^0).$$

Then $e \cdot p = 1$ because

$$(e \cdot p) \cdot \pi'_d = e \cdot p_d = \pi'_p \qquad (d \in D'^0).$$

Since e is a split mono, it is an isomorphism and thus, $\lim \mu$ is in $\mathcal{M}$. □

Corollary. Given $\mathcal{M}$-monos $m_2\colon A'_i \to A_i$ $(i \in I)$, then the product morphism

$$\prod_{i \in I} m_i\colon \prod_{i \in I} A'_i \to \prod_{i \in I} A_i$$

is also an $\mathcal{M}$-mono.

Remark. Let D be a diagram with

$$A = \operatorname{colim} D,$$

let $\varepsilon_d\colon dD \to A$ denote the injections $(d \in \mathcal{D}^o)$. If $\varepsilon_d = e_d \cdot m_d$ are image factorizations, then

$$\bigcup_{d \in \mathcal{D}^o} m_d = 1_A.$$

Indeed, given an $\mathcal{M}$-mono $m\colon B \to A$ with $m_d \subset m (d \in \mathcal{D}^o)$, we prove that m is an isomorphism. For each $d \in \mathcal{D}^o$ we have $f_d\colon dD \to B$ with $m_d = f_d \cdot m$. Since m is a mono, it is easy to check that the family $e_d \cdot f_d (d \in \mathcal{D}^o)$ is compatible. Hence, there exists $f\colon A \to B$ with $e_d \cdot f_d = \varepsilon_d \cdot f$. Then $f \cdot m = 1_A$ (because $\varepsilon_d \cdot f \cdot m = e_d \cdot f_d \cdot m = e_d \cdot m_d = \varepsilon_d$ for each d). Thus, m is a split epi and a mono, hence, m is an isomorphism.

Dually, given a natural $\mathcal{E}$-transformation of two diagrams, the colimit morphism is in $\mathcal{E}$, too. In particular, if $e_i \in \mathcal{E} (i \in I)$, then

$$\coprod e_i \in \mathcal{E}.$$

For directed diagrams, more can be said. Recall that a *directed diagram* is a diagram $D\colon \mathcal{D} \to \mathcal{K}$ such that $\mathcal{D}$ is a directed poset $[\mathcal{D} = (I, \leq)$ such that for all $i, j \in I$ there exists $k \in I$ with $i \leq k$ and $j \leq k]$. Then D consists of objects $D_i (i \in I)$ and morphisms $d_{ij}\colon D_i \to D_j (i \leq j)$. We say that D is a diagram of $\mathcal{E}$-epis if $d_{ij} \in \mathcal{E}$ whenever $i \leq j$.

Proposition. Let D be a directed diagram of $\mathcal{E}$-epis. If D has a colimit, then colimit injections are $\mathcal{E}$-epis, too.

Proof. Denote by $(I, \leq)$ the scheme of D and let $A = \operatorname{colim} D$ with the injections

$$p_i : D_i \to A \qquad (i \in I).$$

For any $i_0 \in I$ we choose an image factorization

$$p_{i_0} = e \cdot m.$$

To prove that $m: A' \to A$ is an isomorphism, it is clearly sufficient to verify that any p_i $(i \in I)$ factors through m.

For each $i \in I$ choose $k \in I$ with $i_0 \leq k$ and $i \leq k$, and use the diagonal fill-in:

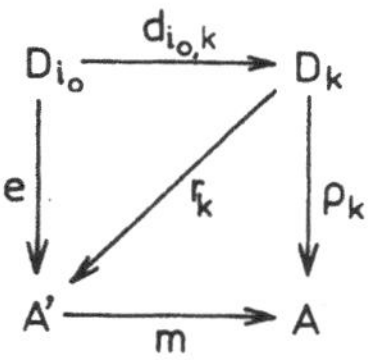

Then $i \leq k$ implies

$$p_i = d_{i,k} \cdot p_k = d_{i,k} \cdot r_m \cdot m.$$

Hence, m is an isomorphism, consequently, $p_{i_0} \in \mathcal{E}$. □

5.6. Lemma. In each factorization system, the class $\mathcal{M}$ determines the class $\mathcal{E}$: a morphism $f: A \to B$ is in $\mathcal{E}$ iff for each commuting triangle

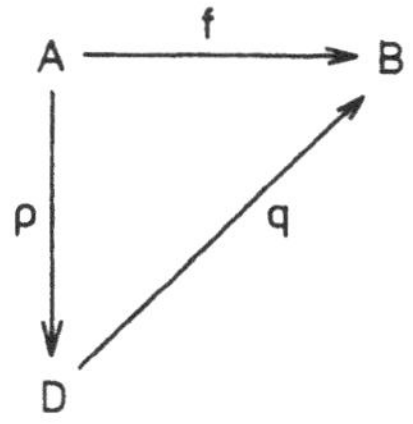

with $q \in \mathcal{M}$, q is an isomorphism.

Proof. Let $f \in \mathcal{E}$. Then we use the diagonal fill-in:

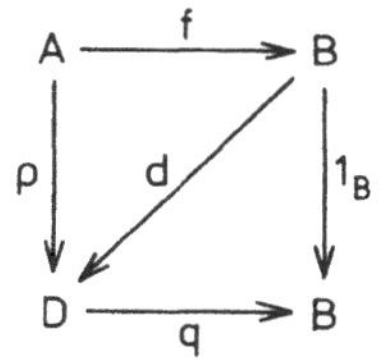

Since q is a split epi and a mono, it is an isomorphism.

Let $f \notin \mathscr{E}$. The image factorization of f forms a triangle as above, with $p \in \mathscr{E}$ and $q \in \mathscr{M}$. Then q is not an isomorphism (because else $f = p \cdot q \in \mathscr{E}$). □

Remarks. (i) If $\mathscr{K}$ is an $(\mathscr{E}, \mathscr{M})$-category with $\mathscr{M}$ = all monos, then $\mathscr{E}$ is the class of all *extremal epis*, i.e., morphisms f such that if $f = p \cdot q$ and q is a mono, then q is an isomorphism.

(ii) If $\mathscr{K}$ is either complete and well-powered or, dually, cocomplete and cowell-powered, then it is an (extremal epi, mono)-category. The proof can be found in Herrlich and Strecker [1979].

If, moreover, regular epis are closed under composition, then extremal = regular. (This is usually fulfilled, for example in **Set**, **Pos**, **Top**, R-**Mod**.) Then we say that $\mathscr{K}$ has *regular factorizations*, which means that it is a (regular epi, mono)-category. The category of semigroups does not have regular factorizations, though it is complete and well-powered and hence, is an (extremal epi, mono)-category.

(iii) Dually, the class $\mathscr{M}$ is determined by the class $\mathscr{E}$. For example, if $\mathscr{K}$ is an (epi, $\mathscr{M}$)-category, then $\mathscr{M}$ is the class of all *extremal monos*, i.e., morphisms f such that if $f = p \cdot q$ and p is an epi, then p is an isomorphism.

Each complete and well-powered (or cocomplete and cowell-powered) category is an (epi, extremal mono)-category.

Exercises III.5

A. Monos and epis. Even without factorization of morphisms, the class of all monos has a lot of properties proved above for the class $\mathscr{M}$ of an $(\mathscr{E}, \mathscr{M})$-category. We spell out the dual properties.

(i) Prove that epis are left cancellative, i.e., if a morphism $e_1 \cdot e_2$ is an epi, then e_2 is an epi.

(ii) Prove that a coproduct $\coprod e_i : \coprod X_i \to \coprod Y_i$ of epis is always an epi.

(iii) Prove that, more generally, for each epitransformation $e: D \to D'$ of two diagrams, the colimit morphism colim e: colim $D \to$ colim D$'$ is an epi.

B. Directed diagrams of epis. (i) Generalizing III.5.5, prove that for each directed diagram D of $\mathscr{E}$-epis, a morphism f: colim $D \to Y$ is in $\mathscr{E}$ iff each component of f is in $\mathscr{E}$.

(ii) Prove the same statement about the class $\mathscr{E}$ of all epis not assuming any factorization of morphisms.

Notes to Chapter III

III.2

Algebras of a "functorial" type were investigated in the category of sets by O. Wyler [1966], V. Trnková and P. Goralčík [1969], V. Koubek and V. Kůrková-Pohlová [1974]. (In those papers, two set functors F and G are given, and algebras are pairs (Q, δ), where Q is a set and $\delta: QF \to QG$ is a map; the corresponding categories are called generalized algebraic categories.) The investigation of set functors (see Notes for III.4 below) was stimulated by those papers. F-algebras over a general category were first introduced by M. Barr [1970].

M. A. Arbib and E. G. Manes presented their model of automata based on a functor $F: \mathcal{K} \to \mathcal{K}$ in a series of papers ([1974a, b; 1975a, b]). All results in III.2 can be found in some form in those papers.

III.3

The fact that basic varieties are categories of F-algebras, and the corresponding examples of set functors, have been a folklore on our seminar. The general problem of characterizing categories of F-algebras over an arbitrary category has been attacked by J. Reiterman [1974].

III.4

Properties of set functors are investigated in a series of papers of V. Koubek, J. Reiterman and V. Trnková. This section presents a selection of those papers. In particular, III.4.2—6 are from V. Trnková [1969, 1971], except for Theorem III.4.5 proved by J. Adámek, V. Koubek and V. Pohlová [1972]. For III.4.7, see V. Koubek [1971], and for Exercise III.4.F, see V. Trnková [1971].

III.5

Axiomatizations of subobjects and quotients appear since the very start of the category theory. The present form is due to J. R. Isbell [1957]; further historical references appear in M. Barr [1971]. A theory of factorization systems is developed by H. Herrlich and G. E. Strecker [1979].

Chapter IV: Construction of Free Algebras

IV.1. Introduction

Free universal algebras are constructed "iteratively": the sets W_n of all trees of depth $\leq n$ are defined by induction, and then the free algebra is W_k for a sufficiently large ordinal k (see II.3.6). In the present chapter we study an iterative construction of free algebras in a category: objects W_n are defined by induction, and if the construction stops, then W_k is the free algebra for a sufficiently large ordinal k. The categorical construction is simple and natural, and it can be applied to a number of situations beyond universal algebra.

We work first with initial algebras, i.e., free algebras on 0 generators. In universal algebra, given a type Σ, the free Σ-algebra on n generators is precisely the initial algebra of the type $\Sigma^{(n)}$ obtained from Σ by adjoining n nullary operation symbols. Also generally, free and initial algebras are closely related. If (Q, δ) is the initial algebra of type $F\colon \mathcal{K} \to \mathcal{K}$, then $\delta\colon QF \to Q$ is an isomorphism and thus, Q is a *fixed point* of F (i.e., an object isomorphic to its F-image). Moreover, Q is the least fixed point in the sense defined below. The initial-algebra construction we study below is a natural generalization of the well-known construction of the least fixed point of an order-preserving map due to Knaster and Tarski. We recall it below.

All constructions studied in the present chapter are transfinite. We define a chain of objects and morphisms in a category $\mathcal{K}$, indexed by all ordinals:

$$W_0 \xrightarrow{w_{0,1}} W_1 \xrightarrow{w_{1,2}} \ldots \to W_\omega \xrightarrow{w_{\omega,\omega+1}} W_{\omega+1} \to \ldots \to W_n \xrightarrow{w_{n,n+1}} W_{n+1} \to \ldots$$

i.e., a functor

$$W\colon \mathbf{Ord} \to \mathcal{K}$$

(where **Ord** is the ordered class of all ordinals). A construction W is said to *stop* after k steps if $w_{k,k+1}$ is an isomorphism; then all $w_{k,n}$ turn out to be isomorphisms for $k \leq n$. The object W_k is then an underlying object of the free algebra.

Free algebras need not exist, in general. A type-functor F for which free algebras exist is called a *varietor*. If F is a varietor preserving monos, we prove that the free-algebra construction must stop eventually (under mild additional hypothesis on the base category $\mathcal{K}$). It follows, e.g., that a non-constant set functor is a varietor iff it has arbitarily large fixed points. If the free-algebra construction stops after ω steps, we call F a *finitary varietor.* Whereas in the category of sets finitary varietors are essentially just the finitary functors, in "suitably" ordered categories a surprising number of functors are finitary varietors. We present a lot of examples illustrating the way in which the free-algebra construction works in the category of sets and in other concrete categories. We also apply our construction to obtain free completions of partial algebras and, more generally, of span algebras. We conclude the chapter by a criterion on a functor F to be a (possibly not "constructive") varietor, in terms of generation of F-algebras.

We are using the *transfinite induction*: in order to define W_n for each ordinal n, it is sufficient to define

(a) W_0;

(b) W_{n+1} for any ordinal n for which W_n has been defined;

(c) W_i for any limit ordinal i for which W_n, $n < i$ have been defined.

Also, to prove a statement concerning W_n for all ordinals n, it is sufficient to prove this statement (a) for W_0, (b) for W_{n+1} if it holds for W_n and (c) for W_i, i a limit ordinal, if it holds for each W_n with $n < i$.

IV.2. Initial-Algebra Construction

2.1. Knaster-Tarski construction. We recall the construction of the least fixed point of an order-preserving map $f: (X, \le) \to (X, \le)$, where $(X, \le)$ is a complete lattice. The first step is the least element $\bot$ of X. The next steps are $(\bot)f$, $(\bot)f^2, \ldots$. The ω-step is

$$(\bot)f^\omega = \bigvee_{n<\omega} (\bot)f^n$$

and $(\bot)f^{\omega+1} = [(\bot)f^\omega]f$, etc. In general,

$$(\bot)f^{n+1} = [(\bot)f^n]f \qquad \text{for each ordinal } n;$$

$$(\bot)f^i = \bigvee_{n<i} (\bot)f^n \qquad \text{for each limit ordinal } i.$$

This construction obviously stops, i.e., there exists an ordinal k with

$$(\bot)f^k = (\bot)f^{k+1}.$$

Then $(\bot)f^k$ is the least fixed point of f.

Complete lattices are special cases of categories; order-preserving maps are then precisely functors. What follows is a natural generalization of Knaster-Tarski construction.

2.2. The initial-algebra construction. Let $\mathscr{K}$ be a chain-cocomplete category (Exercise III.1.F) and let $\perp$ denote its initial object (the colimit of the empty chain).

For each functor $F: \mathscr{K} \to \mathscr{K}$ we define a transfinite chain of objects

$$\perp F^n \qquad (n \in \mathbf{Ord})$$

and morphisms

$$w_{n,m}: \perp F^n \to \perp F^m \qquad (n, m \in \mathbf{Ord},\ n \leq m)$$

by the following transfinite induction:

$$\perp \xrightarrow{w_{0,1}} \perp F \xrightarrow{w_{1,2}} \perp F^2 \to \dots \perp F^\omega = \operatorname*{colim}_{n<\omega} \perp F^n \xrightarrow{w_{\omega,\omega+1}} \perp F^{\omega+1} \dots .$$

(a) First step:

$$\perp F^0 = \perp;$$
$$\perp F^1 = \perp F;$$

$w_{0,1}: \perp \to \perp F$ is the unique morphism.

(b) Isolated step:

$$\perp F^{n+1} = (\perp F^n)F;$$
$$w_{n+1,m+1} = w_{n,m}F: (\perp F^n)F \to (\perp F^m)F$$

for arbitratry ordinals n and m with $n \leq m$.

(c) Limit step:

$$\perp F^i = \operatorname{colim}\,(\perp F^n;\ w_{n,m})_{n \leq m \leq i}$$

for each limit ordinal i, with the colimit injections

$w_{n,i}: \perp F^n \to \perp F^i$ for each $n < i$.

Remark. The fact that the morphisms $w_{n,m}$ are supposed to form a chain fills in all the "missing" morphisms. For example, we need not define explicitly

$$w_{1,3}: \perp F \to \perp F^3$$

because

$$w_{1,3} = w_{1,2} \cdot w_{2,3} = w_{0,1}F \cdot w_{0,1}F^2.$$

Also, $w_{n,n} = 1_{\perp F^n}$. Analogously, we need not define

$$w_{\omega,\omega+1}: W_\omega = \operatorname*{colim}_{n<\omega} \perp F^n \to W_\omega F$$

because it is uniquely determined by the fact that for each $n < \omega$,

$$w_{n+1,\omega} \cdot w_{\omega,\omega+1} = w_{n+1,\omega+1} = w_{n,\omega}F.$$

Therefore, $w_{\omega,\omega+1}$ is the (unique) morphism from $\perp F^{\omega} = \operatorname{colim}_{1 \le n < \omega} \perp F^{n+1}$ with components $w_{n,\omega}F$. Further, $w_{n,\omega+1} = w_{n,\omega} \cdot w_{\omega,\omega+1}$, and $w_{\omega,\omega+2} = w_{\omega,\omega+1} \cdot w_{\omega+1,\omega+2}$, etc.

2.3. Definition. We say that the initial-algebra construction *stops after k steps* if $w_{k,k+1}$ is an isomorphism.

2.4. Proposition. If the initial-algebra construction stops after k steps, then the initial F-algebra is

$$(\perp F^{k}, w_{k,k+1}^{-1}).$$

Observation. Even if the initial-algebra construction does not stop, it has the following universal property: for each algebra (Q, δ) there exists a unique compatible collection

$$f_n : \perp F^{n} \to Q \quad (n \in \mathbf{Ord})$$

with $f_n F \cdot \delta = f_{n+1}$ for each $n \in \mathbf{Ord}$.

Proof. For each F-algebra (Q, δ) we define morphisms $f_n : \perp F^{n} \to Q$ by the following transfinite induction:

(a) $f_0 : \perp \to Q$ is the unique morphism;
(b) $f_{n+1} = f_n F \cdot \delta$, for each ordinal n;
(c) $f_i : \operatorname{colim}_{n<i} \perp F^{n} \to Q$ has components $f_n (n < i)$ for each limit ordinal i.

We must prove that this collection is compatible, i.e.,

$$f_n = w_{n,m} \cdot f_m \qquad (n < m),$$

and this will prove both that (c) is well-defined, and that the observation above is true. We proceed by transfinite induction on m: assume that m_0 is the least ordinal such that the equation above fails (for some n). We prove that (a) $m_0 > 1$ and (b) m_0 is not isolated and (c) m_0 is not a limit ordinal. Thus, m_0 does not exist.

(a) For $m_0 = 1$ we have $f_0 = w_{0,1} \cdot f_1 : \perp \to Q$, since $\perp$ is the initial object.
(b) We prove two auxiliary statements.

(b_1) If $f_n = w_{n,m} \cdot f_m$, then $f_{n+1} = w_{n+1,m+1} \cdot f_{m+1}$. We have

$$f_{n+1} = f_n F \cdot \delta = w_{n,m} F \cdot (f_m F \cdot \delta) = w_{n+1,m+1} \cdot f_{m+1}.$$

(b_2) If n_0 is a limit ordinal with $f_n = w_{n,m_0} \cdot f_{m_0}$ for all $n < n_0$, then $f_{n_0} = w_{n_0,m_0} \cdot f_{m_0}$. This follows easily from $\perp F^{n_0} = \operatorname{colim}_{n<n_0} \perp F^{n}$.

Now assume that m_0 is isolated, and let n_0 be the least ordinal for which

$f_{n_0} \neq w_{n_0, m_0} \cdot f_{m_0}$. Then n_0 cannot be isolated by (b_1), and it cannot be a limit ordinal by (b_2).

(c) If m_0 is a limit ordinal with $f_n = w_{n,m} \cdot f_m$ for all $n \leq m < m_0$, then $f_n = w_{n, m_0} \cdot f_{m_0}$ by the definition of f_{m_0}.

Let the initial-algebra construction stop after k steps. Since

$$f_k = w_{k,k+1} \cdot f_{k+1} = w_{k,k+1} \cdot f_k F \cdot \delta,$$

we have

$$w_{k,k+1}^{-1} \cdot f_k = f_k F \cdot \delta,$$

and thus

$$f_k : (\perp F^k, w_{k,k+1}^{-1}) \to (Q, \delta)$$

is a homomorphism. To prove the uniqueness, we verify that each homomorphism

$$g : (\perp F^k, w_{k,k+1}^{-1}) \to (Q, \delta)$$

fulfis $w_{n,k} \cdot g = f_n$, by induction on $n \leq k$.

(a) $w_{0,k} \cdot g = f_0 : \perp \to Q$ because $\perp$ is initial.

(b) $w_{n,k} \cdot g = f_n$ implies

$$\begin{aligned} w_{n+1,k} \cdot g &= w_{n+1,k+1} \cdot w_{k,k+1}^{-1} \cdot g \\ &= w_{n+1,k+1} \cdot gF \cdot \delta \\ &= (w_{n,k} \cdot g)F \cdot \delta \\ &= f_n F \cdot \delta \\ &= f_{n+1}. \end{aligned}$$

(c) For each limit ordinal i with $w_{n,k} \cdot g = f_n$ $(n < i)$ we have $\perp F^i = \operatorname{colim}_{n<i} \perp F^n$ and for each $i < n$,

$$\begin{aligned} w_{n,i} \cdot (w_{i,k} \cdot g) &= w_{n,k} \cdot g \\ &= f_n \\ &= w_{n,i} \cdot f_i. \end{aligned}$$

Thus, $w_{i,k} \cdot g = f_i$. This concludes the proof. □

Remark. In the above proof we used a type of transfinite induction which will be often encountered below: we want to prove a statement $S_{n,m}$ for all pairs of ordinals with $n < m$, and we know that

$$S_{n,m} \text{ and } S_{m,n} \text{ imply } S_{n,k}$$

for arbitrary $n < m < k$. Then it is sufficient to prove the following:

(a) $S_{0,1}$;

(b_1) $S_{n,m}$ implies $S_{n+1,m+1}$ (for each $n < m$);

(b$_2$) If n_0 is a limit ordinal and $m > n_0$ is an ordinal with $S_{n,m}$ for all $n < n_0$, then $S_{n_0,m}$;

(c) If i is a limit ordinal with $S_{n,m}$ for all $n < m < i$, then $S_{n,i}$ for all $n < i$. The verification of (b$_2$) and (c) will usually be elementary.

Examples. (i) The number k of steps necessary for the stopping of the initial-algebra construction can be arbitrary. Let

$$\mathcal{K} = \textbf{Ord}$$

be the category of ordinals. Define a functor

$$F: \mathcal{K} \to \mathcal{K}$$

by

$$nF = \begin{cases} n + 1 & \text{if } n < k \\ n & \text{if } n \geq k. \end{cases}$$

The the initial-algebra construction proceeds as follows:

$$1, 2, \ldots, k, k, k, \ldots$$

and it stops after k steps.

(ii) For the functor

$$F: \mathcal{K} \to \mathcal{K}$$

defined by

$$nF = n + 1 \quad (n \in \textbf{Ord})$$

the construction never stops. There exists no F-algebra, let alone an initial F-algebra.

(iii) Let $\mathcal{K} = \textbf{Ord} \cup \{\infty\}$ denote the category obtained from the ordered class of all ordinals by adding the largest element ∞. For the functor $G: \mathcal{K} \to \mathcal{K}$ defined by

$$nG = n + 1 \quad (n \in \textbf{Ord}) \quad \text{and} \quad \infty G = \infty,$$

the initial-algebra construction never stops: its k-th step is k. Nevertheless, G has a unique (hence, initial) algebra $(\infty, 1_\infty)$.

2.5. Proposition. (i) If the initial-algebra construction stops after k steps, then each $w_{k,n}$ $(n \geq k)$ is an isomorphism.

(ii) Let $w_{n,m}$ be an isomorphism for a pair of ordinals with $n < m$. Then the initial-algebra construction stops after m steps.

(iii) For each limit ordinal k the initial-algebra construction stops after k steps iff F preserves the colimit

$$\perp F^k = \operatorname*{colim}_{n<k} \perp F^n.$$

Proof. (i) If $w_{k,k+1}$ is an isomorphism, then $w_{k+1,k+2} = w_{k,k+1}F$ is also an isomorphism and so is $w_{k,k+2} = w_{k,k+1} \cdot w_{k+1,k+2}$. Analogously further. For the limit step $k + \omega$ we have

$$\perp F^{k+\omega} = \operatorname*{colim}_{n<\omega} \perp F^{k+n}$$

(see Exercise III.1.F) and a colimit of a chain of isomorphisms is obviously formed by isomorphisms. Etc.

(ii) If $w_{n,m}$ is an isomorphism ($n < m$), then we prove that $w_{m,m+1}$ is inverse to

$$u = w_{n,m}^{-1}F \cdot w_{n+1,m} \colon \perp F^{m+1} \to \perp F^m.$$

Firstly,

$$u \cdot w_{m,m+1} = w_{n,m}^{-1}F \cdot w_{n+1,m+1} = w_{n,m}^{-1}F \cdot w_{n,m}F = \text{id}.$$

Secondly,

$$\begin{aligned} w_{m,m+1} \cdot u &= w_{n,m}^{-1} \cdot w_{n,m} \cdot w_{m,m+1} \cdot u \\ &= w_{n,m}^{-1} \cdot w_{n,m+1} \cdot u \\ &= w_{n,m}^{-1} \cdot w_{n,n+1} \cdot w_{n+1,m+1} \cdot u \\ &= w_{n,m}^{-1} \cdot w_{n,n+1} \cdot w_{n,m}F \cdot w_{n,m}^{-1}F \cdot w_{n+1,m} \\ &= w_{n,m}^{-1} \cdot w_{n,n+1} \cdot w_{n+1,m} \\ &= w_{n,m}^{-1} \cdot w_{n,m} = \text{id}. \end{aligned}$$

(iii) If the construction stops after k steps, then F preserves the colimit $\perp F^k$ because colimits are unique up to isomorphism and so

$$\perp F^{k+1} = \operatorname*{colim}_{n<k} (\perp F^n)F = \operatorname*{colim}_{n<k} \perp F^{n+1} = \operatorname*{colim}_{1 \le n \le k} \perp F^n$$

with injections $w_{n+1,k+1} = w_{n+1,k} \cdot w_{k,k+1}$. Conversely, if F preserves the colimit $\perp F^k$, then $\perp F^{k+1} = \operatorname*{colim}_{n<k} \perp F^{n+1} = \operatorname*{colim}_{1 \le n < k} \perp F^n$ with injections $w_{n,k}F = w_{n+1,k+1} = w_{n+1,k} \cdot w_{k,k+1}$ and hence, $w_{k,k+1}$ is an isomorphism. □

Example: Initial universal algebras. Let

$$\mathcal{K} = \mathbf{Set} \quad \text{and} \quad F = H_\Sigma$$

(III.2.5) for a type Σ. The initial-algebra construction starts as follows:

$$\perp = \emptyset; \ \perp H_\Sigma = \Sigma_0; \ \perp H_\Sigma^2 = \Sigma_0 H_\Sigma; \text{ etc.}$$

If we represent the elements of XH_Σ by trees of the following kind

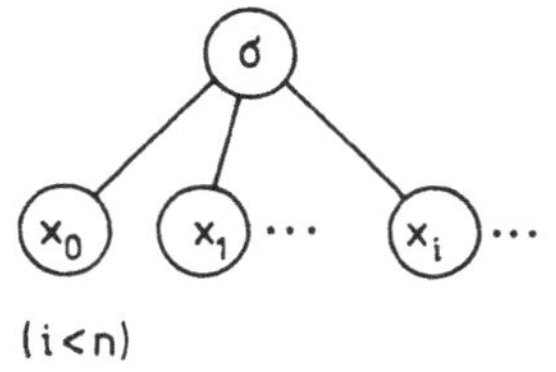

($\sigma \in \Sigma_n$ and $x_i \in X$), then the elements of $\perp H_\Sigma^k$ are the following trees

$\perp$

$\perp H_\Sigma$

$\varphi \in \Sigma_0$

$\perp H_\Sigma^2$

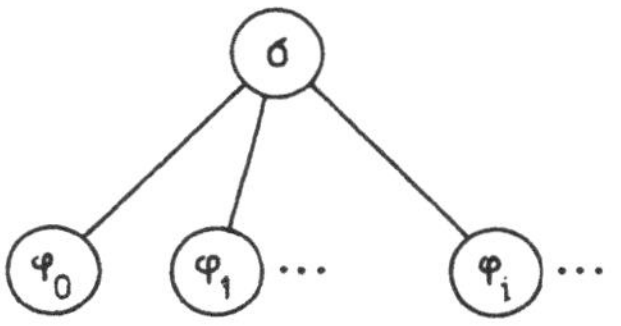

$(i<n)$

$\sigma \in \Sigma_n, \varphi_i \in \Sigma_0$

$\perp H_\Sigma^3$

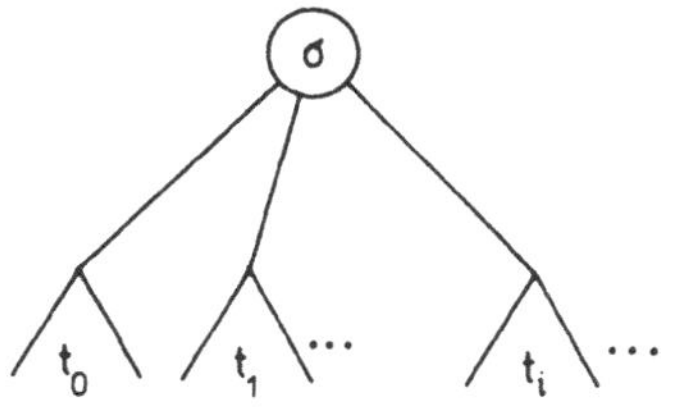

$(i<n)$

$\sigma \in \Sigma_n, t_i \in \perp H_\Sigma^2$

etc.

Since H_Σ preserves inclusion (III.4.5) and since $w_{0,1}$ is the inclusion map of $\emptyset$, it follows that $w_{1,2} = w_{0,1}H_\Sigma$ is the inclusion map too and hence, also $w_{2,3} = w_{1,2}H_\Sigma$, etc. Therefore, we have $\perp \subset \perp H_\Sigma \subset \perp H_\Sigma^2 \ldots$ and the colimit $\perp H_\Sigma^\omega$ is just the union

$$\perp H_\Sigma^\omega = \bigcup_{n<\omega} \perp H_\Sigma^n.$$

Again, $w_{\omega,\omega+1}$ is the inclusion map (because its restrictions to $\perp H_\Sigma^n$ are inclusion maps), etc. On each isolated step we have $\perp H_\Sigma^n \subset \perp H_\Sigma^{n+1}$, and on each limit step, $\perp H_\Sigma^i = \bigcup_{n<i} H_\Sigma^n$.

When does this construction stop? If Σ has no nullary operation, then after 0 steps, since $\perp = \perp H_\Sigma$. Assume $\Sigma_0 \neq \emptyset$. For Σ finitary, the functor H_Σ preserves directed unions (III.4.3 and Exercise III.4.E) and hence, it preserves the colimit $\perp H_\Sigma^\omega = \operatorname{colim}_{n<\omega} \perp H_\Sigma^n$. It follows by the preceding proposition that the construction stops after ω steps. And this corresponds to the results of II.1.5: we see that $\perp H_\Sigma^\omega$ is the set of all finite Σ-trees over the empty set.

For Σ infinitary, the number of steps is larger: it is obvious that the construction does *not* stop after k steps unless k is larger than all arities. On the other hand, if k is a regular cardinal larger than all arities, then H_Σ obviously preserves unions of k-chains, and the constructions stops after k steps.

2.6. Recall that a *fixed point* of a functor $F: \mathcal{K} \to \mathcal{K}$ is an object Q isomorphic to QF. More precisely, a fixed point is an object Q together with an isomorphism $\delta: FQ \to Q$.

Definition. The *least fixed point* of a functor $F: \mathcal{K} \to \mathcal{K}$ is a fixed point (Q_0, δ_0) such that for each fixed point (Q, δ) there exists a unique morphism $f: Q_0 \to Q$ with $fF = \delta_0 \cdot f \cdot \delta^{-1}$.

Proposition. The initial F-algebra, whenever it exists, is the least fixed point of F.

Proof. Let (Q_0, δ_0) be the initial algebra of F. It is clearly sufficient to prove that δ_0 is an isomorphism. Then (Q_0, δ_0) is the least fixed point because for each fixed point (Q, δ) there exists a unique homomorphism $f: (Q_0, \delta_0) \to (Q, \delta)$. The condition $fF = \delta_0 \cdot f \cdot \delta^{-1}$ is equivalent to the fact that f is a homomorphism, i.e., $fF \cdot \delta = \delta_0 \cdot f$.

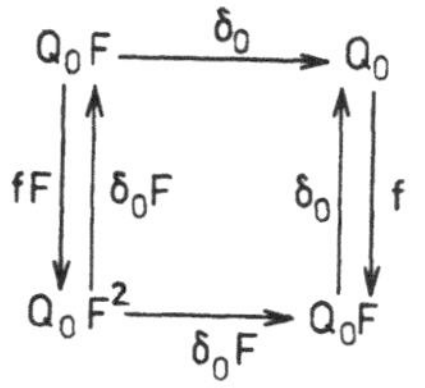

For the F-algebra (Q_0F, δ_0F) there exists a unique homomorphism $f: (Q_0, \delta_0) \to (Q_0F, \delta_0F)$. Then

$$f \cdot \delta_0: (Q_0, \delta_0) \to (Q_0, \delta_0)$$

is a homomorphism, too, because $\delta_0 \cdot f = fF \cdot \delta_0 F$ implies

$$\delta_0 \cdot (f \cdot \delta_0) = fF \cdot \delta_0 F \cdot \delta_0 = (f \cdot \delta_0) F \cdot \delta_0.$$

By the definition of initial object there exists only one homomorphism of (Q_0, δ_0) into itself. Since 1_{Q_0} is a homomorphism, we conclude that

$$f \cdot \delta_0 = 1_{Q_0}.$$

Now, $\delta_0 \cdot f = fF \cdot \delta_0 F$ implies

$$\delta_0 \cdot f = 1_{Q_0} F = 1_{Q_0 F}.$$

Hence,

$$f = \delta_0^{-1}$$

which proves that (Q_0, δ_0) is a fixed point. □

Examples. (i) The power-set functor

$$P: \mathbf{Set} \to \mathbf{Set}$$

(III.3.4) does not have an initial algebra. Indeed, P has no fixed point since for each set X,

$$\operatorname{card} X < \operatorname{card} PX.$$

(ii) For the functor

$$H_2 + C_1: \mathbf{Set} \to \mathbf{Set}$$

each infinite set X is a fixed point since $\operatorname{card} X = \operatorname{card}(X \times X + \{0\})$. The least fixed point is the free groupoid on one generator.

Remark. Interesting problems arise concerning the interrelationship of the notions above:

(i) If a functor has a fixed point, does it have the least one?
(ii) If a functor has the least fixed point, does it have the initial algebra?
(iii) If a functor has the initial algebra, can it be obtained by the initial-algebra construction?

The answers are negative, in general (see the exercises below). But under mild restrictive conditions on the category $\mathscr{K}$, we shall show in Section IV.4 that the answers are affirmative for all monos-preserving functors.

2.7. If the initial-algebra construction stops for F, what does this mean for related functors? We prove that for quotient functors it also stops; see Exercise. IV.2.D below for the analogous result about subfunctors. First, we prove a technical lemma which will be needed on several occasions.

Lemma. Let $\mathscr{E}$ be a class of epis in a category $\mathscr{K}$, which is closed under composition with isomorphisms, and let $\mathscr{K}$ be $\mathscr{E}$-cowell-powered. Given transfinite chains $U, V: \mathbf{Ord} \to \mathscr{K}$ such that U stops (i.e., starting from some ordinal, U consists entirely of isomorphisms) and given a compatible collection of $\mathscr{E}$-epis $e_n: U_n \to V_n$ ($n \in \mathbf{Ord}$), then V also stops.

Proof.

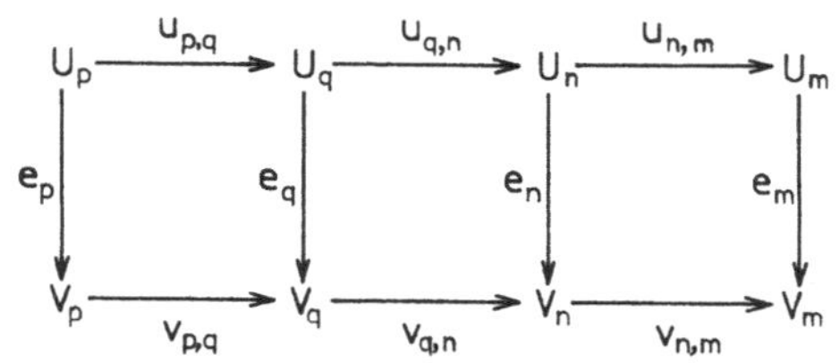

Let p be an ordinal such that each $u_{n,m}$ with $p \leq n < m$ is an isomorphism. Then each

$$u_{p,n} \cdot e_n \colon U_p \to V_n \quad (p \leq n)$$

is an $\mathscr{E}$-epi. Since U_p has only a set of $\mathscr{E}$-quotients, there exists an ordinal q such that all $u_{p,n} \cdot e_n$ with $q \leq n$ represent the same quotient of U_p. Then each $v_{n,m}$ with $q \leq n \leq m$ is an isomorphism: since $u_{p,n} \cdot e_n$ and $u_{p,m} \cdot e_m$ represent the same quotient, there exists an isomorphism

$$i \colon V_n \to V_m$$

with

$$u_{p,n} \cdot e_n \cdot i = u_{p,m} \cdot e_m .$$

We have $u_{p,m} \cdot e_m = u_{p,n} \cdot u_{n,m} \cdot e_m = u_{p,n} \cdot e_n \cdot v_n$, and therefore,

$$(u_{p,n} \cdot e_n) \cdot i = (u_{p,n} \cdot e_n) \cdot v_{n,m} .$$

Since $u_{p,n} \cdot e_n$ is an epi, it follows that $i = v_{n,m}$, thus $v_{n,m}$ is an isomorphism. □

Proposition. Let $\mathscr{K}$ be a cowell-powered category. If the initial-algebra construction stops for $F \colon \mathscr{K} \to \mathscr{K}$, then it stops for each quotient functor of F preserving epis.

Remark. More in general, if $\mathscr{K}$ is an $\mathscr{E}$-cowell-powered $(\mathscr{E}, \mathscr{M})$-category, the same result holds for $\mathscr{E}$-quotients of F preserving $\mathscr{E}$-epis.

Proof. Let $\varepsilon \colon F \to G$ be an epitransformation. We present a collection of epis

$$e_n \colon \perp F^n \to \perp G^n \quad (n \in \mathbf{Ord})$$

which is compatible, i.e., fulfils

$$(*) \qquad w_{n,m} \cdot e_n = e_n \cdot \bar{w}_{n,m} \quad (n < m)$$

(where $w_{n,m} \colon \perp F^n \to \perp F^m$ and $\bar{w}_{n,m} \colon \perp G^n \to \perp G^m$ denote the morphisms of the respective initial-algebra constructions). This will prove the proposition by the preceding lemma.

(a) $e_0 = 1_\perp : \perp \to \perp$;

(b) $e_{n+1} = \varepsilon_{\perp F^n} \cdot e_n G = e_n F \cdot \varepsilon_{\perp G^n} \quad : \quad \perp F^{n+1} \to \perp G^{n+1}$;

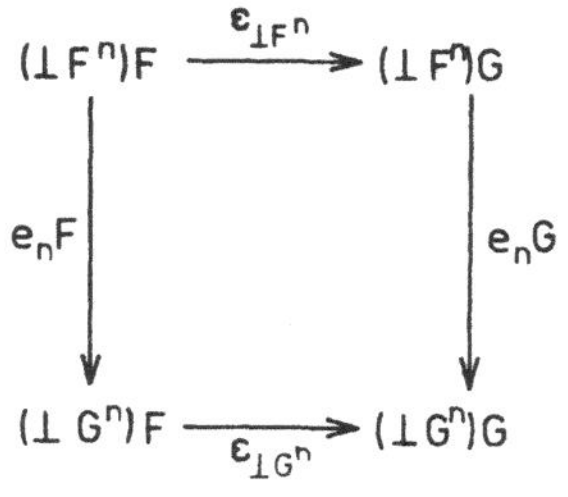

(c) For each limit ordinal i the components of

$$e_i : \operatorname*{colim}_{n<i} \perp F^n \to \operatorname*{colim}_{n<i} \perp G^n$$

are $e_n \cdot \bar{w}_{n,i}$ $(n < i)$. We must prove the compatibility (in order to verify that e_i is well-defined). We use induction in the sense of Remark IV.2.4.

(a) $w_{0,1} \cdot e_1 = e_0 \cdot \bar{w}_{0,1}$ because $\perp$ is the initial object.

(b_1) If (*) holds, then the following diagram

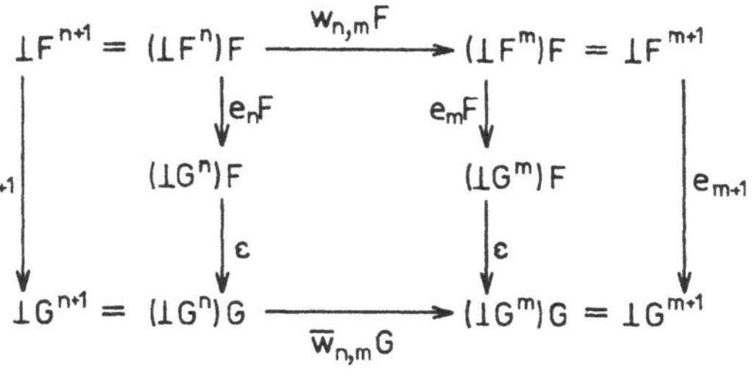

commutes.

(b_2) If (*) holds for all $n < n_0$, where n_0 is a limit ordinal, then it holds for n_0 because $\perp F^{n_0} = \operatorname*{colim}_{n<n_0} \perp F^n$.

(c) If (*) holds for all $n \leq m < i$, where i is a limit ordinal, then it holds for i by the definition of e_i.

It remains to prove that each e_n is an epi.

(a) $n = 0$: this is clear.

(b) If e_n is an epi, then $e_n G$ is an epi and hence, $e_{n+1} = \varepsilon_{\perp F^n} \cdot e_n G$ is an epi.

(c) For the limit step see Exercise III.5.B(ii). □

Exercises IV.2

A. Least fixed point which is not an initial algebra. Let $\mathscr{K}$ be the category of partial groupoids $(X, \cdot)$ (where $\cdot$ is a partial map from $X \times X$ to X) and homomorphisms $f: (X, \cdot) \to (Y, \circ)$ [which are maps such that $f(x_1 \cdot x_2) = f(x_1) \circ f(x_2)$ whenever $x_1 \cdot x_2$ is defined]. Denote by T the singleton (total) groupoid.

Define a functor $F: \mathcal{K} \to \mathcal{K}$ on objects by $(X, \cdot)F = T$ whenever $\cdot$ is non-empty and $(X, \emptyset)F = (PX, \emptyset)$ (where P is the power-set functor). On morphisms $f: (X, \cdot) \to (Y, \circ)$ put $fF = fP$ if $\cdot$ and $\circ$ are empty, else fF is constant.

Prove that T is the least fixed point of F. Using the fact that P has no initial algebra, prove that F also has none.

B. Fixed points exist but none of them is the least one. Find such an example, using posets as categories.

C. Unexpected start of the construction. Let us redefine the power set functor (III.3.4) to obtain the following functor P^* : **Set** $\to$ **Set**.

For each set $X \neq \emptyset$, $XP^* = \{M \subset X;\ M \text{ infinite or } M = \emptyset\}$, $\emptyset P^* = \omega = \{0, 1, 2, \ldots\}$. For each map $f: X \to Y$ either $MfP^* = MfP$ [if $(X)f$ is infinite] or $MfP^* = \emptyset$ [if $(X)f$ is finite and $Y \neq \emptyset$] or $fP^* = 1_\omega$(if $f = 1_\emptyset$).

Prove that the initial-algebra construction stops after ω steps, with $\mathrm{card}(\perp (P^*)^\omega) = 1$. Yet,

$$\mathrm{card} \perp P^* = \aleph_0;\ \mathrm{card} \perp (P^*)^2 = 2^{\aleph_0};\ \mathrm{card} \perp (P^*)^3 = 2^{2^{\aleph_0}}, \ldots.$$

D. Quotient functors and subfunctors. (i) Verify that the functor F: **Ord** $\to$ **Ord** defined by

$$nF = n + 1$$

is a quotient of the identity functor $1_{\mathbf{Ord}}$ and yet, the initial-algebra construction does not stop for F. Why does it not contradict to Proposition IV.2.7?

(ii) Let $\mathcal{K}$ be well-powered. If the initial-algebra construction stops for $F: \mathcal{K} \to \mathcal{K}$, prove that it also stops for each subfunctor of F preserving monos. Hint: Find a compatible collection of monos analogously to the proof of Proposition IV.2.7.

IV.3. Free-Algebra Construction

3.1. Let $\mathcal{K}$ be a category with finite coproducts. We are going to show that the free F-algebra over an object I of $\mathcal{K}$ is precisely the initial algebra of the functor

$$F + C_I: \mathcal{K} \to \mathcal{K}.$$

Here C_I denotes the constant functor of value I and thus,

$$X(F + C_I) = XF + I;$$
$$f(F + C_I) = fF + 1_I.$$

Each $(F + C_I)$-algebra (Q, δ) is given by a morphism

$$\delta: QF + I \to Q$$

or, equivalently, by a pair of morphisms

$$\delta_0: QF \to Q \quad \text{and} \quad \delta_1: I \to Q.$$

Conversely, given an F-algebra (Q, δ_0) and a morphism $\delta_1: I \to Q$, then we obtain a unique $(F + C_I)$-algebra (Q, δ) where δ_0 and δ_1 are components of δ.

Proposition. An F-algebra (Q, δ_0) is freely generated by the object I with the injection morphism $\delta_1: I \to Q$ iff the corresponding $(F + C_I)$-algebra (Q, δ) is initial.

Proof. Denote by $j_X: XF \to XF + I$ and $i_X: I \to XF + I$ the coproduct injections.

I. Let (Q, δ) be the initial $(F + C_I)$-algebra. Then for each F-algebra $(\bar{Q}, \bar{\delta}_0)$ and each morphism $\bar{\delta}_1: I \to \bar{Q}$ we have the corresponding $(F + C_I)$-algebra $(\bar{Q}, \bar{\delta})$. The unique $(F + C_I)$-homomorphism

$$\bar{\delta}_1^{\#}: (Q, \delta) \to (\bar{Q}, \bar{\delta})$$

is an F-homomorphism $\bar{\delta}_1^{\#}: (Q, \delta_0) \to (\bar{Q}, \bar{\delta}_0)$ extending δ_1, because the following diagram

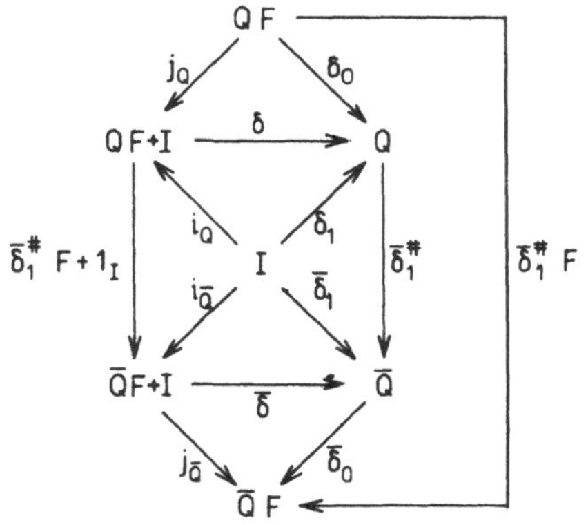

commutes. Conversely, any F-homomorphism extending $\bar{\delta}_1$ defines an $(F + C_I)$-homomorphism from (Q, δ) and thus, it equals to $\bar{\delta}_1^{\#}$.

II. Let (Q, δ_0) be the free algebra with the injection $\delta_1: I \to Q$. For each $(F + C_I)$-algebra $(\bar{Q}, \bar{\delta})$ we have an F-algebra $\bar{\delta}_0: \bar{Q}F \to \bar{Q}$ and a morphism $\bar{\delta}_1: I \to \bar{Q}$. Let $\bar{\delta}_1^{\#}: (Q, \delta_0) \to (\bar{Q}, \bar{\delta}_0)$ be the unique F-homomorphism extending δ_1. Then, again, the diagram above commutes and thus

$$\bar{\delta}_1^{\#}: (Q, \delta) \to (\bar{Q}, \bar{\delta})$$

is an $(F + C_I)$-homomorphism. Conversely, any homomorphism from (Q, δ) to $(\bar{Q}, \bar{\delta})$ is an F-homomorphism extending δ_1; hence, it equals to $\bar{\delta}_1^{\#}$. □

Remark. The free algebra $I^{\#}$ can be constructed by an application of the initial-algebra construction to $F + C_I$. We present a simpler construction (which coincides with the previous one on all infinite steps, as we shall prove): instead of starting with $\perp(F + C_I) = \perp F + I$, we start with I alone.

Since we need both finite coproducts (for the "translation" from F to $F + C_I$ above) and colimits of chains (for the initial-algebra construction), we shall assume that the base category $\mathscr{K}$ is cocomplete. It would be sufficient to assume that $\mathscr{K}$ is chain-cocomplete and has coproducts. (The existence of finite coproducts implies the existence of all coproducts in any chain-cocomplete category.)

3.2. The free-algebra construction. Let $\mathscr{K}$ be a cocomplete category and let $F: \mathscr{K} \to \mathscr{K}$ be a functor. For each object I we define objects $W_n (n \in \mathbf{Ord})$ and morphisms $w_{n,m} (n \leq m)$ which form a functor $W: \mathbf{Ord} \to \mathscr{K}$:

$$I \xrightarrow{w_{0,1}} I + IF \xrightarrow{1_I + w_{0,1}F} I + (I + IF)F \to \dots \to W_\omega \xrightarrow{w_{\omega,\omega+1}} 1_I + W_\omega F \to \dots \to W_n \xrightarrow{w_{n,n+1}} W_{n+1} = I + W_n F \to \dots .$$

We proceed by transfinite induction.

(a) First step:

$$W_0 = I;$$
$$W_1 = I + IF;$$
$$w_{0,1}: I \to I + IF$$

is the first coproduct injection.

(b) Isolated step:

$$W_{n+1} = I + W_n F,$$
$$w_{n+1,m+1} = 1_I + w_{n,m}F: 1_I + W_n F \to 1_I + W_m F$$

for all $n, m \in \mathbf{Ord}$ with $n \leq m$ (see Exercise III.1.C for the concept of a coproduct like $1_I + w_{n,m}F$).

(c) Limit step:

$$W_i = \operatorname*{colim}_{n < i} W_n$$

for each limit ordinal i for which the chain $W_n (n < i)$ and $w_{n,m} (n \leq m < i)$ has been defined:

$$w_{n,i}: W_n \to W_i \qquad (i < n)$$

are the colimit injections.

Remarks. (i) As in the case of the initial-algebra construction, all the "missing" morphisms can be easily filled in. For example,

$$w_{1,3} = w_{1,2} \cdot w_{2,3} = (1_I + w_{0,1}F) \cdot (1_I + w_{1,2}F),$$

and $w_{\omega, \omega+1}$ is the unique morphism with

$$w_{n+1,\omega} \cdot w_{\omega,\omega+1} = 1_I + w_{n,\omega}F \quad \text{for each } n < \omega,$$

etc.

(ii) Denote the coproduct injections of $W_{n+1} = I + W_n F$ by

$$\eta_n : I \to W_{n+1} \quad \text{and} \quad \varphi_n : W_n F \to W_{n+1}.$$

Each W_n is "almost" an F-algebra: φ_n leads from $W_n F$ to W_{n+1} (instead to W_n). If $w_{k,k+1} : W_k \to W_{k+1}$ is an ismorphism for some k, we obtain an F-algebra:

$$\varphi_k \cdot w_{k,k+1}^{-1} : W_k F \to W_k.$$

We are going to prove that this is the free algebra (with the injection η_k).

Definition. The free-algebra construction is said to *stop after k steps* if $w_{k,k+1}$ is an isomorphism.

The functor F is called a *constructive varietor* if the free-algebra construction stops for each I; F is a *finitary varietor* if it always stops after ω steps.

Examples. (i) S_Σ: **Set** → **Set** is a finitary varietor (see III.2.4):

$$W_0 = I,$$
$$W_1 = I + I \times \Sigma = I \times (\{\emptyset\} + \Sigma),$$
$$W_2 = I + (I + I \times \Sigma) \times \Sigma = I + I \times \Sigma + I \times \Sigma^2 = I \times \bigcup_{k=0}^{2} \Sigma^k,$$

in general,

$$W_n = I \times \bigcup_{k=0}^{n} \Sigma^k \qquad (n < \omega)$$

and

$$W_\omega = \operatorname*{colim}_{n<\omega} W_n = \bigcup_{k=0}^{\infty} W_n = I \times \Sigma^*.$$

(ii) More in general, each coadjoint is a finitary varietor (see III.2.11):

$$W_0 = I,$$
$$W_1 = I + IF,$$
$$W_2 = I + (I + IF)F = I + IF + IF^2,$$
$$\dots$$
$$W_\omega = \coprod_{n<\omega} IF^n.$$

In particular, the functor V_Σ: R-**Mod** → R-**Mod** is a finitary varietor.

(iii) The functor S_Σ: R-**Mod** → R-**Mod** is a finitary varietor (see III.2.4):

$$W_0 = I,$$
$$W_1 = I + IS_\Sigma = I + I + \Sigma,$$
$$W_2 = I + (I + I + \Sigma)S_\Sigma = I + I + I + \Sigma + \Sigma,$$

in general, W_{n+1} is the set of all pairs of polynomials $(i_0 + i_1 z + \ldots + i_n z^n, \sigma_0 + \sigma_1 z + \ldots + \sigma_{n-1} z^{n-1})$.

(iv) The functor H_Σ: **Set** → **Set** is a varietor, and it is finitary iff the type Σ is finitary. This can be derived from Exercise IV.2.5 and the following proposition.

3.3. Proposition. The free-algebra construction over an object I coincides on all infinite steps with the initial-algebra construction of the functor $F + C_I$.

Remarks. (i) The individual steps of the free-algebra construction are defined as colimits (coproducts for isolated steps and chain-colimits for limit steps). Since colimits are determined only up to isomorphism, the proposition just states that there exist isomorphisms

$$\alpha_n \colon \bot(F + C_I)^n \to W_n \qquad (n \in \textbf{Ord},\ n \geq \omega)$$

which are compatible with the constructions. [That is, denoting by $\bar{w}_{n,m} \colon \bot(F + C_I)^n \to \bot(F + C_I)^m$ the connecting morphisms, then for all $\omega \leq n \leq m$ we have $\bar{w}_{n,m} \cdot \alpha_m = \alpha_n \cdot w_{n,m}$.]

(ii) The two constructions can differ dramatically on finite steps. For example, let $I \neq \emptyset$ be a finite set and let F be the functor P^* of Exercise IV.2.C. Then

$$\operatorname{card} \bot(P^* + C_I)^n = 2^{2^{\cdot^{\cdot^{\cdot^{2^{\aleph_0}}}}}} \qquad (n < \omega)$$

and yet,

$$W_1 = I + IP^* = I + \{\emptyset\},$$
$$W_2 = I + (I + \{\emptyset\})P^* = I + \{\emptyset\}$$

etc., and hence, the free-algebra construction stops after 1 step.

Proof. We write $C_I + F$ rather than $F + C_I$, and we denote by

$$\bar{w}_{n,m} \colon \bar{W}_n \to \bar{W}_m$$

the objects and morphisms of the initial-algebra construction of $C_I + F$. Thus

$$W = \bot;$$
$$\bar{W}_{n+1} = I + \bar{W}_n F;$$
$$\bar{W}_i = \operatorname*{colim}_{n < i} \bar{W}_n \quad (i \text{ limit ordinal}).$$

We are going to define compatible morphisms

$$\alpha_n : \bar{W}_n \to W_n \qquad\qquad (n \in \mathbf{Ord})$$

and

$$\beta_n : W_n \to \bar{W}_{n+1} \qquad\qquad (n < \omega),$$

and then we prove that α_n are isomorphisms for all $n \geq \omega$.

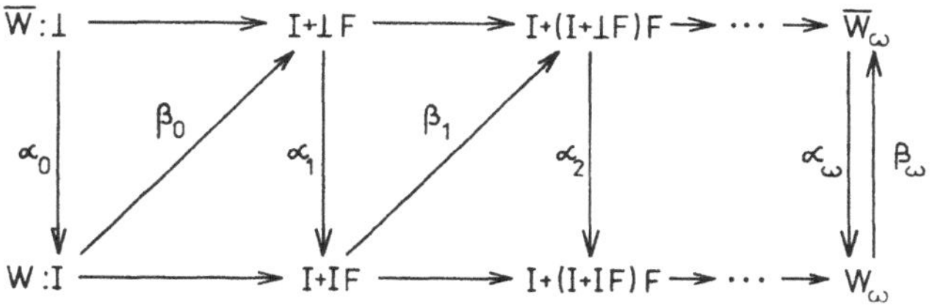

There is a unique morphism

$$\alpha_0 : \bot \to I;$$

given α_n, put

$$\alpha_{n+1} = 1_I + \alpha_n F : I + \bar{W}_n F \to I + W_n F;$$

for each limit ordinal i let

$$\alpha_i : \operatorname*{colim}_{n < i} \bar{W}_n \to \operatorname*{colim}_{n < i} W_n$$

have the components $\alpha_n \cdot w_{n,i}$ $(n < i)$. We prove that these morphisms are compatible, i.e., that the following squares

$$
\begin{array}{ccc}
\bar{W}_n & \xrightarrow{\ \bar{W}_{n,m}\ } & \bar{W}_m \\
\alpha_n \downarrow & & \downarrow \alpha_m \\
W_n & \xrightarrow[\ w_{n,m}\ ]{} & W_m
\end{array}
$$

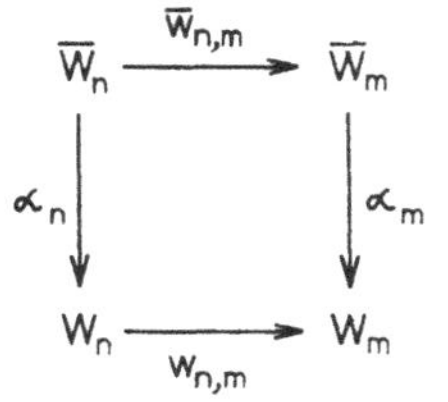

commute for all $n < m$, by induction. See Remark IV.2.4.

(a) $m = 1$: this is clear, since $\bar{W}_0 = \bot$.

(b$_1$) If the square above commutes, then the following one

$$
\begin{array}{ccc}
\bar{W}_{n+1} = I + \bar{W}_n F & \xrightarrow{\ 1_I + \bar{W}_{n,m} F\ } & I + \bar{W}_m F = \bar{W}_{m+1} \\
\alpha_{n+1} \downarrow \qquad \downarrow 1_I + \alpha_n F & \quad 1_I + \alpha_m F \downarrow & \qquad \downarrow \alpha_{m+1} \\
W_{n+1} = I + W_n F & \xrightarrow[\ 1_I + w_{n,m} F\ ]{} & I + W_m F = W_{m+1}
\end{array}
$$

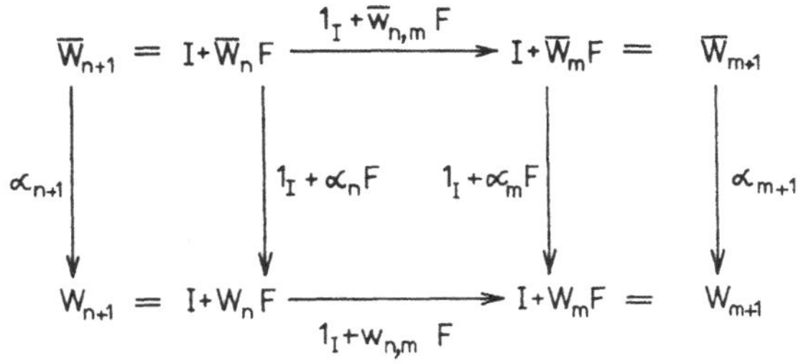

also commutes.

(b$_2$) If the squares above commute for all $n < n_0$ where n_0 is a limit ordinal, then they commute for n_0 because $\bar{W}_{n_0} = \operatorname{colim}_{n < n_0} \bar{W}_n$.

(c) If the squares above commute for all $n \leq m < i$, where i is a limit ordinal, then they commute for i by the definition of α_i.

Further, let us define $\beta_n : W_n \to \bar{W}_{n+1}$ $(n < \omega)$ by the following induction:

$$\beta_0 : I \to I + \perp F$$

is the first coproduct injection, and

$$\beta_{n+1} = 1_I + \beta_n F \qquad (n < \omega).$$

It is easy to prove by induction that

$$\alpha_n \cdot \beta_n = \bar{w}_{n,\, n+1} \qquad (n < \omega),$$
$$\beta_n \cdot \alpha_{n+1} = w_{n,\, n+1} \qquad (n < \omega),$$

and

$$w_{n,\, m} \cdot \beta_m = \beta_n \cdot \bar{w}_{n+1,\, m+1} \qquad (n \leq m < \omega).$$

Hence, the morphisms

$$\beta_n \cdot \bar{w}_{n+1,\, \omega} : W_n \to \bar{W}_\omega \qquad (n < \omega)$$

are compatible, giving rise to the unique morphism

$$\beta_\omega : W_\omega \to \bar{W}_\omega$$

with

$$w_{n,\, \omega} \cdot \beta_\omega = \beta_m \cdot \bar{w}_{n+1,\, \omega} \qquad (n < \omega).$$

We claim that α_ω is an isomorphism with the inverse morphism β_ω: firstly, $\beta_\omega \cdot \alpha_\omega = 1$ because for each $n < \omega$,

$$\begin{aligned} w_{n,\, \omega} \cdot (\beta_\omega \cdot \alpha_\omega) &= \beta_n \cdot \bar{w}_{n+1,\, \omega} \cdot \alpha_\omega \\ &= \beta_n \cdot \alpha_{n+1} \cdot w_{n+1,\, \omega} \\ &= w_{n,\, n+1} \cdot w_{n+1,\, \omega} \\ &= w_{n,\, \omega}. \end{aligned}$$

Secondly, $\alpha_\omega \cdot \beta_\omega = 1$ because for each $n < \omega$,

$$\begin{aligned} \bar{w}_{n,\, \omega} \cdot (\alpha_\omega \cdot \beta_\omega) &= \alpha_n \cdot w_{n,\, \omega} \cdot \beta_\omega \\ &= \alpha_n \cdot \beta_n \cdot \bar{w}_{n+1,\, \omega} \\ &= \bar{w}_{n,\, n+1} \cdot \bar{w}_{n+1,\, \omega} \\ &= \bar{w}_{n,\, \omega}. \end{aligned}$$

It follows by transfinite induction that each α_n, $n \geq \omega$, is an isomorphism. For example, $\alpha_{\omega+1} = 1_I + \alpha_\omega F$ is an isomorphism (inverse to $1_I + \beta_\omega F$) and $\alpha_{\omega+2} = 1_I + \alpha_{\omega+1} F$ is an isomorphism, etc. Also $\alpha_{2\omega} = \operatorname{colim}_{k < \omega} \alpha_{\omega+k}$ is an isomorphism because each $\alpha_{\omega+k}$ is an isomorphism, etc. This completes the proof. □

3.4. Corollary. If the free-algebra construction stops after k steps, then

$$I^{\#} = W_k$$

with $\varphi = \varphi_k \cdot w_{k,k+1}^{-1}: I^{\#}F \to I^{\#}$ and $\eta = \eta_k: I \to I^{\#}$.

For a limit ordinal k, the free-algebra construction stops after k steps iff F preserves the colimit $W_k = \operatorname{colim}_{n<k} W_n$.

For $k \geq \omega$ this follows from the fact that the initial-algebra construction of the functor $F + C_I$ stops after k steps, see Propositions IV.2.4 and IV.3.1 [and for the latter statement also Proposition IV.2.5(c)]. If the free-algebra construction stops after $k < \omega$ steps, then it also stops after ω steps and $W_\omega = W_k$.

Remark. Even if the free-algebra construction does not stop, it has the following universal property: For each F-algebra (Q, δ) and each morphism

$$f: I \to Q$$

there exists a unique compatible collection $f_n: W_n \to Q$ ($n \in \mathbf{Ord}$) such that the components of

$$f_{n+1}: I + W_n F \to Q$$

are $f: I \to Q$ and $f_n F \cdot \delta: W_n F \to Q$. In fact, the first step is $f_0 = f$, the isolated step is given and the limit step is determined by compatibility (the components of $f_i: \operatorname{colim}_{n<i} W_n \to Q$ must be f_n, $n < i$).

Let the free-algebra construction stop after k steps. Then

$$f^{\#} = f_k: (W_k, \varphi_k \cdot w_{k,k+1}^{-1}) \to (Q, \delta).$$

3.5. Corollary. Each functor preserving colimits of k-chains for some infinite cardinal k is a constructive varietor. Each functor preserving colimits of ω-chains is a finitary varietor.

Remark. Let $\mathscr{K}$ be a category with finite coproducts and colimits of k-chains for all $k \leq k_0$. Then we can define the first $k_0 + 2$ members of the free-algebra construction as above. Again, if $w_{k,k+1}$ is an isomorphism for some $k \leq k_0$, then $I^{\#} = W_k$.

Thus, for example, in the category $\mathscr{K}$ of countable sets and maps we can investigate finitary varietors.

Example: Free-many sorted algebras. Let Σ be a type of many-sorted algebras (III.3.7). Say, two-sorted, for simplicity:

$$\Sigma = \langle \Sigma^{(1)}, \Sigma^{(2)} \rangle.$$

Let us apply the free-algebra construction to the functor H_Σ: **Set**$^2 \to$ **Set**^{2} in order to obtain the free algebra generated by an object $I = \langle I_1, I_2 \rangle$. Recall that for each $\langle X_1, X_2 \rangle$ we have

$$\langle X_1, X_2 \rangle H_\Sigma = \left\langle \coprod_{\sigma \in \Sigma^{(1)}_{n_1, n_2}} X_1^{n_1} \times X_2^{n_2}, \coprod_{\sigma \in \Sigma^{(2)}_{m_1, m_2}} X_1^{m_1} \times X_2^{m_2} \right\rangle$$

Let us represent the elements in the σ-th summand of the first sort by the following trees:

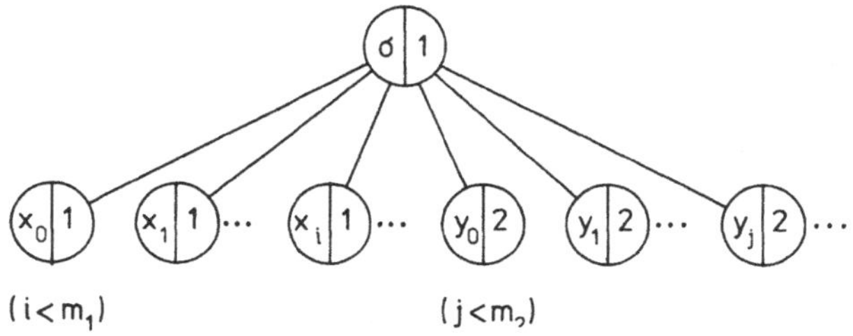

where $\sigma \in \Sigma^{(1)}$ has arity $|\sigma| = (m_1, m_2)$ and $x_i \in X_1$ for $i < m_1$, $y_j \in X_2$ for $j < m_2$. Analogously for the elements of the second sort. The elements of

$$W_0 = \langle I_1, I_2 \rangle$$

are represented by singleton trees with two labels:

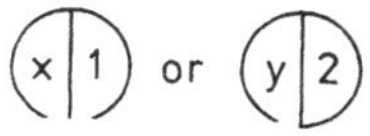

where $x \in I_1$ and $y \in I_2$.
Given $W_n = \langle W^{(1)}_n, W^{(2)}_n \rangle$, then

$$W_{n+1} = I + W_n H_\Sigma = \langle W^{(1)}_{n+1}, W^{(2)}_{n+1} \rangle.$$

The elements of

$$W^{(1)}_{n+1} = I_1 + \coprod_{\sigma \in \Sigma^{(1)}_{m_1, m_2}} (W^{(1)}_n)^{m_1} \times (W^{(2)}_n)^{m_2}$$

are represented (a) by the (singleton) trees of I_1 and (b) by the following trees:

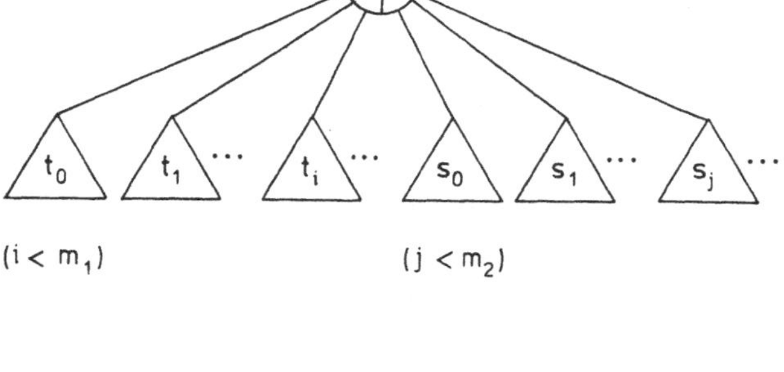

where $t_i \in W_n^{(1)}$ for $i < m_1$ and $s_j \in W_n^{(2)}$ for $j < m_2$. (Analogously with $W_n^{(2)}$.) We see that the elements of each W_n are labelled trees such that

(i) the label of each leaf is (x, k) with $k = 1$ or 2 and $x \in I_k$ or $x \in \Sigma_{0,0}^{(k)}$;

(ii) the other nodes have labels (σ, k) where $k = 1, 2$ and $\sigma \in \Sigma_{m_1, m_2}^{(k)}$ $(m_1 + m_2 > 0)$ and they have m_1 successors of sort 1 [i.e., of label $(—, 1)$ on the root] and m_2 successors of sort 2 [label $(—, 2)$]. The successors of the first sort are depicted left-hand to those of the second sort.

If Σ is a finitary type, then H_Σ preserves colimits of ω-chains and hence, it is a finitary varietor:

$$I^\# = W_\omega.$$

Here, W_ω is the algebra of all finite labelled trees satisfying (i) and (ii) above.

For an infinitary type Σ, let k be a regular infinite cardinal such that given $\sigma \in \Sigma$ with $|\sigma| = (n_1, n_2)$, then $n_1 < k$ and $n_2 < k$. Then H_Σ preserves colimits of k-chains and hence,

$$I^\# = W_k.$$

Analogously to the one-sorted case (II.3.6) it can be proved that W_k is the algebra of all finite-path labelled trees satisfying (i) and (ii) above.

3.6. Example: Free commutative groupoids. We have seen in III.3.1 that commutative groupoids are just P_2-algebras. Let us apply the free-algebra construction to a set I assuming, for simplicity, that I does not contain any element of the type $\{x, y\}$ (i.e., $I \cap XP_2 = \emptyset$ for each set X). Then

$$W_0 = I;$$
$$W_1 = I \cup IP_2 = \{x;\ x \in I\} \cup \{\{x, y\};\ x, y \in I\};$$
$$W_2 = I \cup (I \cup IP_2)P_2 = \{x;\ x \in I\} \cup \{\{x, y\};\ x, y \in W_1\},$$
etc.

The elements of

$$W_\omega = \bigcup_{n < \omega} W_n$$

are (i) the elements of I, (ii) the sets $\{x, y\}$ with $x, y \in I$, (iii) the sets $\{x, y\}$ where x and y are elements of type (i) or (ii), etc. Since P_2 preserves ω-colimits, we have $I^\# = W_\omega$. The operation is defined by $x \cdot y = \{x, y\}$.

Exercises IV.3

A. Non-constructive varietor. Denote by **Gra** the category of *graphs*, i.e., pairs (X, ρ) where X is a set and $\rho \subset X \times X$; the morphisms are *compatible maps* $f\colon (X, \rho) \to (Y, \sigma)$, i.e., maps such that $x_1 \rho x_2$ implies $(x_1)f \sigma (x_2)f$.

(i) For each cardinal n denote by

$$C_n = (n, \rho_n)$$

the *complete graph* of power n: here n is the set of all ordinals $i < n$ and $i\rho_n j$ iff $i \neq j\,(i, j < n)$. The *chromatic number* of a graph (X, ρ) is the least cardinal n for which there exists a morphism from (X, ρ) to C_n; we denote it by $\chi(X, \rho)$. Verify that each graph (X, ρ) without *loops* (i.e., without points $x \in X$ with $x\rho x$) has a chromatic number and that $\chi(X, \rho) \leq \chi(Y, \sigma)$ whenever a compatible map $f\colon (X, \rho) \to (Y, \sigma)$ exists.

(ii) We define a functor F: **Gra** → **Gra** which turns out to be a non-constructive varietor. Let

$$T = (\{t\}, \{(t, t)\})$$

denote the graph which consists of a singleton loop. Put

$$(X, \rho)F = \begin{cases} T & \text{if } X = \emptyset \text{ or } (X, \rho) \text{ has loops} \\ C_{2^n} & \text{if } \chi(X, \rho) = n > 0. \end{cases}$$

For each morphism $f\colon (X, \rho) \to (Y, \sigma)$ let fF be the constant map to t if $(Y, \sigma)F = T$; else, we have $(X, \rho)F = C_{2^n}$ and $(Y, \sigma)F = C_{2^m}$ with $2^n \leq 2^m$ and fF is the inclusion map.

Verify that F is well-defined and that it is a varietor with $I^{\#} = I + T$ for each graph I. [Hint: For each F-algebra (Q, δ) the graph Q has a loop. Thus, $\delta\colon T \to Q$ is only a choice of a loop.]

(iii) Let I be a graph without loops. Prove that the free-algebra construction never stops by verifying that $\chi(W_n) < \chi(W_{n+1})$ for each n.

B. Free ordered Σ-algebras, see III.3.5, are just the free Σ-algebras (of labelled trees) with an apropriate ordering. Describe the ordering, using the free-algebra construction.

C. Free algebras in concrete categories. Let $(\mathcal{K}, U)$ be a concrete category (III.3.8) with products and colimits preserved by U. Generalize the preceding example to prove that free Σ-algebras in $\mathcal{K}$ are just "appropriately structured" algebras of Σ-trees.

Compare this with unary linear algebras (III.2.4).

D. Functorial varieties. Each of the equations below describes a variety of algebras which coincides with some F-**Alg**, see III.3.3. Find F, and describe the free algebras (applying the free-algebra construction).

(i) $\Sigma = \Sigma_2 = \{\cdot\}$; $x \cdot x = y \cdot y$.
(ii) $\Sigma = \{\cdot, 0\}$ with $|\cdot| = 2$, $|0| = 0$; $x \cdot x = 0$.
(iii) $\Sigma = \Sigma_2 = \{\cdot, +\}$; $x \cdot x = x + x$.

IV.4. Characterization Theorem

4.1. Definition. A class $\mathcal{M}$ of monos in a category $\mathcal{K}$ is said to be *chain-cocomplete* if $\mathcal{K}$ is $\mathcal{M}$-well-powered, $\mathcal{M}$ contains all isomorphisms, and each α-chain of $\mathcal{M}$-monos in $\mathcal{K}$

$$p_{n,m}: P_n \to P_m \quad (n \leq m < \alpha)$$

has a colimit $p_n: P_n \to P(n < \alpha)$ with the folloving properties:

(a) $p_n \in \mathcal{M}$ for each $n < \alpha$;

(b) given a compatible collection $q_n: P_n \to Q(n < \alpha)$ in $\mathcal{M}$, then the unique factorization morphism $q: P \to Q$ is in $\mathcal{M}$, too.

Remark. It follows that, furthermore:

(c) $\mathcal{M}$ is closed under composition. [Given $p: P_0 \to P_1$ and $p': P_1 \to P_2$ in $\mathcal{M}$, we obtain a 2-chain the colimit of which is P_2 with injections $p \cdot p'$, p', 1_{P_2}. Thus, by (a) above, $p \cdot p' \in \mathcal{M}$.]

(d) $\mathcal{K}$ has an initial object $\perp$ and the (unique) morphisms $q: \perp \to Q$ are all in $\mathcal{M}$. [Indeed, apply (b) above to $\alpha = 0$.]

We are going to show that the questions asked in Remark IV.2.6 have affirmative answers provided that F preserves $\mathcal{M}$-monos, i.e., given $m: A \to B$ in $\mathcal{M}$, then $mF: AF \to BF$ is also in $\mathcal{M}$.

Theorem. Let $\mathcal{K}$ be a category with a chain-cocomplete class $\mathcal{M}$. For each functor $F: \mathcal{K} \to \mathcal{K}$ preserving $\mathcal{M}$-monos, the following are equivalent:

(i) F has a fixed point;

(ii) F has a least fixed point;

(iii) F has an initial algebra;

(iv) the initial-algebra construction stops.

Proof. By Propositions IV.2.6 and IV.2.4, the implications (iv)$\to$(iii)$\to$(ii)$\to$(i) are clear. We prove (i)$\to$(iv).

We show first that the morphisms $w_{n,m}$ in the initial-algebra construction are $\mathcal{M}$-monos; it follows that the colimits defining $\perp F^i$ for limit ordinals i really exist. We proceed by transfinite induction on m using Remark IV.2.4. The first step is clear: $w_{0,1} \in \mathcal{M}$ by (d) above. (b_1) Let n, m be ordinals with $w_{n,m} \in \mathcal{M}$. Then

$$w_{n+1,m+1} = w_{n,m}F \in \mathcal{M}$$

because the functor F preserves $\mathcal{M}$-monos.

(b_2), (c) These steps follow from (a) and (b) in the definition above.

Let J be a fixed point of F, and let

$$j: JF \to J$$

be an isomorphism. We define compatible $\mathcal{M}$-monos

$$p_n\colon \perp F^n \to J \qquad (n \in \mathbf{Ord})$$

by the following transfinite induction:

$$p_0\colon \perp \to J$$

is the unique morphism [in $\mathcal{M}$, by (d) above].

Given p_n, put

$$p_{n+1} = p_n F \cdot j\colon (\perp F^n)F \to J.$$

Then $p_{n+1} \in \mathcal{M}$ by (c) above.

For each limit ordinal i let

$$p_i\colon \operatorname*{colim}_{n<i} \perp F^n \to J$$

have components

$$w_{n,i} \cdot p_i = p_n \qquad (n < i).$$

Then $p_i \in \mathcal{M}$ by (b) above. Let us prove that the morphisms p_n are compatible, i.e.,

$$p_n = w_{n,m} \cdot p_m$$

for all $n < m$, by induction. See Remark IV.2.4.

(a) $m = 1$: Clearly, $p_0 = w_{0,1} \cdot p_1$.

(b) If $p_n = w_{n,m} \cdot p_m$, then

$$\begin{aligned} p_{n+1} &= p_n F \cdot j \\ &= w_{n,m}F \cdot p_m F \cdot j \\ &= w_{n+1,m+1} \cdot p_{m+1}. \end{aligned}$$

The limit steps (b_2) and (c) are clear.

Since $\mathcal{K}$ is $\mathcal{M}$-well-powered, there exists an ordinal k such that p_k represents the same subobject of J as any p_n, $n \geq k$. In particular, as p_{k+1}. Thus, there is an isomorphism $u\colon \perp F^k \to \perp F^{k+1}$ with $p_k = u \cdot p_{k+1}$. Since also $p_k = w_{k,k+1} \cdot p_{k+1}$ and p_{k+1} is a mono, we conclude that $u = w_{k,k+1}$. Hence, the initial-algebra construction stops after k steps. □

4.2. Definition. A class $\mathcal{M}$ of monos is called *constructive* if it is chain-cocomplete and "additive", i.e., given $p\colon P \to Q$ and $p'\colon P' \to Q'$ in $\mathcal{M}$, then the morphism $p + p'\colon P + P' \to Q + Q'$ is also in $\mathcal{M}$.

Characterization Theorem. Let $\mathcal{K}$ be a cocomplete category with a constructive class $\mathcal{M}$. For each functor $F\colon \mathcal{K} \to \mathcal{K}$ preserving $\mathcal{M}$-monos and each object I, equivalent are:

(i) F has a free-algebra over I;

(ii) the free-algebra construction over I stops;
(iii) there exists an object J isomorphic to $I + JF$.

Proof. Apply the preceding theorem to $F + C_I$ (see IV.3.1). Since $\mathcal{M}$ is "additive" and both F and C_I preserve $\mathcal{M}$-monos, $F + C_I$ also preserves $\mathcal{M}$-monos. □

Examples. (i) $\mathcal{K}$ = **Set** has a (unique) constructive class, viz, all monos. The colimit of a chain of inclusions $p_{n,m}: P_n \to P_m$ $(n \le m < \alpha)$ is the union $P = \bigcup_{n<\alpha} P_n$. The condition (b) above states that each map $q: P \to Q$ which is one-to-one on any P_n, is one-to-one on all of P.

(ii) In a lot of current categories, all monos form a constructive class. For example, in $\mathcal{K}$ = posets, topological spaces, groups, lattices, etc.

Regular monos also often form a constructive class—see Exercise IV.4.A below.

(iii) The category $\mathcal{K}$ of rings and ring homomorphisms fails to have any constructive class of monos. Indeed, $\perp$ is the ring of integers and there are lots of rings Q such that the unique homomorphism $\perp \to Q$ is not one-to-one (= mono).

Corollary. A functor $F: \mathcal{K} \to \mathcal{K}$ preserving monos of a constructive class is a varietor iff for each object I there exists an object $J \cong I + JF$.

Remarks. (i) In the proof of Theorem IV.4.1 we have seen that

$$w_{n,m} \in \mathcal{M} \qquad \text{for each} \qquad n \le m.$$

The same holds for the free-algebra construction (under the hypotheses of the Characterization Theorem). This relates well to our intuition that W_n is the "n-th approximation" of the free algebra $I^\#$. Consequently, $\varphi \in \mathcal{M}$.

(ii) The functor G: **Ord** → **Ord** of Exercise IV.2.4(iii) is a non-constructive varietor though it (trivially) preserves monos. This shows that the hypothesis that $\mathcal{K}$ be $\mathcal{M}$-well-powered is essential.

(iii) The functor of Exercise IV.3.A is a non-constructive varietor (though all monos are constructive in **Gra**). This shows that the hypothesis that F preserve $\mathcal{M}$-monos is essential.

4.3. The Characterization Theorem makes it possible to give a simple full characterization of varietors in **Set** (see below) and in categories of vector spaces (see the next section).

Each constant functor

$$C_M: \mathbf{Set} \to \mathbf{Set}$$

is a finitary varietor. Here

$$W_0 = I;$$

$$W_1 = I + M;$$
$$W_2 = I + M;$$
etc.

The free-algebra construction stops after 1 step with $I^\# = I + M$. Also the constant functor C_M^h (III.4.1) is a finitary varietor for any map $h: M_0 \to M$. For $I \neq \emptyset$, the construction stops after 1 step, and for $I = \emptyset$ we have

$$W_0 = \emptyset:$$
$$W_1 = \emptyset C_M^h = M_0,$$
$$W_2 = M_0 C_M^h = M,$$
$$W_3 = M C_M^h = M,$$

and we see that the construction stops after 2 steps.

Theorem. A non-constant functor

$$F: \mathbf{Set} \to \mathbf{Set}$$

is a (constructive) varietor iff it has arbitrarily large fixed points, i.e., for each cardinal α there is a set X with card X = card $XF \geq \alpha$.

Proof. Since F is non-constant, there is a cardinal γ such that card $XF \geq$ card X whenever card $X \geq \gamma$ (see III.4.7).

(i) Let F be a varietor. For each cardinal α choose a set I of power max $(\alpha, \gamma, \aleph_0)$. The free algebra $I^\#$ is a fixed point of $F + C_I$ (IV.3.1), thus,

$$\text{card } I^\# = \text{card } I^\# F + \text{card } I = \text{card } I^\# F$$

(since card $I^\# F$ is infinite and larger or equal to card $IF \geq$ card I). Hence, $I^\#$ is a fixed point of F of cardinality $\geq \alpha$.

(ii) Let F have arbitrarily large fixed points. Since F need not preserve monos (because of the empty maps), let us redefine F on the empty sets and empty maps—the resulting functor F' is defined by $XF' = XF$ if $X \neq \emptyset$; $\emptyset F' = \emptyset$ and for each non-empty map f, $fF' = fF$.

Then F' preserves monos and has arbitrarily large fixed points. Hence, for each set I there exists an infinite fixed point J of power $\geq$ max(γ, card I). Then

$$J \simeq I + JF'$$

because card $JF \geq$ card $J \geq$ card I implies card $I + JF'$ = card JF' = card J. By Corollary IV.4.2, F' is an constructive varietor.

It follows immediately that for each non-empty set I the free-algebra construction for F stops, too. Let us verify that the free-algebra construction over $\emptyset$ also stops. This is clear if $\emptyset F = \emptyset$; let us assume $\emptyset F \neq \emptyset$ and hence $XF \neq \emptyset$ for any set X(see III.4.2). We have $W_n = \emptyset F^n$. Let us choose a fixed point J of

F with

$$\text{card } J \geq \text{card } \emptyset F$$

and let

$$j: JF \to J$$

be a bijection. As in the proof of Theorem IV.4.1, it is sufficient to exhibit a compatible family of monos $p_n: \emptyset F^n \to J$ (for all ordinals n).

(a) First two steps:

$$p_0: \emptyset \to J \text{ is the (unique) empty map;}$$
$$p_1: \emptyset F \to J \text{ is an arbitrary mono}$$

(we use the fact that card $J \geq$ card $\emptyset F$).

(b) Isolated step:

$$p_{n+1} = p_n F \cdot j: (\emptyset F^n) F \to J$$

for each ordinal $n > 0$. The verification of compatibility is the same induction as in the proof of Theorem IV.4.1. Since $p_n: \emptyset F^n \to J$ is a non-empty map, the assumption that p_n is mono implies that $p_n F$ is mono and hence, so is p_{n+1}.

(c) Limit step is trivial. □

4.4. In universal algebra the number of steps after which the free-algebra construction stops is independent of the generating set I. In general, the number of steps can increase with increasing I without any bound. Let us exhibit such an example.

Example. Given a class C of cardinals, define a functor P_C: **Set** → **Set** as follows: for each set X put

$$XP_C = \{M \subset X;\ \text{card } M \in C \text{ or } M = \emptyset\};$$

for each map $f: X \to Y$ put

$$(M)fP_C = \begin{cases} (M)f & \text{if } f \text{ is one-to-one on } M \\ \emptyset & \text{else.} \end{cases}$$

Whether P_C is a varietor or not depends on C; for example, if C = all cardinals, then P_C obviously has no fixed point, hence no free algebra. On the other hand, P_C is a varietor for each class C with the following property:

(*) There exist arbitrarily large cardinals k such that $k^n = k$ for any $n \in C$, $n \leq k$.

In fact, any infinite set X of power k as above is a fixed point of P_C because

$$\text{card } XP_C = \text{card} \sum_{\substack{n \in C \\ n \leq k}} X^n = \sum_{\substack{n \in C \\ n \leq k}} k^n \leq k \cdot k = k = \text{card} X.$$

It is easy to construct a class C of cardinals which fulfils (*) and has arbitratrily large elements. (For example, assuming the Generalized Continuum Hypothesis, we can choose $C = C_e$, the class of all $\aleph_n$ where n is an even ordinal.) Then the number of steps necessary for the free-algebra construction over P_C increases over all bounds: if card$I = m \in C$, then the construction requires more that m steps. In fact, for each $n \leq m$ we can find a set $A \subset W_n$ of cardinality m such that $A \not\subset W_{n'}$ for any $n' < n$. Then A is an element of $W_n P_C$ with $(A)\varphi_n \notin (W_n)w_{n,n+1}$ and hence, $w_{n,n+1}$ is not surjective.

Example: A coproduct of two varietors which is not a varietor. Let C be a class of cardinals such that both C and $\bar{C}$ (the class of all cardinals not belonging to C) satisfy (*) and contain arbitrarily large cardinals. Then P_C and $P_{\bar{C}}$ are varietors but $P_C + P_{\bar{C}}$ is not a varietor, in fact, $P_C + P_{\bar{C}}$ has no fixed point. Such a class C can be easily constructed (for example, C_e has this property under the Generalized Continuum Hypothesis).

This example is rather surprising in view of the role which coproducts of type functors play (see III.2.5).

A subfunctor of a varietor need not be a varietor (Exercise IV.4.C below) and, dually, a quotient of a varietor need not be a varietor, see IV.2.D.

Exercises IV.4

A. Constructive classes of monos. (i) Prove that in every concrete, cocomplete and well-powered category $\mathscr{K}$ the class

$$\mathscr{M} = \text{all monos}$$

is constructive if (1) monos are just the one-to-one morphisms, (2) they are "additive" and (3) colimits of chains of monos are (suitably structured) unions.

Verify that all this holds in the categories of posets, graphs, topological spaces and modules.

(ii) Verify that all regular monos in **Pos** (= embeddings of subposets) form a constructive class.

In contrast, regular monos in **Top** (= embeddings of subspaces) are not constructive: consider the ω-chain of discrete spaces on $n = \{0, 1, \ldots, n-1\}(n < \omega)$; its colimit is the discrete space on ω. Let Q be the topological space on ω in which a set $M \subset \omega$ is closed iff M is finite or $0 \in M$. Then the embeddings $n \to Q$ are regular monos though the induced map $\mathrm{id}_\omega : \omega \to Q$ is not regular.

(iii) Verify that the category of σ-complete lattices and σ-complete homomorphisms has no constructive class of monos. Consider the ω-chain of lattices $A_n \cup \{T\}$ $(n < \omega)$ where T is the largest element and A_n is the set of all subsets of $\{1, 2 \ldots, n\}$ ordered by inclusion. Its colimit P is uncountable and there exist σ-complete homomorphisms $q: P \to Q$ which are not one-to-one though each restriction to $A_n \cup \{T\}$ is one-to-one.

B. Quotients of a varietor. (i) Let $\mathcal{K}$ be a cocomplete, cowell-powered category. Prove that for each constructive varietor $F: \mathcal{K} \to \mathcal{K}$, all epis-preserving quotients of F are also constructive varietors. (Hint: IV.2.7.)

(ii) Verify that the hypothesis that $\mathcal{K}$ be cowell-powered is essential. (Hint: Exercise IV.2.D.)

C. Subfunctors of a varietor. (i) Let $\mathcal{K}$ satisfy the hypothesis of the Characterization Theorem. Prove that for each constructive varietor $F: \mathcal{K} \to \mathcal{K}$ preserving $\mathcal{M}$-monos, all $\mathcal{M}$-subfunctors of F preserving $\mathcal{M}$-monos are also constructive varietors. (Hint: Let $\mu: G \to F$ be a natural transformation with all μ_X in $\mathcal{M}$. It is sufficient to find compatible $\mathcal{M}$-monos p_n from W_n, the free G-algebra-construction over I, to $\bar{W}_n$, the free F-algebra-construction. Put $p_0 = 1_I$. Given $p_n: W_n \to \bar{W}_n$ put $p_{n+1} = 1_I + p_n G \cdot \mu_{\bar{W}_n}$.)

(ii) Consider the varietor F: **Gra** $\to$ **Gra** of Exercise. IV.3.1. Prove that its subfunctor G, defined as F except that

$$(X, \rho)G = (\emptyset, \emptyset) \quad \text{if } (X, \rho) \text{ has loops or } X = \emptyset$$

is no varietor: it yields no non-empty algebra.

D. Cocompleteness in the Characterization Theorem. (i) Verify that the Characterization Theorem would not be true if instead of cocompleteness of $\mathcal{K}$ we would assume only chain-cocompleteness. (Hint: Use a finite poset as $\mathcal{K}$.)

(ii) Prove that, however, the Characterization Theorem remains true if cocompleteness is weakened to chain-cocompleteness and the existence of finite coproducts. Remark: from this, the existence of all coproducts follows anyway.

IV.5. Algebras in Concrete Categories

5.1. Let $(\mathcal{K}, U)$ be a concrete category (see III.3.8). We prove a powerful criterion for a functor to be a constructive varietor.

Definition. A functor $F: \mathcal{K} \to \mathcal{K}$ is said to be *non-increasing* if there exist arbitrarily large cardinals n such that

$$\text{card } AU \leq n \quad \text{implies card } (AF)U \leq n \qquad \text{for all objects } A \text{ in } \mathcal{K}.$$

Example. A set functor is non-increasing iff it is a (constructive) varietor. In fact, each varietor F is either constant (and then it is clearly non-increasing) or F has arbitrarily large fixed points (IV.4.3). Let $n \geq \text{card } \emptyset F$ be the cardinality of a fixed point of F. Then

$$\text{card } A \leq n \quad \text{implies card } AF \leq n$$

by Proposition III.4.7. (If B is a fixed point of cardinality n, then card $A \leq$ card B implies card $AF \leq$ card $BF = n$ for $A \neq \emptyset$, and we have card $\emptyset F \leq n$).

Conversely, let F be a non-increasing set functor. If F is constant, then F is a finitary varietor. If F is non-constant, we choose γ such that card $AF \geq$ card A for any set A of cardinality $\geq \gamma$ (III.4.7). Then F is a varietor because each cardinal $n \geq \gamma$ with the property above is a cardinality of some fixed point of F: if card $A = n$, then card $AF \leq n$ and, via $n \geq \gamma$, card $AF \geq$ card $A = n$, therefore, card $AF = n =$ card A. □

5.2. Definition. Let $\mathcal{K}$ be a concrete category which has a free object M^* for each set M of generators (III.3.8). We say that $\mathcal{K}$ has *bounded free objects* if there is a cardinal α such that card $M =$ card M^*U for any set M of cardinality $\geq \alpha$.

Examples. (i) Each variety of algebras, considered as a concrete category, has bounded free objects. It is sufficient to choose a regular infinite cardinal α larger than all arities and larger than the number of operations.

(ii) The categories **Top**, **Pos**, **Gra**, etc. have bounded free (= discrete) objects; here $M^*U = M$ for each set M and thus, we can choose $\alpha = 0$.

(iii) The category **Comp** of compact Hausdorff spaces and continuous maps has free objects, but not bounded. The free object generated by an infinite set M is the Čech-Stone compactification βM of the discrete topology on M; it is well-known that card $\beta M >$ card M.

5.3. Theorem. Let $\mathcal{K}$ be a cocomplete and cowell-powered concrete category with bounded free objects. Each non-increasing functor $F: \mathcal{K} \to \mathcal{K}$ which preserves epis is a constructive varietor.

Proof. Denote by Φ: **Set** $\to \mathcal{K}$ the free-object functor (Remark III.3.8).

I. The functor

$$\bar{F} = \bar{\Phi} \cdot F \cdot U: \textbf{Set} \to \textbf{Set}$$

is a constructive varietor.

To prove this, let α be a cardinal with card $M =$ card $(M)\Phi \cdot U$ whenever card $M \geq \alpha$. There exist arbitrarily large cardinals $n \geq \alpha$ such that card $AU \leq n$ implies card $(AF)U \leq n$. For each of these cardinals and each set X,

$$\text{card } X = n \qquad \text{implies} \qquad \text{card } X\bar{F} = \text{card } (X\Phi)F \cdot U \leq n.$$

Thus, $\bar{F}$ is a non-increasing functor, in other words, a constructive varietor.

II. The free-algebra construction stops for each free object $I = X\Phi$.

We shall use the following functor

$$G = F \cdot U \cdot \Phi : \mathscr{K} \to \mathscr{K}.$$

It is sufficient to prove that the free-algebra construction of G stops for each X. In fact, F is a quotient of G: the epitransformations $\varepsilon: U \cdot \Phi \to 1_{\mathscr{K}}$ (see Remark III.3.8) yields an epitransformation $F\varepsilon: F \cdot U \cdot \Phi \to F$. Therefore, $F + C_I$ is a quotient functor of $G + C_I$ (because a coproduct of epis is an epi, Exercise III.5.A). Moreover, since F preserves epis, $F + C_I$ also preserves epis. Thus, by Proposition IV.2.7, if the intial-algebra construction stops for $G + C_I$, then it stops for $F + C_I$. Consequently, by Proposition IV.3.3, if the free-algebra construction over $I = X\Phi$ stops for G, then it stops for F.

Denote by

$$\tilde{W}: \mathbf{Ord} \to \mathbf{Set}$$

the free-algebra construction over X for the functor F. Since Φ preserves colimits (as any coadjoint, III.2.10), the chain

$$\tilde{W} \cdot \Phi: \mathbf{Ord} \to \mathscr{K}$$

is the free-algebra construction over $X\Phi$ for the functor $G: \mathscr{K} \to \mathscr{K}$. In fact

(a) $(\tilde{W} \cdot \Phi)_0 = \tilde{W}_0\Phi = X\Phi$;

(b) Assuming $(W \cdot \Phi)_n$ is the n-th step in the free-algebra construction of $X\Phi$, we have

$$(\tilde{W} \cdot \Phi)_{n+1} = (X + \tilde{W}_n\tilde{F})\Phi = X\Phi + (\tilde{W} \cdot \Phi)_n G$$

because

$$\tilde{F} \cdot \Phi = \Phi \cdot F \cdot U \cdot \Phi = \Phi \cdot G;$$

(c) Assuming $(\tilde{W} \cdot \Phi)_n$ is the n-th step for each $n < i$ where i is a limit ordinal, then

$$(W \cdot \Phi)_i = \tilde{W}_i\Phi = \left(\operatorname*{colim}_{n<i} \tilde{W}_n\right)\Phi = \operatorname*{colim}_{n<i} (\tilde{W} \cdot \Phi)_n .$$

(Analogously with the morphisms $\tilde{w}_{n,m}$.) Since $\tilde{W}$ stops, there is an ordinal k such that $\tilde{w}_{k,k+1}$ is an isomorphism; hence $\tilde{w}_{k,k+1}\Phi$ is an isomorphism, too.

III. The free-algebra construction stops for each object I.

We shall use the fact that the construction stops for $(IU)\Phi$: let

$$W: \mathbf{Ord} \to \mathscr{K} \text{ and } \tilde{W}: \mathbf{Ord} \to \mathscr{K}$$

denote the free-algebra construction for I and $(IU)\Phi$, respectively. By Lemma IV.2.7, it is sufficient to present compatible epis

$$e_n: \tilde{W}_n \to W_n \qquad (n \in \mathbf{Ord}).$$

We proceed by transfinite induction.

(a) $e_0 = \varepsilon_I : (IU)\Phi \to I$;

(b) $e_{n+1} = \varepsilon_I + e_n F : (IU)\Phi + \bar{W}_n F \to I + W_n F$;

(c) $e_i : \operatorname{colim}_{n<i} \bar{W}_n \to \operatorname{colim}_{n<i} W_n$ has components $e_n \cdot w_{n,i}$ $(n < i)$ for each limit ordinal i.

We verify the compatibility, i.e., that the following squares

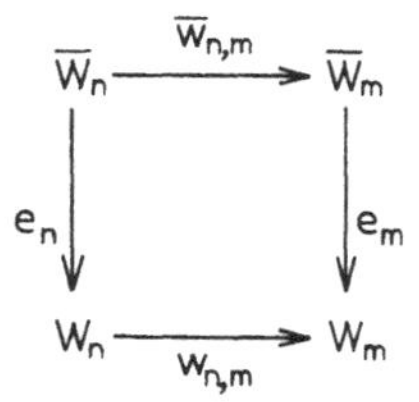

$(n<m)$

commute, by induction using Remark IV.2.4.

(a) For $m = 1$ we have $\bar{w}_{0,1} \cdot e_1 = \varepsilon_I \cdot w_{0,1} = e_0 \cdot w_{0,1}$ because $e_1 = \varepsilon_I + \varepsilon_I F$.

(b_1) If the square above commutes, then the following diagram

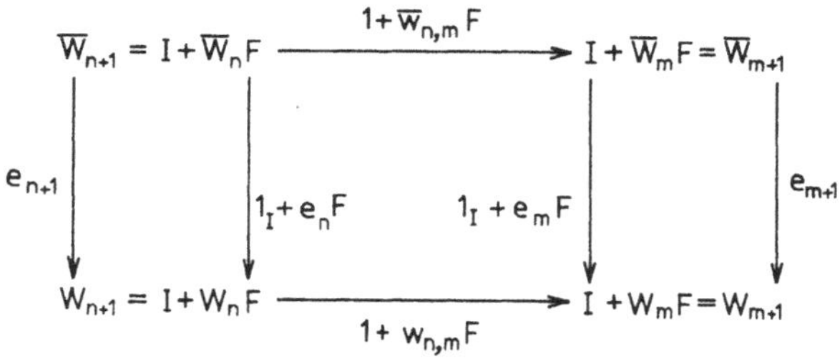

also commutes.

(b_2) If the squares above commute for all $n < n_0$, where n_0 is a limit ordinal, then they commute for n_0 too, because $\bar{W}_{n_0} = \operatorname{colim}_{n<n_0} \bar{W}_n$.

(c) The limit step in m follows from the definition of e_i.

It remains to prove that each e_n is an epi.

(a) $e_0 = \varepsilon_I$ is an epi.

(b) If e_n is an epi, then $e_n F$ is an epi and hence, by Exercise III.5.A, also $e_{n+1} = \varepsilon_I + e_n F$ is an epi.

(c) The limit step follows from Exercise III.5.B.

This concluses the proof that the free-algebra construction W stops. □

5.4. Example. For each type Σ, the functor

$$H_\Sigma : \mathcal{K} \to \mathcal{K} \text{ (III.2.5)}$$

is a constructive varietor, assuming that $\mathcal{K}$ is a concrete category as above and, moreover, $\mathcal{K}$ has concrete products and U preserves epis.

Proof. I. For each cardinal k the "k-th power functor" $H_k : \mathcal{K} \to \mathcal{K}$ is non-increasing. This follows from the fact that U preserves products. For each infinite regular cardinal $n > k$ we have:

$$\operatorname{card} AU = n \quad \text{implies} \quad \operatorname{card}(AH_k)U = \operatorname{card}(AU)^k = n^k = n.$$

Next, H_k preserves epis: given an epi $e: A \to B$, then $eU: AU \to BU$ is onto, hence

$$e^{(k)}U = (eU)^{(k)}: (AU)^k \to (BU)^k$$

is onto and this implies that $e^{(k)} = eH_k$ is epi.

II. It follows immediately that the coproduct

$$H_\Sigma = \coprod_{\sigma \in \Sigma} H_k, \text{ where } k = |\sigma|$$

preserves epis (a coproduct of epis is an epi, see Exercise III.5.A). Let us verify that H_Σ is non-increasing. Let α be the cardinal of Definition IV.5.2, and let β be an infinite cardinal larger than card Σ, α and the arity of any $\sigma \in \Sigma$. We shall prove that each infinite regular cardinal $n \geq \beta$ has the property that

$$\operatorname{card} AU = n \quad \text{implies} \quad \operatorname{card}(AH_\Sigma)U \leq n.$$

We have card $A^kU \leq n$ for any $\sigma \in \Sigma$ with $|\sigma| = k$. Since also card $\Sigma \leq n$, it follows that the coproduct $X = \coprod_{\sigma \in \Sigma_k} (A^kU)$ has power $\leq n$. Then $n \geq \alpha$ implies card $X\Phi U \leq n$. We have $X\Phi = \coprod_{\sigma \in \Sigma_k} (A^kU)\Phi$, and there is an obvious epimorphism $e: X\Phi \to AH_\Sigma$. Since eU is an epi in **Set**, we conclude that card $(AH_\Sigma)U \leq n$. □

5.5. Example: A finitary varietor which is not non-increasing. Let **Ab** be the category of Abelian groups and homomorphisms. We define a functor F: **Ab** → **Ab** as a composition of several "naturally defined" functors.

Let **Abf** denote the full subcategory of torsion-free groups. For each Abelian group A the subgroup of all torsion elements (i.e., elements $a \in A$ such that $n \cdot a = a + a + \ldots + a$ is zero for some $n > 0$) is denoted by **Tor** (A). Let

$$R: \mathbf{Ab} \to \mathbf{Abf}$$

be the reflector, assigning to each Abelian group A its quotient group

$$AR = A/\mathbf{Tor}\,(A)$$

(and analogously on morphisms). Denote by

$$U\colon \mathbf{Abf} \to \mathbf{Set}$$

the forgetful functor and by

$$P\colon \mathbf{Set} \to \mathbf{Set}$$

the power-set functor (III.3.4). Finally, let

$$\Phi_2\colon \mathbf{Set} \to \mathbf{Ab}$$

be the functor assigning to each set M the free Z_2-module generated by M (i.e., the elements of $M\Phi_2$ are all maps $t\colon M \to \{0, 1\}$ of finite support and the addition is defined by $(m)(t_1 + t_2) = 0$ iff $(m)t_1 = (m)t_2$ for each $t_1, t_2 \in M\Phi_2$ and $m \in M$) and to each map $f\colon M \to M'$ the unique homomorphism extending f [if each $m \in M$ is considered as $t\colon M \to \{0, 1\}$ where $(m)t = 1$ and $(m')t = 0$ for all $m' \neq m$].

The functor

$$F = R \cdot U \cdot P \cdot \Phi_2 : \mathbf{Ab} \to \mathbf{Ab}$$

is a finitary varietor. Indeed, starting the free-algebra construction, we have

$$W_0 = I,$$
$$W_1 = I + IF$$

and since IF is a torsion group [hence, $(IF)R = 0$], clearly $W_1F = IF$;

$$W_2 = I + IF;$$
$$W_3 = I + IF;$$

etc. The construction stops after one step.

Nevertheless, F fails to be non-increasing: let I be the free group on n generators, $n \geq \aleph_0$. Then

$$\operatorname{card} I = n.$$

Since I is torsion-free, we have $IR = I$, thus,

$$\operatorname{card} IF = 2^n.$$

Remark. In the preceding example we had $\mathscr{K} = \mathbf{Ab}$, the category of modules over the ring of integers. Nevertheless, in categories of modules over *fields*, non-increasing functors are precisely the varietors, as we prove now. Hence, the situation is analogous to $\mathscr{K} =$ Set for R-**Vect** (but not for R-**Mod**, in general).

Theorem. Let R be a commutative field. The following conditions are equivalent for each functor

$$F: R\text{-}\mathbf{Vect} \to R\text{-}\mathbf{Vect}:$$

(i) F is a varietor;
(ii) F is a constructive varietor;
(iii) F is non-increasing;
(iv) F has arbitrarily large fixed points or it is a constant functor.

Proof. In the category R-**Vect** all monos and all epis split. Thus, F trivially preserves monos and epis.

(iv) → (iii) This is clear.

(iii) → (ii) This follows from Theorem IV.5.3, all assumptions of which are fulfilled by R-**Vect**.

(ii) → (i) This is clear.

(i) → (iv) Let us apply Characterization Theorem IV.4.2. Here $\mathcal{M}$ = all monos. If F is non-constant, then there exists a cardinal γ such that $\dim XF \geq \dim X$ wherever $\dim X \geq \gamma$ [see Exercise III.4.A(iv)]. For each cardinal α choose a vector space I of dimension $\max(\alpha, \gamma, \aleph_0)$. The free algebra $I^\#$ is a fixed point of $F + C_I$ (IV.3.1), thus

$$\dim I^\# = \dim I^\# F + \dim I = \dim I^\# F.$$

Hence, $I^\#$ is a fixed point of F and $\dim I^\# \geq \alpha$. □

Exercises IV.5

A. Varietors and fixed points. The notions varietor, non-increasing functor and arbitrarily large fixed points are relatively independent. We illustrate this on the category **Gra** (IV.4.A); we denote by $T = (\{t\}, \{(t, t)\})$ the terminal object. In each case, verify that F is a well-defined functor and that it has the properties claimed.

(i) A non-increasing varietor without fixed points (except, of course, the initial algebra): on objects (X, R) put $(X, R)F = T$ if $R \neq \emptyset$ and $(X, \emptyset)F = (X, \emptyset) + T$; on morphisms $f: (X, R) \to (Y, S)$ let $(t)fF = t$ (in T) and in case $R = S = \emptyset$, $fF = f$ on X.

(ii) An increasing varietor without fixed points: See IV.3.A.

(iii) An increasing varietor with arbitrarily large fixed points: define F analogously to IV.3.A except that $(X, \emptyset)F = (X, \emptyset)$ and for morphisms $f: (X, R) \to (Y, S)$, $fF = f$ if $S = \emptyset$ and fF is constant if $R = \emptyset \neq S$.

B. Ordered types. Let Σ be an ordered type, i.e., a type of algebras with an order on each of the sets Σ_k. An ordered Σ-algebra is then a Σ-algebra (Q, δ), ordered in the usual sense (III.3.5) and such that, moreover, if $\sigma, \tau \in \Sigma_k$ and $\sigma \leq \tau$, then

$$(x_i)\sigma \leq (x_i)\tau \qquad \text{for all } (x_i) \in Q^k.$$

Define a functor F: **Pos** → **Pos** by appropriately enriching the order of H_Σ: if $Q = (X, \leq)$, then $QF = \coprod_{\sigma \in \Sigma_k} X^k$ where $(x_i)\sigma \leq (y_j)\tau$ iff $|\sigma| = |\tau|$ $(= k)$, $\sigma \leq \tau$ in Σ_k and $x_i \leq y_i$ for each i.

Verify that F is a constructive varietor and the free algebras are appropriately ordered algebras of finite-path Σ-trees (see II.3.6).

C. Concrete categories with concrete coproducts. (i) Verify that the categories **Pos, Top, Metr, Gra** fulfil the assumptions of Exercise IV.5.4 and, moreover, U preserves coproducts.

(ii) Generalize functors $H_\Sigma: \mathscr{K} \to \mathscr{K}$, where $\mathscr{K}$ has all the properties mentioned in (i), to obtain algebras of "structured types" as follows. For each cardinal k with $\Sigma_k \neq \emptyset$ choose an object A_k with $A_k U = \Sigma_k$. Define $F: \mathscr{K} \to \mathscr{K}$ by

$$XF = \coprod_{\substack{k \text{ cardinal} \\ \Sigma_k \neq \emptyset}} A_k \times X^k.$$

Note that $H_\Sigma \cdot U = F \cdot U$ (and $H_\Sigma = F$ if A_k are discrete objects). Prove that F is a varietor and that the free algebras are the Σ-tree algebras with an apropriate structure.

(iii) If $\mathscr{K}$ = **Pos** we described F-algebras in Exercise B above. Describe them if $\mathscr{K}$ = **Top**.

D. Varietors in Vect. (i) Prove that each of the functors F in Exercise C (ii) above is a varietor in R-**Vect**.

(ii) For each finitary type Σ define

$$\tilde{H}_\Sigma: R\text{-}\mathbf{Vect} \to R\text{-}\mathbf{Vect}$$

using the tensor product ⊗ instead of product: $\tilde{H}_n$ is defined by $X\tilde{H}_n = X \otimes X \otimes \ldots \otimes X$ (n times) and $\tilde{H}_\Sigma = \coprod_{\sigma \in \Sigma_n} \tilde{H}_n$. Prove that $\tilde{H}_\Sigma$ is a finitary varietor. [Hint: Prove that each $\tilde{H}_n$ preserves ω-colimits—then so does each $\tilde{H}_\Sigma$. To prove that, say, $\tilde{H}_2$ preserves the colimit of any chain $f_{n,m}: X_n \to X_m (n \leq m < \omega)$ choose a basis $B_n \subset X_n$ such that $f_{n,m}(B_n) \subset B_m \cup \{0\}$ for each $n \leq m < \omega$. Then $B_n \otimes B_n$ is a basis of $X_n \otimes X_n$ and, again, $f_{n,m} \otimes f_{n,m}(B_n \otimes B_n) \subset B_m \otimes B_m \cup \{0\}$.]

(iii) Find two varietors in R-**Vect** the coproduct (= product) of which is not a varietor. (Hint: For each functor F: **Set** → **Set** we have a functor $\tilde{F} = U \cdot F \cdot \Phi$: R-**Vect** → R-**Vect** which is a varietor iff F is. Use Example IV.4.4).

(iv) Biduals do not form a varietor. The hom-functor

$$\hom(-, R): R\text{-}\mathbf{Vect} \to R\text{-}\mathbf{Vect}^{\mathrm{op}}$$

defined on object X by hom (X, R) (with the usual addition and scalar multiplication of linear maps) is a well-known duality functor. Composing it with itself we obtain a functor F: R-**Vect** → R-**Vect**. Verify that F is not a varietor: indeed, it has no free algebra except the (trivial) initial algebra. (Hint: If dim X is infinite, then dim $XF >$ dim X. Use Theorem IV.5.5.)

IV.6. Finitary Varietors

6.1. In the present section we study finitary varietors, i.e., functors for which the free-algebra construction stops after ω steps. In the category of sets, these are essentially just the finitary functors and this "essentially" is a question of the axioms of set theory. In contrast, we shall prove that in suitably ordered categories the class of finitary varietors is extremely large, including a lot of infinitary functors.

6.2. Recall that each set functor is "almost" standard (III.4.5). A standard set functor F is said to *preserve unions of ω-chains of subsets*, provided that

$$X = \bigcup_{n<\omega} X_n \quad \text{implies} \quad XF = \bigcup_{n<\omega} X_n F$$

for each ω-chain $X_0 \subset X_1 \subset X_2 \ldots$.

Theorem. A standard set functor is a finitary varietor iff it preserves unions of ω-chains of subsets.

Proof. I. Let F preserve unions of ω-chains. Since F preserves monos (see III.4.7), and the coproduct injection $w_{0,1}: I \to I + IF$ is a mono, it is easy to verify by induction that $w_{n,m}$ is a mono for each $n \leq m < \omega$. Therefore we can assume that $W_0 \subset W_1 \subset W_2 \ldots$ and that $w_{n,m}$ are the inclusion maps. Then the colimit $W_\omega = \operatorname{colim} W_n$ is just the union $W_\omega = \bigcup W_n$. Since F preserves this union, and since $w_{n,m}F$ are inclusion maps too, it follows that F preserves the colimit W_ω. This means that the free-algebra construction stops after ω steps.

II. Let F be a finitary varietor. We can assume that F is non-constant. Given sets $X_0 \subset X_1 \subset X_2 \subset \ldots$, we put $X = \bigcup_{n<\omega} X_n$ and we prove $XF = \bigcup_{n<\omega} (X_n F)$.

By III.4.7, there exists a cardinal γ such that card $YF \geq$ card Y for any set Y with card $Y \geq \gamma$. Let I be an infinite set with $X \subset I$ and card $I \leq \gamma$. We use the fact that the free-algebra construction for I stops after ω steps. We are going to present one-to-one maps

$$t_n : X_n \to W_n \qquad (n < \omega)$$

which fulfil

$$(X_{n+1} - X_n)t_{n+1} \subset W_{n+1} - (W_n)w_{n,n+1} \qquad (n < \omega)$$

and which are compatible, i.e.,

$$t_n \cdot w_{n,n+1} = j_n \cdot t_{n+1} \qquad (n < \omega),$$

where $j_n : X_n \to X_{n+1}$ denotes the inclusion map.

Let

$$t_0 : X_0 \to I$$

be the inclusion map (we have $X_0 \subset X \subset I$). Then

$$\text{card } (X_1 - X_0) \leq \text{card } (W_1 - (W_0)w_{0,1})$$

because $W_1 - (W_0)w_{0,1} = IF$ and card $IF \geq$ card $I \geq$ card $X \geq$ card X_1. Therefore, we can extend the map $t_0 \cdot w_{0,1} : X_0 \to W_1$ to a one-to-one map $t_1 : X_1 \to W_1$ satisfying $(X_1 - X_0)t_1 \subset W_1 - (W_0)w_{0,1}$. Analogously, since

$$\text{card } (X_2 - X_1) \leq \text{card } (W_2 - (W_1)w_{1,2}),$$

we can extend $t_1 \cdot w_{1,2}$ to t_2, etc. Denote by

$$t_\omega : X \to W_\omega$$

the (unique) map extending each $t_n \cdot w_{n,\omega} : X_n \to W_\omega (n < \omega)$. This map is one-to-one and fulfils

$$[(W_n)w_{n,\omega}]t_\omega^{-1} = X_n \qquad (n < \omega).$$

Since F preserves preimages for one-to-one maps (III.4.7), this implies

$$[(W_nF)w_{n,\omega}F](t_\omega F)^{-1} = X_nF \qquad (n < \omega).$$

Finally, F preserves the colimit $W_\omega = \operatorname{colim}_{n<\omega} W_n$ (IV.3.4) and hence, by Remark III.5.4 we get

$$W_\omega F = \bigcup_{n<\omega} (W_{n,\omega}F)w_{n,\omega}F.$$

Thus,

$$XF = (W_\omega F)(t_\omega F)^{-1} = \bigcup_{n<\omega} [(W_nF)w_{n,\omega}F](t_\omega F)^{-1} = \bigcup_{n<\omega} X_nF.$$

This concludes the proof. □

6.3. Recall that a set functor F is finitary iff for each set X and each point $a \in XF$ there exists a finite set Y and a map $f: Y \to X$ with $a \in (YF)fF$. What is the relation between the properties "finitary functor", equivalent to the preservation of directed unions (Exercise III.4.E) and "finitary varietor", equivalent to the preservation of ω-unions? The answer depends on the axioms of set theory.

An infinite cardinal n is said to be *measurable* if on each set X of power n there exists a non-trivial σ-additive measure $\mu: X \to \{0, 1\}$ [i.e., a map with $(X)\mu = 1$, $(M)\mu = 0$ for each finite $M \subset X$, and $(\bigcup_{n<\omega} M_n)\mu = \sum_{n<\omega} (M_n)\mu$ if $M_n \subset X$ are pairwise disjoint]. We use a formulation based on the concept of filter, i.e., a collection $\mathcal{F}$ of non-empty subsets of X closed under finite intersections and super-sets (i.e., $M_1, M_2 \in \mathcal{F}$ implies $M_1 \cap M_2 \in \mathcal{F}$ and $M_3 \in \mathcal{F}$ for any $M_3 \supset M_1$). Recall that maximal filters are called *ultrafilters*; they are characterized by the property that for each $M \subset X$ either $M \in \mathcal{F}$ or $X - M \in \mathcal{F}$. For example, the collection $\mathcal{F}_x$ of all subsets containing a given element $x \in X$ is an ultrafilter. A cardinal n is measurable iff for each set X of power n there exists an ultrafilter $\mathcal{F}$ which is non-trivial, i.e., $\mathcal{F} \neq \mathcal{F}_x$ for each $x \in X$, and is closed under countable intersections. In fact, each such ultrafilter yields a non-trivial measure μ defined by $\mu(M) = 1$ iff $M \in \mathcal{F}$ (and vice versa).

The assumption "there exists no measurable cardinal" is well-known to be consistent with the theory of sets.

Theorem. Assume that no cardinal is measurable. A standard set functor is a finitary varietor iff it is a finitary functor.

Proof. Each finitary set functor preserves ω-unions (Exercise III.4.E) and hence, it is a finitary varietor (IV.6.2). Conversely, let F be a finitary varietor. Then F preserves ω-unions: for each set X and each point $a \in XF$ we prove that there exists a finite set $Y \subset X$ with $a \in YF$. Put

$$Y^* = \bigcap \{Y;\ Y \subset X \text{ and } a \in YF\}.$$

(i) Let $a \in Y^*F$. If Y^* is finite, the proof is concluded. If it is infinite, we choose pairwise distinct elements $y_0, y_1, y_2, \ldots$ in Y^* and put $Z_n = Y^* - \{y_{n+1}, y_{n+2}, \ldots\}$. Then $Z_0 \subset Z_1 \subset Z_2 \ldots$ and $\bigcup_{n<\omega} Z_n = Y^*$; we have

$$a \in Y^*F = (\bigcup_{n<\omega} Z_n)F.$$

Since F preserves unions of ω-chains, there exists n with $a \in Z_nF$. This contradicts to the definition of Y^* because $Y^* \not\subset Z_n$.

(ii) Let $a \notin Y^*F$. Put $X' = X - Y^*$ and define a collection of subsets of X',

$$\mathcal{F} = \{Z \subset X';\ a \in (Z \cup Y^*)F\}.$$

Since F preserves inclusion and finite intersections (III.4.6), $\mathcal{F}$ is a filter on the set X'. By Zorn's Lemma (applied to the set of all filters on X', ordered by inclusion), there exists an ultrafilter $\mathcal{G}$ on X' with $\mathcal{F} \subset \mathcal{G}$. By definition of Y^*, clearly $\bigcap_{Z \in \mathcal{F}} Z = \emptyset$ and hence, $\bigcap_{Z \in \mathcal{G}} Z = \emptyset$. It follows that each set in $\mathcal{G}$ is infinite (because any ultrafilter $\mathcal{G}$ containing finite sets equals $\mathcal{F}_x$ for some x). Since card X' is not measurable, the ultrafilter $\mathcal{G}$ is not closed under countable intersections, thus, we can choose sets $Y_k \in \mathcal{G}$ with $\bigcap_{k<\omega} Y_k \notin \mathcal{G}$. Put

$$Z_0 = \bigcap_{k<\omega} Y_k$$

and define sets $Z_n \subset X'$ by induction:

$$Z_{n+1} = Z_n \cup (X' - Y_n).$$

Then clearly $Z_0 \subset Z_1 \subset Z_2 \ldots$ and $\bigcup_{n<\omega} Z_n = X'$. Since F preserves unions of ω-chains and

$$a \in XF = (\bigcup_{n<\omega} (Z_n \cup Y^*))F,$$

there exists n with $a \in (Z_n \cup Y^*)F$, i.e., with $Z_n \in \mathcal{F}$. Let n_0 be the least number with $Z_{n_0} \in \mathcal{F} \subset \mathcal{G}$. Then $n_0 \neq 0$, hence $Z_{n_0} = Z_{n_0-1} \cup (X' - Y_{n_0-1})$. This is a contradiction: any ultrafilter $\mathcal{G}$ has the property that $A \cup B \in \mathcal{G}$ implies either $A \in \mathcal{G}$ or $B \in \mathcal{G}$. Yet, neither Z_{n_0-1} nor $X' - Y_{n_0-1}$ is an element of $\mathcal{G}$.

□

Corollary. The following statements are equivalent:
(i) no cardinal is measurable;
(ii) each standard set functor which is a finitary varietor is a finitary functor.

The proof of(i) → (ii) was presented above. For the converse assume that there exist measurable cardinals and define a set functor M as follows. For each set X let XM be the set of all σ-additive measures on X. For each map $f: X \to Y$ and each measure $\mu: X \to \{0, 1\}$ define the measure $(\mu)fM = \bar{\mu}$ by

$$(A)\bar{\mu} = ((A)f^{-1})\,\mu \quad \text{for each } A \subset Y.$$

It is easy to check that $\bar{\mu}$ is a σ-additive measure and that M is a well-defined functor. Moreover, the existence of measurable cardinals clearly implies that M is not a finitary functor. On the other hand, M preserves countable unions: for each σ-additive measure $\mu: \bigcup_{n<\omega} X_n \to \{0, 1\}$ there exists n_0 with $(X_{n_0})\mu = 1$, and then the restriction μ' of μ to X_{n_0} fulfils

$$\mu = (\mu')i_{n_0}F$$

where $i_{n0}: X_{n0} \to \bigcup_{n<\omega} X_n$ is the inclusion map. Thus, M is a finitary varietor. By III.4.5, there exists a standard functor F naturally isomorphic to M. Then F is a counterexample to (ii).

6.4. In contrast to **Set**, in the full subcategory

$$\mathbf{Set}_\omega$$

of all countable sets, each functor is a finitary varietor. In fact, a finitary functor, which is defined as in **Set** (III.4.3).

Theorem. Each functor

$$F: \mathbf{Set}_\omega \to \mathbf{Set}_\omega$$

is finitary and hence, a finitary varietor.

Proof. I. Let us first observe that for each infinite set X there exists an uncountable collection $(M_j)_{j \in J}$ (i.e., card $J > \aleph_0$) of infinite subsets $M_j \subset X$ such that $M_j \cap M_{j'}$ is finite for each $j, j' \in J$ with $j \neq j'$.

We can assume that X contains the set of all rational numbers. Let J be the set of all irrational numbers. For each $j \in J$ choose a sequence x_n of rational numbers with $\lim_{n\to\infty} x_n = j$ and put $M_j = \{x_n; n < \omega\}$.

II. Let $F: \mathbf{Set}_\omega \to \mathbf{Set}_\omega$ be an arbitrary functor. We can assume that F is a standard functor—this is quite analogous to the category **Set**, see III.4.5. Let us extend F to a standard functor

$$G: \mathbf{Set} \to \mathbf{Set}$$

as follows. For each set X let

$$XG = \bigcup X'F$$

where the union ranges over all countable subsets X' of X (and hence, $XG = XF$ if X is countable); for each map $f: X \to Y$ define $fG: XG \to YG$ as follows: given $a \in XG$, there is a countable set $X' \subset X$ with $a \in X'G = X'F$, and we denote by $f': X' \to (X')f$ the restriction of f, and put

$$(a)fG = (a)f'F'.$$

Since F is standard, G is obviously a well-defined extension of F. It is sufficient to prove that G is a finitary functor. Then G is a finitary varietor, and since F coincides with G on the category $\mathbf{Set}_\omega$, it follows that F is also both a finitary functor and a finitary varietor.

For each set X and each point $a \in XG$ we present a finite set $Y \subset X$ with $a \in YG$. Let $X' \subset X$ be a countable set with $a \in X'G$. If X' is finite, the proof is concluded. If X' is infinite, we have a collection $M_j \subset X'$ $(j \in J)$ as in I. above. For each j there exists a one-to-one map $f_j: X' \to X'$ with

$$M_j = (X')f_j.$$

The set of all

$$(a)f_jF(\in X'G) \qquad j \in J,$$

is countable simply because $X'G = X'F$ is countable. Since the set J is uncountable, there exist distinct $j, j' \in J$ with

$$(a)f_jG = (a)f_{j'}G = b.$$

Put $\bar{Y} = M_j \cap M_{j'}$. Since $(X'G)(f_jG) = M_jG$ (see III.4.7) and analogously with j', and since G preserves finite intersections (III.4.6), we have

$$b \in M_jG \cap M_{j'}G = \bar{Y}G.$$

The set $Y = (\bar{Y})f_j^{-1}$ is finite, since f_j is one-to-one, and

$$a \in YG$$

because the fact that f_j is one-to-one implies that $YG = (\bar{Y}G)(f_jG)^{-1}$ by III.4.7. This proves that G, and hence F, is a finitary functor. □

6.5. For a certain type of ordered categories we shall exhibit a mild criterion for a functor to be a finitary varietor.

By an *ordered category* is meant a category together with a partial ordering $\leq_{A,B}$ (or just $\leq$) on each set hom (A, B) of morphisms from A to B, compatible with the composition in the sense that

$$f \leq_{A,B} f' \text{ and } g \leq_{B,C} g' \quad \text{imply} \quad f \cdot g \leq_{A,C} f' \cdot g'.$$

Recall that a poset $(X, \leq)$ is said to be *ω-complete* if it has a least element 0 and each increasing ω-sequence has a join. Maps preserving the least element and joins of increasing ω-sequences are said to be *ω-continuous.*

Defintion. An *ω-category* is an ordered category in which
(i) each hom (A, B) is an ω-complete poset;
(ii) composition is ω-continuous, i.e., given $f_0 \leq f_1 \leq f_2 \ldots : A \to B$ and $g_0 \leq g_1 \leq g_2 \ldots : B \to C$, then

$$\left(\bigvee_{n=0}^{\infty} f_n\right) \cdot \left(\bigvee_{n=0}^{\infty} g_n\right) = \bigvee_{n=0}^{\infty} f_n \cdot g_n : A \to C,$$

and for the least elements,

$$0_{A,B} \cdot 0_{B,C} = 0_{A,C}.$$

Examples. (i) **Pos**$_\omega$, the category of ω-complete posets and ω-continuous maps, is an ω-category. Here, hom (A, B) is naturally ordered by $f \leq g$ iff $(x)f \leq (x)g$ for each $x \in A$.

(ii) **Pfn**, the category of sets and partial functions, is an ω-category. The ordering of hom(X, Y) is defined "by extension": given $f, g: X \to Y$, then

$$f \leq g \quad \text{iff} \quad (x)f = (x)g \quad \text{for all } x \text{ with } (x)f \text{ defined.}$$

The least map is the empty (nowhere defined) map. The join of a sequence $f_0 \le f_1 \le f_2 \le \dots : X \to Y$ is the map $f: X \to Y$ defined by $(x)f = (x)f_n$ whenever $(x)f_n$ is defined.

(iii) The category R-**Mod** becomes trivially an ω-category when we put $f \le g$ iff either $f = g$ or f is the zero map.

6.6. Definition. A diagram $D: \mathscr{D} \to \mathscr{K}$ in an ordered category $\mathscr{K}$ is said to have an *isotone limit* if it has a limit $\pi_d: X \to dD (d \in \mathscr{D}^\circ)$ such that for arbitrary $p, q: Y \to X$ we have

$$p \le q \quad \text{iff} \quad p \cdot \pi_d \le q \cdot \pi_d \quad \text{for each } d \in \mathscr{D}^\circ.$$

Remarks. (i) If $\mathscr{K}$ is an ω-category, then each isotone limit is in fact ω-continuous in the sense that (1) given $p, q_n: Y \to X$ with $q_0 \le q_1 \le q_2 \le \dots$, then

$$p = \bigvee_{n<\omega} q_n \quad \text{iff} \quad p \cdot \pi_d = \bigvee_{n<\omega} q_n \cdot \pi_d \qquad \text{for each } d \in \mathscr{D}^\circ,$$

and (2)

$$p = 0_{Y,X} \quad \text{iff} \quad p \cdot \pi_d = 0_{Y,dD} \qquad \text{for each } d \in \mathscr{D}^\circ.$$

This follows easily from the ω-continuity of the composition map $- \cdot \pi_d: \hom(Y, X) \to \hom(Y, dD)$.

(ii) In an ordered category $\mathscr{K}$, all limits are isotone iff

(a) products are isotone, i.e., $\prod_{i \in I} f_i \le \prod_{i \in I} g_i$ whenever $f_i \le g_i$ for each $i \in I$;

(b) equalizers are isotone, i.e., $\mathscr{K}$ has equalizers and for each regular mono $m: X \to X'$ and for arbitrary $p, g: Y \to X$,

$$p \cdot m \le q \cdot m \quad \text{implies} \quad p \le q.$$

The proof that (a) and (b) imply that each limit is isotone is analogous to the non-ordered variant (Exercise III.1.E).

(iii) Isotone colimits are defined dually. An ordered category has isotone colimits iff it has isotone coproducts and isotone coequalizers.

6.7. Definition. Let $\mathscr{K}$ be an ω-category. By an *ω-functor* we mean a functor $F: \mathscr{K} \to \mathscr{K}$ which preserves the join of any ω-sequence $p_0 \le p_1 \le p_2 \le \dots : A \to B$, i.e., it fulfils

$$\left(\bigvee_{n<\omega} p_n\right) F = \bigvee_{n<\omega} p_n F: AF \to BF.$$

Example. For each type Σ the functor

$$H_\Sigma: \mathscr{K} \to \mathscr{K}$$

is an ω-functor, assuming that $\mathscr{K}$ has isotone products and coproducts.

Remark. Each ω-functor is clearly "locally isotone", i.e., if $p \leq q$ in hom(A, B) then $pF \leq qF$ in hom(AF, BF).

Note that, however, F is not supposed to preserve the least element of hom(A, B).

Theorem. Let $\mathscr{K}$ be an ω-category with isotone ω-colimits. For each ω-functor $F: \mathscr{K} \to \mathscr{K}$, the initial-algebra construction stops after ω steps.

Proof. Let us define morphisms

$$t_n: \perp F^{n+1} \to \perp F^n \qquad (n < \omega)$$

by the following induction:

$$t_0: \perp F \to \perp$$

is the least element of hom$(\perp F, \perp)$;

$$t_{n+1} = t_n F: \quad (\perp F^{n+1})F \to (\perp F^n)F.$$

Then

(1) $$w_{n,n+1} \cdot t_n = 1_{\perp F^n} \quad \text{and} \quad t_n \cdot w_{n,n+1} \leq 1_{\perp F^{n+1}} \qquad (n < \omega).$$

This is obvious if $n = 0$: $w_{0,1} \cdot t_0 = 1$ because hom$(\perp, \perp) = \{1_\perp\}$, and $t_0 \cdot w_0 \leq 1_{\perp F}$ because t_0 is the least element of hom$(\perp F, \perp)$ and hence, $t_0 \cdot w_{0,1}$ is the least element of hom$(\perp F, \perp F)$ (because the composition $- \cdot w_{0,1}$ is ω-continuous). Since F is "locally isotone", the induction step is obvious, too. Given $n < m$, put

$$t_{n,m} = t_{m-1} \cdot t_{m-2} \cdot \ldots \cdot t_n: \perp F^m \to \perp F^n.$$

Then (1) clearly implies

(2) $$w_{n,m} \cdot t_{n,m} = 1_{\perp F^n} \quad \text{and} \quad t_{n,m} \cdot w_{n,m} \leq 1_{\perp F^m} \qquad (n < m < \omega).$$

For a fixed $k = 0, 1, 2, \ldots$, define a family of morphisms from $\perp F^n$ $(n < \omega)$ to $\perp F^k$ by

$$w_{n,k}: \perp F^n \to \perp F^k \qquad \text{for all } n = 0, 1, \ldots, k;$$
$$t_{n,k}: \perp F^n \to \perp F^k \qquad \text{for all } n = k+1, k+2, \ldots .$$

Using (2), it is easy to check that this family is compatible with the initial-algebra construction. Hence, there exists a unique morphism

$$s_k: \perp F^\omega \to \perp F^k$$

with

(3) $$w_{n,\omega} \cdot s_k = \begin{cases} w_{n,k} & (n \leq k) \\ t_{n,k} & (n > k). \end{cases}$$

Let us verify that the morphisms $s_k \cdot w_{k,\omega}: \perp F^\omega \to \perp F^\omega$ form an ω-chain

with

$$(4) \qquad \bigvee_{k=0}^{\infty} s_k \cdot w_{k,\omega} = 1_{\perp F\omega}.$$

On the one hand, for each $k < \omega$ we have

$$s_k \cdot w_{k,\omega} \leq 1_{\perp F\omega}$$

because the colimit $\perp F^{\omega} = \operatorname{colim}_{n<\omega} \perp F^n$ is isotone, and for each $n = k+1$, $k+2, \ldots$ we have

$$\begin{aligned} w_{n,\omega} \cdot (s_k \cdot w_{k,\omega}) &= t_{n,k} \cdot w_{k,\omega} \\ &= (t_{n,k} \cdot w_{k,n}) \cdot w_{k,\omega} \\ &\leq w_{n,\omega}. \end{aligned}$$

Analogously with the inequality

$$s_k \cdot w_{k,\omega} \leq s_{k+1} \cdot w_{k+1,\omega} \qquad (k < \omega).$$

On the other hand, let $f\colon \perp F^{\omega} \to \perp F^{\omega}$ be a morphism with

$$f \geq s_k \cdot w_{k,\omega} \qquad \text{for each } k < \omega.$$

For each $k < \omega$ we have $w_{k,\omega} \cdot s_k = w_{k,k} = 1$, hence

$$w_{k,\omega} \cdot f \geq w_{k,\omega} \cdot s_k \cdot w_{k,\omega} = w_{k,\omega}.$$

This implies that $f \geq 1_{\perp F\omega}$, and the proof of (4) is concluded.

We shall prove that $w_{\omega,\omega+1}$ is an isomorphism with the inverse

$$s = \bigvee_{k=0}^{\infty} s_k F \cdot w_{k+1,\omega}\colon \perp F^{\omega+1} \to \perp F^{\omega}.$$

First, to show that s is well-defined we must check that $s_k F \cdot w_{k+1,\omega} \leq s_{k+1} F \cdot w_{k+2,\omega}$ for each $k < \omega$. We have

$$s_k = s_{k+1} \cdot t_k$$

(because for each $n > k+1$ clearly $w_{n,\omega} \cdot s_k = t_{k,n} = w_{n,\omega} \cdot s_{k+1} \cdot t_k$), and we conclude that

$$\begin{aligned} s_k F \cdot w_{k+1,\omega} &= s_{k+1} F \cdot t_k F \cdot w_{k+1,k+2} \cdot w_{k+2,\omega} \\ &= s_{k+1} F \cdot (t_k \cdot w_{k,k+1}) F \cdot w_{k+2,\omega} \\ &= s_{k+1} F \cdot w_{k+2,\omega} \end{aligned}$$

[by (i) above]. Next

$$w_{\omega,\omega+1} \cdot s = 1_{\perp F\omega}$$

beacuse for each $n < \omega$ we have

$$w_{n+1,\omega} \cdot (w_{\omega,\omega+1} \cdot s) = w_{n+1,\omega} \cdot s$$

$$= w_{n,\omega}F \cdot \bigvee_{k=0}^{\infty} s_k F \cdot w_{k+1,\omega}$$

$$= \bigvee_{k=0}^{\infty} (w_{n,\omega} \cdot s_k)F \cdot w_{k+1,\omega}$$

$$= \bigvee_{k=0}^{\infty} w_{n+1,k+1} \cdot w_{k+1,\omega}$$

$$= w_{n+1,\omega}.$$

Finally,

$$s \cdot w_{\omega,\omega+1} = 1_{\perp F\omega+1}$$

because F is an ω-functor and hence, $\bigvee_{k=0}^{\infty} (s_k \cdot w_{k,\omega})F = \left(\bigvee_{k=0}^{\infty} s_k \cdot w_{k,\omega}\right)F$. Therefore,

$$s \cdot w_{\omega,\omega+1} = \bigvee_{k=0}^{\infty} s_k F \cdot w_{k+1,\omega} \cdot w_{\omega,\omega+1}$$

$$= \bigvee_{k=0}^{\infty} (s_k \cdot w_{k,\omega})F$$

$$= \left(\bigvee_{k=0}^{\infty} s_k \cdot w_{k,\omega}\right)F$$

$$= 1_{\perp F\omega+1}.$$

This conclude the proof. □

Corollary. Let $\mathscr{K}$ be an ω-category with isotone countable colimits. Then each ω-functor $F: \mathscr{K} \to \mathscr{K}$ is a finitary varietor.

Indeed, since finite coproducts are isotone (hence, ω-continuous, see Remark IV.6.6) and since both F and each constant functor C_I are ω-functors, $F + C_I$ is clearly an ω-functor, too. It follows from Proposition IV.3.3 and the preceding theorem that F is a finitary varietor.

Examples. (i) The ω-category $\mathbf{Pos}_\omega$ has isotone limits and colimits, as we verify presently. Thus, for each type Σ, the functors

$$H_\Sigma, H_\Sigma^0: \mathbf{Pos}_\omega \to \mathbf{Pos}_\omega$$

(see III.3.6) are finitary varietors. And this holds even for infinitary types Σ. The inner reason will be seen below.

(a) Limits. The product $\prod X_i$ in $\mathbf{Pos}_\omega$ is the cartesian product, ordered coordinate-wise. Given morphisms $p, q: P \to \prod X_i$ such that each projection π_i fulfils $p \cdot \pi_i \leq q \cdot \pi_i$, then $p \leq q$ (because the ordering is coordinate-wise).

The equalizer $e: X \to Y$ of two morphisms $f_1, f_2: Y \to Z$ is the embedding of the subposet of all $y \in Y$ with $(y)f_1 = (y)f_2$. Given $p, q: P \to X$ with $p \cdot e \leq q \cdot e$, then, obviously, $p \leq q$.

(b) Colimits. The coproduct $\coprod_{i \in I} X_i$ in $\mathbf{Pos}_\omega$ is the disjoint union with the least elements merged. Given morphisms $p, q: \coprod_{i \in I} X_i \to P$ such that $p/X_i \leq q/X_i$ for each $i \leq I$, then $p \leq q$.

The coequalizers in $\mathbf{Pos}_\omega$ are not so easy to describe. It is clear, however, that each coequalizer $c: X \to Y$ has the following property: if Y' is a subposet of Y containing $(X)c$ and closed under joins of increasing ω-sequences, then $Y' = Y$. Given morphisms $p, q: Y \to P$ with $c \cdot p \leq c \cdot q$, let Y' be the subposet of Y of all $y \in Y$ with $(y)p \leq (y)q$. Then Y' has the properties mentioned above and hence, $Y' = Y$. Thus $p \leq q$.

(ii) The ω-category **Pfn** has isotone limits and colimits—see Exercise IV.6.B below. However, the product in **Pfn** (which we denote by $\hat{\times}$ and $\hat{\prod}$) is not the cartesian product. We have

$$X \hat{\times} Y = (X \times Y) + X + Y$$

and, in general,

$$\hat{\prod_{i \in I}} X_i = \coprod_{\emptyset \neq J \subset I} \prod_{i \in J} X_i.$$

[Given partial maps $f_i: T \to X_i$, $i \in I$, then the unique factorizing map $f: T \to \hat{\prod} X_i$ is defined by $(t)f = \{(t)f_i\}_{i \in J}$ where $J = \{i \in I;\ (t)f_i \text{ is defined}\}$; $(t)f$ is undefined iff each $(t)f_i$ is undefined.]

Thus, the functors

$$H_\Sigma: \mathbf{Pfn} \to \mathbf{Pfn}$$

are finitary varietors for each (even infinitary) type Σ, but there seems to be no intuition what the H_Σ-algebras are like.

On the other hand, the functors defined by the cartesian product are not ω-functors, in general. For each cardinal n let us denote by

$$H_n^*: \mathbf{Pfn} \to \mathbf{Pfn}$$

the functor of the n-th cartesian power:

$$XH_n^* = X^n \quad \text{and} \quad fH_n^* = f^{(n)}.$$

Then H_n^* is an ω-functor iff n is finite. As an example, let $X_0 \subsetneq X_1 \subsetneq X_2 \ldots$ be arbitrary sets and put $X = \bigcup_{k=0}^{\infty} X_k$. Let

$$f_k: X \to X \qquad (k < \omega)$$

be defined by

$$(x)f_k = x \qquad \text{if } x \in X_k, \text{ else undefined.}$$

Then

$$\bigvee_{k=0}^{\infty} f_k = 1_X$$

but, for any infinite n,

$$\bigvee_{n=0}^{\infty} f_k^{(n)} \neq 1_{X^n}.$$

Indeed, let $(x_i)_{i<n} \in X^n$ be an element such that $x_0 \in X_0$, $x_1 \in X_1 - X_0$, $x_2 \in X_2 - X_1, \ldots$; then 1_{X^n} is defined in $(x_i)_{i<n}$ but $\vee f_k^{(n)}$ is not.

(iii) The category R-**Mod**, with its trivial order, has neither products nor coproducts isotone. For example, consider

$$f\colon X \times Y \to X \times Y$$

defined by

$$f \cdot \pi_X = \pi_X \qquad \text{and} \qquad f \cdot \pi_Y = 0.$$

Though $f \not\leq 1_{X \times Y}$, we have

$$f \cdot \pi_X \leq 1_{X \times Y} \cdot \pi_X \qquad \text{and} \qquad f \cdot \pi_Y \leq 1_{X \times Y} \cdot \pi_Y.$$

6.8. Free ω-continuous algebras. For each type Σ we are going to describe the free ω-continuous Σ-algebras, i.e., the free algebras of the functor

$$H_\Sigma^0\colon \mathbf{Pos}_\omega \to \mathbf{Pos}_\omega.$$

(See Exercise IV.6.A below for H_Σ.) We use the description of Σ-trees as maps $t\colon n^* \to I \cup \Sigma$ $(n = \bigvee_{\sigma \in \Sigma} |\sigma|)$ introduced in II.3.

Let I be an ω-complete poset (with $I \cap \Sigma = \emptyset$, for convenience). In the first step of the free-algebra construction we have

$$W_1 = I + H_\Sigma^0.$$

Here, $+$ is the disjoint union with the least element $0 \in I$ merged with the (formal!) least element of IH_Σ^0. Thus, as a set, $W_1 = I \cup IH_\Sigma^0$, where the elements of IH_Σ^0 are represented (as usual) by the following trees:

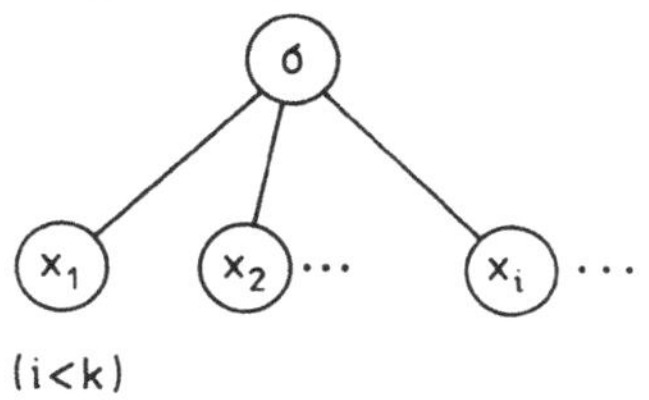

$(\sigma \in \Sigma_k \quad \text{and} \quad x_i \in X)$.

The ordering is such that 0 is smaller then each such tree, while any $x \in I - \{0\}$ is incompatible with each such tree.

In general,

$$W_{m+1} = I + W_m H_\Sigma^0$$

is (as a set) the disjoint union of I and of $W_m H_\Sigma^0$. The elements of the latter set are represented by trees

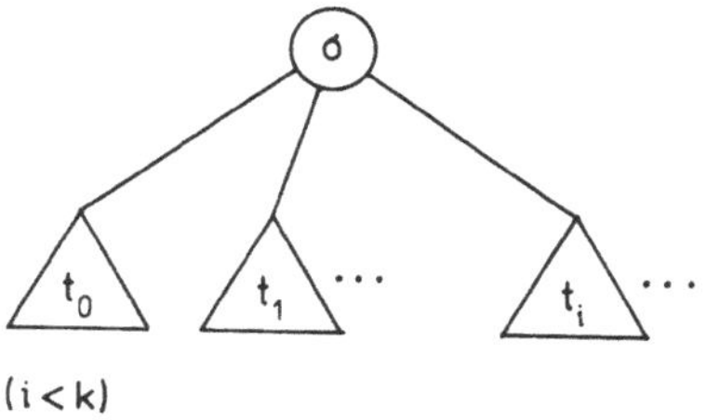

($\sigma \in \Sigma_k$ and $t_i \in W_m$). Hence, W_{m+1} is the set of all trees $t: n^* \to I \cup \Sigma$ of depth $\leq m + 1$. The ordering of trees $t, t': n^* \to I \cup \Sigma$ in W_m is defined as follows:

(*) $t \leq t'$ iff for each $z \in n^*$
(i) $(z)t \in \Sigma$ implies $(z)t = (z)t'$;
(ii) $(z)t \in I - \{0\}$ implies $(z)t \leq (z)t' \in I - \{0\}$.

This is easy to verify by induction on m. Example:

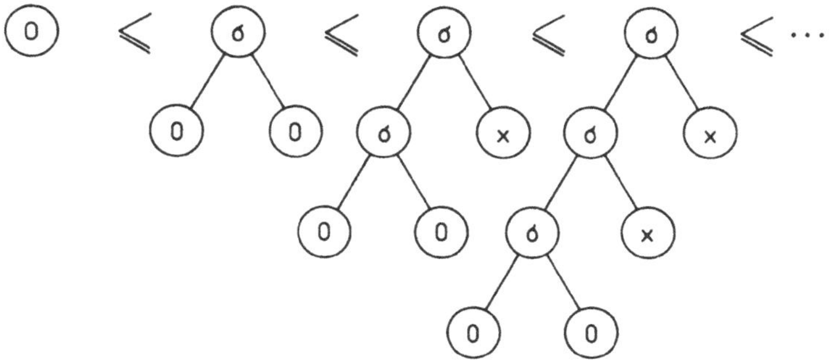

Thus

$$\bigcup_{m=0}^{\infty} W_m$$

is the poset of all Σ-trees of finite depth, ordered by (*). This is not W_ω; indeed, the poset $\bigcup_{m=0}^{\infty} W_m$ is not ω-complete, as indicated by the example above.

Observation. Let $I^\#$ denote the poset of all (finite and infinite) Σ-trees over I with the ordering (*) above. Then $I^\# = \operatorname{colim}_{m<\omega} W_m$ with respect to the inclusion maps

$$w_{m,\omega}: W_m \to I^\# \qquad (m < \omega).$$

Proof. (i) The poset $I^\#$ is ω-complete. Its least element is $0 \in I \subset I^\#$. The join of a sequence $t_0 \leq t_1 \leq t_2 \ldots$ of Σ-trees is the following Σ-tree $t: n^* \to i \cup \Sigma$: for each $z \in n^*$,

$(z)t = x \in I$ iff $(z)t_m \in I$ for each $m < \omega$, and $x = \bigvee_{m<\omega} (z)t_m$ in I;

$(z)t = \sigma \in \Sigma$ iff $(z)t_m = \sigma$ for all but finitely many $m < \omega$.

(ii) The inclusion maps are clearly ω-continuous.

(iii) Each tree $t \in I$ is a join of an ω-sequence $t^{(0)} \leq t^{(1)} \leq t^{(2)} \leq \ldots$ with $t^{(m)} \in W_m$ defined by

$$(i_1 \ldots i_r)t^{(m)} = \begin{cases} (i_1 \ldots i_r)t & \text{if } r < m \text{ and } (i_1 \ldots i_r)t \text{ is defined;} \\ 0 & \text{if } r = m \text{ and } (i_1 \ldots i_r)t \text{ is defined} \end{cases}$$

and undefined else. It is easy to prove by induction that $t^{(m)} \in W_m$; using the description of ω-joins in (i), we see that

$$t = \bigvee_{m<\omega} t^{(m)} \text{ for each } t \in I^\#.$$

(iv) Let P be an ω-complete poset and let $p_m: W_m \to P\ (m < \omega)$ be a compatible family, i.e., a family such that p_{m+1} extends $p_m\ (m < \omega)$. Define $p: I^\# \to P$ by

$$(t)p = \bigvee_{m<\omega} (t^{(m)})p_m.$$

Then p clearly extends **each** p_m, and it is sufficient to show that p is ω-continuous [the uniqueness follows immediately from (iii)]. For each $m < \omega$ we have $t^{(m)} = \bigvee_{k<\omega} t_k^{(m)}$ — this is easy to check using the description of ω-joins presented in (i) above. Hence,

$$\begin{aligned}(t)p &= \bigvee_{m<\omega} (t_k^{(m)})p_m \\ &= \bigvee_{m<\omega} \bigvee_{k<\omega} (t_k^{(m)})p_m \\ &= \bigvee_{k<\omega} \bigvee_{m<\omega} (t_k^{(m)})p_m \\ &= \bigvee_{k<\omega} (t_k)p. \end{aligned}$$

□

Corollary. The free ω-continuous Σ-algebra generated by an ω-complete poset I is the algebra $I^\#$ of all Σ-trees over I ordered as in (*) above and with the operations $\varphi_\sigma: (I^\#)^k \to I^\#\ (\sigma \in \Sigma_k)$ defined by

$$(t_j)_{j<k}\varphi_\sigma = t \quad \text{iff} \quad (\emptyset)t = \sigma \quad \text{and} \quad \partial_j t = t_j \qquad (j < k).$$

This follows from the fact that H^0_Σ is a finitary varietor (hence, $I^\# = W_\omega$) and from the preceding observation. The definition of φ_σ above is correct if all the trees $t_j, j < k$, are from one W_m for some $m < \omega$: then φ_σ is the coproduct injection of W^k_m into W_{m+1}; in general, it follows from the ω-continuity of φ_σ, since

$$(t_j)_{j<k}\varphi_\sigma = \bigvee_{m<\omega} (t_j^{(m)})_{j<k}\varphi_\sigma.$$

Exercises IV.6

A. Free strictly ω-continuous Σ-algebras. These are algebras for the functor H_Σ: **Pos**$_\omega$ → **Pos**$_\omega$; they differ from the (non-strict) ω-continuous algebras by the requirement that the operations preserve the least element, i.e., by the equations

$$(0, 0, \ldots, 0, \ldots)\varphi_\sigma = 0 \text{ for each } \sigma \in \Sigma - \Sigma_0.$$

Verify that the free H_Σ-algebra is obtained from the free H^0_Σ-algebra by the least congruence $\sim$ such that $(0, 0, \ldots, 0, \ldots)\varphi_\sigma \sim 0$ for each $\sigma \in \Sigma - \Sigma_0$. Its elements can be represented by all trees t such that under each node there lies a leaf not labelled by 0. (Formally: for each $p_1 \ldots p_r \in n^*$ in the domain of t there exists $p_1 \ldots p_r p_{r+1} \ldots p_s \in n^*$ with $(p_1 \ldots p_s)t \in (I - \{0\}) \cup \Sigma_0$.)

B. Limits and colimits in Pfn. (i) Verify that coproducts in **Pfn** are those in **Set**, i.e., the disjoint unions. Verify that each coequalizer in **Pfn** is a partial map onto.

Conclude that **Pfn** has isotone colimits.

(ii) Verify that the products in **Pfn** (described in Example IV.6.7) are isotone. Describe equalizers in **Pfn**. Conclude that **Pfn** has isotone limits.

C. Free algebras in Pfn. Denote by H^*_Σ: **Pfn** → **Pfn** the coproduct of the functors $H^*_{|\sigma|}$, $\sigma \in \Sigma$ (IV.6.7). Verify that the free-algebra construction coincides with that in **Set**. Hence, H^*_Σ is a finitary varietor iff Σ is finitary.

H^*_Σ-algebras are precisely partial Σ-algebras but the homomorphisms here are not "natural". Describe H^*_Σ-homomorphisms in **Pfn** and explain why free algebras are just the free universal Σ-algebras.

D. Finitary varietors on R-Vect. (i) Assuming that no cardinal is measurable, prove that a standard functor F: R-**Vect** → R-**Vect** is a finitary varietor iff it is a finitary functor in the sense of III.4.3. [Hint: The proof is completely analogous to that of IV.6.3; use the properties of vector functors, indicated in Exercises III.4.A, D, E.]

(ii) Denote by R-**Vect**$_\omega$ the full subcategory of R-**Vect** over countably-dimensional spaces. Prove that each functor F: R-**Vect**$_\omega$ → R-**Vect**$_\omega$ is finitary. (Hint: Proceed as in IV.6.4.)

IV.7. Free-Completion Construction

7.1. Partial algebras. Throughout this section, $\mathcal{K}$ denotes a category with a fixed factorization system $(\mathcal{E}, \mathcal{M})$. By a *partial morphism* from an object A to an object B in $\mathcal{K}$ is meant a morphism from a subobject of A into B. We denote partial morphisms by $\rightharpoonup$. Thus, a partial morphism $\delta: A \rightharpoonup B$ is represented by a span

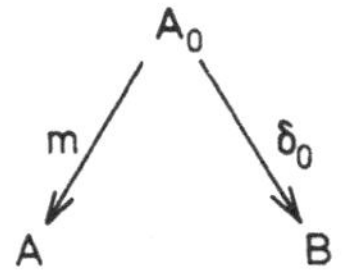

where $m \in \mathcal{M}$. (For each isomorphism $i: A_0' \to A$, the span $i \cdot \delta_0 : A_0' \to B$ and $i \cdot m: A_0' \to A$ represents the same partial morphism.)

Let $F: \mathcal{K} \to \mathcal{K}$ be a functor. A *partial F-algebra* is a pair (Q, δ) consisting of an object Q and a partial morphism $\delta: QF \rightharpoonup Q$.

Example. A partial groupoid (i.e., a partial H_2-algebra) consists of a set Q and a partial binary operation, i.e., a map

$$\Box : D \to Q$$

where

$$D \subset Q \times Q$$

denotes the domain of $\Box$.

Given partial groupoids $(Q, \Box)$ and $(Q', \circ)$, a map $f: Q \to Q'$ is a homomorphism iff

(*) $\qquad (q_1 \Box q_2)f = (q_1)f \circ (q_2)f$

holds for arbitrary $q_1, q_2 \in Q$ for which $q_1 \Box q_2$ is defined.

Remark. There are several natural concepts of homomorphism of partial groupoids (and partial F-algebras in general). All are characterized by (*), and the differences are only in the question which side is to be defined. In Chapters VII and VIII, devoted to partial and nondeterministic automata, we use so-called state morphisms (characterized, in case of groupoids, by (*) whenever $(q_1) f \circ (q_2)f$ is defined).

Notation. A partial F-algebra can be represented by a quadruple (Q, D, δ_0, m), where

$$m: D \to QF$$

is an $\mathcal{M}$-mono (the domain of the operation) and

$$\delta_0 : D \to Q$$

is an arbitrary morphism. All other such representations of the same partial algebra are then

$$(Q, \bar{D}, i \cdot \delta_0, i \cdot m),$$

where $i: \bar{D} \to D$ is an arbitrary isomorphism. Given partial algebras $A = (Q, D, \delta_0, m)$ and $A' = (Q', D', \delta_0', m')$, a morphism $f: Q \to Q'$ is a *homomorphism* iff there exists a (unique) $f_0: D \to D'$ such that the following diagram

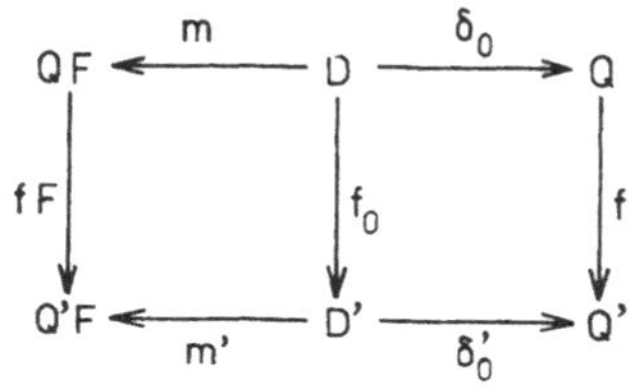

commutes. This is clearly independent of the choice of representatives of A and A'.

7.2. One extreme case of partial F-algebras are the *discrete* F-algebras, i.e., those where δ is nowhere defined. Formally, let $\perp$ be the initial object. A discrete F-algebra on an object Q is represented by

$$(Q, \perp, \delta_0, m)$$

where $\delta_0: \perp \to Q$ and $m: \perp \to QF$ are the unique morphisms. But here we must assume that $\mathcal{M}$ contains all the morphisms from $\perp$.

If a partial F-algebra (Q, δ) is discrete, each morphism from Q is a homomorphism. More precisely, for each partial F-algebra (Q', δ'), all morphisms $f: Q \to Q'$ are homomorphisms $f: (Q, \delta) \to (Q', \delta')$.

The opposite extreme are the (*total*) *F-algebras*, represented by those quadruples (Q, D, δ_0, m) for which m is an isomorphism. [One of the representatives of an F-algebra (Q, δ) is $(Q, QF, \delta, 1_{QF})$, of course.]

7.3. The *free completion* of a partial F-algebra (Q, δ) is a (total) F-algebra $(Q^\#, \delta^\#)$ together with a homomorphism

$$\eta: (Q, \delta) \to (Q^\#, \delta^\#)$$

which has the following universal property:

> For each (total) F-algebra (Q', δ') and each homomorphism $f: (Q, \delta) \to (Q', \delta')$, there exists a unique homomorphism

$f^{\#}: (\mathbf{Q}^{\#}, \delta^{\#}) \to (Q', \delta')$ such that $f = \eta \cdot f^{\#}$.

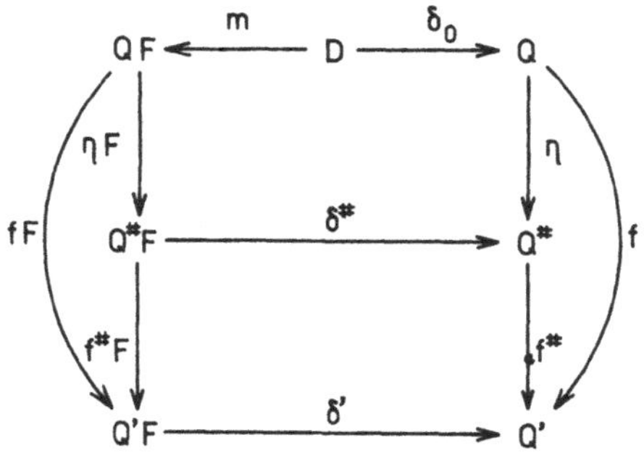

Examples. If (Q, δ) is the discrete algebra, then $(Q^{\#}, \delta^{\#})$ is precisely the free algebra generated by Q. (Indeed, here f is simply a morphism $f: Q \to Q'$.)

On the other hand, if (Q, δ) is a total F-algebra then $(Q^{\#}, \delta^{\#}) = (Q, \delta)$.

7.4. We are going to present a construction of free completion, generalizing the free-algebra construction. Before that, we extend the concept of a partial algebra, no more insisting that the morphism $m: D \to Q$ be a mono. On the one hand, this will simplify some considerations (for example we shall be able to work with discrete algebras without unnecessary assumptions on $\mathcal{M}$); on the other hand, in the construction of free completion we study below, there is no advantage in restricting ourselves to partial algebras.

Let $F: \mathcal{K} \to \mathcal{K}$ be a functor. By a *span F-algebra* is meant a diagram of the following type in $\mathcal{K}$:

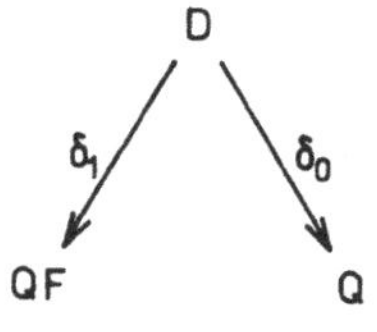

A *span-algebra morphism* from a span algebra $(Q, D, \delta_0, \delta_1)$ to a span algebra $(Q', D', \delta_0', \delta_1')$ is a pair (f, f_0) of $\mathcal{K}$-morphisms such that the following diagram

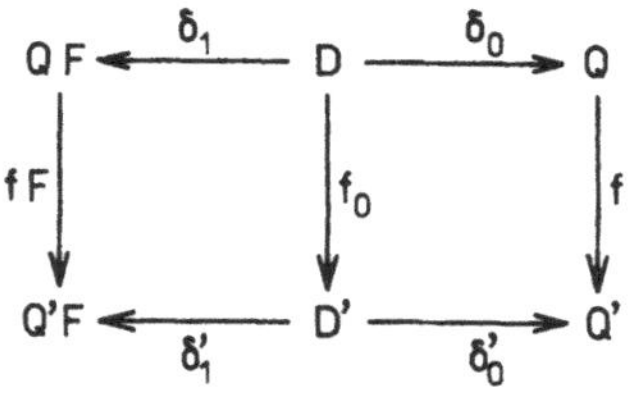

commutes.

Remark. By the above notation, each partial algebra is represented by a span algebra with $\delta_1 \in \mathcal{M}$, and we shall simply identify partial algebras with (any of the) representing span algebras. In this sense homomorphisms of partial algebras are exactly the span-algebra morphisms.

On the other hand, span algebras are much more general than partial algebras—we return to this after introducing relations in a category (see V.2.8).

Each F-algebra (Q, δ) will be identified with the span algebra $(Q, QF, \delta, 1_{QF})$ and each homomorphism f with the span-algebra morphism (f, fF). We often write simply $f: (Q, D, \delta_0, \delta_1) \to (Q', \delta')$ instead of $(f, \delta_1 \cdot fF): (Q, D, \delta_0, \delta_1) \to (Q', Q'F, \delta', 1_{Q'F})$.

Definition. The *free completion* of a span F-algebra A is a (total) F-algebra $A^{\#}$ together with a span-algebra morphism $\eta: A \to A^{\#}$ which has the following universal property:

> For each F-algebra B and each span-algebra morphism $f: A \to B$ there exists a unique homomorphism $f^{\#}: A^{\#} \to B$ with $f = \eta \cdot f^{\#}$.

7.5. Free-completion construction. Let $\mathcal{K}$ be a cocomplete category and let $F: \mathcal{K} \to \mathcal{K}$ be a functor. For each span F-algebra

$$A = (Q, D, \delta_0, \delta_1)$$

we define two transfinite chains

$$V, W: \mathbf{Ord} \to \mathcal{K}$$

and a natural transformation

$$\rho: V \to W$$

by the following transfinite induction.

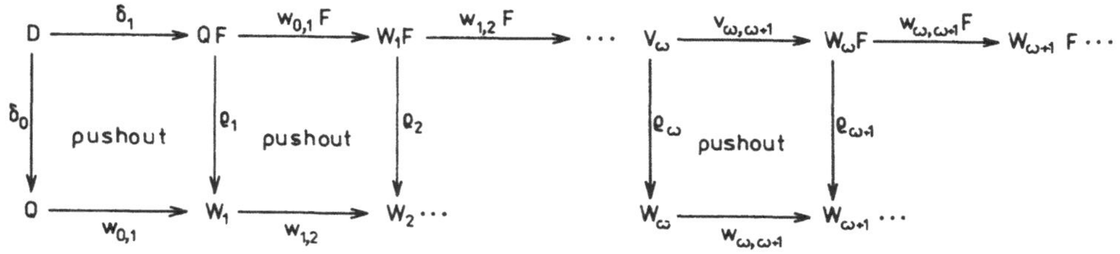

(a) First step:

$$V_0 = D;\ V_1 = QF \text{ and } v_{0,1} = \delta_1;$$
$$W_0 = Q;\ \rho_0 = \delta_0.$$

(b) Isolated step. For each ordinal n put

$$V_{n+1} = W_n F,$$
$$v_{n+1,n+2} = w_{n,n+1} F\colon W_n F \to W_{n+1} F,$$

and let the following be a pushout (defining $w_{n,n+1}$ and ρ_{n+1}):

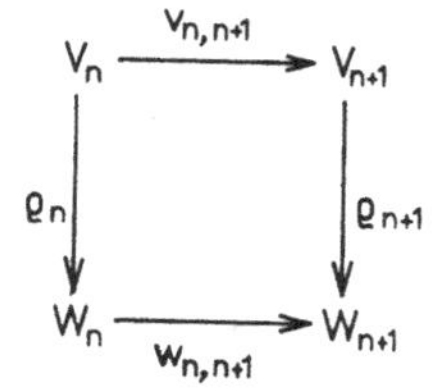

(c) Limit step. For each limit ordinal i:

$$V_i = \operatorname{colim}\, (V_n;\ v_{n,m})_{n \le m < i};$$
$$W_i = \operatorname{colim}\, (W_n;\ w_{n,m})_{n \le m < i};$$
$$\rho_i = \operatorname*{colim}_{n<i} \rho_n.$$

Remark. The above definition is "complete". In the first step, ρ_0 and $v_{0,1}$ are given. By (b), their pushout defines

$$\rho_1,\ w_{0,1} \text{ and } v_{1,2} = w_{0,1}F.$$

Next, the pushout of ρ_1 and $v_{1,2}$ defines

$$\rho_2,\ w_{1,2} \text{ and } v_{1,2} = w_{1,2}F,$$

etc. Moreover $v_{0,2} = v_{0,1} \cdot v_{1,2}$, etc. Hence, we obtain all ρ_n, $v_{n,m}$ and $w_{n,m}$ with $n \le m < \omega$. This yields, by colimit,

$$V_\omega,\ W_\omega \text{ and } \rho_\omega\colon V_\omega \to W_\omega.$$

Next, to obtain

$$w_{\omega,\omega+1}\colon V_\omega \to W_\omega F,$$

we use the fact that $V_\omega = \operatorname{colim} V_n$, and for each $n < \omega$, $v_{n+1,\omega}$ $v_{\omega,\omega+1} = v_{n+1,\omega+1} = w_{n,\omega}F\colon W_n F \to W_\omega F$. Then the pushout of ρ_ω and $v_{\omega,\omega+1}$ defines

$$\rho_{\omega+1},\ w_{\omega,\omega+1} \quad \text{and} \quad v_{\omega+1,\omega+2} = w_{\omega,\omega+1}F,$$

etc.

7.6. Example. Let $(Q, \bot, \delta_0, m)$ be the discrete algebra (IV.7.2). The first pushout is a coproduct of Q and QF:

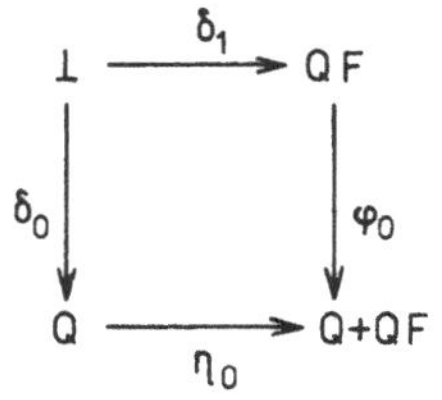

where η_0 and φ_0 are the coproduct injections, see Remark IV.3.2. (Proof: Any square with $\bot$ in the upper left-hand corner commutes, and hence the universal property of pushouts coincides with that of coproducts.) Also the next pushouts are obtained from the free-algebra construction, since the following square

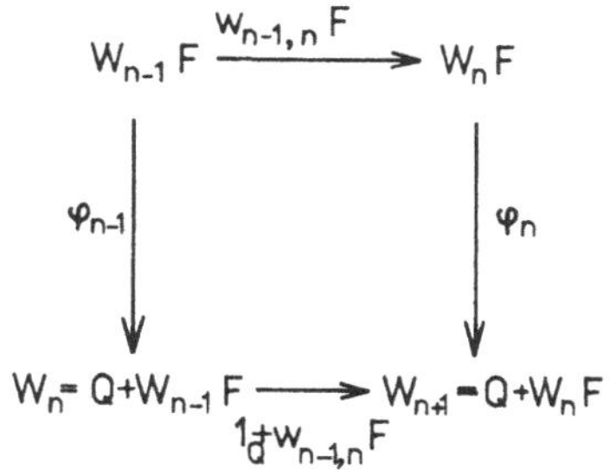

is a pushout. [Proof: Let $p: W_n \to P$ and $q: W_nF \to P$ be morphisms with $\varphi_{n-1} \cdot p = w_{n-1,n}F \cdot q$. The unique morphism $k: W_{n+1} \to P$ with $p = (1_Q + w_{n-1,n}F) \cdot k$ and $q = \varphi_n \cdot k$ has components

$$w_{0,n} \cdot p: Q \to P \quad \text{and} \quad q: W_nF \to P.]$$

Finally, the limit step is obtained by a colimit, analogously as in the free-algebra construction.

Thus, the chain W in the free-completion construction of a discrete algebra is just the free-algebra construction.

Definition. The free-completion construction is said to *stop after k steps* if both $v_{k,k+1}$ and $w_{k,k+1}$ are isomorphisms.

Remark. (i) If $v_{k,k+1}$ is an isomorphism, then the construction stops after k steps because the pushout of ρ_k and $v_{k,k+1}$ is then formed by $\rho_{k+1} = v_{k,k+1}^{-1} \cdot \rho_k$ and $w_{k,k+1} = 1_{W_k}$.

If $w_{k,k+1}$ is an isomorphism, then the construction stops after $k+1$ steps because $v_{k+1,k+2} = w_{k,k+1}F$ is an isomorphism. This is in particular the case whenever k is a limit ordinal such that F preserves the colimit $W_k = \operatorname{colim}_{n<k} W_n$.

(ii) We are going to prove that in case the free-completion construction stops after k steps, the free completion of the given span algebra is $(W_k, \nu_{k,k+1}^{-1} \cdot \rho_k)$. This can be proved directly by induction. We are going to prove this by translating the present construction as the initial-algebra construction.

7.7. Free completion as the least fixed point. Let $A = (Q, D, \delta_0, \delta_1)$ be a span F-algebra. We define a functor G whose least fixed point is the free completion of A. Denote by $\mathscr{R}_A$ the following category:

Objects are pairs (B, f) where $B = (X, Y, x_0, x_1)$ is a span F-algebra (i.e., $x_0 : Y \to X$ and $x_1 : Y \to XF$) and $f = (f_0, f_1)$ is a span-algebra morphism $f: A \to B$, i.e., the following diagram

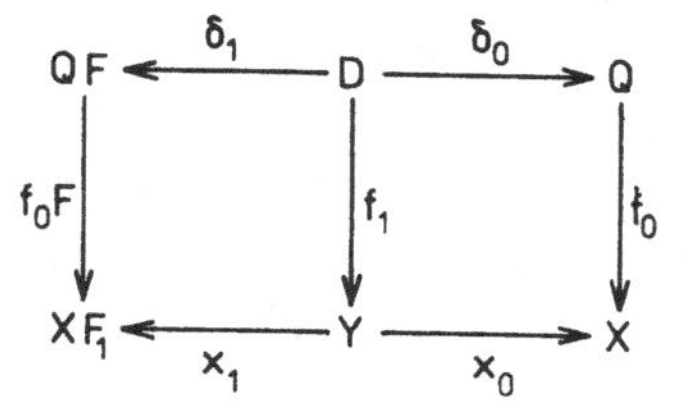

commutes;

Morphisms from (B, f) to (B', f') are such morphisms $g = (g_0, g_1) : B \to B'$ of span algebras which fulfil $f' = f \cdot g$.

The formation of pushouts which is used in the free-completion construction gives rise to the following functor

$$G: \mathscr{R}_A \to \mathscr{R}_A.$$

For each object $(B, f) = (X, Y, x_0, x_1, f_0, f_1)$ we form the pushout of x_0 and x_1:

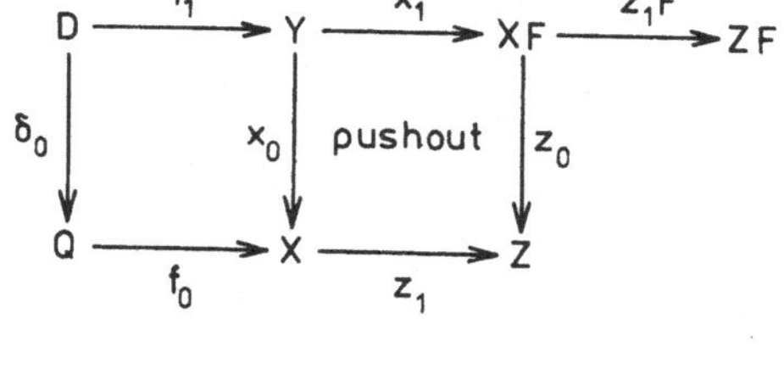

and we put

$$(B, f)G = (Z, XF, z_0, z_1F, f_0 \cdot z_1, f_1 \cdot x_1).$$

For each morphism

$$h = (h_0, h_1) : (B, f) \to (B', f')$$

there exists a unique morphism $k: Z \to Z'$ in $\mathscr{K}$ for which the following diagram

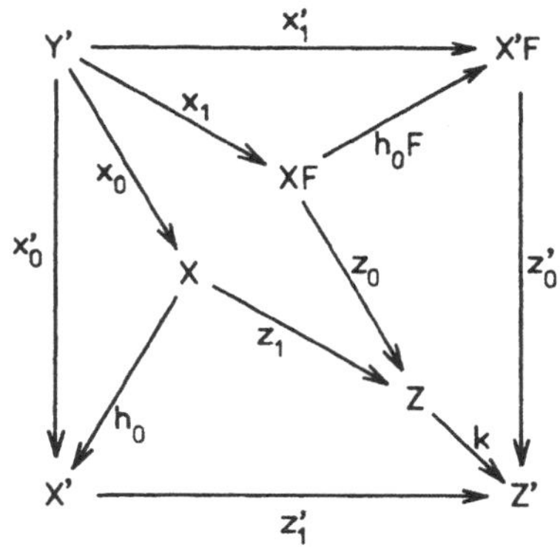

commutes (where z_0' and z_1' are morphisms forming the pushout of x_0' and x_1'). Put

$$hG = (k, h_0F).$$

It is easy to verify that G is a well-defined functor.

Proposition. Let (B, f) be the least fixed point of the functor G. Then B is a (total) F-algebra and the free F-completion of A is

$$f: A \to B.$$

Proof. (i) We prove that B is an F-algebra. Let

$$\alpha: (B, f)G \to (B, f)$$

be the isomorphism for which $((B, f), \alpha)$ is the least fixed point of G. Put

$$(B, f) = (X, Y, x_0, x_1, f_0, f_1)$$

and define

$$\dot{h} = (z_1 \cdot \alpha_0, x_1 \cdot \alpha_1).$$

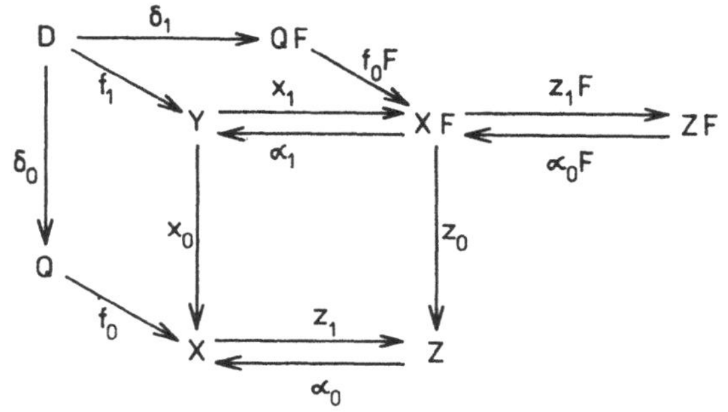

We are going to prove that

$$h:(B, f) \to (B, f)$$

is a morphism of $\mathscr{R}_A$ such that

(1) $\alpha \cdot h = hG \cdot \alpha.$

By the definition of the least fixed point, such h is unique—since also $1_{(B,f)}$ has these properties, we shall conclude that

$$h = 1_{(B,f)}.$$

This means that $z_1 \cdot \alpha_0 = 1_X$ and $x_1 \cdot \alpha_1 = 1_Y$. Since α_0 and α_1 are morphisms, we shall have

(2) $z_1 = \alpha_0^{-1}$ and $x_1 = \alpha_1^{-1}$.

Thus, we shall conclude that x_1 is an isomorphism and B is a total F-algebra (by Remark IV.7.4).

To prove that

$$h: (B, f) \to (B, f)$$

is a morphism in $\mathscr{R}_A$, it is sufficient to inspect the following diagram in $\mathscr{K}$ (which commutes, since α is a morphism in $\mathscr{R}_A$):

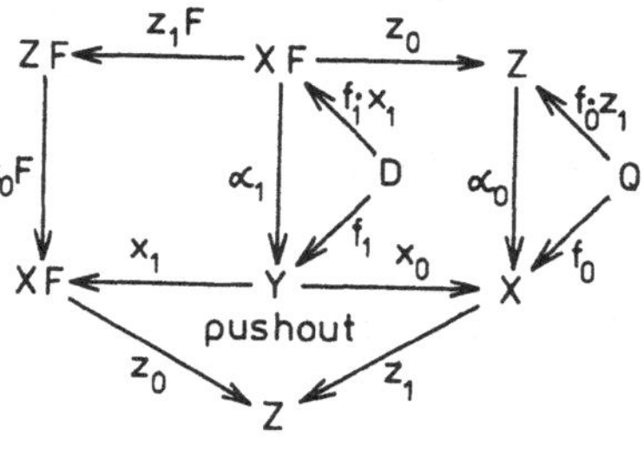

Further, by the definition of G we have

$$hG = (k, h_0F)$$

where $k: Z \to Z$ is defined by

$$z_0 \cdot k = h_0F \cdot z_0 = z_1F \cdot \alpha_0F \cdot z_0 = z_0 \cdot (\alpha_0 \cdot z_1)$$

and

$$z_1 \cdot k = h_0 \cdot z_1 = z_1 \cdot (\alpha_0 \cdot z_1),$$

which implies

$$k = \alpha_0 \cdot z_1.$$

Thus, $hG = (\alpha_0 \cdot z_1,\ z_1F \cdot \alpha_0 F)$, and hence (1) means that

$$\alpha_0 \cdot h_0 = \alpha_0 \cdot z_1 \cdot \alpha_0 \quad \text{and} \quad \alpha_1 \cdot h_1 = z_1F \cdot \alpha_0 F \cdot \alpha_1,$$

both of which are easy to check.

(ii) For each F-algebra $\bar{B}$ and each span-F-algebra morphism $g: A \to \bar{B}$, we prove that there is a unique F-homomorphism $g^{\#}: B \to \bar{B}$ with $g = f \cdot g^{\#}$. Representing $\bar{B}$ as

$$\bar{B} = (\bar{Q}, \bar{Q}, \bar{\delta}, 1_{\bar{Q}})$$

and considering the following pushout

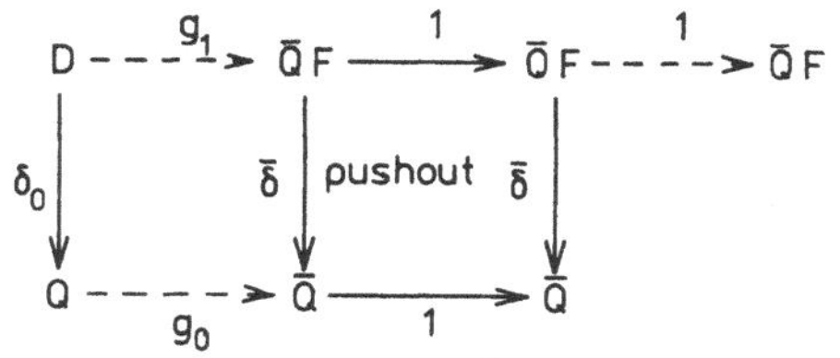

we conclude that the object $(\bar{B}, g)$ of $\mathscr{R}_A$ fulfils

$$(\bar{B}, g)G = (\bar{B}, g).$$

Thus, $(\bar{B}, g)$ is a fixed point of G with the corresponding morphism $1_{(\bar{B}, g)}$. Therefore, there exists a unique G-homomorphism $g^{\#}: ((B, f), \alpha) \to ((\bar{B}, g), 1)$, i.e., a unique F-homomorphism

$$g^{\#}: B \to \bar{B}$$

such that

$$g = f \cdot g^{\#} \quad \text{and} \quad \alpha \cdot g^{\#} = g^{\#}G.$$

To conclude the proof (of the unicity of g), it is sufficient to show that the last equation is superfluous: each F-homomorphism $g^{\#}: B \to \bar{B}$ fulfils

$$\alpha \cdot g^{\#} = g^{\#}G.$$

By remark IV.7.4, we have $g^{\#} = (g_0^{\#},\ x_1 \cdot g_0^{\#}F)$. Then $g^{\#}G = (k,\ g_0 F)$, where k is defined by the commutativity of the following diagram:

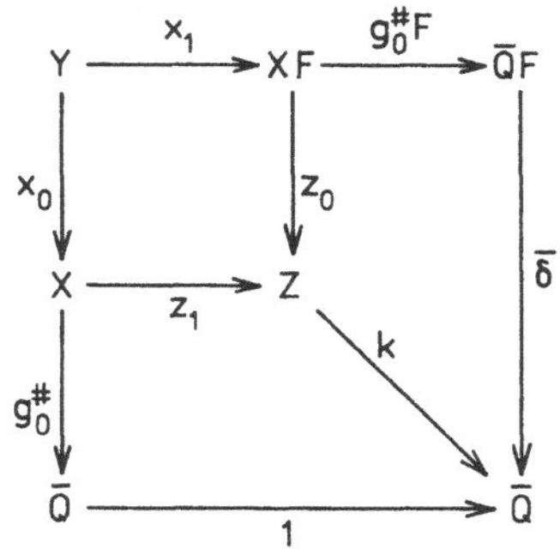

Using (2), we see that

$$k = z_1^{-1} \cdot g_0^{\#} = \alpha_0 \cdot g_0^{\#} \quad \text{and} \quad g_0^{\#} F = \alpha_1 \cdot x_1 \cdot g_0^{\#} F,$$

which implies

$$g^{\#} G = (\alpha_0^{\#} \cdot g_0, \alpha_1 \cdot x_1 \cdot g_0^{\#} F) = \alpha \cdot g^{\#}. \qquad \square$$

7.8. The initial-algebra construction over G is precisely the free-completion construction for A over F. In fact, the initial object of $\mathscr{R}_A$ is clearly

$$\perp \, = (A, 1_A).$$

The next step $\perp G$ is given by the pushout of δ_0 and δ_1

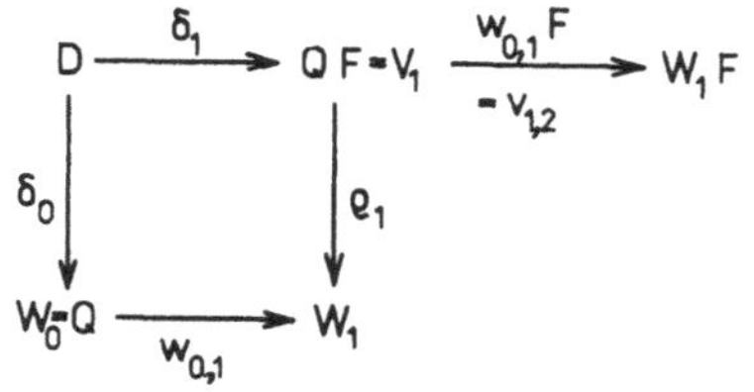

as

$$\perp G = (W_1, V_1, \rho_1, v_{1,2};\ w_{0,1}, v_{0,1})$$

with the connecting morphism

$$(w_{0,1}, v_{0,1})\colon \perp \rightarrow \perp G.$$

Analogously, $\perp G^2$ is obtained from the following pushout

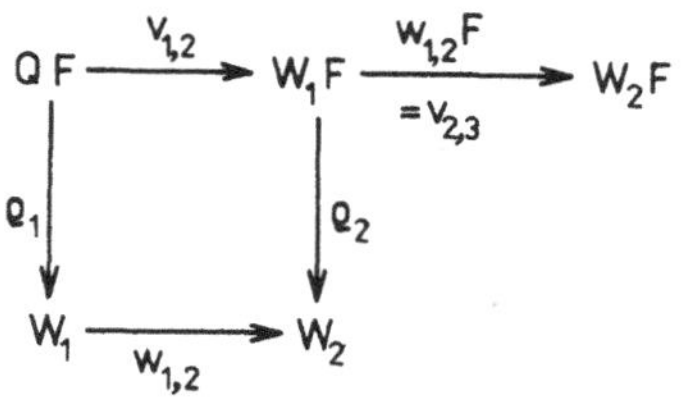

to get

$$(w_{1,2}, v_{1,2})\colon \perp G \rightarrow \perp G^2 = (W_2, V_2, \rho_2, v_{2,3};\ w_{0,2}, v_{0,2}),$$

etc. In order to verify that $\perp G^{\omega}$ also represents the ω-step in the free-completion construction (and, more generally, $\perp G^n$ the n-th step for any ordinal n), we need a characterization of chain-colimits in $\mathscr{R}_A$:

Observation. $\mathscr{R}_A$ is a chain-cocomplete category with coordinate-wise formation of α-colimits (for any limit ordinal α). That is, given an α-chain $h_{n,m}\colon (B_n, f_n) \rightarrow (B_m, f_m)$ $(n \leq m < \alpha)$ with $B_n = (X_n, Y_n, x_{0,n}, x_{1,n})$, then the colimit in $\mathscr{R}_A$ is

$$B = (\operatorname*{colim}_{n<\alpha} X_n,\ \operatorname*{colim}_{n<\alpha} Y_n,\ x_0,\ x_1;\ \bar{f}_0,\ \bar{f}_1)$$

where the morphisms in B are determined by the fact that the following diagram

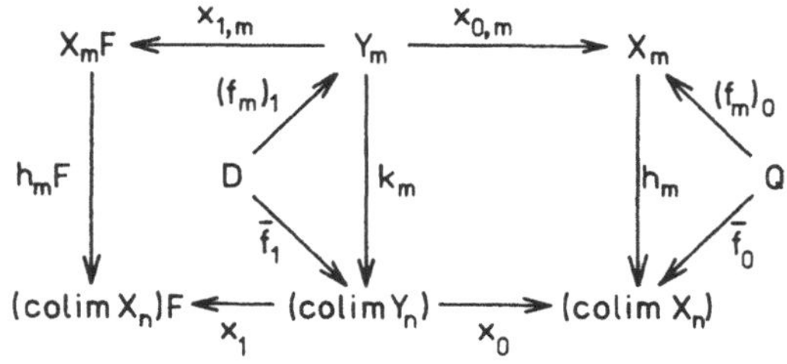

commutes for each $m < \alpha$. Here h_m and k_m denote the respective colimit injections in $\mathscr{K}$, and the colimit injections in $\mathscr{R}_A$ are

$$(h_m, k_m)\colon B_m \to B\ (m < \alpha).$$

Corollary. If the free-completion constructions stops after k steps, then the F-algebra $A^\# = (W_k, \rho_k \cdot w_{k,k+1}^{-1})$ is the free completion of A with respect to

$$w_{0,k}\colon A \to A^\#.$$

In fact, if the free-completion construction stops after k steps, so does the initial-algebra construction over G and hence, $\perp G^k = (A^\#, w_{0,k})$ is the initial G-algebra.

7.9. We say that a functor F has *constructive free completions* if the free-completion construction stops for each span F-algebra.

Theorem. Let $\mathscr{K}$ be a cocomplete, $\mathscr{E}$-cowell-powered $(\mathscr{E}, \mathscr{M})$-category. A functor $F\colon \mathscr{K} \to \mathscr{K}$ preserving $\mathscr{E}$-epis has constructive free completions iff F is a constructive varietor.

Proof. By Example IV.7.6, if F has constructive free completions, then F is a constructive varietor. Conversely, let F be a constructive varietor preserving $\mathscr{E}$-epis. For each span F-algebra $(Q, D, \delta_0, \delta_1)$ denote by

$$\rho\colon V \to W$$

the free-completion construction and, for distinction, by

$$\bar{W}$$

the free-algebra construction over Q. We define a compatible collection of $\mathscr{E}$-epis

$$e_n\colon \bar{W}_n \to W_n;$$

since $\bar{W}$ stops, W also stops by Lemma IV.2.7.

(a) First step: $e_0 = 1_Q : Q \to Q$.

(b) Isolated step: the components of

$$e_{n+1} : Q + \bar{W}_n F \to W_{n+1}$$

are

$$w_{0,n+1} : Q \to W_{n+1} \quad \text{and} \quad e_n F \cdot \rho_{n+1} : \bar{W}_n F \to W_{n+1}.$$

(Here $\rho_{n+1} : V_{n+1} = W_n F \to W_{n+1}$.)

(c) Limit step: the components of

$$e_i : \lim_{n<i} \bar{W}_n \to \lim_{n<i} W_n$$

are $e_n \cdot w_{n,i}\,(i < n)$.

We prove the compatibility. Using Remark IV.2.4, we prove that the following squares

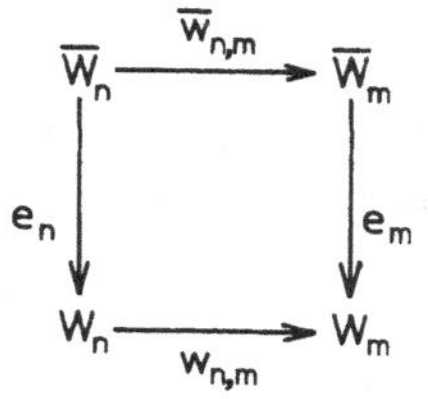

commute for all $n < m$.

(a) $m = 1$: clearly $\bar{w}_{0,1} \cdot e_1 = w_{0,1}$.

(b$_1$) If the square above commutes, then the following diagram

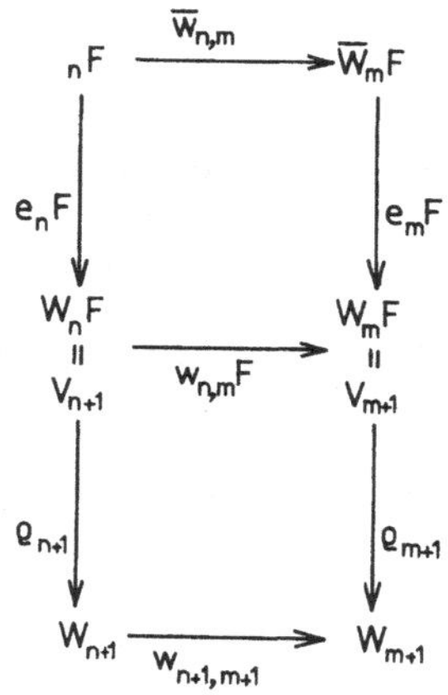

also commutes. This shows that the morphisms

$$\bar{w}_{n+1,m+1} \cdot e_{m+1};\; e_{n+1} \cdot w_{n+1,m+1} : Q + \bar{W}_n F \to W_{m+1}$$

have identical second components. The first components of these two morphisms are also identical (viz, $w_{0,\,m+1}$).

(b_2) If the squares above commute for all $n < n_0$, where n_0 is a limit ordinal, then they commute for n_0 too, because $\bar{W}_{n_0} = \operatorname{colim}_{n<n_0} \bar{W}_n$.

(c) The limit step in m follows from the definition of e_i.

It remains to prove that each e_n is in $\mathscr{E}$:

(a) $e_0 = 1_Q$.

(b) Let $e_n \in \mathscr{E}$, and let

$$e_{n+1} = \bar{e} \cdot \bar{m}$$

be an image factorization, $\bar{e}\colon \bar{W}_{n+1} \to Z$ in $\mathscr{E}$ and $\bar{m}\colon Z \to W_{n+1}$ in $\mathscr{M}$. We have

$$e_n F \cdot \rho_{n+1} = \varphi_n \cdot e_{n+1},$$

where $\varphi_n\colon \bar{W}_n F \to \bar{W}_{n+1} = Q + \bar{W}_n F$ is the coproduct injection. Since e_n and $e_n F$ are $\mathscr{E}$-epis, we can use the diagonal fill-in twice:

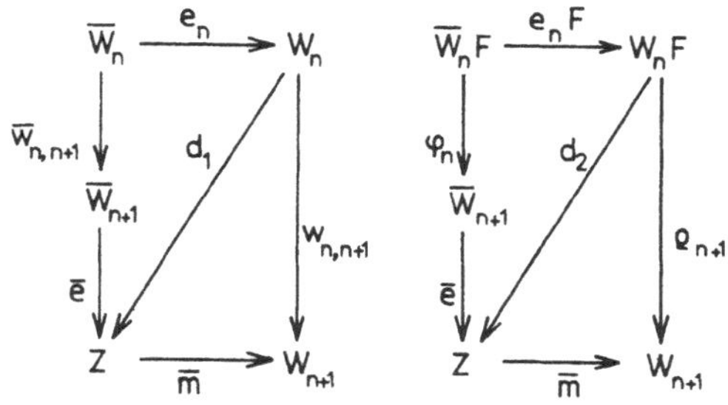

Thus, both $w_{n,\,n+1}$ and ρ_{n+1} factor through the subobject $\bar{m}$. Since these two morphisms form a pushout (of $v_{n,\,n+1}$ and ρ_n), we have $\operatorname{im} w_{n,\,n+1} \cup \operatorname{im} \rho_{n+1} = 1$ by Remark III.5.4. Therefore, $\bar{m}$ is an isomorphism, and this proves $e_{n+1} \in \mathscr{E}$.

(c) The limit step follows from Exercise III.5.B. This concludes the proof. □

7.10. We prove now an important criterion for a functor to have constructive free completions: the preservation of colimits of α-chains of subobjects (for some infinite cardinal α). We know that the preservation of all α-colimits is sufficient for the stopping of the initial-algebra construction (IV.3.5), and it can be easily seen that this is sufficient for the free-completion construction too. Nevertheless, the preservation of colimits of α-chains of $\mathscr{M}$-monos is a considerably milder condition which is often rather easy to verify (because colimits of chains of monos are usually unions).

Definition. Let α be an infinite cardinal. A functor *has rank* α iff it preserves colimits of α-chains of $\mathscr{M}$-monos.

Examples. (i) The functor H_n: **Set** $\rightarrow$ **Set** has rank α iff the cofinality of α is larger than n. In fact, if $\mathrm{cof}(\alpha) > n$, then for each union

$$M = \bigcup_{t<\alpha} M_t$$

of an α-chain $M_0 \subset M_1 \subset \ldots$ and for each element $\{x_i\}_{i<n} \in MH_n = M^n$ there exists $t_0 < \alpha$ such that $x_i \in M_{t_0}$ for all $i < n$. Thus, $\{x_i\} \in M^n_{t_0}$ and this proves

$$MH_n = \bigcup_{t<\alpha} M_t H_n.$$

Conversely, if $\mathrm{cof}(\alpha) \leq n$ and if the union above is strictly increasing, we can find elements $x_i \in M$ for $i < n$, such that for each $t < \alpha$ there exists i with $x_i \notin M_t$. Then $\{x_i\} \in M^n - \bigcup_{t<\alpha} M^n_t$ and hence, H_n does not preserve the union.

(ii) For each finitary type Σ, the rank of H_Σ: **Set** $\rightarrow$ **Set** is ω. If Σ is infinitary, the rank of H_Σ is any infinite regular cardinal larger than all arities.

(iii) Let $\mathcal{K}$ be a concrete category (III.3.8) with both chain-colimits and products preserved by the forgetful functor U and with $\mathcal{M} = \{m; mU \text{ is one-to-one}\}$. (These hypotheses on $\mathcal{K}$ are rather mild, e.g., all varieties of finitary algebras and **Pos, Top**, **Gra** fulfil them.) Then $H_n: \mathcal{K} \rightarrow \mathcal{K}$ has rank α iff $\mathrm{cof}(\alpha) > n$. This follows from (i) above.

7.11. Theorem. Let $\mathcal{K}$ be a cocomplete, $\mathcal{E}$-cowell-powered $(\mathcal{E}, \mathcal{M})$-category. Each functor $F: \mathcal{K} \rightarrow \mathcal{K}$ with a rank has constructive free completions.

Proof. I. For each span algebra A we form the free-completion construction $\rho: V \rightarrow W$. Let us consider the image factorizations

$$w_{n,m} = e_{n,m} \cdot j_{n,m} \qquad (n \leq m)$$

with $e_{n,m}: W_n \rightarrow J_{n,m}$ in $\mathcal{E}$ and $j_{n,m}: J_{n,m} \rightarrow W_n$ in $\mathcal{M}$. Since $\mathcal{K}$ is $\mathcal{E}$-cowell-powered, for each ordinal n there exists an ordinal n^* such that e_{m,n^*} represents the same quotient of W_n as any $e_{n,m}$ with $m \geq n^*$. Thus, there are isomorphism $s_{n,m}: J_{n,n^*} \rightarrow J_{n,m}$ with

$$e_{n,m} = e_{n,n^*} \cdot s_{n,m} \qquad (m \geq n^*).$$

Define a map t: **Ord** $\rightarrow$ **Ord** by transfinite induction as follows:

$$(0)t = 0;$$
$$(n+1)t = (n)t^*;$$
$$(i)t = \bigvee_{n<i} (n)t \text{ for a limit ordinal } i.$$

We shall prove that F preserves the colimit of the chain

$$\bar{W}: \alpha \to \mathscr{K}$$

defined by

$$\bar{W}_n = W_{(n)t} \quad \text{and} \quad \bar{w}_{n,m} = w_{(n)t,(m)t}.$$

Then F preserves the colimit $W_{(\alpha)t} = \operatorname{colim}_{n < (\alpha)t} W_n$: the set of all $(n)t$ with $n < \alpha$ is cofinal in $(\alpha)t$ and hence, by Exercise III.1.F,

$$W_{(\alpha)t} = \operatorname{colim}_{n < \alpha} W_{(n)t} = \operatorname{colim} \bar{W}.$$

Then the free-completion construction stops after $(\alpha)t + 1$ steps by Remark IV.7.6.

II. To prove that F preserves colim $\bar{W}$, we define an α-chain of $\mathscr{M}$-monos $\bar{J}$ as follows. For each ordinal n put

$$\bar{J}_n = J_{(n)t,(n)t^*}$$

and put, for short,

$$\bar{e}_n = e_{(n)t,(n)t^*}: \bar{W}_n \to \bar{J}_n;$$
$$\bar{j}_n = j_{(n)t,(n)t^*}: \bar{J}_n \to \bar{W}_{n^*} = W_{(n)t^*}.$$

By the diagonal fill-in, for each $n \leq m$ there is a (unique)

$$\bar{j}_{n,m}: \bar{J}_n \to \bar{J}_m$$

such that the following diagram

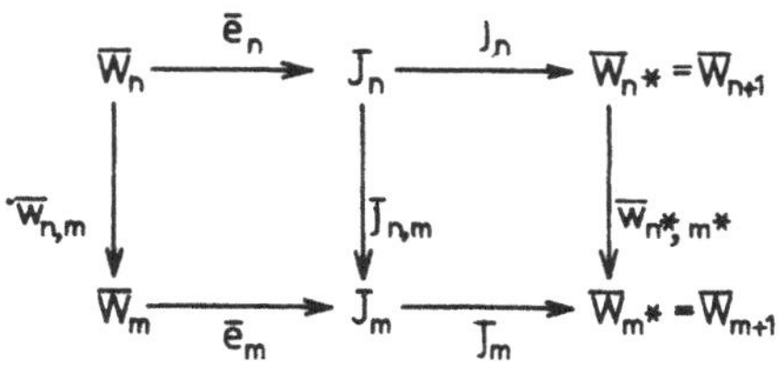

commutes. To prove that

$$\bar{j}_{n,m} \in \mathscr{M} \qquad (n \leq m),$$

it is sufficient to show that

$$\bar{j}_{n,m} \cdot \bar{j}_m = s_{(n)t,(m)t^*} \cdot j_{(n)t,(m)t^*}$$

(the right-hand side morphism is clearly in $\mathscr{M}$, and $\mathscr{M}$ is right cancellative by III.5.1). This equation follows from the fact that $\bar{e}_n$ is an epi, since we have

$$\begin{aligned}\bar{e}_n \cdot \bar{j}_{n,m} \cdot \bar{j}_m &= \bar{w}_{n,m} \cdot \bar{e}_m \cdot \bar{j}_m \\ &= w_{(n)t,(m)t} \cdot w_{(m)t,(m)t^*} \\ &= w_{(n)t,(m)t^*} \\ &= e_{(n)t,(m)t^*} \cdot j_{(n)t,(n)t^*} \\ &= \bar{e}_n \cdot s_{(n)t,(m)t^*} \cdot j_{(n)t,(m)t^*} .\end{aligned}$$

Therefore, $\bar{J}$ is an α-chain of $\mathcal{M}$-monos and hence, F preserves the colimit of $\bar{J}$.

The morphisms $\bar{e}_n$ form a natural transformation

$$\bar{e}: \bar{W} \to \bar{J}$$

(this follows from the definition of $\bar{j}_{n,m}$). Also, if we define $\bar{W}^+ : \alpha \to \mathcal{K}$ by

$$\bar{W}_n^+ = \bar{W}_{n+1} \quad \text{and} \quad \bar{w}_{n,m}^+ = \bar{w}_{n+1,m+1},$$

then the morphisms $\bar{j}_n$ from a natural transformation

$$\bar{j}: \bar{J} \to \bar{W}^+.$$

It follows that $\bar{W}$ and $\bar{J}$ have the same compatible collections (and therefore, the same colimits): if

$$f_n : \bar{W}_n \to X \qquad (n < \alpha)$$

is compatible for $\bar{W}$, then

$$\bar{j}_n \cdot f_{n+1} : \bar{J}_n \to X \qquad (n < \alpha)$$

is compatible for $\bar{J}$; conversely, if

$$g_n : \bar{J}_n \to X \qquad (n < \alpha)$$

is compatible for $\bar{J}$, then

$$e_n \cdot g_n : \bar{W}_n \to X \qquad (n < \alpha)$$

is compatible for $\bar{W}$. It follows that

$$(\operatorname{colim} \bar{W})F = (\operatorname{colim} \bar{J})F = \operatorname{colim} \bar{J} \cdot F.$$

Also the diagrams $\bar{W} \cdot F$ and $\bar{J} \cdot F$ are in the same interrelationship: we have natural transformations

$$\bar{e}F: \bar{W} \cdot F \to \bar{J} \cdot F$$

and

$$\bar{j}F: \bar{J} \cdot F \to \bar{W}^+ \cdot F = (\bar{W} \cdot F)^+.$$

Therefore, $\bar{J} \cdot F$ and $\bar{W} \cdot F$ have the same colimit, and we conclude that

$$(\operatorname{colim} \bar{W})F = \operatorname{colim} \bar{J} \cdot F = \operatorname{colim} \bar{W} \cdot F. \qquad \square$$

Corollary. For $\mathcal{K}$ as above, each functor with a rank is a constructive varietor.

7.12. Returning to the *partial* algebras, we are interested to know whether the free-completion $\eta: A \to A^{\#}$ of a partial F-algebra A is an "extension", i.e., $\eta \in \mathcal{M}$. We say that the *pushout axiom* holds if in each pushout

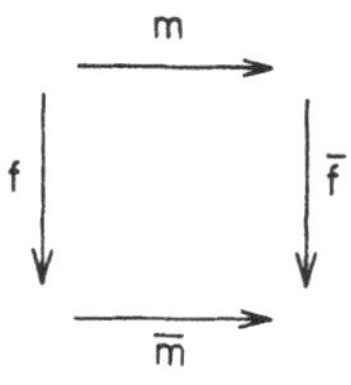

$m \in \mathcal{M}$ implies $\bar{m} \in \mathcal{M}$.

Proposition. Let $\mathcal{K}$ be an $(\mathcal{E}, \mathcal{M})$-category such that $\mathcal{M}$ is a chain-cocomplete class and the pushout axiom holds. Let $F: \mathcal{K} \to \mathcal{K}$ have constructive free completions and preserve $\mathcal{M}$-monos. Then for each partial F-algebra A the free completion $\eta: A \to A^{\#}$ fulfils $\eta \in \mathcal{M}$.

Proof. It is sufficient to prove that in the free-completion construction of A all $w_{n,m}$ and $v_{n,m}$ are $\mathcal{M}$-monos (which also implies that the colimits defining V_i and W_i for limit ordinals i exist). Since A is a partial algebra, we have $v_{0,1} \in \mathcal{M}$. By the pushout axiom, also $w_{0,1} \in \mathcal{M}$, hence, $v_{1,2} = w_{0,1}F \in \mathcal{M}$, etc. For each ordinal n, assuming $w_{n,n+1} \in \mathcal{M}$, we get $v_{n+1,n+2} = w_{n,n+1}F \in \mathcal{M}$ and hence $w_{n+1,n+2} \in \mathcal{M}$. Using the chain-cocompleteness of $\mathcal{M}$, we can easily check that $w_{n,m} \in \mathcal{M}$. Since F has constructive free completions, the chain W stops, say, after k steps. Then $\eta = w_{0,k} \in \mathcal{M}$. □

Exercises IV.7

A. Free completion of partial groupoids. (i) Let $(X, *)$ be a partial groupoid. Put $W_0 = X$; $W_1 = X + X^2/\sim$ where $\sim$ is the least equivalence with $(x_1, x_2) \sim x$ whenever $x_1 * x_2 = x$; etc. Verify that $\bigcup_{n=0}^{\infty} W_n$ is the free completion of $(X, *)$. Compare this with the free-completion construction.

(ii) Do the same with a relational groupoid, i.e, a pair $(X, *)$ where $*$ is a ternary relation on X.

B. Free completion of many-sorted algebras. For each type Σ of many-sorted algebras verify that H_Σ has a rank, and conclude that each partial Σ-algebra has a free completion which is its extension.

C. Free completions in Set and R-Vect. Characterize functors F: **Set** → **Set** and F: R-**Vect** → R-**Vect** which have constructive free completions. (Hint: IV.7.9.)

D. Functors with a rank. (i) Prove that if a functor $F: \mathcal{K} \to \mathcal{K}$ is a coproduct of functors with rank α, then F has rank α, too.

(ii) Prove that each H_Σ has rank assuming that $\mathcal{K}$ is a concrete category with both chain-colimits and products preserved by the forgetful functor. [Hint: See IV.7.10(iii).]

IV.8. Categories of Algebras

8.1. In the present section we investigate the basic properties of the category

$$F\text{-}\mathbf{Alg}$$

of F-algebras. Recall that

$$U: F\text{-}\mathbf{Alg} \to \mathcal{K}$$

denotes the forgetful functor (III.2.B) defined by $(Q, \delta)U = Q$ and $fU = f$.

Limits of F-algebras are always formed on the level of $\mathcal{K}$-objects. More precisely, if $\mathcal{K}$ has a certain type of limits, then F-**Alg** also has these limits and the forgetful functor U preserves them. For example, let $\mathcal{K}$ have products. Then for an arbitrary collection of F-algebras

$$(Q_i, \delta_i),\ i \in I,$$

we can form the product

$$Q = \prod_{i \in I} Q_i$$

in $\mathcal{K}$ with projections $\pi_i: Q \to Q_i$. By definition of product, there exists a unique morphism

$$\delta: QF \to Q$$

with components $\pi_i F \cdot \delta_i$, i.e., with

$$\delta \cdot \pi_i = \pi_i F \cdot \delta_i \qquad \text{for each } i \in I.$$

The last equation states that

$$\pi_i: (Q, \delta) \to (Q_i, \delta_i) \qquad (i \in I)$$

are homomorphisms. It follows that we obtain the product

$$(Q, \delta) = \prod_{i \in I} (Q_i, \delta_i)$$

in the category F-**Alg**. More generally:

Proposition. Let $D: \mathcal{D} \to F\text{-}\mathbf{Alg}$ be a diagram of algebras and let $Q = \lim D \cdot U$ be a limit in $\mathcal{K}$ with projections $\pi_d: Q \to d(D \cdot U)(d \in \mathcal{D}^0)$. There exists a unique morphism $\delta: QF \to Q$ such that $(Q, \delta) = \lim D$ in F-**Alg** with the same projections.

Proof. For each $d \in \mathscr{D}^0$ put $dD = (Q_d, \delta_d)$. The family

$$\pi_d F \cdot \delta_d : QF \to Q_d \qquad (d \in \mathscr{D}^0)$$

is compatible with the diagram $D \cdot U$: for each $f: d \to d'$ in $\mathscr{D}$ we know that $fD: (Q_d, \delta_d) \to (Q'_d, \delta'_d)$ is a homomorphism, i.e.,

$$\delta_d \cdot fD = (fD \cdot F) \cdot \delta_{d'},$$

and that $\pi_{d'} = \pi_d \cdot fD$. It follows that

$$\pi_{d'} F \cdot \delta_{d'} = \pi_d F \cdot (fD \cdot F) \cdot \delta_{d'} = (\pi_d F \cdot \delta_d) \cdot fD.$$

Therefore, there exists a unique morphism

$$\delta: QF \to Q$$

with

$$\delta \cdot \pi_d = \pi_d F \cdot \delta_d \qquad (d \in \mathscr{D}^0).$$

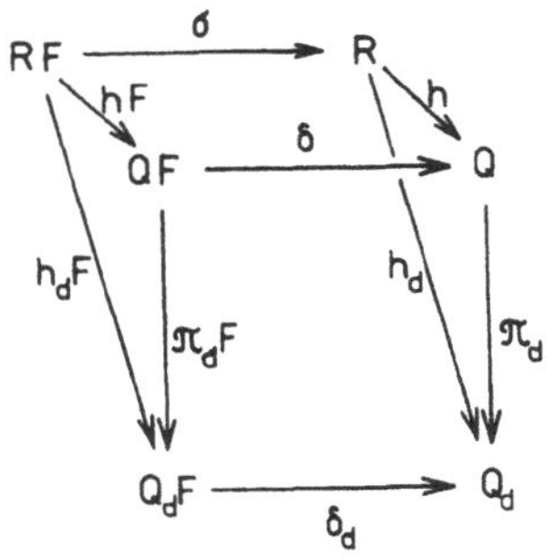

Thus,

$$\pi_d: (Q, \delta) \to (Q_d, \delta_d)$$

are homomorphisms.

Let

$$h_d: (R, \sigma) \to (Q_d, \delta_d) \qquad (d \in \mathscr{D}^0)$$

be a compatible family for D. Then the morphisms $h_d: R \to Q_d$ form a compatible family for $D \cdot U$, and hence, there is a unique

$$h: R \to Q$$

with

$$h \cdot \pi_d = h_d \qquad (d \in \mathscr{D}^0).$$

It remains to prove that h is a homomorphism, i.e.,

$$\sigma \cdot h = hF \cdot \delta.$$

This follows from the fact that each h_d is a homomorphism:

$$(\sigma \cdot h) \cdot \pi_d = \sigma \cdot h_d = h_dF \cdot \delta_d = hF \cdot (\pi_dF \cdot \delta_d) = (hF \cdot \delta) \cdot \pi_d .$$

This concludes the proof. □

Corollary. If $\mathcal{K}$ is complete, then F-**Alg** is complete, and the fortgeful functor preserves limits.

8.2. Colimits of algebras are closely related to free completions.

For example, let $\mathcal{K}$ have coproducts. Given a collection of algebras

$$(Q_i, \delta_i) \qquad\qquad (i \in I),$$

we form the coproducts $\coprod_{i \in I} Q_i$ (with injections p_i) and $\coprod_{i \in I} Q_iF$ (with injections r_i) in $\mathcal{K}$. We have canonical morphisms:

$$\delta_0 \colon \coprod Q_iF \to \coprod Q_i,$$
$$\delta_1 \colon \coprod Q_iF \to (\coprod Q_i)F,$$

i.e., morphisms such that the following diagrams

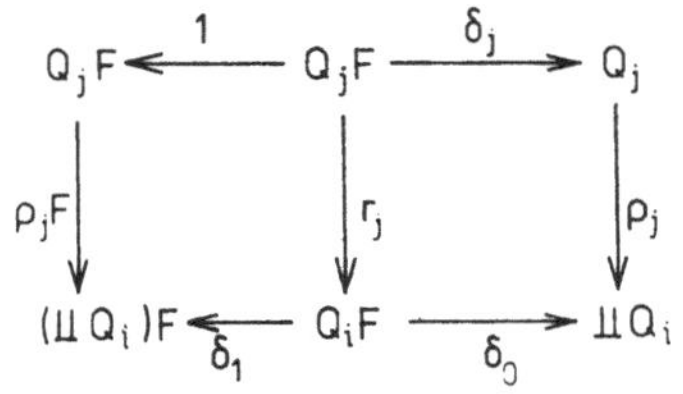

commute for all $j \in J$. Then

$$(\coprod_{i \in I} Q_i, \coprod_{i \in I} Q_iF, \delta_0, \delta_1)$$

is a span F-algebra (IV.7.4).

Proposition. Let

$$\eta \colon (\coprod Q_i, \coprod Q_iF, \delta_0, \delta_1) \to (Q^{\#}, \delta^{\#})$$

be the free completion of the above span algebra. Then

$$(Q^{\#}, \delta^{\#}) = \coprod_{i \in I} (Q_i, \delta_i)$$

is a coproduct in F-**Alg** with injections

$$p_i \cdot \eta \colon (Q_i, \delta_i) \to (Q^{\#}, \delta^{\#}) \qquad\qquad (i \in I).$$

Proof. The above diagrams show that

$$(p_j, r_j): (Q_j, \delta_j) \to (\coprod Q_i, \coprod Q_i F, \delta_0, \delta_1)$$

are morphisms of span algebras. It follows that $p_j \cdot \eta$ are also morphisms $(j \in J)$.

Let

$$f_i: (Q_i, \delta_i) \to (\bar{Q}, \bar{\delta}) \qquad (i \in I)$$

be a family of homomorphisms. There exists a unique morphism

$$g_0: \coprod Q_i \to \bar{Q}$$

with

$$p_j \cdot g_0 = f_j \qquad \text{for each } j \in I,$$

and a unique morphism

$$g_1: Q_j F \to \bar{Q}F$$

with

$$r_j \cdot g_1 = p_j F \cdot g_0 F \qquad \text{for each } j \in I.$$

Then

$$(g_0, g_1): (\coprod Q_i, \coprod Q_i F, \delta_0, \delta_1) \to (\bar{Q}, \bar{\delta})$$

is a span-algebra morphism because each of the following diagrams $(j \in J)$

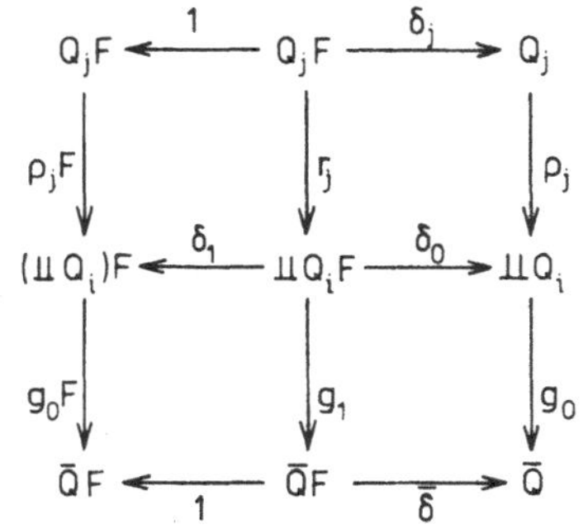

commutes. There exists a unique homomorphism

$$g^{\#}: (Q^{\#}, \delta^{\#}) \to (\bar{Q}, \bar{\delta})$$

with

$$(g_0, g_1) = \eta \cdot g^{\#}.$$

Then

$$f_j = p_j \cdot g_0 = (p_j \cdot \eta) \cdot g^* \qquad \text{for each } j \in J,$$

and such g^* is obviously unique. □

Remark. The same can be easily proved for each diagram D in F-**Alg**: we obtain a natural span algebra

$$(\operatorname{colim} D \cdot U, \operatorname{colim} D \cdot U \cdot F, \delta_0, \delta_1)$$

and its free completion, if it exists, is the colimit of D in the category F-**Alg**.

Corollary. Let $\mathscr{K}$ be a cocomplete category. For each functor $F: \mathscr{K} \to \mathscr{K}$ with constructive free completions the category F-**Alg** is also cocomplete.

Corollary. Let $\mathscr{K}$ be a cocomplete, $\mathscr{E}$-cowell-powered $(\mathscr{E}, \mathscr{M})$-category. Then F-**Alg** is cocomplete whenever F is

(i) a functor with a rank, or
(ii) a constructive varietor preserving $\mathscr{E}$-epis.

See IV.7.11 for (i) and IV.7.9 for (ii).

8.3. The construction of a colimit of F-algebras becomes easy in case the functor F preserves the corresponding type of colimits. For example, let F preserve coproducts. Then the coproduct of F-algebras

$$(Q_i, \delta_i) \qquad (i \in I)$$

is the algebra $(\coprod_{i \in I} Q_i, \delta)$ where δ is defined by

$$\delta = \coprod_{i \in I} \delta_i : (\coprod Q_i)F \to \coprod Q_i$$

[which is meaningful since $(\coprod Q_i)F = \coprod Q_iF$]. More generally:

Proposition. Let $\mathscr{D}$ be a diagram scheme such that $\mathscr{K}$ has colimits of diagrams $\mathscr{D} \to \mathscr{K}$, and F preserves these colimits. Then the category F-**Alg** has colimits of all diagrams $\mathscr{D} \to F$-**Alg**, and they are preserved by the forgetful functor.

Proof. For each diagram $D: \mathscr{D} \to F$-**Alg** with objects $dD = (Q_d, \delta_d)$ for $d \in \mathscr{D}^\circ$ we form the colimit $Q = \operatorname{colim} D \cdot U$ in $\mathscr{K}$ with injections $\varepsilon_d: Q_d \to Q$ $(d \in \mathscr{D}^\circ)$. The collection

$$\delta_d \cdot \varepsilon_d : Q_dF \to Q \qquad (d \in \mathscr{D}^\circ)$$

is clearly compatible for $D \cdot U \cdot F$ and hence, there exists a unique morphism $\delta: QF = \operatorname{colim} D \cdot U \cdot F \to Q$ with

$$\delta_d \cdot \varepsilon_d = \varepsilon_d F \cdot \delta \qquad (d \in \mathscr{D}^\circ),$$

i.e., such that $\varepsilon_d: (Q_d, \delta_d) \to (Q, \delta)$ are homomorphisms. Then $(Q, \delta) = \operatorname{colim} D$ in the category F-**Alg** with injections ε_d. In fact, given a compatible collection

$$f_d: (Q_d, \delta_d) \to (Q', \delta') \qquad (d \in \mathscr{D}^o),$$

there exists a unique morphism $f: Q \to Q'$ in $\mathscr{K}$ with components f_d. Then f is a homomorphism, i.e.,

$$\delta \cdot f = fF \cdot \delta' : QF \to Q'$$

because $QF = \operatorname{colim} Q_dF$ and for each $d \in \mathscr{D}^o$ we have

$$\begin{aligned} \varepsilon_d F \cdot (\delta \cdot f) &= \delta_d \cdot \varepsilon_d \cdot f \\ &= \delta_d \cdot f_d \\ &= f_dF \cdot \delta' \\ &= \varepsilon_d F \cdot (fF \cdot \delta'). \end{aligned}$$ □

Corollary. Let $\mathscr{K}$ be cocomplete and let F be a co-adjoint. Then F-**Alg** is cocomplete, with colimits preserved by U.

Example. Unary algebras have colimits preserved by the forgetful functor: this is the case of $\mathscr{K}$ = **Set** and of the co-adjoint $F = S_\Sigma$. In Exercise D below we show that no other algebras in **Set** have colimits preserved by U. In fact, not even finite colimits.

8.4. Counterexample. We now present a finitary varietor

$$F: \mathbf{Pos} \to \mathbf{Pos}$$

for which F-**Alg** is not cocomplete. In fact, we will find an algebra (Q, δ) such that the coproduct $(Q, \delta) + (Q, \delta)$ fails to exist.

We denote by P the power-set functor (III.3.4).

For each poset A denote by $A^{(3)}$ the set of all 3-chains in A, plus an additional point ξ (which is not any 3-chain). That is,

$$A^{(3)} = \{(x, y, z) \in A^3;\ x < y < z\} \cup \{\xi\}.$$

On the set $A^{(3)}P$ of all subsets of $A^{(3)}$ we consider the following (trivial) order:

$$X \preceq Y \quad \text{iff} \quad X = Y \text{ or } X = \emptyset \qquad \text{for all } X, Y \subset A^{(3)}.$$

This defines a functor F: **Pos** → **Pos** on objects:

$$AF = A^{(3)}P.$$

Define it on morphisms $f: A \to B$ by

$$fF = f^{(3)}P$$

where the map $f^{(3)}: A^{(3)} \to B^{(3)}$ is defined by

$$(x, y, z)f^{(3)} = \begin{cases} ((x)f, (y)f, (z)f) & \text{if } (x)f < (y)f < (z)f, \\ \xi & \text{else.} \end{cases}$$

It is easy to check that F is a well-defined functor.

F is a finitary varietor: indeed, the free-algebra construction stops after one step. We have

$$W_0 = I;$$
$$W_1 = I + I^{(3)}P,$$

and thus, W_1 is a disjoint union of I and $I^{(3)}P$. Since the latter poset has no 3-chains, clearly $W_1^{(3)} = I^{(3)}$, hence,

$$W_2 = I + W_1^{(3)}P = I + I^{(3)}P = W_1$$

with $w_{1,2} = 1_{W_1}$, etc. Here, W_1 is the free F-algebra.

We exhibit an F-algebra (Q, δ) such that the coproduct $(Q, \delta) + (Q, \delta)$ does not exist in F-**Alg**. Let Q be a 3-chain $Q = \{x, y, z\}$ with $x < y < z$, thus,

$$Q^{(3)} = \{t, \xi\} \text{ where } t = (x, y, z).$$

Let

$$\delta: \{t, \xi\} P \to \{x, y, z\}$$

be the constant map with value z.

Lemma. There does not exist the coproduct

$$(Q, \delta) + (Q, \delta)$$

in the category F-**Alg**.

Proof. Assuming that an F-algebra (R, σ) is the coproduct

$$(R, \sigma) = (Q, \delta) + (Q, \delta)$$

with injections $v_1, v_2: (Q, \delta) \to (R, \sigma)$, we shall derive a contradiction.

We have a pair of homomorphisms $1_Q, 1_Q: (Q, \delta) \to (Q, \delta)$, and by the definition of coproduct, we get a homomorphism $d: (R, \delta) \to (Q, \delta)$ with $1_Q = v_1 \cdot d = v_2 \cdot d$. It follows that v_1 and v_2 are one-to-one maps.

Put

$$z^* = (\emptyset)\sigma \in R.$$

Then

$$(z)v_1 = (z)v_2 = z^*$$

because v_i $(i = 1, 2)$ is a homomorphism, and we have

$$(z)v_i = (\emptyset)\delta \cdot v_i = ((\emptyset)v_i F)\sigma = (\emptyset)\sigma = z^*.$$

It follows that

$$(x)v_1 < (y)v_1 < z^*$$

is a 3-chain in R (because v_1 is one-to-one), and we put

$$t_1 = ((x)v_1, (y)v_1, z^*) \in R^{(3)}.$$

Analogously,

$$t_2 = ((x)v_2, (y)v_2, z^*) \in R^{(3)}.$$

Let us prove that

$$t_1 \neq t_2.$$

Let R_0 be the following poset:

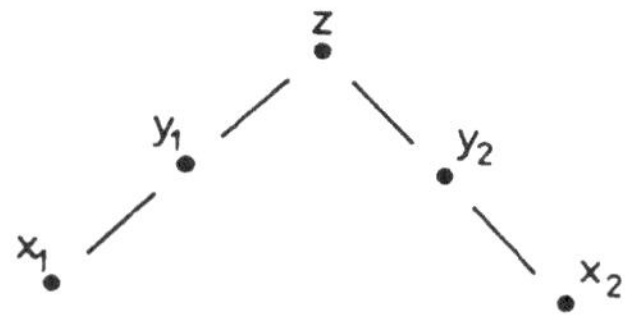

and let $\sigma_0 : R_0F \to R_0$ be the constant map with value z. The two obvious embeddings $w_1, w_2 : Q \to R_0$ are homomorphisms $w_1, w_2 : (Q, \delta) \to (R_0, \sigma_0)$. There exists a homomorphism $w : (R, \sigma) \to (R_0, \delta_0)$ with $w_1 = v_1 \cdot w$ and $w_2 = v_2 \cdot w$. Then

$$(x)v_1 \neq (x)v_2$$

because $((x)v_1)w \neq ((x)v_2)w$, and hence, $t_1 \neq t_2$.

Let us prove that for distinct sets $X, X' \in R^{(3)}P$,

(*) $\quad X \notin \{t_1, \xi\}P \cup \{t_2, \xi\}P$ implies $(X)\sigma \neq (X')\sigma$.

Assuming $(X)\sigma = (X')\sigma = r \in R$, we denote by $\bar{R}$ the poset obtained from R by splitting r to two points. That is

$$\bar{R} = (R - \{r\}) \cup \{r_1, r_2\},$$

where r_1 and r_2 are incompatible points, the ordering of $R - \{r\}$ is as in R, and for each $x \in R - \{r\}$ we have $r_1 \leq x$ iff $r \leq x$ (analogously with $r_2 \leq x$ and $r_1 \geq x$, $r_2 \geq x$). We define order-preserving maps

$$h : \bar{R} \to R \quad \text{and} \quad k : R \to \bar{R}$$

by

$$(x)h = (x)k = x \qquad \text{for each } x \in R - \{r\};$$
$$(r)k = r_1 \quad \text{and} \quad (r_1)h = (r_2)h = r.$$

Note that

$$k \cdot h = 1_R.$$

Define an operation $\bar{\sigma}: \bar{R}^{(3)}P \to \bar{R}$ by

$$(Y)\bar{\sigma} = \begin{cases} r_2 & \text{if } (Y)h^{(3)}P = X \\ (Y)h^{(3)}P \cdot \sigma \cdot k & \text{else.} \end{cases} \qquad (Y \subset \bar{R}^{(3)}).$$

Then

$$v_1 \cdot k, v_2 \cdot k : (Q, \delta) \to (\bar{R}, \bar{\sigma})$$

are homomorphisms: for each $Z \subset Q^{(3)}$ clearly $(Z)v_1^{(3)}P \subset \{t_1, \xi\}$, thus

$$X \neq (Z)v_1^{(3)}P = [(Z)v_1^{(3)}P \cdot k^{(3)}P]h^{(3)}P.$$

Therefore, if $Y = (Z)v_1^{(3)}P \cdot k^{(3)}P$, then

$$\begin{aligned} (Y)\bar{\sigma} &= (Y)h^{(3)}P \cdot \sigma \cdot k \\ &= (Z)v_1^{(3)}P \cdot ((k^{(3)}P \cdot h^{(3)}P) \cdot \sigma \cdot k && [k \cdot h = 1] \\ &= (Z)v_1^{(3)}P \cdot \sigma \cdot k \\ &= (Z)\delta \cdot v_1 \cdot k && [v_1 \text{ a homomorphism}]. \end{aligned}$$

This proves that $(v_1 \cdot k)^{(3)}P \cdot \bar{\sigma} = \delta \cdot (v_1 \cdot k)$, i.e., that $v_1 \cdot k$ is a homomorphism; analogously with $v_2 \cdot k$.

By the definition of coproduct, there exists a homomorphism

$$\bar{k}: (R, \sigma) \to (\bar{R}, \bar{\sigma})$$

with $v_1 \cdot \bar{k} = v_1 \cdot k$ and $v_2 \cdot k = v_2 \cdot \bar{k}$. Also, $h: (\bar{R}, \bar{\sigma}) \to (R, \sigma)$ is a homomorphism, i.e., $\bar{\sigma} \cdot h = h^{(3)}P \cdot \sigma$: given $Y \subset \bar{R}^{(3)}$, then $(Y)h^{(3)}P = X$ implies

$$(Y)\bar{\sigma} \cdot h = (r_2)h = r = (X)\sigma = (Y)h^{(3)}P \cdot \sigma$$

and $(Y)h^{(3)}P \neq X$ implies

$$(Y)\bar{\sigma} \cdot h = (Y)h^{(3)}P \cdot \sigma \cdot k \cdot h = (Y)h^{(3)}P \cdot \sigma.$$

Thus,

$$\bar{k} \cdot h : (R, \sigma) \to (R, \sigma)$$

is an endomorphism with

$$v_i \cdot (\bar{k} \cdot h) = v_i \cdot k \cdot h = v_i \qquad \text{for } i = 1, 2,$$

hence,

$$\bar{k} \cdot h = 1_R.$$

Therefore, $\bar{k}^{(3)}P \cdot h^{(3)}P = 1$ and hence, $[(X)\bar{k}^{(3)}P]h^{(3)}P = X$. It follows that $[(X)\bar{k}^{(3)}P]\bar{\sigma} = r_2$. Since $\bar{k}$ is a homomorphism, i.e., $\sigma \cdot \bar{k} = \bar{k}^{(3)}P \cdot \bar{\sigma}$, we get

$$\begin{aligned} r_2 &= ((X)\bar{k}^{(3)}P)\bar{\sigma} \\ &= (X)\sigma \cdot \bar{k} \\ &= (r)k \end{aligned}$$

$$= (X')\sigma \cdot \bar{k}$$
$$= [(X')\bar{k}^{(3)}P]\sigma.$$

This implies that $Y = (X')\bar{k}^{(3)}P$ fulfils $(Y)h^{(3)}P = X$ [else $(Y)\bar{\sigma} = (Y)h^{(3)}P \cdot \sigma \cdot k \neq r_2$]. Thus,

$$X = (Y)h^{(3)}P = (X')k^{(3)}P \cdot h^{(3)}P = X',$$

which proves (*).

The required contradiction will be obtained by proving that the set

$$K = \{r \in R;\ z^* \leq r\}$$

admits a one-to-one map from $KP - \{\emptyset\}$ into K, and yet K has more than one point. [This is obviously impossible: for any set K with card $K > 1$ we have card $(KP - \{\emptyset\}) = 2^{\text{card } K} - 1 >$ card K.] For each non-empty set $M \subset K$ put

$$\bar{M} = \{(x_1, y_1, m);\ m \in M\} \cup \{(x_2, y_2, z'\} \in R^{(3)}P.$$

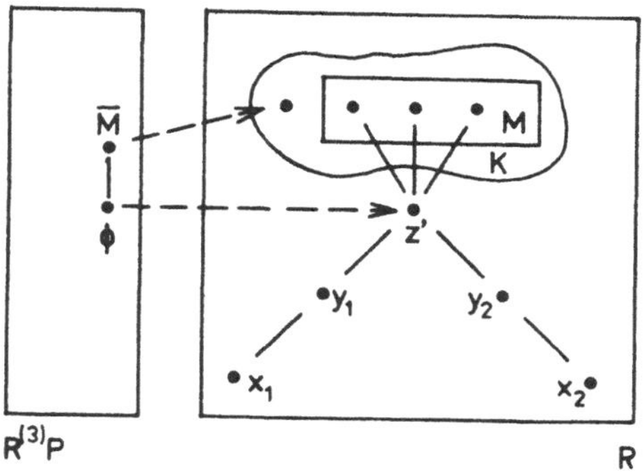

Note that $M \notin \{t_1, \xi\}P \cup \{t_2, \xi\}P$ and $M \neq M'$ implies $\bar{M} \neq \bar{M}'$ for all M, $M' \subset K$. By (*) we conclude that $(\bar{M})\sigma \neq (\bar{M}')\sigma$, and since $\emptyset \preceq M$ in $R^{(3)}P$ implies $z^* = (\emptyset)\sigma \leq (\bar{M})\sigma$, also

$$(\bar{M})\sigma \in K \qquad \text{for each } M \subset K,\ M \neq \emptyset.$$

Thus, $M \mapsto (\bar{M})\sigma$ defines a one to-one map as required. Finally, to prove that K has more than one point, put

$$z_1 = (\{t_1, t_2\})\sigma.$$

Then $z_1 \in K$ because $\emptyset \preceq \{t_1, t_2\}$ in $R^{(3)}P$, and by (*) we have $(\emptyset)\sigma \neq z_1$. Analogously, for

$$z_2 = (\{z_1, t_1, t_2\})\sigma$$

we have $z_2 \in K$ and by (*) again, $z_1 \neq z_2$. This concludes the proof. □

8.5. Next, let us consider image factorizations (III.5) of homomorphisms of F-algebras. Let $F: \mathcal{K} \to \mathcal{K}$ be an arbitrary functor. Let $(\mathcal{E}, \mathcal{M})$ be a factorization system in $\mathcal{K}$ such that F preserves $\mathcal{E}$-epis, i.e.,

$$e \in \mathcal{E} \quad \text{implies} \quad eF \in \mathcal{E}.$$

Then $(\mathcal{E}, \mathcal{M})$ is also a factorization system in the category

F-Alg

of algebras. More precisely, each homomorphism

$$f: (Q, \delta) \to (Q', \delta')$$

factors as a homomorphism in $\mathcal{E}$ followed by a homomorphism in $\mathcal{M}$. To prove this, let $f = e \cdot m$ be an image factorization of f,

$$e: Q \to R \text{ in } \mathcal{E} \quad \text{and} \quad m: R \to Q' \text{ in } \mathcal{M}.$$

We use diagonal fill-in:

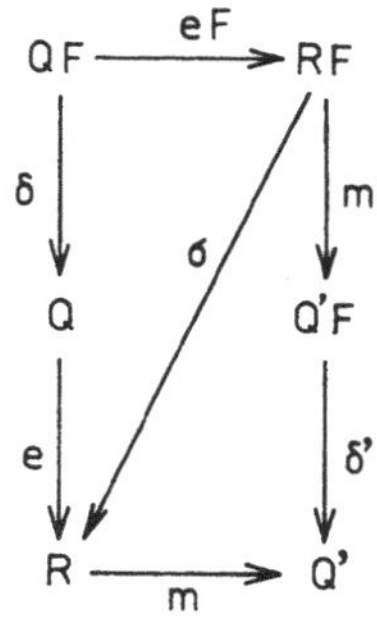

Then (R, σ) is an F-algebra such that

$$e: (Q, \delta) \to (R, \sigma) \quad \text{and} \quad m: (R, \sigma) \to (Q', \delta')$$

are homomorphisms.

$\mathcal{M}$-subobjects in **F-Alg** are called *subalgebras*. Thus, a subalgebra of an F-algebra (Q, δ) is represented by a homomorphism $m: (R, \sigma) \to (Q, \delta)$ with $m \in \mathcal{M}$. Let $m_0: R_0 \to Q$ be an $\mathcal{M}$-subobject of Q. We say that a subalgebra $m: (R, \sigma) \to (Q, \delta)$ is *generated by the subobject* m_0 if m is the least subalgebra with $m_0 \subset m$. We say that m_0 *generates* the algebra (Q, δ) if no proper subalgebra contains m_0, i.e., for each $m: (R, \sigma) \to (Q, \delta)$ in $\mathcal{M}$,

$$m_0 \subset m \text{ implies } m \text{ is an isomorphism.}$$

$\mathcal{E}$-quotients in **F-Alg** are called *quotient algebras*. A quotient algebra of (Q, δ) is represented by a homomorphism $e: (Q, \delta) \to (R, \sigma)$ with $e \in \mathcal{E}$.

Lemma. Let $\mathcal{K}$ be an $(\mathcal{E}, \mathcal{M})$-category with intersections of $\mathcal{M}$-subobjects, and let $F: \mathcal{K} \to \mathcal{K}$ preserve $\mathcal{E}$-epis. Then each $\mathcal{M}$-subobject of each F-algebra (Q, δ) generates a subalgebra of (Q, δ).

Proof. Let $\bar{m}: \bar{R} \to Q$ be an $\mathcal{M}$-mono. Let

$$m_i: (R_i, \sigma_i) \to (Q, \sigma) \qquad (i \in I)$$

be the (possibly large) collection of all subalgebras of (Q, δ) containing $\bar{m}$. Let

$$m^*: R^* \to Q$$

by the intersection of all m_i, $i \in I$. For each $i \in I$ we have a morphism $f_i: R^* \to R_i$ with $m^* = f_i \cdot m_i$. Since $\bar{m} \subset m_i$ $(i \in I)$ implies $\bar{m} \subset m^*$, it is sufficient to exhibit an operation morphism

$$\sigma^*: R^*F \to R^*$$

such that $m^*: (R^*, \sigma^*) \to (Q, \sigma)$ is a subalgebra. Then m^* is clearly generated by $\bar{m}$.

Let

$$m^*F \cdot \delta = \hat{e} \cdot \hat{m}$$

be an image factorization, $\hat{e}: R^*F \to T$ in $\in \mathcal{E}$ and $\hat{m}: T \to Q$ in $\mathcal{M}$. For each $i \in I$ we have

$$\hat{e} \cdot \hat{m} = f_iF \cdot m_iF \cdot \delta = f_iF \cdot \sigma_i \cdot m_i,$$

and we can use the diagonal fill-in:

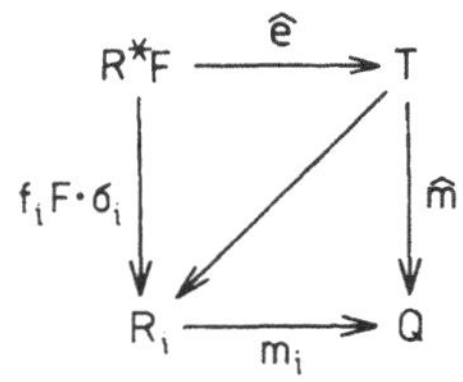

We see that $\hat{m} \subset m_i$ $(i \in I)$ and hence, $\hat{m} \subset m^*$. Let $f: T \to R^*$ fulfil $\hat{m} = f \cdot m^*$, and put $\sigma^* = \hat{e} \cdot f$. Then

$$\sigma^* \cdot m^* = \hat{e} \cdot f \cdot m^* = \hat{e} \cdot \hat{m} = m^*F \cdot \delta,$$

i.e., m^* is a homomorphism. This concludes the proof. □

8.6. We conclude this chapter by an important criterion for varietors which is expressed by generation. Let us say that a functor $F: \mathcal{K} \to \mathcal{K}$ has *bounded generation* if each object R in $\mathcal{K}$ generates only a (small) set of F-algebras. That is, there exists a set (Q_i, δ_i), $i \in I$, of F-algebras such that

(i) each (Q_i, δ_i) is generated by some $\mathcal{M}$-mono $R \to Q_i$;

(ii) if an algebra (Q, δ) is generated by some $\mathcal{M}$-mono $R \to Q$, then (Q, δ) is isomorphic to (Q_i, δ_i) for some $i \in I$.

Theorem. Let $\mathcal{K}$ be a complete $(\mathcal{E}, \mathcal{M})$-category which is $\mathcal{E}$-cowell-powered and $\mathcal{M}$-well-powered. A functor $F: \mathcal{K} \to \mathcal{K}$ preserving $\mathcal{E}$-epis is a varietor iff F has bounded generation.

Proof. (i) Let F have bounded generation. We shall prove that the forgetful functor U: F-**Alg** $\to \mathcal{K}$ is an adjoint (Exercise III.2.B) by verifying the hypotheses of the Adjoint Functor Theorem (III.2.10). We know that F-**Alg** is complete and U preserves limits. It remains to show that (i) R has only a set of $\mathcal{E}$-quotients R', (ii) each R' generates only a set of F-algebras (Q, δ), and (iii) for each (Q, δ) there is only a set of morphisms from R to Q. Therefore, we can find a (small) representative set of morphisms

$$f: R \to (Q, \delta)$$

where (Q, δ) is an algebra generated by a quotient of R, and $f: R \to Q$ is an arbitrary morphism. This is a solution set: each morphism

$$g: R \to (\bar{Q}, \bar{\delta})U = \bar{Q}$$

factors as

$$g = f \cdot hU$$

where $h: (Q, \delta) \to (\bar{Q}, \bar{\delta})$ is a homomorphism and (Q, δ) is generated by a quotient of Q. To prove this, let $g = e_0 \cdot m_0$ be an image factorization,

$$e_0: R \to R' \text{ in } \mathcal{E} \text{ and } m_0: R' \to \bar{Q} \text{ in } \mathcal{M}.$$

The subobjects m_0 generates a subalgebra $\tilde{m}: (\tilde{Q}, \tilde{\delta}) \to (\bar{Q}, \bar{\delta})$; then $(\tilde{Q}, \tilde{\delta})$ is generated by the quotient R' of R, and we have $m_0 \subset \tilde{m}$, i.e., $m_0 = p \cdot \tilde{m}$ for some $p: R' \to Q$. Thus,

$$g = (e_0 \cdot p) \cdot \tilde{m}U.$$

(ii) Let F be a varietor, and let R be an object of $\mathcal{K}$. We shall prove that whenever an algebra (Q, δ) is generated by R, then Q is an $\mathcal{E}$-quotient of $R^{\#}$. Since $R^{\#}$ has only a set of $\mathcal{E}$-quotients, and since for each of these quotients Q we have only a set of morphisms $\delta: QF \to Q$, this will prove that R generates only a set of algebras.

Let $m: R \to Q$ be an $\mathcal{M}$-mono generating (Q, δ). The homomorphism $m^{\#}: (R^{\#}, \varphi) \to (Q, \delta)$ can be factored as $m^{\#} = e \cdot \bar{m}$, where $e: (R^{\#}, \varphi) \to (\bar{Q}, \bar{\delta})$ ia a quotient algebra and $\bar{m}: (\bar{Q}, \bar{\delta}) \to (Q, \delta)$ is a subalgebra. Then $m \subset m$ [because $m = \eta \cdot m^{\#} = (\eta \cdot e) \cdot \bar{m} \subset \bar{m}$] and (Q, δ) is generated by m; hence, $\bar{m}$ is an isomorphism. Therefore, $m^{\#} \in \mathcal{E}$ and Q is an $\mathcal{E}$-quotient of $R^{\#}$. □

Remark. Let $\mathscr{K}$ be a concrete category (III.3.8) such that for each cardinal n there is, up to isomorphism, only a set of objects R with card $(RU) \leq n$. Then bounded generation can be expressed more concisely as follows: For each cardinal n there is a cardinal n^* such that any F-algebra on n generators [i.e., generated by an object R with card $(RU) = n$] has at most n^* points [i.e., it is an algebra (Q, δ) with card $(QU) \leq n^*$].

Exercises IV.8

A. Colimits without free completions. Verify that the following functor F: **Ord** $\to$ **Ord**

$$0F = 0;\ nF = n + 1 \text{ if } n \neq 0$$

does not have free completions of partial algebras and yet, F-**Alg** is cocomplete.

B. Coequalizers often exist. (i) Let $\mathscr{K}$ be $\mathscr{E}$-cowell-powered and let F preserve $\mathscr{E}$-epis. Prove that the free-completion construction stops for each span algebra $(Q, D, \delta_0, \delta_1)$ with $\delta_1 \in \mathscr{E}$. (Hint: Prove that $w_{0,n} \in \mathscr{E}$ and $v_{0,n} \in \mathscr{E}$ for each n by induction, and use the fact that Q and D have only a set of quotients.)

(ii) Prove that for each $\mathscr{E}$-epis preserving functor F in a cocomplete, $\mathscr{E}$-cowell-powered category, F-**Alg** has coequalizers. (Hint: The span algebra of Remark IV.8.2 fulfils $\delta_1 \in \mathscr{E}$.)

C. Finite colimits preserved by the forgetful functor. For each standard set-functor F, prove the equivalence of the following statements:

(a) F-**Alg** has finite colimits preserved by U;
(b) F-**Alg** has countable colimits preserved by U;
(c) F preserves finite colimits.

[Hint: (c) implies that F preserves countable colimits (see Exercise III.4.F) and hence (b) follows from IV.8.3. It is sufficient to prove that (a) implies (c). Since $\emptyset F = \emptyset$ [for F-**Alg** has an initial object $(\emptyset, \delta)$], we only have to prove that F preserves the pushout of arbitrary maps $f : X \to Y$ and $g: X \to Z$. By III.4.6, F preserves finite intersection and hence, it is clearly sufficient to prove that F preserves the pushout of the extended maps

$$f' : X \cup \{a\} \to Y \cup \{a\} \text{ and } g' : X \cup \{a\} \to Z \cup \{a\}$$

where a is a new element and $(a)f' = a = (a)g'$. Define an operation $\delta_X : (X \cup \{a\})F \to X \cup \{a\}$ as the constant to a, analogously with δ_Y and δ_Z. Then f' and g' become homomorphisms. Since U preserves the pushout of f' and g', it is easy to show that F preserves the pushout in **Set**.]

D. Colimits preserved by the forgetful functor. Let F be a standard set functor.

(i) If F-**Alg** has colimits preserved by U, then prove that F-**Alg** is the category of unary algebras. [Hint: By III.4.8, it is sufficient to prove that F preserves colimits. For finite colimits see Exercise C above, for coproducts $\coprod_{i \in I} X_i$ proceed analogously: find a new element a and consider the coproduct of the algebras $(X_i \cup \{a\}, \delta_i)$, where δ_i is constant to a.]

(ii) If F-**Alg** has finite colimits preserved by U and if no cardinal is measurable (IV. 6.3), prove that F-**Alg** is the category of unary algebras.

[Hint: By Exercise C above, F preserves countable colimits. To prove that F preserves any coproduct $X = \coprod_{i \in I} X_i$, it is sufficient to show that each $x \in XF$ lies in some X_iF, $i \in I$ (since the sets $X_iF \subset XF$ are pairwise disjoint because F has no distinguished point). The set $\mathscr{F} = \{T \subset X; x \in TF\}$ is an ultrafilter closed under countable intersections because F preserves finite unions and countable intersections. Hence, $\mathscr{F}$ is trivial, $\mathscr{F} = \mathscr{F}_y$ for some $y \in X$ (say, $y \in X_{i0}$). Then $x \in X_{i_0}F$.]

Notes to Chapter IV

IV.2

The initial-algebra construction is a special case of the free-algebra construction of Section IV.3 which was first investigated for set functors by V. Pohlová [1973] and V. Kůrková-Pohlová and V. Koubek [1974], and for general functors by J. Adámek [1974a]. A number of authors study the restriction of the initial-algebra construction to the first ω steps, see M. A. Arbib [1977], M. B. Smyth [1976], M. Wand [1979] and references there. The least-fixed-point construction is from B. Knaster [1928] and A. Tarski [1955].

Proposition IV.2.6 is due to M. Barr [1970] and Proposition IV.2.7 due to J. Reiterman [1977b].

IV.3

The term varietor has been introduced by the present authors; M. A. Arbib and E. G. Manes use the term input process or recursive process. Corollary IV.3.4 is from J. Adámek [1974a]. Exercise IV.3.A was suggested by J. Reiterman a V. Rödl.

IV.4

The Characterization Theorem has been proved by V. Trnková, J. Adámek, V. Koubek and J. Reiterman [1975]. V. Kůrková-Pohlová and. V. Koubek [1974] described varietors in **Set**. Their proof was much more complicated than the present one, based on the Characterization Theorem, but they covered the more general situation of F-G-algebras (see the notes to III.2 above).

IV.5

Theorem IV.5.3 and Example IV.5.5 are due to J. Reiterman [1977b], and Theorem IV.5.5 due to V. Trnková, J. Adámek, V. Koubek and J. Reiterman [1975].

IV.6

J. Adámek and V. Koubek [1979] characterized finitary varietors in Set. They also proved that a category $\mathcal{K}$ which has the fixed-point property (i.e., each functor $F: \mathcal{K} \to \mathcal{K}$ has a fixed point) can be neither complete nor cocomplete. An example of a category with the fixed-point property is the category $\mathbf{Set}_\omega$; this result, in the form of Theorem IV.6.4, is due to V. Trnková [1974].

We use the term ω-category for the enriched categories over the cartesian closed category $\mathbf{Pos}_\omega$. A number of authors study the finitary initial-algebra construction in an ω-category, notably M. Wand [1979] whose technique was used in the proof of Theorem IV.6.7. The notion of isotone colimits and the general form of IV.6.7 are new; they were announced in J. Adámek [1978]. The surprising fact that even for Σ infinitary the functor H_Σ is a finitary varietor in $\mathbf{Pos}_\omega$ was established by G. Jarzembski [1982]. Free ω-continuous algebras were described by J. A. Goguen *et al.* [1977] and E. Nelson [1981], free strictly ω-continuous algebras by J. Adámek, E. Nelson and J. Reiterman [1982].

IV.7

Free completion of span algebras was investigated by V. Koubek and J. Reiterman [1979] who proved Theorems IV.7.9 and IV.7.11. The interpretation of the free completion as the least fixed point (see IV.7.7) is due to G. M. Kelly's survey [1982] of transfinite constructions.

The fact that a functor with a rank is a varietor has been established already by M. Barr [1970]. His proof is existential, and his additional hypotheses are somewhat more restrictive than those of Corollary IV.7.11.

IV.8

Colimits of F-algebras were studied by J. Adámek and V. Koubek [1977b, 1978]. The earlier paper is the source of the results in IV.8.2, of Proposition IV.8.3 and Theorem IV.8.5, the latter of Counterexample IV.8.4 (which is a modification of an example of J. Adámek [1977b]). Theorem IV.8.3. is from J. Adámek [1974b]. The Exercises C and D are new.

Chapter V: Minimal Realization and Reduction

V.1. Minimal Reduction

1.1. The present chapter is devoted to the problem which varietors have minimal realization (III.2.9) for all behavior morphisms $b: I^{\#} \to \Gamma$, and to consequences of this property. We consider first non-initial F-automata and their minimal reduction. This can be formulated without the hypothesis that F be a varietor. It turns out that, in suitable categories, functors with minimal reduction are finitary varietors anyway.

1.2. Non-initial automata. Let $F: \mathscr{K} \to \mathscr{K}$ be an arbitrary functor in an $(\mathscr{E}, \mathscr{M})$-category $\mathscr{K}$ (see III.2.7). A non-initial F-automaton is a quadruple

$$(Q, \delta, \Gamma, \gamma)$$

consisting of a F-algebra (Q, δ) and a morphism $\gamma: Q \to \Gamma$. A morphism $f: (Q, \delta, \Gamma, \gamma) \to Q', \delta', \Gamma, \gamma')$ of non-initial F-automata is a morphism $f: Q \to Q'$ of $\mathscr{K}$ ch that the following diagram

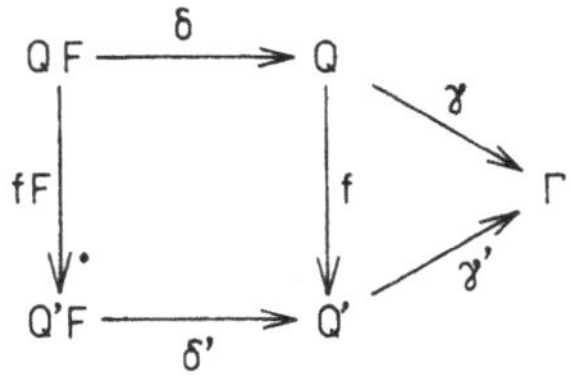

commutes (i.e., f is a homomorphism commuting with the outputs).

Definition. Let A be a non-initial F-automaton. A *reduction* of A is a non-initial F-automaton A' together with a morphism $f: A \to A'$ such that $f \in \mathscr{E}$.

A reduction $r: A \to A_0$ is *minimal* if for each reduction $f: A \to A'$ there exists a unique morphism $g: A' \to A_0$ with

$$r = f \cdot g.$$

Remark. Since $f \cdot g = r \in \mathscr{E}$, we conclude that $g \in \mathscr{E}$ (III.5.1). Hence, the minimal reduction A_0 is characterized by the property that any reduction of A can be further reduced to A_0.

Proposition. Let $\mathcal{K}$ be an $(\mathcal{E}, \mathcal{M})$-category. For each varietor F, equivalent are:

(i) every behavior has a minimal realization;
(ii) every non-initial automaton has a minimal reduction.

Proof. (ii) $\to$ (i). Let

$$\beta: I^\# \to \Gamma$$

be a behavior morphism. Put

$$A = (I^\#, \varphi, \Gamma, \beta).$$

This is a non-initial F-automaton. Let

$$r: A \to A_0 = (Q_0, \delta_0, \Gamma, \gamma_0)$$

be its minimal reduction.

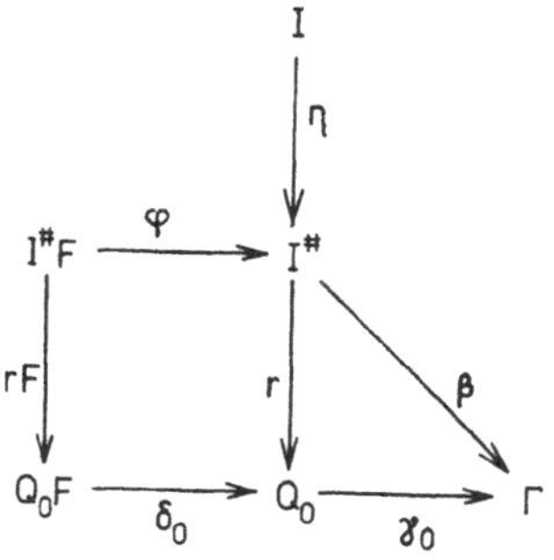

Then the following (initial) F-automaton

$$\bar{A}_0 = (Q_0, \delta_0, \Gamma, \gamma_0, I, \eta \cdot r)$$

is the minimal realization of β. Indeed, the run morphism of $\bar{A}_0$ is r [because $r: (I^\#, \varphi) \to (Q, \delta)$ is a homomorphism and $\eta \cdot r$ is the initialization of $\bar{A}_0$]. Hence, the behavior of $\bar{A}_0$ is

$$r \cdot \gamma_0 = \beta$$

and $\bar{A}_0$ is reachable, since $r \in \mathcal{E}$. Further, for each reachable realization of β:

$$A' = (Q', \delta', \Gamma, \gamma', I, \lambda'),$$

the run morphism

$$\rho': (I^\#, \varphi) \to (Q', \delta')$$

is a reduction of A:

$$\rho': A \to (Q', \delta', \Gamma, \gamma'),$$

because $\beta = \rho' \cdot \gamma'$ (for A realizes β) and $\rho' \in \mathscr{E}$. Thus, there exists a morphism of non-initial automata

$$g: (Q', \delta', \Gamma, \gamma') \to A_0$$

with

$$r = \rho' \cdot g.$$

The last implies $\eta \cdot r = \lambda' \cdot g$, hence,

$$g: A' \to \bar{A}_0$$

is a morphism of initial automata as well.

(i) $\to$ (ii). Let

$$A = (Q, \delta, \Gamma, \gamma)$$

be a non-initial automaton. Consider the initial automaton $A^{\mathrm{i}} = (Q, \delta, \Gamma, \gamma, Q, 1_Q)$. The run map of A^{i} is the unique homomorphism

$$\rho: (Q^{\#}, \varphi) \to (Q, \delta)$$

with

$$\eta \cdot \rho = 1_Q.$$

The behavior of A^{i},

$$\beta = \rho \cdot \gamma: Q^{\#} \to \Gamma$$

has a minimal realization

$$A_0 = (Q_0, \delta_0, \Gamma, \gamma_0, Q, \lambda_0).$$

We prove that

$$\lambda_0: A \to \bar{A} = (Q_0, \delta_0, \Gamma, \gamma_0)$$

is a minimal reduction of A. First, the initial automaton A^{i} is a reachable realization of β (for $\eta \cdot \rho = 1$ implies $\rho \in \mathscr{E}$ by III.5.1). Thus, there exists a morphism

$$r: A^{\mathrm{i}} \to A_0.$$

Then $r = \lambda_0$ (since r preserves the initialization) and

$$r: A \to \bar{A}_0$$

is a reduction of A. (Indeed, $\rho \cdot r$ is the run morphism of A_0 by III.2.5, and $\rho \cdot r \in \mathscr{E}$ implies $r \in \mathscr{E}$ by III.5.1.) Next, given a reduction

$$f: A \to (Q', \delta', \Gamma, \gamma')$$

of A, we form the initial automaton

$$A' = (Q', \delta', \Gamma, \gamma', I, f).$$

Its run morphism is $\rho \cdot f$, because this is a homomorphism with $\eta \cdot (\rho \cdot f) = f$. Hence, A' is reachable and it realizes $\rho \cdot f \cdot \gamma' = \rho \cdot \gamma = \beta$. Therefore, there exists a morphism

$$g: A' \to A_0.$$

Then

$$g: (Q', \delta', \Gamma, \gamma') \to \bar{A}_0$$

is a morphism with

$$r = f \cdot g$$

(because $r = \lambda_0$ and g preserves the initialization). This concludes the proof. □

1.3. Cointersections. Recall the concept of intersection (III.5.2); the dual concept is cointersection. All $\mathscr{E}$-quotients $e: X \to Y$ ($\in \mathscr{E}$) of an object X are ordered as follows: $e \leq e'$ iff there exists a commuting traingle

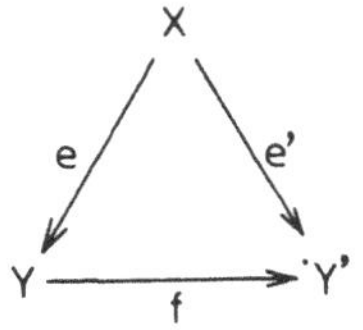

The meets in this ordering are called $\mathscr{E}$-cointersections. Each $\mathscr{E}$-cowell-powered category which is either complete or cocomplete has (possibly large) cointersections—this is the dual of III.5.3. Also, if $e: X \to Y$ is a cointersection of $\mathscr{E}$-quotients $e_j: X \to Y_j$ ($j \in J$) and $f_j: Y_j \to Y$ are morphisms with $e = e_j \cdot f_j$, then the following diagram

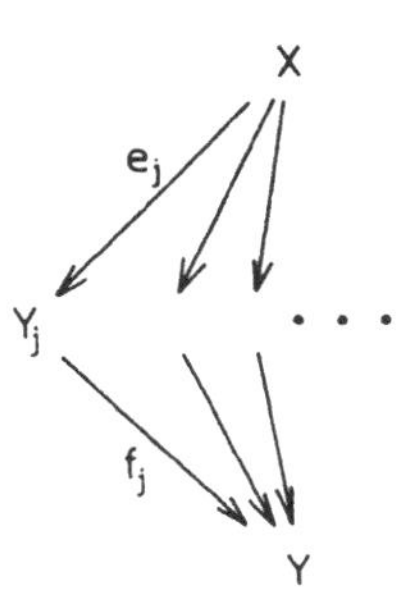

is a multiple pushout. This is dual to III.5.2.

A functor $F: \mathscr{K} \to \mathscr{K}$ preserving $\mathscr{E}$-epis (i.e., such that $e \in \mathscr{E}$ implies $eF \in \mathscr{E}$) is said to *preserve cointersections* if for each cointersection $e: A \to B$ of $\mathscr{E}$-quotients $e_j: A \to B_j$ $(j \in J)$ eF is a cointersection of the $\mathscr{E}$-quotients e_jF $(j \in J)$. This is a central concept of the present chapter. There is a close interrelationship between preservation of cointersections and the existence of minimal realizations. We present below various characterizations of functors preserving cointersections. For example if $\mathscr{K} =$ **Set** or $\mathscr{K} = R$-**Vect**, this characterizes the finitary functors.

Theorem. Let $\mathscr{K}$ be an $(\mathscr{E}, \mathscr{M})$-category with cointersections. Each functor $F: \mathscr{K} \to \mathscr{K}$ preserving cointersections has minimal reductions.

Remark. We prove a slightly more general result. Let us say that *F weakly preserves cointersections* if it preserves the cointersection of any collection of $\mathscr{E}$-quotients, $e_j: A \to B_j$ $(j \in J)$ for which there exist "operation" morphisms

$$\delta: AF \to A$$

and

$$\delta_j: B_jF \to B_j \qquad (j \in J)$$

turning each e_j into a homomorphism $e_j: (A, \delta) \to (B_j, \delta_j)$. We prove that also this weaker condition is sufficient for minimal reductions.

Proof. Let

$$A = (Q, \delta, \Gamma, \gamma)$$

be a non-initial F-automaton and let

$$e_j: A \to A_j = (Q_j, \delta_j, \Gamma, \gamma_j) \qquad (j \in J)$$

be a collection of all reductions of A.

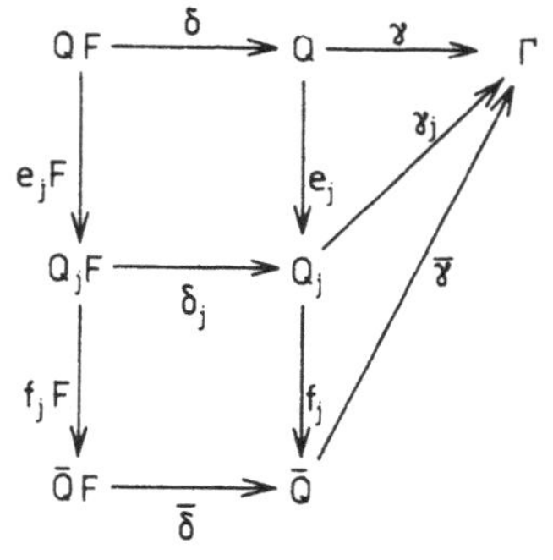

We form the cointersection

$$\bar{e}: Q \to \bar{Q}$$

of the collection e_j $(j \in J)$, and we obtain a multiple pushout

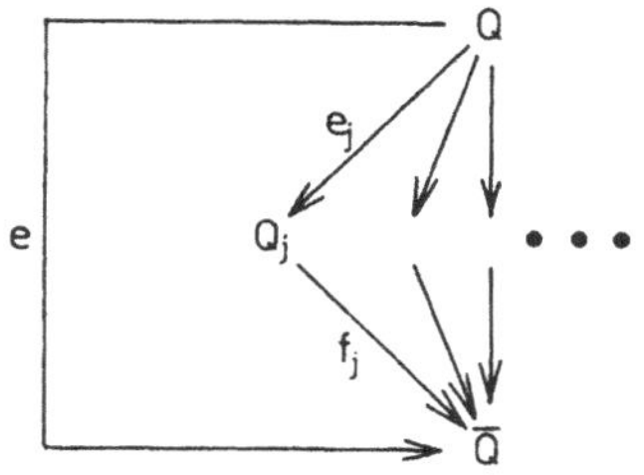

Since each e_j preserves the initializations, we have

$$e_j \cdot \gamma_j = \gamma \quad \text{for each } j \in J$$

and hence, there exists a unique $\bar{\gamma}: \bar{Q} \to \Gamma$ with

(1) $\qquad \gamma_j = f_j \cdot \bar{\gamma} \quad (j \in J).$

Furthermore, if F (weakly) preserves cointersections, then $\bar{e}F$ is the cointersection of e_jF and hence, the following diagram

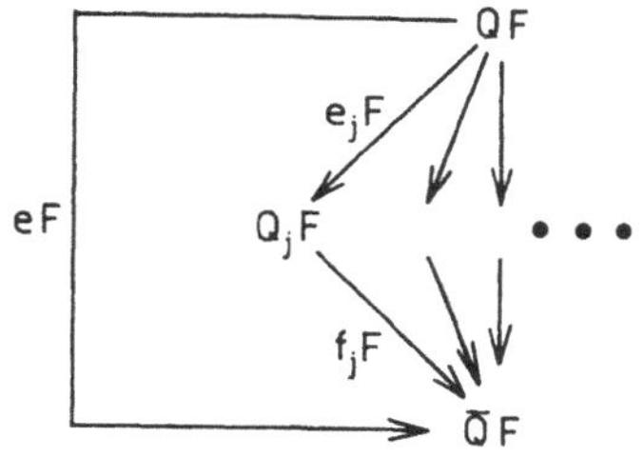

is a multiple pushout. Since each e_j is a homomorphism, we have

$$e_jF \cdot (\delta_j \cdot f_j) = \delta \cdot e_j \cdot f_k = \delta \cdot e.$$

Thus, there exists $\bar{\delta}: \bar{Q}F \to \bar{Q}$ with

(2) $\qquad \delta_j \cdot f_j = f_jF \cdot \bar{\delta} \quad (j \in J).$

We claim that

$$\bar{e}: A \to \bar{A} = (\bar{Q}, \bar{\delta}, \Gamma, \bar{\gamma})$$

is the minimal reduction of A.

First, $\bar{e}$ is a morphism because (for any $j \in J$)

$$\begin{aligned} \delta \cdot \bar{e} &= \delta \cdot e_j \cdot f_j \\ &= e_jF \cdot \delta_j \cdot f_j && \text{by (2)} \\ &= e_jF \cdot f_jF \cdot \bar{\delta} \\ &= \bar{e}F \cdot \bar{\delta} \end{aligned}$$

and

$$\begin{aligned}\bar{e} \cdot \bar{\gamma} &= e_j \cdot f_j \cdot \bar{\gamma} && \text{by (1)} \\ &= e_j \cdot \gamma_j \\ &= \gamma.\end{aligned}$$

Given an arbitrary reduction of A,

$$e_{j0}\colon A \to A_{j0},$$

then (1) and (2) imply that

$$f_{j0}\colon A_{j0} \to \bar{A}$$

is a morphism, and

$$e = e_{j0} \cdot f_{j0}.$$

This concludes the proof. □

1.4. For the converse of the preceding result we need a certain (mild) additional condition of the $(\mathscr{E}, \mathscr{M})$-category $\mathscr{K}$: we say that $\mathscr{K}$ has *regular finite coproducts* if it has finite coproducts and the coproduct injections

$$A \to A + B$$

are $\mathscr{M}$-monos (for arbitrary objects A and B).

Theorem. Let $\mathscr{K}$ be an $(\mathscr{E}, \mathscr{M})$-category with cointersections and regular finite coproducts. An $\mathscr{E}$-epi preserving functor $F\colon \mathscr{K} \to \mathscr{K}$ has minimal reductions iff F weakly preserves cointersections.

Remark. For further reference, we denote by (*) the (single) argument in the following proof in which the regularity of finite coproducts is needed.

Proof. By Remark V.1.3, it is sufficient to prove that F weakly preserves cointersections assuming the existence of minimal reductions.

(i) Let $e_j\colon (Q, \delta) \to (Q_j, \delta_j)$ $(j \in J)$ be homomorphisms with $e_j \in \mathscr{E}$ and let $\hat{e}\colon Q \to \hat{Q}$ be the cointersection of this collection in $\mathscr{K}$. We are going to prove that then the $\mathscr{E}$-epis e_jF $(j \in J)$ have the cointersection $\hat{e}F$. Denote by

$$f_j\colon Q_j \to \hat{Q}$$

the morphisms with

$$\hat{e} = e_j \cdot f_j \quad (j \in J).$$

To prove that $\hat{e}F$ is the cointersection of e_jF $(j \in J)$, let

$$p\colon QF \to Y$$

be an arbitrary $\mathscr{E}$-epi with $e_jF \leq p$, i.e., for suitable $g_j: Q_jF \to Y$,

$$e_jF \cdot g_j = p \quad (j \in J).$$

We are going to prove that

$$\hat{e}F \leq p,$$

i.e., that there exists $g: \hat{Q}F \to Y$ with

$$p = \hat{e}F \cdot g.$$

(ii) For each F-algebra (A, α) we define an F-algebra $(\bar{A}, \bar{\alpha})$ as follows. Put

$$\bar{A} = A + AF$$

with injections $v_A: A \to \bar{A}$ and $w_A: AF \to \bar{A}$. Define

$$\bar{\alpha}_0: A + AF \to A$$

by

$$v_A \cdot \bar{\alpha}_0 = 1_A \quad \text{and} \quad w_A \cdot \bar{\alpha}_0 = \alpha.$$

Finally, put

$$\bar{\alpha} = \bar{\alpha}_0 F \cdot w_A: \bar{A}F \to \bar{A}.$$

For each homomorphism

$$f: (A, \alpha) \to (B, \beta),$$

also

$$f + fF: (\bar{A}, \bar{\alpha}) \to (\bar{B}, \bar{\beta})$$

is a homomorphism, i.e., the following diagram

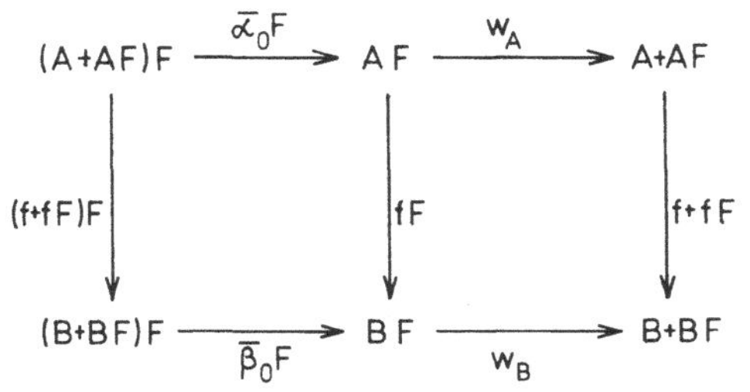

commutes. In fact

$$\bar{\alpha}_0 \cdot f = (f + fF) \cdot \bar{\beta}_0: A + AF \to B$$

because both

$$v_A \cdot (\bar{\alpha}_0 \cdot f) = f = f \cdot v_B \cdot \bar{\beta}_0 = v_A[(f + fF) \cdot \bar{\beta}_0]$$

and

$$w_A \cdot (\bar{\alpha}_0 \cdot f) = \alpha \cdot f = fF \cdot \beta = fF \cdot w_B \cdot \bar{\beta}_0 = w_A[(f + fF) \cdot \bar{\beta}_0].$$

(iii) To find g as above, we use the following automaton

$$A = (\bar{Q}, \bar{\delta}, \hat{Q} + Y, \hat{e} + p),$$

the output morphism of which is

$$\hat{e} + p: QF \to \hat{Q} + Y.$$

Let

$$r: A \to A^* = (Q^*, \delta^*, \hat{Q} + Y, \gamma^*)$$

be the minimal reduction of A. For each $j \in J$ we have $e_jF \in \mathscr{E}$ (since F preserves $\mathscr{E}$-epis), hence, $e_j + e_jF: \bar{Q} \to \bar{Q}_j$ is in $\mathscr{E}$, see III.5.5. Then

$$e_j + e_jF: A \to A_j = (\bar{Q}_j, \bar{\delta}_j, \hat{Q} + Y, f_j + g_j)$$

is a reduction of A: it is a homomorphism by (ii), and it preserves the outputs, since

$$(e_j + e_jF) \cdot (f_j + g_j) = (e_j \cdot f_j) + (e_jF \cdot g_j) = \hat{e} + p.$$

Therefore, for each $j \in J$ there exists a morphism

$$h_j : A_j \to A^*$$

$$r = (e_j + e_jF) \cdot h_j .$$

Since

$$v_Q \cdot r = v_Q \cdot (e_j + e_jF) \cdot h_j = e_j \cdot v_{Q_j} \cdot h_j ,$$

and since $e_j \cdot f_j = \hat{e}\,(j \in J)$ is a multiple pushout, there exists $t: \hat{Q} \to Q^*$ with $v_{Q_j} \cdot h_j = f_j \cdot t$. Hence,

$$v_Q \cdot r = e_j \cdot f_j \cdot t = \hat{e} \cdot t.$$

Since r is a morphism of automata, the following diagram

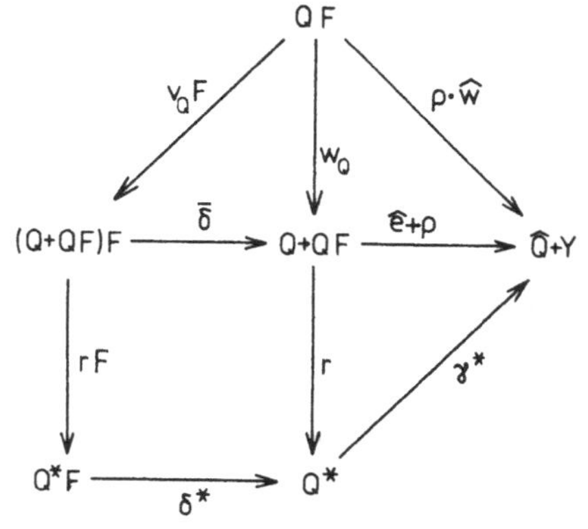

commutes, where $\hat{w}: Y \to \hat{Q} + Y$ denotes the coproduct injection. We conclude that

$$\begin{aligned} p \cdot \hat{w} &= (v_Q \cdot r)F \cdot \delta^* \cdot \gamma^* \\ &= \hat{e}F \cdot (tF \cdot \delta^* \cdot \gamma^*). \end{aligned}$$

Moreover

(*) $\hat{w}$ is an $\mathcal{M}$-mono because of the regularity of finite coproducts.

Hence, the morphism g we are looking for is obtained by the diagonal fill-in (Lemma III.2.7):

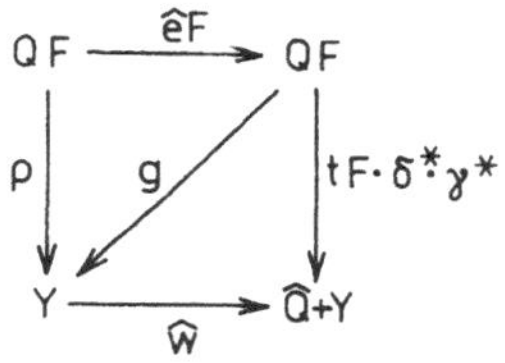

This concludes the proof. □

1.5. A category is said to be *connected* if from each object there leads a morphism into any non-initial object, i.e.,

$$\hom(A, B) \neq \emptyset \text{ whenever } B \neq \bot.$$

In a connected category, a more satisfactory result holds, with preservation of cointersections, not the weak preservation. Lots of categories are connected: **Set**, *R*-**Mod** (because of the zero maps), **Pos** and **Top** (because of the constant maps), etc. But there are natural categories which are non-connected (graphs, semigroups, etc.).

Theorem. Let $\mathcal{K}$ be a connected, finitely cocomplete $(\mathcal{E}, \mathcal{M})$-category with cointersections. An $\mathcal{E}$-epi preserving functor $F: \mathcal{K} \to \mathcal{K}$ has minimal reductions iff F preserves cointersections.

Remark. We are going to use the terminal object of $\mathcal{K}$; under the present hypotheses, $\mathcal{K}$ has one. Indeed, let A be an arbitrary object, $A \neq \bot$. Let $e: A \to T$ denote the cointersection of *all* quotients of A. Then

(a) for each object X there exists a morphism $t_X: X \to T$ (because $A \neq \bot$ implies that there exists a morphism $f: X \to A$; put $t_X = f \cdot e$);

(b) this morphism is unique [given $t'_X: X \to T$, we form the coequalizer $e_0: T \to T_0$ of t_X and t'_X; since $e_0 \in \mathcal{E}$ (III.5.1), we have $e \cdot e_0 \in \mathcal{E}$ and $e \cdot e_0: A \to T_0$ is a quotient of A—hence, $e \cdot e_0 \leq e$ and this proves that e_0 is an isomorphism, thus, $t_X = t'_X$].

Thus, T is a terminal object. If $\mathscr{K}$ has only one object $\perp$, then $\perp$ is terminal.

Proof. Assuming the existence of minimal reductions, we prove first the weak preservation and then the preservation of cointersections. The converse implication has been proved above.

(i) Weak preservation. We distinguish two cases.

(A) For some object $A_0 \neq \perp$ we have $\hom(A_0, \perp) \neq \emptyset$. It follows immediately that $\hom(A, B) \neq \emptyset$ for arbitrary objects A, B. [Indeed, if $B \neq \perp$, this is the connectedness, and if $B = \perp$, then we use the fact that $\hom(A, A_0) \neq \emptyset$.] Then $\mathscr{K}$ has regular finite coproducts, in fact, the coproduct injections

$$i: A \to A + B$$

are all split monos. It is sufficient to choose a morphism $f: B \to A$, then 1_A and f determine a morphism $\bar{f}: A + B \to A$ with $i \cdot \bar{f} = 1_A$. Thus, $i \in \mathscr{M}$ by III.5.1. Hence, we can apply the preceding theorem.

(B) For all objects $A_0 \neq \perp$ we have $\hom(A_0, \perp) = \emptyset$. In this case, the coproduct injections are split monos, too, with the exception of

$$\perp \to \perp + B = B \quad \text{for } B \neq \perp .$$

Let us inspect the proof of the preceding theorem. The only application of the regularity of coproducts is denoted by (*) (see the remark preceding the proof). It concerns the injection

$$Y \to Q + Y.$$

Thus, we only have to check that the proof becomes trivial if $Y = \perp$. Since we have morphisms $p: QF \to Y$ and $g_j: Q_jF \to Y$, we conclude that

$$QF = \perp \quad \text{and} \quad Q_jF = \perp \quad (j \in J).$$

Assuming $Q \neq \perp$, F is, necessarily, the constant functor $C_\perp$ of value $\perp$ (in which case the preservation of cointersections is obvious). Indeed, for each object A we have a morphism $f: A \to Q$, hence, a morphism $fF: AF \to QF = \perp$, and we conclude $AF = \perp$. Analogously, assuming $Q_j \neq \perp$ for any $j \in J$, we have $F = C_\perp$. Thus, we can assume

$$Q = \perp \quad \text{and} \quad Q_j = \perp \quad (j \in J)$$

and the proof is trivial because each e_j is $1_\perp$ $(j \in J)$.

(ii) Preservation. Let $e_j: Q \to Q_j$ $(j \in J)$ be a collection of $\mathscr{E}$-epis. If $Q \neq \perp$, then we exhibit morphisms $\delta: QF \to Q$ and $\delta_j: Q_jF \to Q_j$ such that $e_j: (Q, \delta) \to (Q_j, \delta_j)$ are homomorphisms; if $Q = \perp$, then we show that the proof is trivial. Then clearly (i) implies (ii). We denote by T the terminal object of $\mathscr{K}$ (see the preceding remark).

Assuming $Q \neq \perp$, we choose an arbitrary morphism $f: T \to Q$. Put

$$\delta = t_{QF} \cdot f: QF \to Q$$

and

$$\delta_j = t_{Q_jF} \cdot f \cdot e_j : Q_jF \to Q_j \quad (j \in J).$$

Then each e_j is a homomorphism since the following diagram

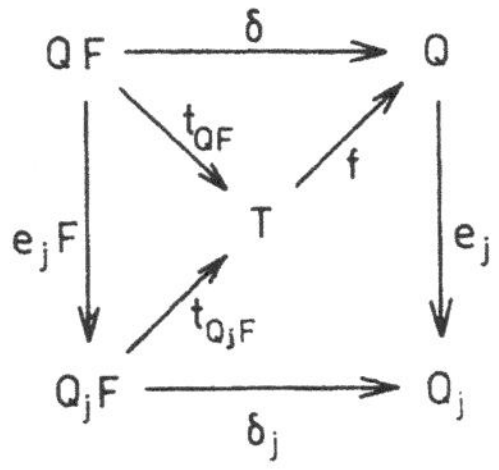

commutes.

If $Q = \perp$, the situation is trivial. We can clearly assume that no $e_j, j \in J$, is an isomorphism (since isomorphisms do not influence the resulting cointersection). Then all e_j represent the same quotient of Q: given $j, j' \in J$, then $Q_j \neq \perp \neq Q_{j'}$ implies that there are morphisms

$$f: Q_j \to Q_{j'} \quad \text{and} \quad g: Q_{j'} \to Q_j.$$

The the fact that $Q = \perp$ implies

$$e_j \cdot f = e_{j'} \quad \text{and} \quad e_{j'} \cdot g = e_j,$$

hence,

$$e_j \cdot (f \cdot g) = e_j \quad \text{and} \quad e_j' \cdot (g \cdot f) = e_j'.$$

Since both e_j and $e_{j'}$ are epis, we conclude that $f = g^{-1}$ and hence, e_j and $e_j \cdot f = e_j'$ represent the same quotient. The cointersection of a single quotient is the quotient itself; each functor preserves such a cointersection. □

Example. The functor

$$H_\Sigma: \textbf{Set} \to \textbf{Set}$$

(III.2.5) preserves cointersections iff Σ is a finitary type.

Indeed, H_Σ is a varietor and hence, it has minimal reductions iff it has minimal realizations. This holds iff Σ is finitary (II.3.8). On the other hand, **Set** is an (epi, mono)-category which is cocomplete, connected and has cointersections (because equivalence lattices are complete). Hence, H_Σ has minimal reductions iff it preserves cointersections.

We prove below that a set functor preserves cointersections iff it is finitary, i.e., a quotient of some H_Σ with Σ finitary.

Remark. We characterize functors preserving cointersections in Section V.4. But first we prove certain consequences of the theorems above in V.3, using relations which we introduce in V.2.

Exercises V.1

A. Preservation of cointersections is not necessary. We present a finitary varietor F: **Gra** $\to$ **Gra** (the category of graphs, see Exercise IV.3.A) which does not preserve cointersections and yet, has minimal reductions. We consider the factorization system (epi, embedding) in **Gra**.

(i) Verify that **Gra** fulfils the hypotheses of Theorem V.1.4.

(ii) Denote by T the terminal graph $T = (\{t\}, \{t, t\})$. Define a functor

$$F: \mathbf{Gra} \to \mathbf{Gra}$$

on objects by

$$(X, R)F = T \quad \text{if} \quad R \neq \emptyset, \quad (X, \emptyset)F = (X^\omega, X^\omega \times X^\omega);$$

on morphisms $f: (X, R) \to (Y, S)$ by

$$fF = \text{const } t \quad \text{if } S \neq \emptyset, \quad fF = f^{(\omega)} = fH_\omega \quad \text{if } S = \emptyset.$$

Verify that F is a well-defined functor preserving epis.

(iii) Since H_ω: **Set** $\to$ **Set** fails to preserve cointersections (see Example V.1.5), conclude that F: **Gra** $\to$ **Gra** fails, too. [Hint: Given epis $e_i: A \to B_i$ in **Set**, consider the epis $e_i: (A, \emptyset) \to (B_i, \emptyset)$ in **Gra**.]

(iv) Verify that F is a finitary varietor: for each graph I we have

$$I^\# = W_2 = I + T$$

with $\varphi: T \to I^\#$ and $\eta: I \to I^\#$ the coproduct injections.

(v) Prove that each behavior $\beta: I + T \to \Gamma$ has a minimal realization A, defined by the image factorization of $\beta = e \cdot m$, with $e: I + T \to Q$ and $m: Q \to \Gamma$:

$$A = (Q, \delta, \Gamma, m, I, \eta \cdot e)$$

where $\delta: QF = T \to Q$ sends t to $(t)e \in Q$.

(iv) Conclude that F weakly preserves cointersections. Prove this also directly.

B. Sequential topological automata. (i) Verify that if Σ is a compact Hausdorff space, the functor S_Σ: **Top** $\to$ **Top** preserves cointersection [with $(\mathscr{E}, \mathscr{M})$ either (regular epi, mono) or (epi, regular mono)] and hence, every sequential Σ-automaton has a minimal reduction (V.1.3). [Hint: A cointersection of epis $e_i: X \to Y_i$ in **Top** is created on the level of sets: if $f_i: Y_i \to Y$ form the cointersection in **Set**, then a set $M \subset Y$ is open iff $(M)f_i^{-1}$ is open for each i. Prove that if Σ is a compact Hausdorff space, then again a set $M \subset X \times \Sigma$ is open iff $(M)(f_i \times 1_\Sigma)^{-1}$ is open for each i.]

(ii) For non-compact Σ, the functor S_Σ need not have minimal reductions. Let Σ be the set of all rational numbers with the usual topology. Prove that the following S_Σ-automaton $(Q, \delta, \Gamma, \gamma)$ has no minimal reduction.

Let Q' be the subspace of the real plane $R \times R$ consisting of the points (x, y) with x an integer and $y = 0$ or $y = \frac{1}{n}$ $(n = 1, 2, 3, \ldots)$, put

$$Q = Q' + (Q' \times \Sigma) + \{a\} \quad (a \notin Q', a \notin Q' \times \Sigma)$$

with the topology of topological sum (disjoint union). Define $\delta\colon Q \times S \to Q$ by

$$(q', \sigma)\delta = (q', \sigma) \text{ for } q' \in Q', \quad (q, \sigma)\delta = a \text{ for } q \in Q - Q'.$$

Denote by Γ the quotient space of Q under the least equivalence $\sim$ with $(x, 0) \sim (\bar{x}, 0)$ for all $(x, 0), (\bar{x}, 0) \in Q'$ and $(x, 0, \sigma) \sim (\bar{x}, 0, \sigma)$ for all $(x, 0, \sigma)$, $(\bar{x}, 0, \sigma) \in Q' \times \Sigma$. Finally, γ is the quotient map. [Hint: For each pair $p = (x', x'')$ of integers we have a reduction A_p obtained by merging $(x', 0)$ with $(x'', 0)$ as well as $(x', 0, \sigma)$ with $(x'', 0, \sigma)$. If A is a minimal reduction, then its state object is a quotient $Q/\approx$ with $\approx$ larger or equal to $\sim$, in fact, equal to $\sim$ because of γ. Use the fact that the topology of $(Q'/\sim) \times \Sigma$ is distinct from that of $(Q' \times \Sigma)/\sim$.]

C. Finitary varietors in Set^{op}. The category **Set**^{op}, which is well-known to be equivalent to that of complete atomic Boolean algebras, has the following property: H_Σ (III.2.5) is a finitary varietor for each, possibly infinitary, type Σ. Prove it. [Hint: In **Set**, prove that coproducts commute with limits of ω-chains. Hence, in **Set**^{op}, each H_n preserves colimits of ω-chains, and then so does each H_Σ.]

V.2. Relations in a Category

We introduce here the concept of relation in an $(\mathscr{E}, \mathscr{M})$-category, and the basic theory concerning relations. The reader can skip this section without breaking the continuity of the text; the only concepts needed for Chapter V below are that of relation, and of equivalence relation. We shall use relations in Chapters VI and VII.

A relation from an object X to an object Y is a subobject of $X \times Y$. Thus we assume throughout the section that a finitely complete $(\mathscr{E}, \mathscr{M})$-category $\mathscr{K}$ is given.

2.1. A relation

$$r\colon X \rightharpoonup Y$$

from X to Y is a subobject r of $X \times Y$. This can be represented by an $\mathscr{M}$-mono

$$r\colon R \to X \times Y,$$

or by the pair

$$r_{(1)}: R \to X \quad \text{and} \quad r_{(2)}: R \to Y$$

of components of r. We also say that a pair of morphisms

$$f_1 : R \to X \quad \text{and} \quad f_2 : R \to Y$$

represents the relation r if the induced morphism $R \to X \times Y$ has image r (i.e., if the morphism $f: R \to X \times Y$ with components f_1 and f_2 has an image factorization $f = e \cdot r$). We write

$$r = [f_1, f_2].$$

In particular,

$$r = [r_{(1)}, r_{(2)}].$$

For each $e: \bar{R} \to R$ in $\mathscr{E}$,

$$r = [e \cdot r_{(1)}, e \cdot r_{(2)}]$$

(and conversely, any pair $[f_1, f_2]$ representing r has the form $f_1 = e \cdot r_{(1)}$ and $f_2 = e \cdot r_{(2)}$ with $e \in \mathscr{E}$).

Each morphism $f: X \to Y$ is considered as a special case of relation:

$$[1_X, f]: X \rightharpoonup Y.$$

A relation $r: X \rightharpoonup Y$ is called a *partial morphism* if $r_{(1)}: R \to X$ is in $\mathscr{M}$. Then R is a subobject of X (the domain of r) and hence, r is a morphism from a subobject of X into Y. (See IV.7.1) A relation

$$[f_1, f_2]: X \rightharpoonup Y$$

is a morphism f iff $f_1 \in \mathscr{E}$ and $f_2 = f_1 \cdot f$; it is a partial morphism $g: R \rightharpoonup Y$ (for a subobject $m: R \to X$) iff $f_1 = e \cdot m$ and $f_2 = e \cdot g$ for some $e \in \mathscr{E}$.

Examples. (i) **Set**. A relation $X \rightharpoonup Y$ is a subset $R \subset X \times Y$. It is represented by any pair $f_1 : \bar{R} \to X$ and $f_2 : \bar{R} \to X$ of maps with

$$R = \{((x)f_1, (x)f_2);\ x \in \bar{R}\}.$$

In particular, by the pair of projections $r_{(1)}: R \to X$ and $r_{(2)}: R \to Y$.

(ii) **Pos**. In the (epi, embedding)-category **Pos**, a relation from $(X, \leq)$ to $(Y, \preceq)$ is any subset $R \subset X \times Y$, ordered component-wise:

$$(x, y) \sqsubseteq^* (x', y') \quad \text{iff} \quad x \leq x' \text{ and } y \preceq y'$$

for all $x, x' \in X$ and $y, y' \in Y$.

In the (quotient, mono)-category **Pos**, a relation $(R, \sqsubseteq)$ from $(X, \leq)$ to $(Y, \preceq)$ carries any ordering $\sqsubseteq$, contained in $\sqsubseteq^*$. For example, R can be discretely ordered (no matter how X and Y are ordered).

2.2. We are going to define the composition of relations. In **Set**, given

$$r: X \rightharpoonup Y \quad \text{and} \quad s: Y \rightharpoonup Z$$

then

$$r \circ s: X \rightharpoonup Z$$

is the relation of all $(x, z) \in X \times Z$ for which there exists $y \in Y$ with

$$(x, y) \in R \quad \text{and} \quad (y, z) \in S.$$

This can be expressed by the pullback of the projections $r_{(i)}$ and $s_{(i)}$ $(i = 1, 2)$, representing r and s, as follows. Let us form the pullback of $r_{(2)}$ and $s_{(1)}$:

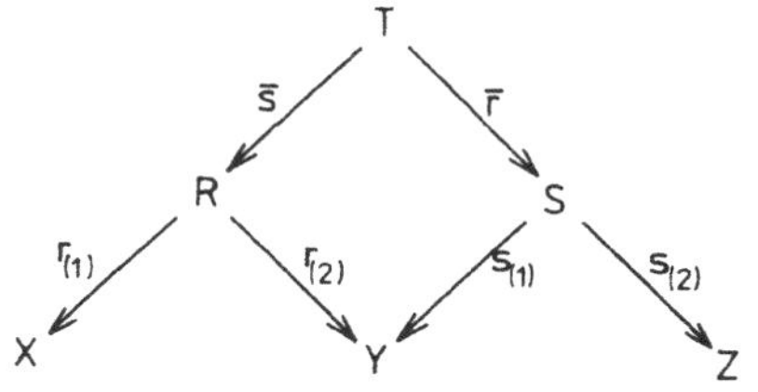

Here

$$\begin{aligned} T &= \{(a, b) \in R \times S;\ (a)r_{(2)} = (b)s_{(1)}\} \\ &= \{(a_1, a_2, b_1, b_2) \in R \times S;\ a_2 = b_1\} \end{aligned}$$

and s, r are the projections,

$$(a_1, a_2, b_1, b_2)s = (a_1, a_2); \quad (a_1, a_2, b_1, b_2)r = (b_1, b_2).$$

The pair of maps

$$\bar{s} \cdot r_{(1)}: T \to X \quad \text{and} \quad \bar{r} \cdot s_{(2)}: T \to Y$$

clearly represents $R \circ S$. We use this for the general definition.

Definition. The *composition* of relations

$$r: X \rightharpoonup Y \quad \text{and} \quad s: Y \rightharpoonup Z$$

is the relation

$$r \circ s = [\bar{s} \cdot r_{(1)}, \bar{r} \cdot S_{(2)}]: X \rightharpoonup Z,$$

where $\bar{s}$ and $\bar{r}$ are defined by the following pullback:

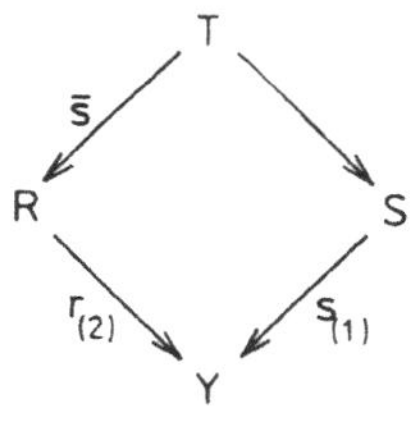

Remarks. (i) This definition is independent of the representatives: if $\mathcal{M}$-monos r and r' represent the same relation $X \rightharpoonup Y$ and if s and s' represent the same relation $Y \rightharpoonup Z$, then also $r \circ s$ and $r' \circ s'$ represent the same relation — this is easy to verify.

Note, however, that we formed the pullback above using $r_{(i)}$ and $s_{(i)}$, not arbitrary pairs $r = [f_1, f_2]$ and $s = [g_1, g_2]$! We return to this problem below when discussing the pullback axiom.

(ii) The composition of morphisms agrees with that of relations. Indeed, given morphisms $f: X \to Y$ and $g: Y \to Z$, the diagram above defining $f \circ g$ has the following form:

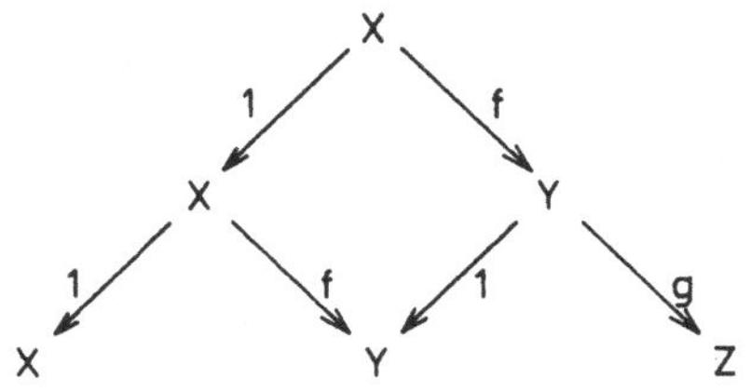

(iii) If $r: X \rightharpoonup Y$ and $s: Y \rightharpoonup Z$ are partial morphisms and $\bar{s}$, $\bar{r}$ are as above, then $\bar{s} \in \mathcal{M}$, so that $\bar{s} \cdot r_{(1)} \in \mathcal{M}$, consequently

$$(r \circ s)_{(1)} = \bar{s} \cdot r_{(1)} \quad \text{and} \quad (r \circ s)_{(2)} = \bar{r} \cdot s_{(2)}.$$

It follows that partial morphisms are closed under composition.

2.3. There is another way of defining the composition of relations:

Lemma. Given relations, represented by $\mathcal{M}$-monos

$$r: R \to X \times Y \quad \text{and} \quad s: S \to Y \times Z,$$

let us form the pullback of $r \times 1_Z$ and $1_X \times s$:

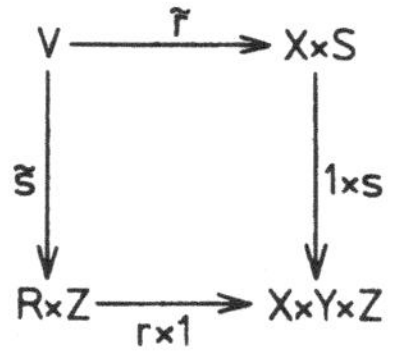

Put $v = \tilde{s} \cdot (r \times 1) = \tilde{r} \cdot (1 \times s)$ and denote by π_1, π_2, π_3 the projections of $X \times Y \times Z$. Then

$$r \circ s = [v \cdot \pi_1, v \cdot \pi_3]: X \to Z$$

Proof. Denote by

$$\pi_1': R \times Z \to R \quad \text{and} \quad \pi_2': R \times Z \to Z$$

the projections; analogously,

$$\pi_1'': X \times S \to X \quad \text{and} \quad \pi_2'': X \times S \to S.$$

Obviously,

$$v \cdot \pi_1 = \tilde{s} \cdot (r \times 1) \cdot \pi_1 = \tilde{s} \cdot \pi_1' \cdot r_{(1)}$$

and

$$v \cdot \pi_3 = \tilde{r} \cdot (1 \times s) \cdot \pi_3 = \tilde{r} \cdot \pi_2'' \cdot s_{(2)}.$$

To prove that $r \circ s = [v \cdot \pi_1, v \cdot \pi_3]$, it is sufficient to check that the following square

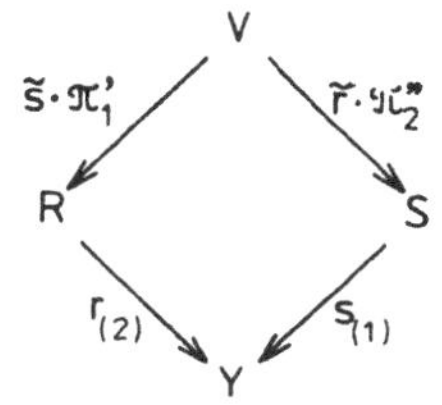

is a pullback. This square obviously commutes. Let

$$p: P \to R \quad \text{and} \quad q: P \to S$$

be morphisms with

$$p \cdot r_{(2)} = q \cdot s_{(1)}.$$

Let

$$\tilde{p}: P \to R \times Z$$

have components p and $q \cdot s_{(2)}$, and let

$$\tilde{q}: P \to X \times S$$

have components $p \cdot r_{(1)}$ and s. Then the morphisms

$$\tilde{p} \cdot (r \times 1_Z), \tilde{q} \cdot (1_X \times s): P \to X \times Y \times Z$$

both have components

$$p \cdot r_{(1)};\ p \cdot r_{(2)} = q \cdot s_{(1)};\ q \cdot s_{(2)}.$$

Hence, the two morphisms above are equal and there exists a unique h for

which the following diagram

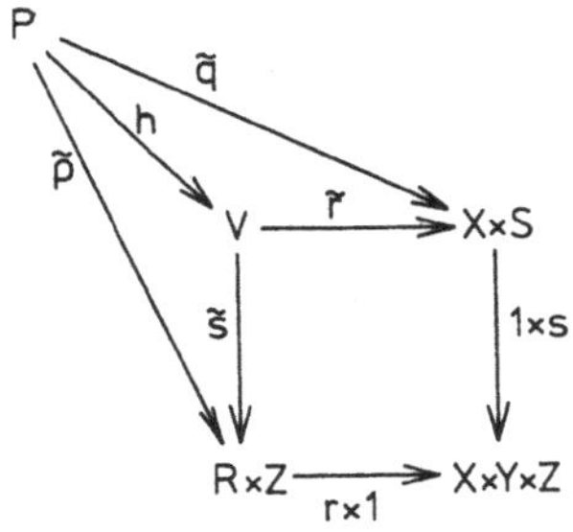

commutes. This is the unique morphism with

$$p = h \cdot \tilde{s} \cdot \pi_1' \quad \text{and} \quad q = h \cdot \tilde{r} \cdot \pi_2''.$$

This conclude the proof. □

2.4. Relations from X to Y are naturally ordered, as subobjects of $X \times Y$ (III.5.2): given

$$r, s : X \rightharpoonup Y$$

represented by $\mathcal{M}$-monos $r: R \to X \times Y$ and $s: S \to X \times Y$, then

$$r \subset s \quad \text{iff} \quad r = f \cdot s$$

for some morphism $f: R \to S$. Note that for each relation $r = [f_1, f_2]$, where $f_1: R \to X$ and $f_2: R \to Y$ are morphisms, and for each morphism $h: S \to R$, we have

$$[h \cdot f_1, h \cdot f_2] \subset r.$$

Moreover,

$$[h \cdot f_1, h \cdot f_2] = r \quad \text{iff} \quad h \in \mathcal{E}.$$

And for the canonical representation $r = [r_{(1)}, r_{(2)}]$ also conversely:

$$\text{if } [h \cdot r_{(1)}, h \cdot r_{(2)}] = r, \quad \text{then } h \in \mathcal{E}.$$

(Proof: if $[h \cdot r_{(1)}, h \cdot r_{(2)}] = r$, then $h \cdot r: S \to X \times Y$ has the image r, i.e., there exists $e \in \mathcal{E}$ with $h \cdot r = e \cdot r$ and since r is a mono, this implies $h = e \in \mathcal{E}$.)

Remark. We shall often use the following consequence of the definition of the order of relations: if $r = [g_1, g_2] \subset s$, then there exists h such that

$$g_1 = h \cdot s_{(1)} \quad \text{and} \quad g_2 = h \cdot s_{(2)}.$$

(Indeed, if $f: R \to S$ fulfils $r = f \cdot s$, then $f \cdot s_{(1)} = r_{(1)}$ and $f \cdot s_{(2)} = s_{(2)}$; since

$[g_1, g_2]$ is a representation of r, there exists $e \in \mathscr{E}$ with

$$e \cdot r_{(1)} = g_1 \quad \text{and} \quad e \cdot r_{(2)} = g_2.$$

Hence, $h = e \cdot f$ fulfils the equations.)

Proposition. Composition of relations is order-preserving, i.e.

$$r \subset r' \quad \text{and} \quad s \subset s' \quad \text{imply} \quad r \circ s \subset r' \circ s'$$

for arbitrary relations $r, r': X \rightharpoonup Y$ and $s, s': Y \rightharpoonup Z$.

Proof. Let $r = f \cdot r'$ and $s = g \cdot s'$, then $r \times 1_Z = (f \times 1_Z) \cdot (r' \times 1_Z)$, analogously with $1_X \times s$, and we form the corresponding pullbacks as in Lemma V.2.3:

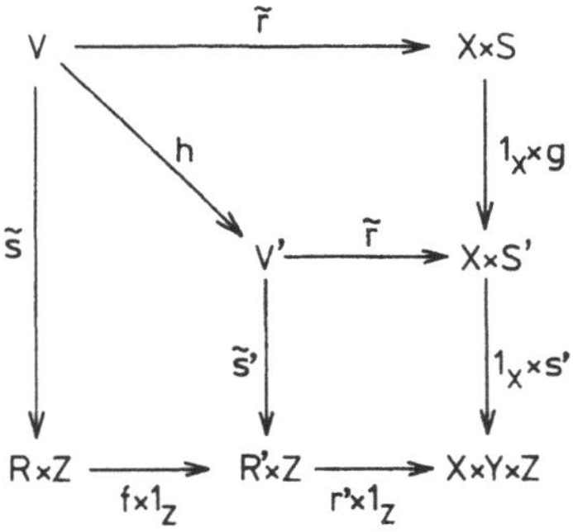

There exists a unique $h: V \to V'$ such that the diagram above commutes. The morphisms

$$v = \tilde{r} \cdot (1 \times s) \quad \text{and} \quad v' = \tilde{r}' \cdot (1 \times s')$$

then fulfil

$$v = h \cdot v'.$$

Hence,

$$r \circ s = [v \cdot \pi_1, v \cdot \pi_3] \subset [v' \cdot \pi_1, v' \cdot \pi_3] = r' \circ s'. \qquad \square$$

2.5. The *inverse relation* to $r: X \rightharpoonup Y$ is the relation

$$r^{-1} = [r_{(2)}, r_{(1)}]: Y \rightharpoonup X.$$

For any representation $r = [f_1, f_2]$ obviously

$$[f_1, f_2]^{-1} = [f_2, f_1].$$

To express this by $\mathscr{M}$-monos, denote by

$$\xi: X \times Y \to Y \times X$$

the canonical isomorphism, defined by

$$\xi \cdot \pi_1 = \pi_2 \quad \text{and} \quad \xi \cdot \pi_2 = \pi_1.$$

Then for each $\mathcal{M}$-mono $r: R \to X \times Y$ we have

$$r^{-1} = r \cdot \xi.$$

Proposition. The operation of inverse is an order-isomorphism on the class of all relations (i.e., a bijective map such that $r \subset s$ iff $r^{-1} \subset s^{-1}$) which inverts the composition [i.e., $(r \circ s)^{-1} = s^{-1} \circ r^{-1}$] and is an involution [$r = (r^{-1})^{-1}$].

Proof. (i) The isomorphism $\xi: X \times Y \to Y \times X$ above fulfils $\xi \cdot \xi = 1_{X \times Y}$ and hence, for each relation r,

$$(r^{-1})^{-1} = (r \cdot \xi) \cdot \xi = r.$$

Thus, the operation of inverse is onto; it is also one-to-one, since

$$r \neq s \quad \text{implies} \quad r^{-1} = r \cdot \xi \neq s \cdot \xi = s^{-1}.$$

(ii) Let $r, s: X \rightharpoonup Y$ be two relations. If $r \subset s$, then $r = f \cdot s$ for some f and this implies

$$r^{-1} = [r_{(2)}, r_{(1)}] = [f \cdot s_{(2)}, f \cdot s_{(1)}] \subset [s_{(2)}, s_{(1)}] = s^{-1}.$$

If $r^{-1} \subset s^{-1}$, then $r = (r^{-1})^{-1} \subset (s^{-1})^{-1} = s$.

(iii) The composition $r \circ s$ is defined by the following pullback of $r_{(2)}$ and $s_{(1)}$

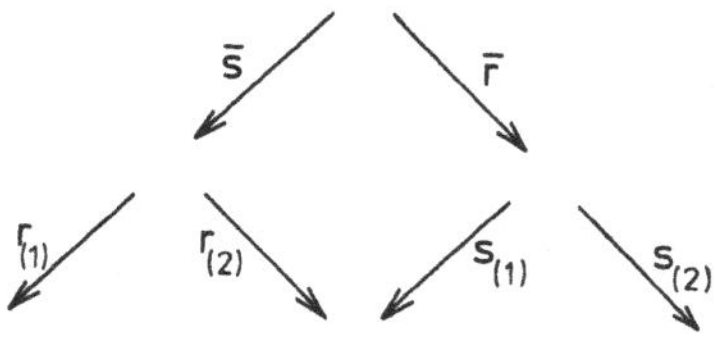

as $r \circ s = [\bar{s} \cdot r_{(1)}, \bar{r} \cdot s_{(2)}]$. Hence,

$$(r \circ s)^{-1} = [\bar{r} \cdot s_{(2)}, \bar{s} \cdot r_{(1)}].$$

Since the composition of s^{-1} and r^{-1} is defined by the pullback of $s_{(1)}$ and $r_{(2)}$, which is $\bar{r}$ and $\bar{s}$, we have

$$s^{-1} \circ s^{-1} = [\bar{r} \cdot s_{(2)}, \bar{s} \cdot r_{(1)}]. \qquad \square$$

Remarks. (i) A relation

$$r: X \rightharpoonup X$$

is said to be *symmetric* if

$$r = r^{-1}.$$

Symmetric relations are closed under unions and intersections—this follows from the preceding proposition.

(ii) For each object X we define the *diagonal relation*

$$\Delta_X = [1_X, 1_X] : X \rightharpoonup X.$$

A relation $r: X \rightharpoonup X$ is said to be *reflexive* if $\Delta_X \subset r$, i.e., if there exists $f: X \to R$ with

$$f \cdot r_{(1)} = f \cdot r_{(2)} = 1_X.$$

(iii) Each $\mathscr{E}$-epi $e: X \to Y$ "preserves" reflexive relations: if $r: X \rightharpoonup X$ is reflexive, so is $[r_{(1)} \cdot e, r_{(2)} \cdot e] : Y \to Y$. Indeed, if

$$[1_X, 1_X] \subset [r_{(1)}, r_{(2)}],$$

then

$$[e, e] \subset [r_{(1)} \cdot e, r_{(2)} \cdot e]$$

and since $e \in \mathscr{E}$ implies $[e, e] = [1_Y, 1_Y]$, we conclude

$$\Delta_Y \subset [r_{(1)} \cdot e, r_{(2)} \cdot e].$$

(iv) A relation

$$r: X \rightharpoonup X$$

is said to be *transitive* if

$$r \circ r \subset r.$$

A reflexive, symmetric and transitive relation is called an *equivalence*.

2.6. Proposition. For each morphism $f: X \to Y$ let us form the pullback of f and f:

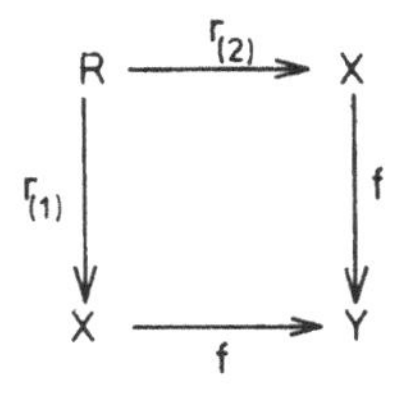

Then $[r_{(1)}, r_{(2)}] : X \rightharpoonup X$ is an equivalence, called the *kernel equivalence* of f.

Proof. (i) Since $1_X \cdot f = 1_X \cdot f$, there exists a unique $h\colon X \to R$ for which the following diagram

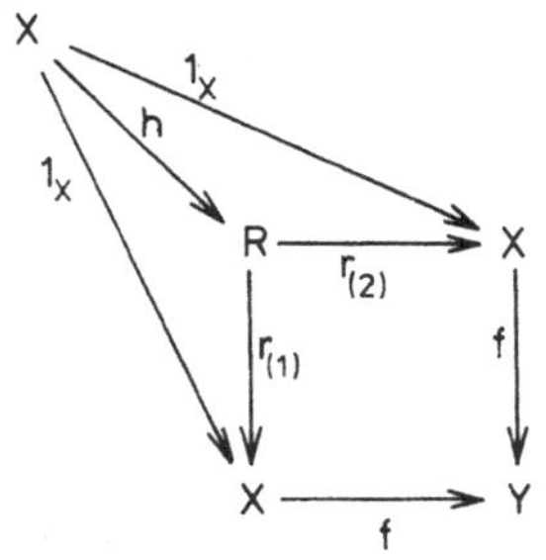

commutes. Hence,

$$[1_X, 1_X] \subset [r_{(1)}, r_{(2)}]$$

which proves the reflexivity.

(ii) Since $r_{(2)} \cdot f = r_{(1)} \cdot f$, there exists a unique $h\colon R \to R$ for which the following diagram

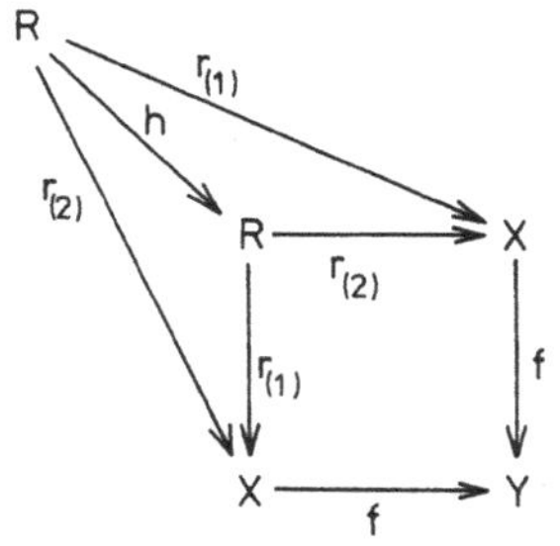

commutes. Since, obviously, $h \cdot h = 1_R$, h is an isomorphism and we see that

$$[r_{(2)}, r_{(1)}] = [h \cdot r_{(1)}, h \cdot \mathrm{r}_{(2)}] = [r_{(1)}, r_{(2)}].$$

Thus, the symmetry is proved.

(iii) To prove the transitivity, we first verify that the morphism

$$r\colon R \to X \times X$$

with components $r_{(1)}$ and $r_{(2)}$ is an $\mathcal{M}$-mono (which then represents the kernel equivalence). Indeed, let $r = e \cdot m$ be its image factorization. Since the projections π_1 and π_2 of $X \times X$ fulfil

$$e \cdot m \cdot \pi_1 \cdot f = r_{(1)} \cdot f = r_{(2)} \cdot f = e \cdot m \cdot \pi_2 \cdot f$$

and since e is an epi, we have

$$m \cdot \pi_1 \cdot f = m \cdot \pi_2 \cdot f$$

and there is a unique h for which the following diagram

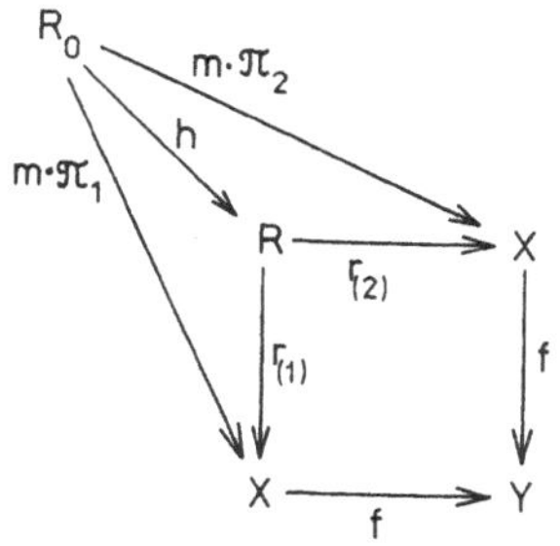

commutes. Then $e \cdot h = 1$ because both

$$(e \cdot h) \cdot r_{(1)} = e \cdot m \cdot \pi_1 = r \cdot \pi_1 = r_{(1)}$$

and, analogously,

$$(e \cdot h) \cdot r_{(2)} = r_{(2)}.$$

Since e is a split mono and an epi, it is an isomorphism, therefore, $r \in \mathcal{M}$.

It follows that $r \circ r$ is defined by the pullback of $r_{(2)}$ and $r_{(1)}$:

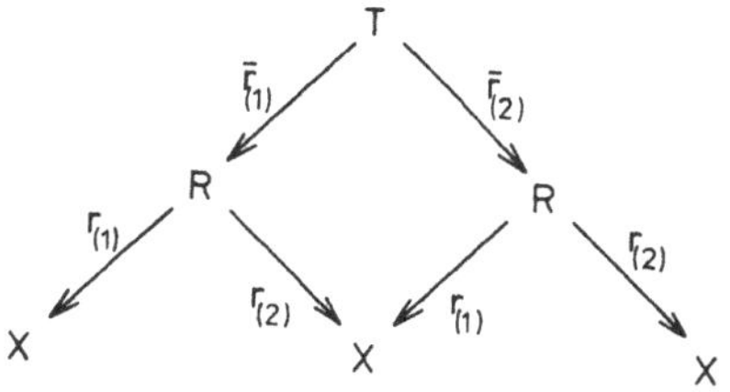

as

$$r \circ r = [\bar{r}_{(1)} \cdot r_{(1)}, \bar{r}_{(2)} \cdot r_{(2)}].$$

To prove $r \circ r \subset r$, it is sufficient to verify that f merges $\bar{r}_{(1)} \cdot r_{(1)}$ and $\bar{r}_{(2)} \cdot r_{(2)}$. And this is clear:

$$\bar{r}_{(1)} \cdot r_{(1)} \cdot f = \bar{r}_{(1)} \cdot r_{(2)} \cdot f = \bar{r}_{(2)} \cdot r_{(1)} \cdot f \underline{=} \bar{r}_{(2)} \cdot r_{(2)} \cdot f.$$

This proves the transitivity. □

Examples. (i) **Set**: Equivalence relations have their usual meaning. Each equivalence relation $R \subset X \times X$ is the kernel equivalence of some morphism (e.g., of the canonical morphism $f: X \to X/R$, assigning to each element its equivalence class).

(ii) **Pos**: In the (epi, embedding)-category **Pos**, an equivalence relation $R \subset X \times X$ is ordered component-wise. This need not be a kernel equivalence of any morphism: consider, as an example, the three-element chain $X = \{0, 1, 2\}$ and the equivalence relation R with two classes:

$$\{0, 2\} \quad \text{and} \quad \{1\}.$$

This is no kernel equivalence (note that the quotient set $X/R = \{[0], [1]\}$ is not ordered, since $0 \leq 1$ implies $[0] \leq [1]$, but $1 \leq 2$ implies $[1] \leq [0]$).

2.7. Definition. An $(\mathscr{E}, \mathscr{M})$-category is said to satisfy the ***pullback axiom*** if in each pullback

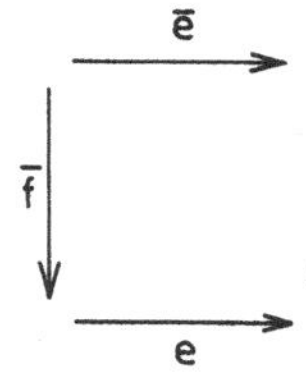

with $e \in \mathscr{E}$ we have $\bar{e} \in \mathscr{E}$.

Remark. The pullback axiom (which is the dual of the pushout axiom used in IV.7.12) makes the work with relations much simpler. For example, the composition of two relations

$$[f_1, f_2] \circ [g_1, g_2]$$

can be defined (for *arbitrary* representing pairs) by the pullback of f_2 and g_1:

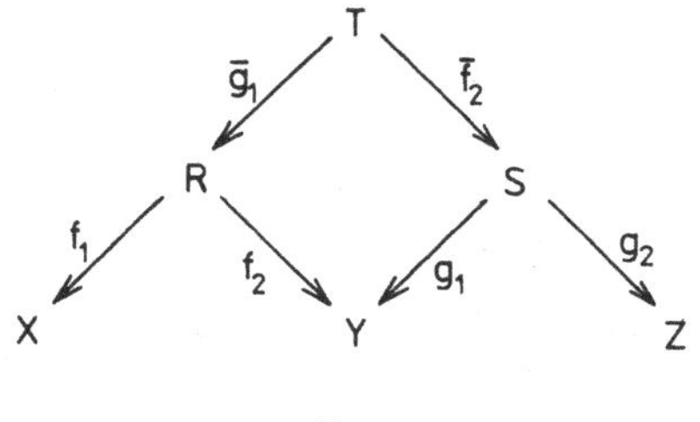

as the relation $[\bar{g}_1 \cdot f_1, \bar{f}_2 \cdot g_2]$.

Proof. If $[f_1, f_2] = r: R_0 \to X \times Y$ then, by definition, r is the image of the morphism $R \to X \times Y$ with the components f_1 and f_2. Thus, there is $e \in \mathscr{E}$ with

$$f_1 = e \cdot r_{(1)} \quad \text{and} \quad f_2 = e \cdot r_{(2)}.$$

Analogously, if $[g_1, g_2] = s: S \to Y \times Z$, then there is an $\mathscr{E}$-epi $\bar{e}: S \to S_0$ with

$$g_1 = \bar{e} \cdot s_{(1)} \quad \text{and} \quad g_2 = \bar{e} \cdot s_{(2)}.$$

We form the following four pullbacks:

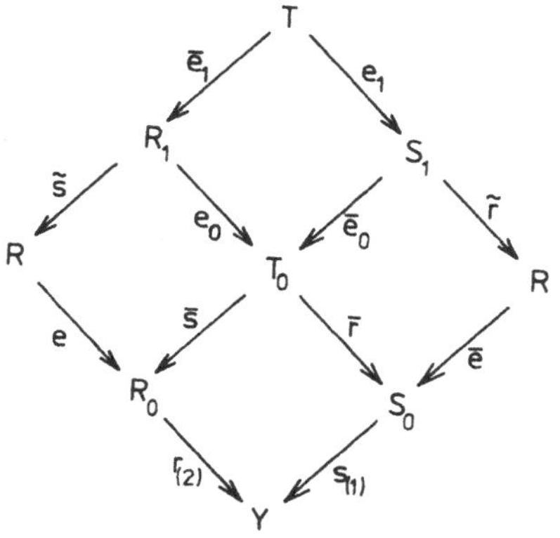

By definition,

$$r \circ s = [\bar{s} \cdot r_{(1)}, \bar{r} \cdot s_{(2)}].$$

Since adjacent pullbacks form a new pullback, the pullback of

$$f_2 = e \cdot r_{(2)} \quad \text{and} \quad g_1 = \bar{e} \cdot s_{(1)}$$

is formed by $\bar{e}_1 \cdot \tilde{s}$ and $e_1 \cdot \tilde{r}$ above. Thus, what we want to prove is that $r \circ s$ is represented by

$$[\bar{e}_1 \cdot \tilde{s} \cdot f_1, e_1 \cdot \tilde{r} \cdot g_2].$$

We use the fact that

$$\tilde{s} \cdot f_1 = \tilde{s} \cdot e \cdot r_{(1)} = e_0 \cdot \bar{s} \cdot r_{(1)}$$

and

$$\tilde{r} \cdot g_2 = \tilde{r} \cdot \bar{e} \cdot s_{(2)} = \bar{e}_0 \cdot \bar{r} \cdot s_{(2)}.$$

Thus, denoting $\hat{e} = \bar{e}_1 \cdot e_0 = e_1 \cdot \bar{e}_0$, we have

$$[e_1 \cdot \tilde{s} \cdot f_1, e_1 \cdot \tilde{r} \cdot g_2] = [\hat{e} \cdot \bar{s} \cdot r_{(1)}, \hat{e} \cdot \bar{r} \cdot s_{(2)}].$$

Finally, the pullback axiom guarantees that

$$e_0 \in \mathscr{E} \quad \text{and} \quad \bar{e}_0 \in \mathscr{E}$$

and, applied again, yields

$$\bar{e}_1 \in \mathscr{E} \quad \text{and} \quad e_1 \in \mathscr{E};$$

consequently,

$$\hat{e} = \bar{e}_1 \cdot e \in \mathscr{E}.$$

Hence, $[s \cdot r_{(1)}, \ \bar{r} \cdot s_{(2)}]$ represents the same relation (which is $r \circ s$) as $[\hat{e} \cdot \bar{s} \cdot r_{(1)}, \ \hat{e} \cdot \bar{r} \cdot s_{(2)}]$. Hence,

$$r \circ s = [\bar{e}_1 \cdot \tilde{s} \cdot f_1, e_1 \cdot \tilde{r} \cdot g_2]. \qquad \square$$

Proposition. The pullback axiom guarantees that the composition of relations is associative:

$$r \circ (s \circ t) = (r \circ s) \circ t$$

(for arbitrary relations $r: X \rightharpoonup Y$, $s: Y \rightharpoonup Z$ and $t: Z \rightharpoonup V$).

Proof. Let us form the following three pullbacks:

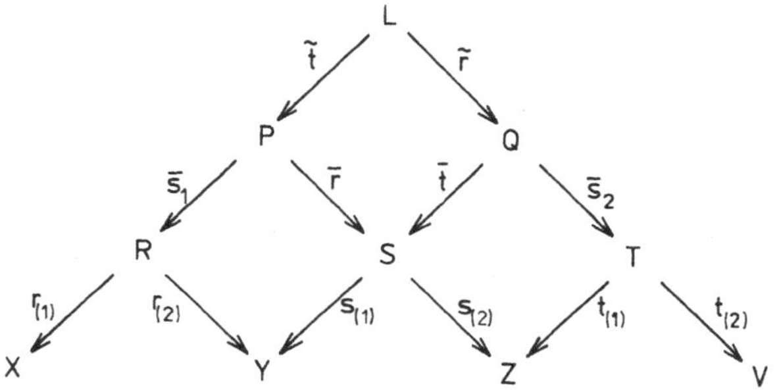

By definition,

$$r \circ s = [\bar{s}_1 \cdot r_{(1)}, \bar{r} \cdot s_{(2)}].$$

Since adjacent pullbacks form a pullback, the pullback of $\bar{r} \cdot s_{(2)}$ and $t_{(1)}$ is $\tilde{t}$ and $\tilde{r} \cdot \bar{s}_2$. Hence, by the preceding remark,

$$(r \circ s) \circ t = [\tilde{t} \cdot \bar{s}_1 \cdot r_{(1)}, \tilde{r} \cdot \bar{s}_2 \cdot t_{(2)}].$$

Analogously,

$$s \circ t = [\bar{t} \cdot s_{(1)}, \bar{s}_2 \cdot t_{(2)}]$$

and, by the preceding remark,

$$r \circ (s \circ t) = [\tilde{t} \cdot \bar{s}_1 \cdot r_{(1)}, \tilde{r} \cdot \bar{s}_2 \cdot t_{(2)}]. \qquad \square$$

Examples. (i) **Set** fulfils the pullback axiom: the pullback

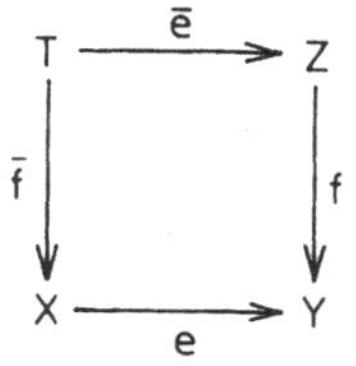

in **Set** can be described as the set

$$T = \{(x, z) \in X \times Z;\ (x)e = (z)f\}$$

with the projections $\bar{e}$ and $\bar{f}$. If e is onto, then so is $\bar{e}$ because for each $z \in Z$ the point $(z)f \in Y = (X)e$ can be expressed as $(z)f = (x_0)e$, $x_0 \in X$, and then

$$(x_0, z) \in T$$

fulfils $z = (x_0, z)\bar{e}$.

(ii) Let $\mathcal{K}$ be a concrete (III.3.8) $(\mathcal{E}, \mathcal{M})$-category with $\mathcal{E}$ = all morphisms which are surjective maps. Assume that finite limits are preserved by the forgetful functor (which they are in all current categories). Then $\mathcal{K}$ satisfies the pullback axiom. This applies to

R-**Mod**,

in fact, to any variety of universal algebras, as well as to

Pos, **Top**, **Gra**, etc.

considered as (epi, embedding)-categories.

(iii) Alas!, for other factorization systems the pullback axiom can fail. For example, the (quotient, mono)-category **Top** does not satisfy it, see Exercise A below.

Therefore, we try to work without the pullback axiom, whenever possible. We must be careful with the composition, however, because it can fail to be associative.

Remark. Composition of partial morphisms is always associative (even if the pullback axiom fails). In the above proof, assume $s_{(1)} \in \mathcal{M}$ and $t_{(1)} \in \mathcal{M}$. Then $\bar{s}_1$, $\bar{t}$ (and hence $\tilde{t}$) are all $\mathcal{M}$-monos. Therefore,

$$r \circ s = [\bar{s}_1 \cdot r_{(1)}, \bar{r} \cdot s_{(2)}] \text{ and } s \circ t = [t \cdot s_{(1)}, \bar{s}_{(2)} \cdot t_{(2)}]$$

and we obtain $(r \circ s) \circ t = r \circ (s \circ t)$.

2.8. Remark. The span algebras introduced in IV.7.4 can be considered as representations of "relational" F-algebras. By a *relational F-algebra* is meant a pair (Q, δ) consisting of a relation $\delta\colon FQ \rightharpoonup Q$. In this sense, the span algebra morphisms $(f, f_0)\colon (Q, D, \delta_0, \delta_1) \to (Q', D', \delta_0', \delta_1')$ correspond exactly to morphisms $f\colon Q \to Q'$ satisfying

$$f \circ fF \subset fF \circ \delta'\colon QF \rightharpoonup Q'$$

[where δ is represented by (δ_0, δ_1) and δ' is represented by (δ_0', δ_1')]. More in detail, for each span algebra morphism (f, f_0), the first component f satisfies the inclusion above. Conversely, representing $\delta\colon QF \rightharpoonup Q$ by (δ_0, δ_1) such that the induced morphism $D \to Q \times QF$ is in $\mathcal{M}$, and analogously with $\delta'\colon Q'F \rightharpoonup Q'$, then for each f satisfying the above inclusion there exists a unique f_0 such that $(f, f_0): (Q, D, \delta_0, \delta_1) \to (Q', D', \delta_0', \delta_1')$ is a span algebra morphism.

The free-completion construction IV.7.6 can also be performed on relations rather than spans: it is easy to verify that the result of this construction is independent of the choice of the representing span.

2.9. Functors applied to relations. For each functor $F: \mathscr{K} \to \mathscr{K}$ and each relation $r: X \rightharpoonup Y$ we define the relation

$$rF: XF \rightharpoonup YF$$

by

$$rF = [r_{(1)}F, r_{(2)}F].$$

Examples. (i) H_2: **Set** → **Set**. Each relation $r: X \rightharpoonup Y$ is represented by the inclusion map of a set $R \subset X \times Y$; then $r_{(1)}$ and $r_{(2)}$ are the projections. The relation $rH_2: X^2 \rightharpoonup Y^2$ is then the inclusion map of the set $R^{(2)} \subset X^2 \times Y^2$ of all pairs $((x_1, x_2), (y_1, y_2))$ for which there exists $(a, b) \in R \times R$ with $(x_1, x_2) = ((a)r_1, (b)r_{(1)})$ and $(y_1, y_2) = ((a)r_{(2)}, (b)r_{(2)})$, i.e., with $a = (x_1, y_1)$ and $b = (x_2, y_2)$. Thus,

$$(x_1, x_2)R^{(2)}(y_1, y_2) \quad \text{iff} \quad x_1Ry_1 \quad \text{and} \quad x_2Ry_2.$$

As a concrete example, consider the relation $r: \{x, y\} \rightharpoonup \{x, y, z\}$, given by the following graph

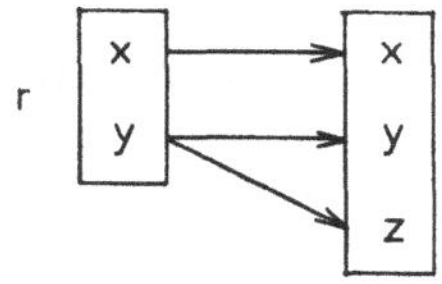

Then rH_2 is the following relation

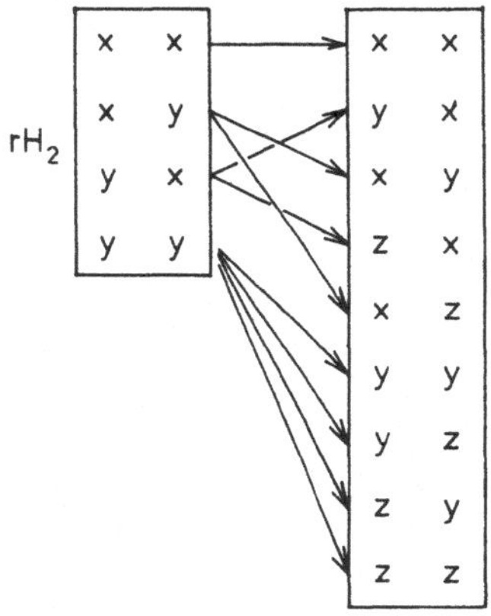

(ii) P: **Set** $\to$ **Set**, the power set functor (III.3.4). Let r be the relation above. Then rP is the following relation

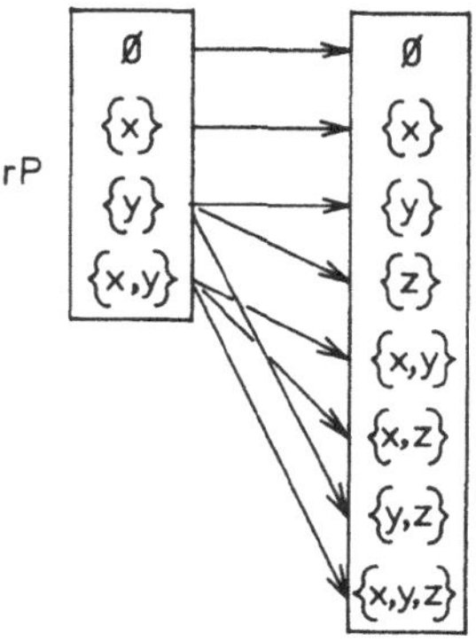

Remarks. (i) If F *preserves* $\mathscr{E}$*-epis* (i.e., if eF is an $\mathscr{E}$-epi for each $\mathscr{E}$-epi e), a more natural definition is the following:

$$\text{if } r = [f_1, f_2], \text{ then } rF = [f_1F, f_2F].$$

Indeed, $r = [f_1, f_2]$ means that there exists an $\mathscr{E}$-epi e with

$$e \cdot r_{(1)} = f_1 \quad \text{and} \quad e \cdot r_{(2)} = f_2.$$

This implies

$$eF \cdot r_{(1)}F = f_1F \quad \text{and} \quad eF \cdot r_{(2)}F = f_2F;$$

since $eF \in \mathscr{E}$,

$$[r_{(1)}F, r_{(2)}F] = [eF \cdot r_{(1)}F, eF \cdot r_{(2)}F].$$

(ii) If F *preserves* $\mathscr{M}$*-monos* (i.e., mF is an $\mathscr{M}$-mono for each $\mathscr{M}$-mono m), then F preserves partial morphisms, and for each partial morphism $r = [r_1, r_2]$, $r_1 \in \mathscr{M}$, we have

$$rF = [r_1F, r_2F].$$

(iii) Each functor F preserves the ordering of relations:

$$r \subset s \quad \text{implies} \quad rF \subset sF$$

for arbitrary relations $r, s: X \rightharpoonup Y$. F also preserves the inverses,

$$(r^{-1})F = (rF)^{-1}.$$

Both are easy to verify.

Proposition. Let $F: \mathscr{K} \to \mathscr{K}$ be a functor preserving $\mathscr{E}$-epis. For arbitrary relations

$$r: X \rightharpoonup Y \quad \text{and} \quad s: Y \rightharpoonup Z$$

we have

$$(r \circ s)F \subset rF \circ sF: XF \rightharpoonup YF.$$

Proof. We form the pullback of $r_{(2)}$ and $s_{(1)}$:

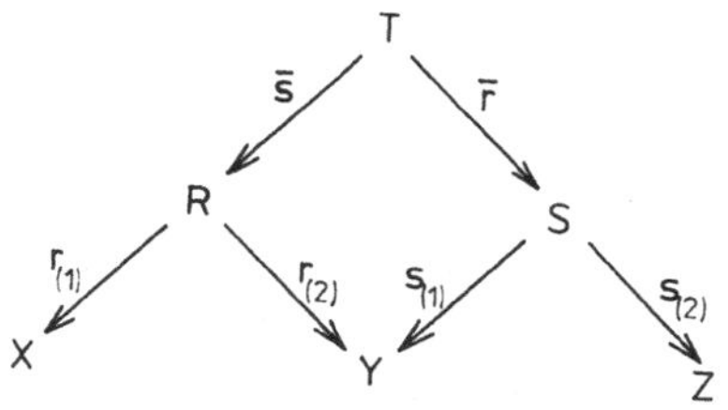

to get

$$(r \circ s)F = [\bar{s}F \cdot r_{(1)}F, \bar{r}F \cdot s_{(2)}F].$$

Let $\hat{r}: \hat{R} \to XF \times YF$ be an $\mathscr{M}$-mono representing rF, and let $e_r: RF \to \hat{R}$ be the $\mathscr{E}$-epi with

$$r_1F = e_r \cdot \hat{r}_{(1)} \quad \text{and} \quad r_2F = e_r \cdot \hat{r}_{(2)}.$$

Analogously, define $\hat{s}: \hat{S} \to YF \times ZF$ in $\mathscr{M}$ and $e_s: SF \to \hat{S}$ in $\mathscr{E}$. Now we form the pullback of $\hat{r}_{(2)}$ and $\hat{s}_{(1)}$:

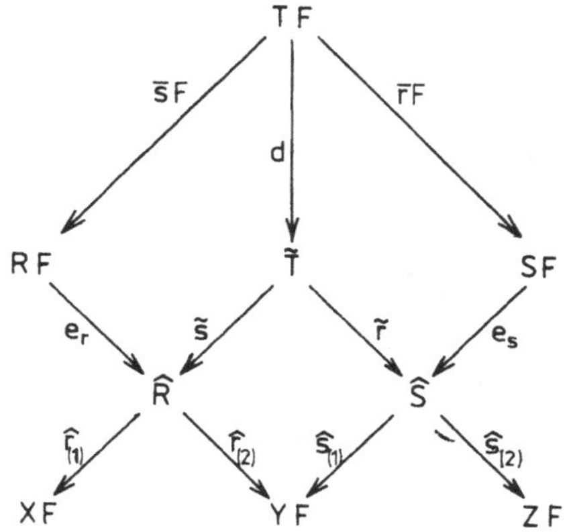

There exists a unique morphism $d: TF \to \tilde{T}$ for which the diagram above commutes. We have

$$rF \circ sF = \hat{r} \circ \hat{s} = [\tilde{s} \cdot \hat{r}_{(1)}, \tilde{r} \cdot s_{(2)}].$$

Therefore,

$$(r \circ s)F = [d \cdot \tilde{s} \cdot \hat{r}_{(1)},\ d \cdot \tilde{r} \cdot \hat{s}_{(2)}] \subset [\tilde{s} \cdot \hat{r}_{(1)},\ \tilde{r} \cdot \hat{s}_{(2)}] = rF \circ sF.$$

This concludes the proof. □

Remark. If $F: \mathcal{K} \to \mathcal{K}$ preserves $\mathcal{M}$-monos (but not necessarily $\mathcal{E}$-epis), then for arbitrary partial morphisms $r: X \rightharpoonup Y$ and $s: Y \rightharpoonup Z$ we have, again, $(r \circ s)F \subset rF \circ sF$. The proof is the same.

Even reasonable functors in reasonable categories can fail to preserve the composition of relations:

Example. The functor P_3: **Set** → **Set** (III.4.1) does not preserve composition of relations. For example, let

$$f: \{x, y, z\} \to \{x, y\}$$

be the map defined by

$$(x)f = x, (y)f = x, \quad (z)f = y,$$

and let $r: \{x, y\} \rightharpoonup \{x, y, z\}$ be the relation of Example V.2.8. Then

$$(f \circ r)P_3 \neq fP_3 \circ rP_3.$$

Inspecting the graph of rP above, we see that rP_3 is the restriction of rP to all non-empty sets, and that the graph of $fP_3 \circ rP_3$ is a subset of $\exp\{x, y, z\} \times \exp\{x, y, z\}$ containing for example the pair $(\{x, y, z\}, \{x, y, z\})$. On the other hand, the relation $f \circ r$ has the following graph

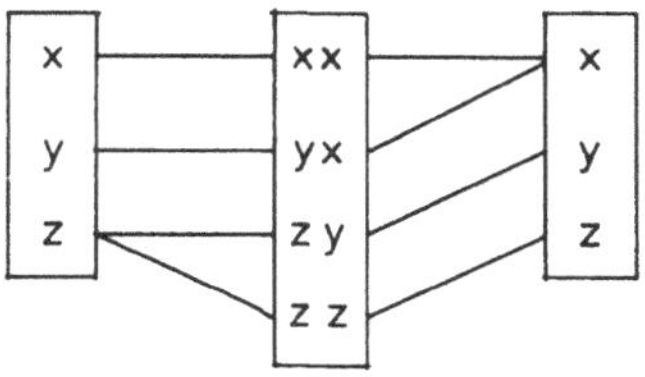

f∘r

There is no three-point subset of $f \circ r$ which is mapped onto $\{x, y, z\}$ simultaneously by both projections. Hence, the pair $(\{x, y, z\}, \{x, y, z\})$ is not contained in the graph of $(f \circ r)P_3$.

2.10. We are going to characterize functors preserving composition of relations.

Given a pullback $\bar{P}$ in the category $\mathcal{K}$:

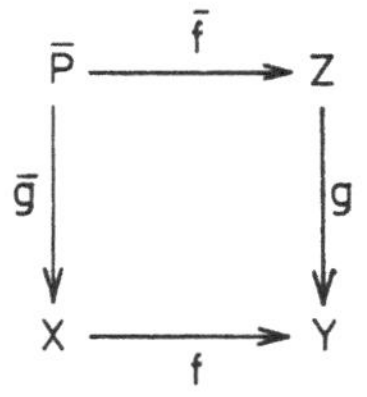

the image of this square factors through the pullback $\tilde{P}$ of fF and gF by a *canonical morphism* p: $\bar{P}F \to \tilde{P}$, i.e., the unique morphism such that the following diagram

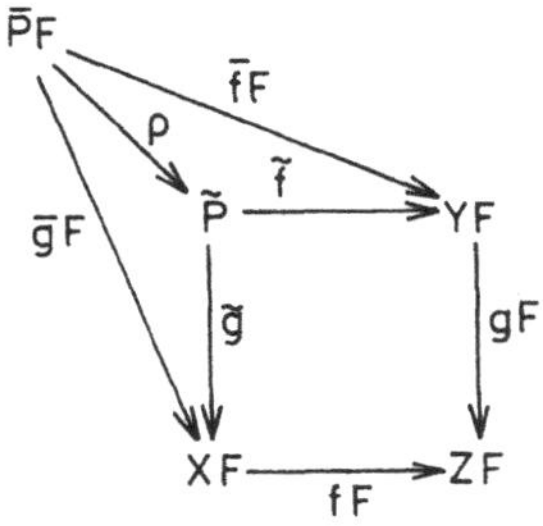

commutes. (Then F preserves the given pullback iff p is an isomorphism.)

Definition. A functor F: $\mathscr{K} \to \mathscr{K}$ is said to *cover pullbacks* if the canonical morphism of each pullback is an $\mathscr{E}$-epi.

Theorem. Let $\mathscr{K}$ satisfy the pullback axiom, and let F: $\mathscr{K} \to \mathscr{K}$ preserve $\mathscr{E}$-epis. Then F preserves the composition of relations iff F covers pullbacks.

Proof. (i) Let F cover pullbacks. To prove that

$$(r \circ s)F = rF \circ sF,$$

we form the pullback of $r_{(2)}$ and $s_{(1)}$:

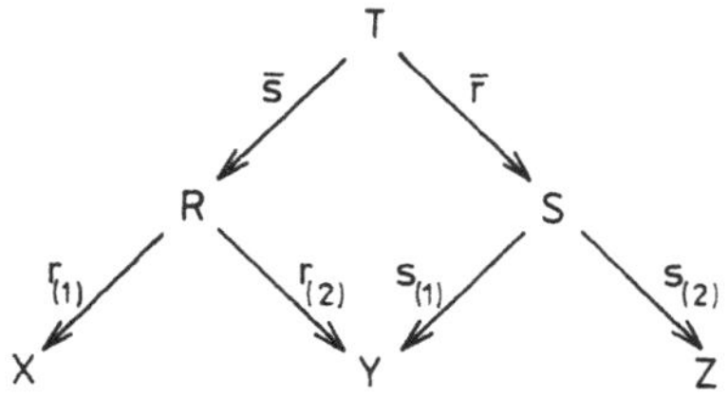

Then

$$(r \circ s)F = [\bar{s}F \cdot r_{(1)}F, \bar{r}F \cdot s_{(2)}F].$$

By Remark V.2.8, the composition of rF and sF is defined by the pullback of $r_{(2)}F$ and $s_{(1)}F$

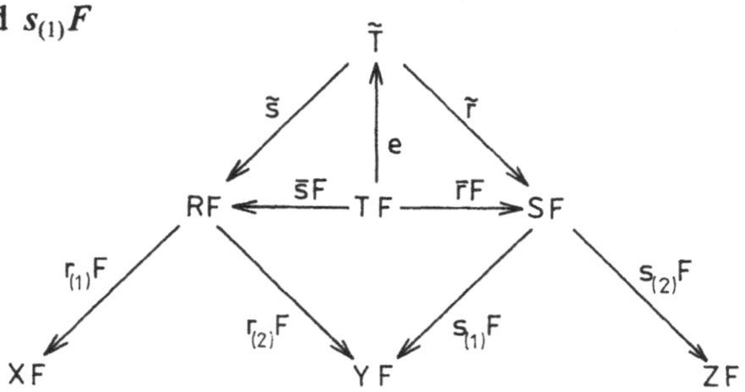

as

$$rF \circ sF = [\tilde{s} \cdot r_{(1)}F,\ \tilde{r} \cdot s_{(2)}F].$$

By hypothesis, there is an $\mathscr{E}$-epi e with

$$\bar{s}F = e \cdot \tilde{s} \quad \text{and} \quad \bar{r}F = e \cdot \tilde{r}.$$

Then $[\bar{s}F \cdot r_{(1)}F,\ \bar{r}F \cdot s_{(2)}F]$ is the same relation as $[\tilde{s} \cdot r_{(1)}F,\ \tilde{r} \cdot s_{(2)}F]$, i.e.,

$$(r \circ s)F = rF \circ sF.$$

(ii) Let F preserve the composition of relations. Given morphisms

$$f: X \to Y \quad \text{and} \quad g: Z \to Y,$$

we use the fact that F preserves the composition of

$$f = [1_X, f]: X \rightharpoonup Y \quad \text{and} \quad g^{-1} = [g, 1_Z]: Y \rightharpoonup Z.$$

This composition $f \circ g^{-1}$ is defined by the pullback of f and g:

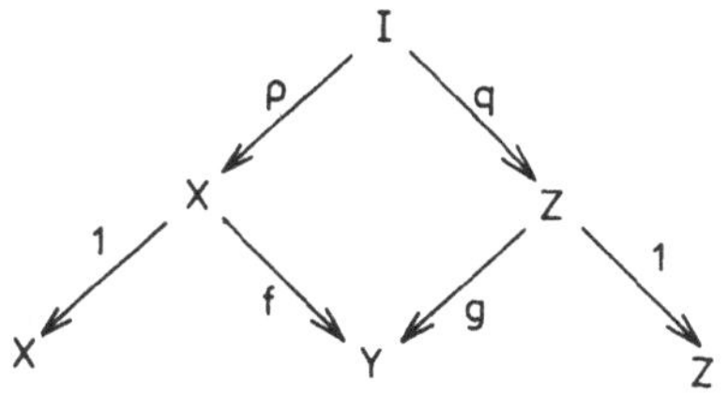

as

$$f \circ g^{-1} = [p, q].$$

We have

$$(f \circ g^{-1})F = fF \circ (gF)^{-1}$$

where $fF \circ (gF)^{-1}$ is defined by the pullback of fF and gF:

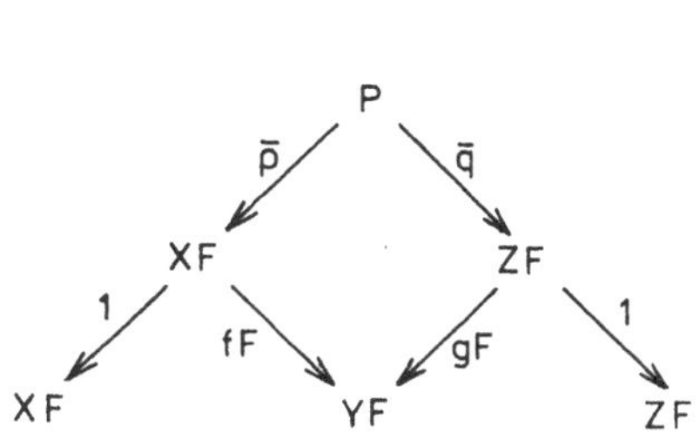

Thus,

$$[pF, qF] = [\bar{p}, \bar{q}]: XF \rightharpoonup ZF.$$

Let e be the canonical morphism with $pF = e \cdot \bar{p}$ and $qF = e \cdot \bar{q}$. Since

$$[e \cdot \bar{p}, e \cdot \bar{q}] = [\bar{p}, \bar{q}],$$

we conclude that $e \in \mathscr{E}$. □

Examples. (i) The functors $H_n: \mathscr{K} \to \mathscr{K}$ preserve composition of relations because they preserve limits. For $\mathscr{K} =$ **Set** (and for a number of current categories), coproducts of functors covering pullbacks also cover them. Hence, each H_Σ: **Set** → **Set** preserves composition of relations.

(ii) The power-set functor P: **Set** → **Set** preserves composition of relations—it is easy to verify that P covers pullbacks. But the subfunctor P_3 of P does not cover pullbacks.

Remark. If $\mathscr{K}$ satisfies the pullback axiom and if $F: \mathscr{K} \to \mathscr{K}$ preserves $\mathscr{M}$-monos, then F preserves the composition of partial morphisms iff F *preserves preimages* (i.e., preserves the pullback of f and g whenever $f \in \mathscr{M}$). This can be proved precisely as the preceding theorem, since the preservation of preimages is equivalent to the covering of preimages. [In fact, for each preimage the canonical morphism p: $\bar{P}F \to \bar{P}$ is an $\mathscr{M}$-mono: $f \in \mathscr{M}$ implies $\bar{f} \in \mathscr{M}$ and therefore $\bar{f}F = p \cdot \bar{f} \in \mathscr{M}$; hence, $p \in \mathscr{M}$ by III.5.1(i). Thus, p is an isomorphism iff p is an $\mathscr{E}$-epi.)

Examples. (iii) The functor P_3 preserves preimages: given a map $g: Z \to Y$ and a subset $X \subset Y$, then

$$(ZP_3)(gP_3)^{-1} = ((Z)g^{-1})P_3.$$

(iv) The following functor D_2: **Set** → **Set** does not preserves preimages. For each set X put

$$XD_2 = \{(x, y) \in X \times X;\ x \neq y\} \cup \{*\}$$

where $*$ is any element outside of $X \times X$; for each map $f: X \to Y$,

$$(x, y)fD_2 = \begin{cases} ((x)f, (y)f) & \text{if } (x)f \neq (y)f; \\ * & \text{if } (x)f = (y)f; \end{cases}$$
$$(*)fD_2 = *.$$

This functor does not preserve the composition of the map f: $\{x, y, z\} \to \{x, y\}$ defined by

$$(x)f = x;\ (y)f = x;\ (z)f = y$$

with the partial map $g: \{x, y\} \rightharpoonup y$, defined by

$$(y)g = y.$$

The relation $f \circ g$ is defined only in z and hence, $(f \circ g)D_2$ is defined only in $*$. On the other hand, $fD_2 \circ gD_2$ is defined in (x, y) and (y, x).

Exercises V.2

A. The pullback axiom in Top. Verify that the (regular epi, mono)-category **Top** does not satisfy the pullback axiom (a regular epi is a surjective continuous map $f: A \twoheadrightarrow B$ such that each open set in A has the form $(U)f^{-1}$ for $U \subset B$ open). Use the following pullback

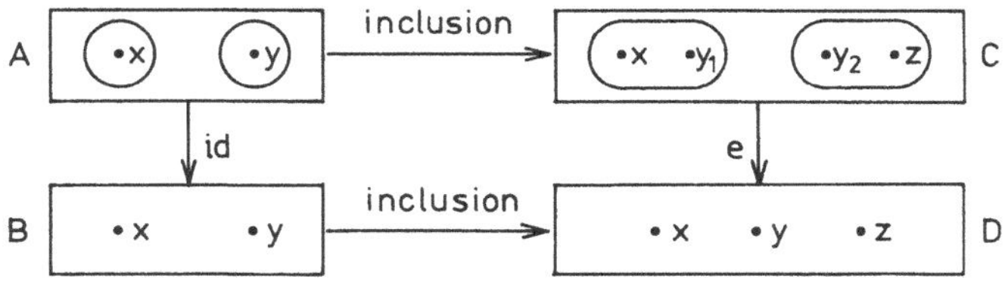

where e is the map merging y_1 and y_2 to y, and the topology on each set is indicated in the picture: A is discrete (each set is open), B and D are indiscrete (only $\emptyset$ and the whole set are open) and C has open sets $\emptyset$, $\{x, y_1\}$, $\{y_2, z\}$ and C.

B. Preservation of composition. (i) Verify that the set functors P, P_f and P_2 cover pullbacks and thus, preserve composition of relations (in contrast to P_3). Verify that P_n does not preserve composition of relations if $3 \leq n < \omega$.

(ii) Verify that the quotient of H_3 given by the equation $(x, x, y)\sigma = (x, x, z)\sigma$ does not preserve preimages.

C. Partial morphisms. For each partial morphism $[f_1, f_2]: X \rightharpoonup Y$, $f_1 \in \mathcal{M}$, prove that a pair $[g_1, g_2]$ with $g_1 \in \mathcal{M}$ represents the same relation iff there is an isomorphism i with $g_1 = i \cdot f_1$ and $g_2 = i \cdot f_2$.

V.3. Finitary Functors

3.1. The existence of minimal reductions is characterized by the preservation of cointersections. But this in itself is a condition difficult to verify. We prove that under additional hypotheses, a functor preserves cointersections iff it is finitary (i.e., preserves directed unions). The latter condition is much easier to check in a concrete situation, and has a clear intuitive meaning. For the implication

finitary $\Rightarrow$ preserves cointersections

we have to restrict the factorization systems to $\mathcal{E}$ = regular epis and $\mathcal{M}$ = monos, and then we obtain quite a general result. For the converse implication, some further requirements are needed. All these are fulfilled by every set functor and every functor on R-**Vect** (R any commutative field). Hence, in these categories, particularly satisfactory results are obtained.

Definition. Let $\mathcal{K}$ be an $(\mathcal{E}, \mathcal{M})$-category. A functor $F: \mathcal{K} \to \mathcal{K}$ is said to be *finitary* if it preserves directed unions.

Remarks. (i) Explicitly, F is finitary iff for each union

$$m = \bigcup_{j \in J} m_j$$

with J directed (i.e., given $j_1, j_2 \in J$ there exists $j \in J$ such that $m_{j_1} \subset m_j$ and $m_{j_2} \subset m_j$) we have

$$\operatorname{im}(mF) = \bigcup_{j \in J} \operatorname{im}(m_j F).$$

(ii) For the categories **Set** and R-**Mod**, the definition above is equivalent to that given in Chapter III: a functor F is finitary iff for each set (module) X and each point $x \in XF$ there exists a morphism $f: Y \to X$ with Y finite (finite-dimensional) and $x \in (YF)fF$.

[Proof. If F is finitary in the present sense, we can use the fact that each object X is a directed union of its finite (or finite-dimensional) subobjects. Conversely, if F is finitary in the sense of Chapter III and if $X = \bigcup_{j \in J} X_j$ is a directed union, then for each $x \in XF$ we find $f: Y \to X$ with $x \in (YF)fF$ and Y finite (or finite-dimensional). Since the union is directed, there exists $j_0 \in J$ with $(Y)f \subset X_{j_0}$. This proves that $XF \subset \bigcup (X_j F) m_j F$, where $m_j: X_j \to X$ are the inclusion maps, and the converse inclusion is clear.]

(iii) Some authors define finitary functors as those preserving directed colimits. For our purposes, this does not make much difference—see the Characterization Theorem below.

Lemma. Let $\mathcal{K}$ have pullbacks. A functor F is finitary iff for each directed family m_j ($j \in J$) of $\mathcal{M}$-subobjects of an object A,

$$\bigcup_{j \in J} m_j = 1_A \quad \text{implies} \quad \bigcup_{j \in J} \operatorname{im}(m_j F) = 1_{AF}.$$

Proof. Let $\bar{m}: A_j \to B$ ($j \in J$) be a directed family of $\mathcal{M}$-monos with union $\bar{m}: A \to B$. We have $\mathcal{M}$-monos $m_j: A_j \to A$ with $\bar{m}_j = m_j \cdot m$ ($j \in J$), and then

$$\bigcup_{j \in J} m_j = 1_A.$$

(In fact, if a subobject $\tilde{m}$ of A contains each m_j, then $\tilde{m} \cdot m$ contains each $m_j \cdot m = \bar{m}_j$ and hence, $\tilde{m} \cdot m$ contains m. It follows that $\tilde{m}$ is an isomorphism.) Therefore,

$$\bigcup_{j \in J} \operatorname{im}(m_j F) = 1_{AF}.$$

Let m^* be an $\mathcal{M}$-subobject of BF containing each $\mathrm{im}(\bar{m}_j F)$. This means that for each $j \in J$ we have

$$\bar{m}_j F = \bar{e}_j \cdot \bar{u}_j \cdot m^*,$$

with $\bar{e}_j \in \mathcal{E}$ and $\bar{u}_j \in \mathcal{M}$. Let

$$m_j F = e_j \cdot u_j$$

be the image factorization. Then we can use the diagonal fill-in:

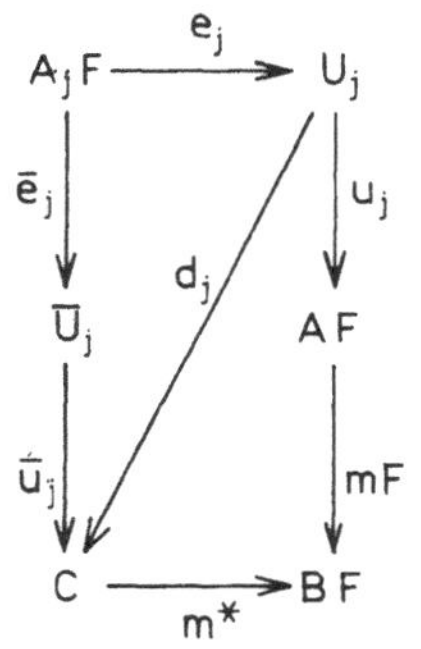

Let p and q form the pullback of m^* and $\bar{m}F$, then for each $j \in J$ there exists a unique r_j such that the following diagram

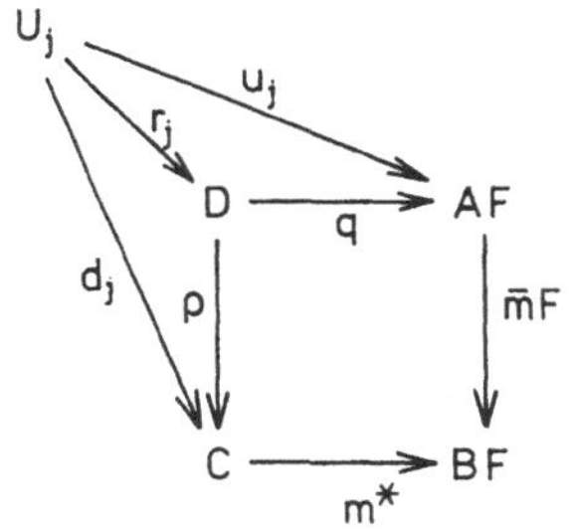

commutes. Thus, q is an $\mathcal{M}$-mono (opposite to $m^* \in \mathcal{M}$ in a pullback, see III.5.1) containing each u_j and hence, q is an isomorphism. Therefore,

$$\bar{m}F = (q^{-1} \cdot p) \cdot m^*,$$

which proves that $\mathrm{im}(\bar{m}F) \subset m^*$.

3.2. Let us assume that $\mathcal{K}$ is a finitely complete category with *regular factorizations*, i.e., an $(\mathcal{E}, \mathcal{M})$-category where

$$\mathcal{E} = \text{regular epis} \quad \text{and} \quad \mathcal{M} = \text{monos}.$$

For each morphism $e: X \twoheadrightarrow Y$ we can form the pullback of e with itself:

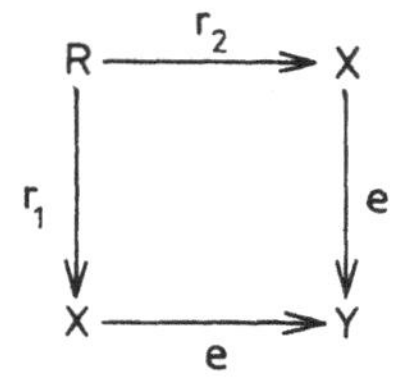

The pair $[r_1, r_2]$ represents an equivalence relation $r: X \rightharpoonup X$ called the kernel equivalence of e, see V.2.6.

Kernel equivalences and regular epis are closely related:

(a) If $e: X \twoheadrightarrow Y$ is a regular epi and $[r_1, r_2]: X \rightharpoonup X$ is its kernel equivalence, then e is the coequalizer of r_1 and r_2.

Proof. Let $p_1, p_2: P \twoheadrightarrow X$ be morphisms such that e is the coequalizer of p_1 and p_2.

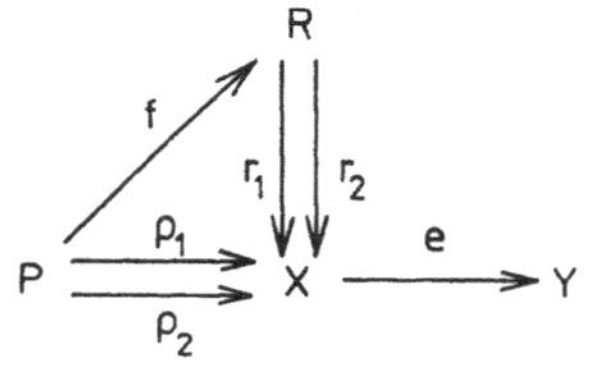

Since $p_1 \cdot e = p_2 \cdot e$, there exists a unique morphism $f: P \twoheadrightarrow R$ with $p_1 = f \cdot r_1$ and $p_2 = f \cdot r_2$. Given a morphism $\bar{e}: X \twoheadrightarrow \bar{Y}$ with $r_1 \cdot \bar{e} = r_2 \cdot \bar{e}$, we have also

$$p_1 \cdot \bar{e} = f \cdot r_1 \cdot \bar{e} = f \cdot r_2 \cdot \bar{e} = p_2 \cdot \bar{e}$$

and hence, $\bar{e}$ factors uniquely through e. Therefore, e is a coequalizer of r_1 and r_2. □

(b) Let e and e' be regular quotients of an object X, and let $r \subset X \times X$ be the kernel equivalence of e and r' the kernel equivalence of e'. Then

$$r \subset r' \quad \text{iff} \quad e \leq e'.$$

Proof. If $r \subset r'$, then $r'_{(1)} \cdot e' = r'_{(2)} \cdot e'$ implies $r_{(1)} \cdot e' = r_{(2)} \cdot e'$. Since e is the coequalizer of $r_{(1)}$ and $r_{(2)}$, it follows that e' factorizes through e, i.e., $e \leq e'$. Conversely, if $e \leq e'$ then $r_{(1)} \cdot e = r_{(2)} \cdot e$ implies $r_{(1)} \cdot e' = r_{(2)} \cdot e'$. Since $r'_{(1)}$ and $r'_{(2)}$ form the pullback of e' and e', there exists a unique morphism f with $r_{(1)} = f \cdot r'_{(1)}$ and $r_{(2)} = f \cdot r'_{(2)}$, i.e., $r \subset r'$. □

(c) Let $e_j: X \twoheadrightarrow Y_j$ $(j \in J)$ be quotients of an object X, and let $r_j \subset X \times X$ be the kernel equivalence of e_j $(j \in J)$. Put

$$r = \bigcup_{j \in J} r_j \subset X \times X.$$

The coequalizer of $r_{(1)}$ and $r_{(2)}$ is the cointersection of e_j, $j \in J$. Hence, if $\mathcal{K}$ has unions and coequalizers, it has cointersections. This follows immediately from (b).

3.3. Remark. We want to characterize functors preserving cointersections. We first consider finite cointersections, i.e., pushouts of regular epis.

Let us say that a pair of morphisms $p_1, p_2 : X \to Y$ is *reflexive* if it represents a reflexive relation (V.2.5) $[p_1, p_2] : Y \rightharpoonup Y$. A functor is said to *preserve reflexive coequalizers* if it preserves the coequalizer of any reflexive pair.

Construction of the least equivalence containing a given relation $r: X \rightharpoonup X$.

Let $\mathcal{K}$ be a complete, well-powered category with regular factorizations. We define relations

$$r_n : X \rightharpoonup X \quad (n \in \mathbf{Ord})$$

by transfinite induction.

(a) Denote by r_0 the least reflexive and symmetric relation containing r, i.e.,

$$r_0 = (r \cup \Delta_X) \cup (r \cup \Delta_X)^{-1}.$$

(b) Given r_n, denote by A_n the least set of relations $X \rightharpoonup X$ containing r_n and closed under composition and the formation of inverses. Put

$$r_{n+1} = \bigcup_{s \in A_n} s.$$

(c) Given a limit ordinal i, put

$$r_i = \bigcup_{n < i} r_n.$$

The unions in (b) and (c) exist because $\mathcal{K}$ is complete and well-powered, and hence, it has intersections, which implies that the poset of all subobjects of any object is a complete lattice. Put

$$r^* = r_k$$

where k is an ordinal with $r_k = r_n$ for all $n \geq k$; such an ordinal exists because $\mathcal{K}$ is well-powered. Then

(i) r^* is the least equivalence containing r;

(ii) each of the sets A_n above is directed;

(iii) r and r^* have the same coequalizer (i.e., a morphism is a coequalizer of $r_{(1)}$ and $r_{(2)}$ iff it is a coequalizer of $r^*_{(1)}$ and $r^*_{(2)}$).

Proof. (i) The relation r^* is

—reflexive: $\Delta \subset r_0 \subset r$;

—symmetric: for each n we have

$$r_{n+1} = \bigcup_{s \in A_n} s \cup s^{-1}$$

because A_n is closed under the formation of inverses. Since a union of symmetric relations is obviously symetric, it can be easily proved by induction that each r_n is symmetric;

—transitive: since $r_k \circ r_k \in A_k$, we have

$$r^* \circ r^* = r_k \circ r_k \subset r_{k+1} = r_k = r^*.$$

Let s be an arbitrary equivalence relation containing r. Let B be the set of all subrelations of s—note that B is clearly closed under composition. To prove $r^* \in B$, we verify that $r_n \in B$ by induction in n. First step: $r_0 = (r \cup \Delta) \circ (r \cup \Delta)^{-1} \subset s \circ s^{-1} = s \in B$. Isolated step: if $r_n \in B$ then $A_n \subset B$, hence, $r_{n+1} \subset s$. Limit step is clear.

(ii) Since each r_n is reflexive, all relations in A_n are, obviously, reflexive. Given $s_1, s_2 \in A_n$ then $s_1 \circ s_2 \in A_n$ and

$$s_1 = s_1 \circ \Delta \subset s_1 \circ s_2; \quad s_2 = \Delta \circ s_2 \subset s_1 \circ s_2.$$

(iii) For each morphism $f: X \to Y$ the statement $r_{(1)} \circ f = r_{(2)} \circ f$ means that r is contained in the kernel equivalence ker f; by (i) this is equivalent to $r^* \subset$ ker f, i.e., to $r^*_{(1)} \circ f = r^*_{(2)} \circ f$. Hence, r and r^* have the same coequalizer. □

3.4. Definition. A functor is said to be *right exact* if it preserves coequalizers of equivalence relations.

Remarks. (i) This is a much weaker condition than the preservation of reflexive coequalizers. For example, we shall prove below that each functor on the category **Set** or R-**Vect** is right exact.

(ii) The terminology here comes from homological algebra: right exact functors are those which preserve short right-exact sequences (which are just the coequalizers of equivalence relations in those categories).

(iii) A functor preserving finite cointersections is right exact. In fact, for each equivalence relation r ($: R \to X \times X$ a mono) there exists $d: X \to R$ with

$$d \cdot r_{(1)} = d \cdot r_{(2)} = 1_X,$$

because r is reflexive. Forming the pushout of $r_{(1)}$ and $r_{(2)}$

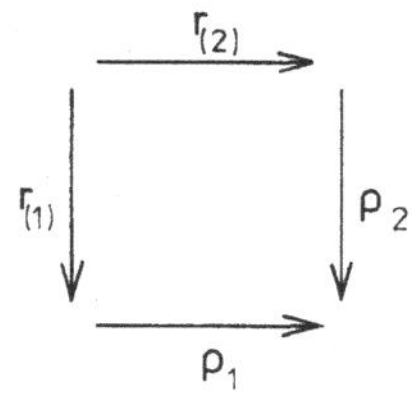

we have

$$p_1 = d \cdot r_1 \cdot p_1 = d \cdot r_2 \cdot p_2 = p_2$$

and hence, $p_1 = p_2$ is the coequalizer of $r_{(1)}$ and $r_{(2)}$. Since $r_{(1)}$ and $r_{(2)}$ are split epis, the pushout above is a cointersection. The preservation of this cointersection is equivalent to the preservation of the coequalizer of $r_{(1)}$ and $r_{(2)}$.

3.5. Theorem. Let $\mathscr{K}$ be a complete, well-powered category with regular factorizations. Each right exact, finitary functor $F: \mathscr{K} \to \mathscr{K}$ preserves cointersections.

Observation. Slightly weaker hypotheses will be needed in the proof below:
(i) $\mathscr{K}$ be finitely complete and well-powered;
(ii) $\mathscr{K}$ have unions and cointersections;
(iii) $\mathscr{K}$ have coequalizers.
Note that the present hypotheses imply that $\mathscr{K}$ has the properties (i)—(iii): $\mathscr{K}$ has intersections and hence also unions. The existence of cointersections follows from III.5.3: $\mathscr{K}$ is regularly cowell-powered because it is well-powered and we can apply V.3.2.b, therefore, $\mathscr{K}$ has cointersections. This implies that for each object Y, all regular quotients form a (small) complete lattice. Consequently, each pair of morphisms $f_1, f_2: X \to Y$ has a coequalizer, viz., the least regular quotient $e: Y \to Y'$ of Y such that $f_1 \cdot e = f_2 \cdot e$.

Proof. I. F preserves the coequalizer of any pair $f_1, f_2: X \to Y$ which represents a reflexive relation $r: Y \rightharpoonup Y$.

To prove this, it is sufficient to show that the least equivalence r^* containing r fulfils

$$rF \subset r^*F \subset (rF)^*$$

(where rF is defined as $[r_{(1)}F, r_{(2)}F]$, see V.2.9). Since rF and $(rF)^*$ have the same coequalizer by V.3.3(iii), it is obvious that also rF and r^*F have the same coequalizer. And F is right exact, therefore, it preserves the coequalizer of r^* and hence, also of r. Finally, $r = [f_1, f_2]$ means that there exists a regular epi e with $f_1 = e \cdot r_{(1)}$ and $f_2 = e \cdot r_{(2)}$; it follows that the coequalizers of f_1, f_2 and $r_{(1)}$, $r_{(2)}$ coincide, too, and hence, F preserves the coequalizer of f_1 and f_2.

The first inclusion follows from III.2.9(iii), since $r \subset r^*$. To prove $r^*F \subset (rF)^*$, we proceed by induction: let r_n denote the steps of the Construction V.3.3 applied to r, and $\bar{r}_n$ the steps for rF; we prove that

$$r_nF \subset \bar{r}_n \quad (n \in \mathbf{Ord}).$$

(a) Since r is a reflexive relation, we have $r_0 = r \circ r^{-1}$ and hence, by V.2.9,

$$r_0F = (r \circ r^{-1})F \subset rF \circ (rF)^{-1} \subset \bar{r}_0.$$

(b) F is finitary, i.e. it preserves the directed unions and A_n is directed, hence

$$r_{n+1}F = \bigcup_{s \in A_n} sF.$$

Also, we have

$$\bar{r}_{n+1} = \bigcup_{s \in \bar{A}_n} \bar{s}$$

where $\bar{A}_n$ is the least set of relations containig r_n and closed under composition and inverses. The set

$$B = \{s : X \rightharpoonup X; \quad sF \subset \bar{s} \text{ for some } \bar{s} \in \bar{A}_n\}$$

is obviously closed under composition and inverses. Therefore, $A_n \subset B$, and we get

$$r_{n+1}F = \bigcup_{s \in A_n} sF \subset \bigcup_{s \in B} sF \subset \bigcup_{\bar{s} \in \bar{A}_n} \bar{s} = \bar{r}_{n+1}.$$

(c) The limit step is clear, since F preserves unions of chains:

$$r_iF = \bigcup_{n<i} r_nF \subset \bigcup_{n<i} \bar{r}_n = \bar{r}_i.$$

II. F preserves finite cointersections, i.e., pushouts

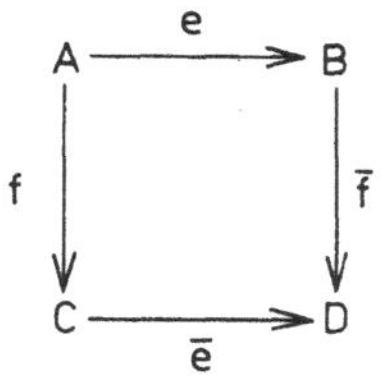

of regular epis.

To prove this, let $r_1, r_2 : R \to A$ be the kernel equivalence of e. The relation $r = [r_1, r_2]$ is reflexive and hence, also $[r_1 \cdot f, r_2 \cdot f] : C \rightharpoonup C$ is a reflexive relation: we have $\Delta_A \subset r$, thus $[f, f] \subset [r_1 \cdot f, r_2 \cdot f]$ and since f is a regular epi, $[f, f] = [1_C, 1_C] = \Delta_C$.

The coequalizer of $r_1 \cdot f$ and $r_2 \cdot f$ is $\bar{e}$. To verify this, consider a morphism $k : C \to X$ with

$$r_1 \cdot f \cdot k = r_2 \cdot f \cdot k.$$

Since e is a coequalizer of r_1 and r_2 [V.3.2(a)], there is a unique $h : B \to X$ with

$$f \cdot k = e \cdot h.$$

Using the universal property of the pushout above, we conclude that k factors through e and hence, e is a coequalizer of $r_1 \cdot f$ and $r_2 \cdot f$.

By I, the functor F preserves this coequalizer: $\bar{e}F$ is the coequalizer of $(r_1 \cdot f)F$ and $(r_2 \cdot f)F$. We are prepared to prove that F preserves the pushout above. Let p and q be morphisms with $fF \cdot p = eF \cdot q$:

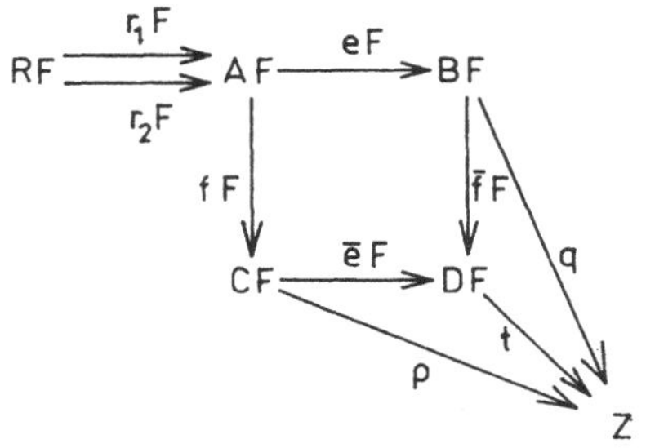

Then

$$(r_1 \cdot f)F \cdot p = r_1F \cdot eF \cdot q = r_2F \cdot eF \cdot q = (r_2 \cdot f)F \cdot p,$$

and because $\bar{e}F$ is the coequalizer of $(r_1 \cdot f)F$ and $(r_2 \cdot f)F$, there is a morphism $t: DF \to Z$ with $p = \bar{e}F \cdot t$. Then also $q = \bar{f}F \cdot t$ because eF is a (regular) epi, and

$$eF \cdot q = fF \cdot p = fF \cdot \bar{e}F \cdot t = eF \cdot \bar{f}F \cdot t.$$

This proves that $\bar{e}F$ and $\bar{f}F$ form the pushout of eF and fF.

III. F preserves cointersections. It is sufficient to prove that F preserves directed cointersections because each cointersection of quotients e_j ($j \in J$) can be obtained as a directed cointersections of finite "subcointersections". That is, consider the collection M_i, $i \in I$, of all finite subsets of J, and for each i let $\bar{e}_i$ be the (finite) cointersection of all e_j, $j \in M_i$. The collection $\bar{e}_i$, $i \in I$, is directed because for arbitrary $i_1, i_2 \in I$ we have a finite subset $M_{i_1} \cup M_{i_2} \subset J$, thus, $M_{i_1} \cup M_{i_2} = M_{i_3}$ for some $i_3 \in I$, and then $\bar{e}_{i_1} \leq \bar{e}_{i_3}$ and $\bar{e}_{i_2} \leq \bar{e}_{i_3}$. It is obvious that the collections e_j ($j \in J$) and $\bar{e}_i$ ($i \in I$) have the same cointersection.

Let

$$e_j: X \to Y_j \quad (j \in J)$$

be a directed collection of regular quotients. For each $j \in J$, let r_j be the kernel equivalence of e_j. Then r_j ($j \in J$) is a directed collection of subobjects of $X \times X$ (see V.3.2). The coequalizer e of the relation

$$r = \bigcup_{j \in J} r_j$$

is the cointersection of e_j ($j \in J$), and the relation r is reflexive (because each r_j is reflexive). By I above, F preserves the coequalizer of r. Hence, eF is the coequalizer of

$$rF = \bigcup_{j \in J} r_jF.$$

Let us prove that eF is the cointersection of e_jF ($j \in J$). For each regular epi

$$\bar{e}: XF \to Z$$

with $e_jF \leq \bar{e}$ $(j \in J)$ the kernel equivalence $\bar{r}$ fulfils

$$r_jF \subset \bar{r}.$$

(Indeed, if $r_j = [p_1, p_2]$ then $p_1 \cdot e_j = p_2 \cdot e_j$ implies $p_1F \cdot e_jF = p_2F \cdot e_jF$ and thus, $p_1F \cdot \bar{e} = p_2F \cdot \bar{e}$. We conclude that $r_jF = [p_1F, p_2F] \subset \ker \bar{e} = \bar{r}$.) Thus,

$$rF \subset \bar{r}.$$

Since eF is the coequalizer of rF and $\bar{e}$ the coequalizer of $\bar{r}$, this implies $eF \leq \bar{e}$. Thus, eF is the cointersection of e_jF $(j \in J)$. □

3.6. Next, we turn to the reverse implication:

$$F \text{ preserves cointersections} \Rightarrow F \text{ is finitary.}$$

Here, we work with an arbitrary factorization system $(\mathscr{E}, \mathscr{M})$ but we need rather extensive additional hypotheses. They include the following:

$$F \text{ preserves finite intersections}$$

(i.e., pullbacks of $\mathscr{M}$-monos). This is no restriction if $\mathscr{K}$ = **Set** (III.4.6) or $\mathscr{K}$ = **Vect** [III.4.A.(iii)] but generally, this is a burden, unlike the right exactness in the preceding theorem (of which we know that it follows from the preservation of cointersections).

Let us discuss the hypotheses on $\mathscr{K}$. First, assume that a *directed collection of pullbacks* is given:

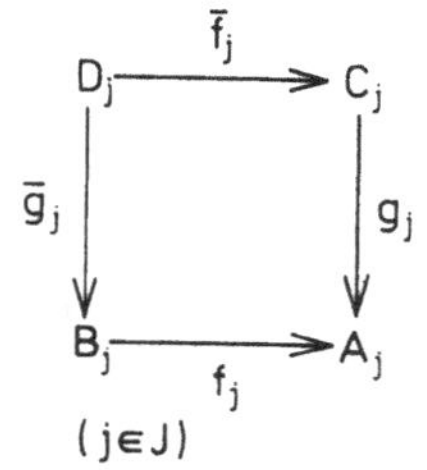

This means that $(J, \leq)$ is a directed poset and that A, B, C and D are diagrams from $(J, \leq)$ into $\mathscr{K}$ (A has objects A_j, $j \in J$, and morphisms a_{jk}, $j \leq k$, analogously B, C and D) and $f, \bar{f}, g$ and $\bar{g}$ are natural transformations. Let us form the colimits of the diagram above:

$$A_0 = \operatorname{colim} A$$

with injections $a_{j0}: A_j \to A_0$ $(j \in J)$, analogously B_0, C_0 and D_0. Here we assume $0 \notin J$ for simplicity. The four transformations above yield four morphisms forming a commuting square:

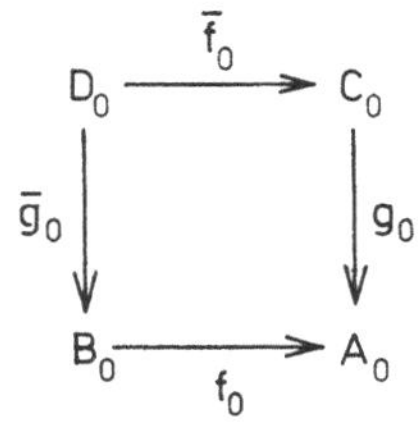

There is, in general, no reason why this colimit square should be a pullback again. We shall postulate this as a condition on $\mathscr{K}$.

Definition. A category with pullbacks and directed colimits is said to have *stable pullbacks* if for each directed collection of pullbacks also the colimit square is a pullback.

Remarks. (i) Given a directed poset $(J, \leq)$, denote by $\mathscr{K}^{(J, \leq)}$ the category of all diagrams over $(J, \leq)$ in $\mathscr{K}$ (as objects) and all natural transformations (as morphisms). The formation of colimit defines a functor

$$\text{colim}: \mathscr{K}^{(J, \leq)} \to \mathscr{K}.$$

Then $\mathscr{K}$ has stable pullbacks iff colim preserves pullbacks for each directed poset.

(ii) It is easy to check that in each category pushouts are stable (i.e., the colimit square of a directed family of pushouts is a pushout). This also follows from the general principle that colimits commute with colimits.

Examples. (i) **Set** has stable pullbacks. Indeed, the functor

$$\text{colim}: \mathbf{Set}^{(J, \leq)} \to \mathbf{Set}$$

above preserves all finite limits. This means that colim preserves both finite products and equalizers, and both are easily verified.

(ii) R-**Mod** has stable pullbacks for any commutative ring R. This follows from the fact that **Set** has this property and that both pullbacks and directed colimits are "created" by the corresponding constructs in **Set**. (I.e., a commuting square of linear maps is a pullback in R-**Mod** iff it is a pullback in **Set**, analogously with the directed colimits.)

Further examples are discussed in the Exercises below.

3.7. Another condition we need concerns the amalgams of subobjects $m: A \to B$, i.e., the objects obtained from the coproduct $B + B$ by the identification of the two copies of A. They are expressed by the pushout of m with itself.

Definition. An $(\mathscr{E}, \mathscr{M})$-category $\mathscr{K}$ is said to have the *amalgamation property* if each pushout

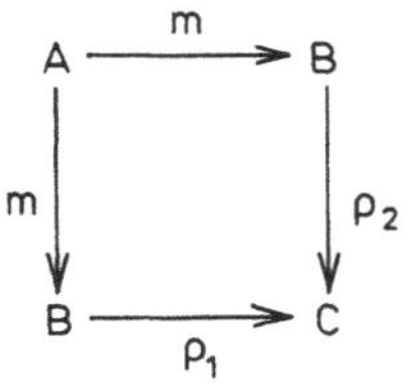

with $m \in \mathscr{M}$ is also a pullback of monos $p_1, p_2 \in \mathscr{M}$.

Examples. (i) **Set** has the amalgamation property. Given $A \subset B$ then

$$C = B + B/\sim \; = B \times \{1, 2\}/\sim,$$

where the equivalence $\sim$ has the following classes: $[a] = \{(a, 1), (a, 2)\}$ for each $a \in A$; $[b, 1] = \{(b, 1)\}$ and $[b, 2] = \{(b, 2)\}$ for each $b \in B - A$.

Here, $p_1 : B \to C$ is the natural injection defined by $(a)p_1 = [a]$ for $a \in A$ and $(b)p_1 = [b, 1]$ for $b \in B—A$. Analogously with p_2. Then

$$(B)p_1 \cap (B)p_2 = (A)m \cdot p_1,$$

thus, the square above is a pullback.

(ii) R-**Mod** has the amalgamation property. Given a module B and its submodule A, then

$$C = B + B/\hat{A} = B \times B/\hat{A}$$

is the quotient module of $B \times B$ under the subspace

$$\hat{A} = \{(a, -a);\ a \in A\}$$

[because the equation $(a, 0) = (0, a)$ is equivalent to $(a, -a) = (0, 0)$].

The two natural embeddings $p_1, p_2 \colon B \to B \times B/\hat{A}$ fulfil $(B)p_1 \cap (B)p_2 = (A)m \cdot p_1$, thus, the square above is a pullback.

3.8. Theorem. Let $\mathscr{K}$ be a cocomplete $(\mathscr{E}, \mathscr{M})$-category with stable pullbacks and the amalgamation property. If an $\mathscr{E}$-epis preserving functor $F \colon \mathscr{K} \to \mathscr{K}$ preserves directed cointersections and finite intersections, then F is finitary.

Proof. (i) Let $m_j \colon X_j \to X$ $(j \in J)$ be a directed family of $\mathscr{M}$-monos. Assuming $\bigcup m_j = 1_X$, we are to prove that $\bigcup \operatorname{im}(m_j F) = 1_{XF}$ (see Lemma V.3.1).

For each $j, k \in J$ with $j \leq k$ we denote by

$$m_{j,k} \colon X_j \to X_k$$

the $\mathscr{M}$-mono with $m_j = m_{j,k} \cdot m_k$. This defines a directed diagram in $\mathscr{K}$. Put

$$Y = \operatorname*{colim}_{j \in J} X_j \quad \text{and} \quad \bar{Y} = \operatorname*{colim}_{j \in J} X_j F$$

with the colimit injections $y_j: X_j \to Y$ and $\bar{y}_j: X_jF \to \bar{Y}$ $(j \in J)$. Since $\bigcup m_j = 1$, the morphism

$$e: Y \to X,$$

defined by $y_j \cdot e = m_j$ $(j \in J)$ is an $\mathscr{E}$-epi (because each m_j factors through im e). To prove that $\bigcup$ im$(m_jF) = 1$, it is sufficient to verify that the morphism

defined by $\bar{y}_j \cdot h = y_jF\,(j \in J)$ is an isomorphism. Indeed, then the fact that $\bigcup \bar{y}_j = 1_{\bar{Y}}$ (see III.5.4) implies

$$\bigcup \operatorname{im}(y_jF) = \operatorname{im} h = 1_{YF}.$$

Since F preserves $\mathscr{E}$-epis, we have then

$$\bigcup \operatorname{im}(m_jF) = \bigcup \operatorname{im}(y_j \cdot e)F = \operatorname{im}(eF) = 1_{XF}.$$

Without loss of generality, $(J, \leq)$ is supposed to have a least element j_0.
(ii) For each $j \in J$ we form the following pushout

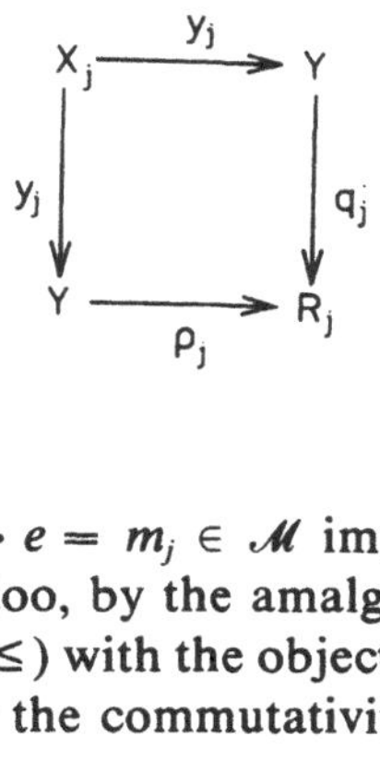

Since $y_j \cdot e = m_j \in \mathscr{M}$ implies $y_j \in \mathscr{M}$ (see III.5.1), this is a pullback of $\mathscr{M}$-monos, too, by the amalgamation property. We define a directed diagram R over $(J, \leq)$ with the objects R_j above and the morphisms $r_{j,k}: R_j \to R_k$ $(j \leq k)$, given by the commutativity of the following diagram:

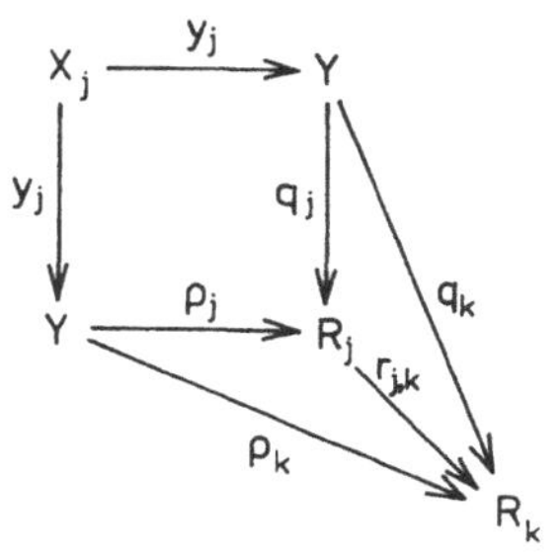

This gives rise to a directed family of pushouts. The colimit of these pushouts is the following square

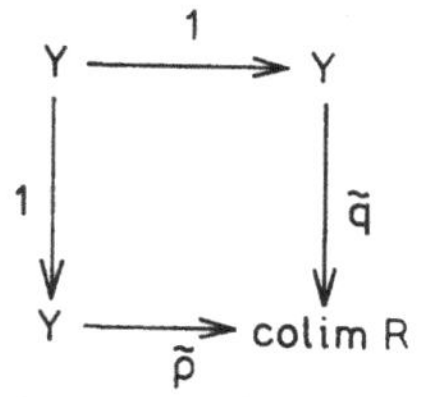

.

This is a pushout (see Remark V.3.6), hence,

$$\operatorname{colim} R = Y \quad \text{and} \quad \tilde{p} = \tilde{q} = 1_Y.$$

Moreover, the given pushouts are pullbacks of $\mathscr{M}$-monos, i.e., finite intersections, and therefore, by the hypotheses on F, also the following squares

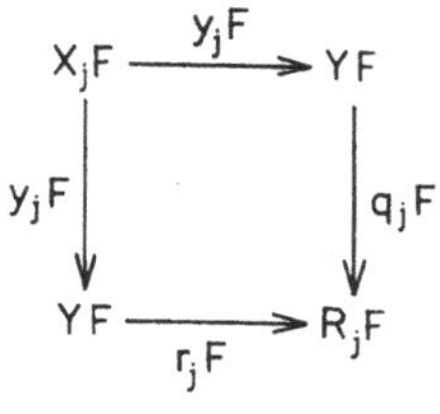

are pullbacks. Since pullbacks are stable, the colimit square

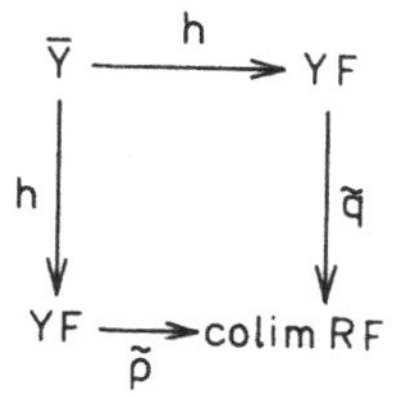

is a pullback, too. To prove that h is an isomorphism, it is sufficient to prove that F preserves the colimit of R. Then, obviously, $\tilde{p} = \tilde{q} = 1_{YF}$ and the pullback of 1 and 1 is formed by an isomorphism, of course.

(iii) F preserves $\operatorname{colim} R$. Indeed, each $r_{j,k}$ ($j, k \in J$, $j \leq k$) is an $\mathscr{E}$-epi, because it is a coequalizer of

$$y_k \cdot p_j;\ y_k \cdot q_j : X_k \to R_j.$$

Therefore, the colimit of R is precisely the cointersection of $r_{j_0,k}$ ($k \in J$), where j_0 is the least element of J. Since F preserves cointersections, it preserves the colimit of R, too. □

Corollary. Let $\mathcal{K}$ be a connected, cocomplete $(\mathcal{E}, \mathcal{M})$-category with cointersections, stable pullbacks and the amalgamation property. Let $F: \mathcal{K} \to \mathcal{K}$ preserve $\mathcal{E}$-epis and finite intersections.

If F has minimal reductions, then F is finitary.

This follows from V.1.5.

3.9. Characterization Theorem. Let $\mathcal{K}$ be a cocomplete, finitely complete and well-powered category with regular factorizations, stable pullbacks, and the amalgamation property.

For each right exact functor $F: \mathcal{K} \to \mathcal{K}$ preserving finite intersections, equivalent are:

(i) F preserves cointersections;
(ii) F preserves directed cointersections;
(iii) F preserves directed colimits;
(iv) F is finitary.

If $\mathcal{K}$ is connected, further equivalent conditions are

(v) F has minimal reductions;
(vi) F is a finitary varietor with minimal realizations.

Proof. (a) We prove below

$$\text{(ii) \& (iv)} \to \text{(iii)}.$$

All the rest follows from the results above—let us first explain how. The hypotheses on $\mathcal{K}$ and F imply (see Observation V.3.5): $\mathcal{K}$ is cowell-powered and has unions and cointersections, and F preserves regular epis.

Thus, we can use Theorem V.3.5 to derive that

$$\text{(iv)} \to \text{(i)} \to \text{(ii)}.$$

By Theorem V.3.8, also

$$\text{(ii)} \to \text{(iv)}.$$

Hence, the conditions (i), (ii) and (iv) are equivalent. After the implication above has been proved, it will be clear that (iii) is also an equivalent condition [since it evidently implies (ii)].

Let $\mathcal{K}$ be connected. Then (v) is equivalent to (i) by Theorem V.1.6. And since (iii) implies that F is a finitary varietor (by Corollary IV.3.5), also (vi) is equivalent to (i) by Proposition V.1.2.

(b) Assuming that F preserves directed cointersections and directed unions, we are going to prove that it preserves directed colimits.

Let us first remark that given two directed diagrams X and Y and a mono-transformation

$$m: X \to Y,$$

then the colimit morphism

$$\operatorname{colim} m \colon \operatorname{colim} X \to \operatorname{colim} Y$$

is also a monomorphism. Indeed, let $(J, \leq)$ be the (joint) scheme of X and Y. For each $j \in J$, the morphism $m_j \colon X_j \to Y_j$ is a mono, i.e., the following square

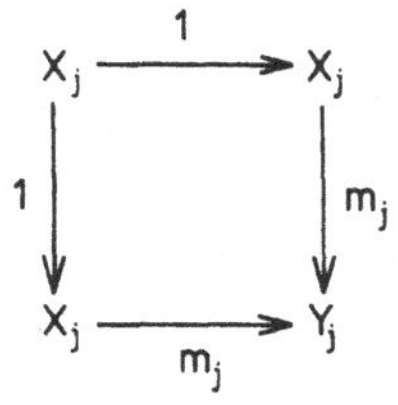

is a pullback. Since pullbacks are stable, the colimit square:

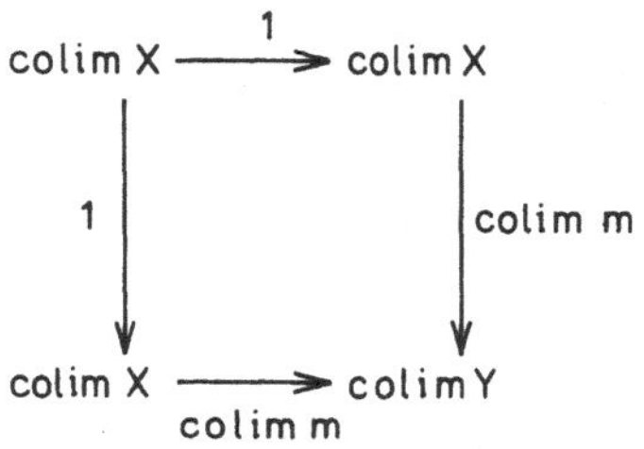

is also a pullback. Hence, colim m is a mono.

(c) Let X be a directed diagram with a scheme $(J, \leq)$; denote by X_j $(j \in J)$ its objects and by $x_{j,k}$ $(j \leq k)$ its morphisms. Let us form the image factorizations

$$x_{j,k} = e_{j,k} \cdot m_{j,k} \qquad (j, k \in J; j \leq k)$$

where

$$e_{j,k} \colon X_j \to T_{j,k} \in \mathscr{E}; \qquad m_{j,k} \colon T_{j,k} \to X_k \in \mathscr{M}.$$

In particular, $T_{j,j} = X_j$ and $e_{j,j} = m_{j,j} = 1$.

For each $j \in J$ we form a directed diagram T^j of regular epis as follows. The objects of T^j are all $T_{j,p}$ with $p \in J$, $j \leq p$. If $j \leq p \leq q$, then the morphism $t^j_{p,q}$ of T^j is the regular epi obtained from the following diagonal fill-in:

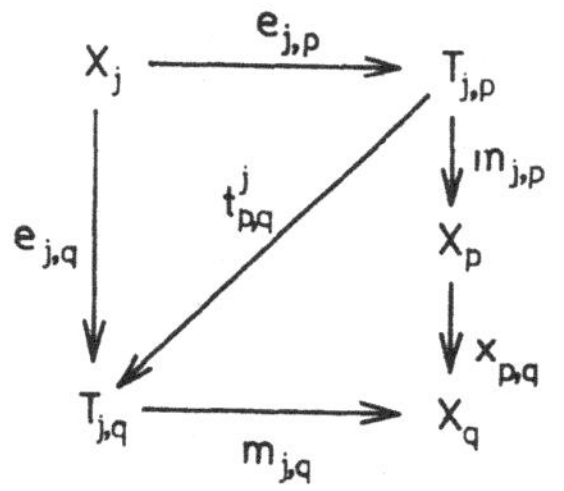

Then colim T^j is the cointersection of the quotients $e_{j,q}: X_j \twoheadrightarrow T_{j,q}$ ($q \in J$, $j \le q$). More precisely, let

$$\bar{e}_j: X_j \twoheadrightarrow \bar{X}_j$$

denote this cointersection. Then for each $q \ge j$ we have $\bar{e}_j = e_{j,q} \cdot f_{j,q}$ and

$$\operatorname{colim} T^j = \bar{X}_j$$

with colimit injections $f_{j,q}: T_{j,q} \twoheadrightarrow \bar{X}_j$.

Since F preserves directed cointersections, also $\bar{e}_j F$ is the cointersection of $e_{j,q}F$ ($q \in J$, $j \le q$), and

$$\operatorname{colim} T^j F = \bar{X}_j F.$$

(d) Given $j \le k \le p$, we use the diagonal fill-in again:

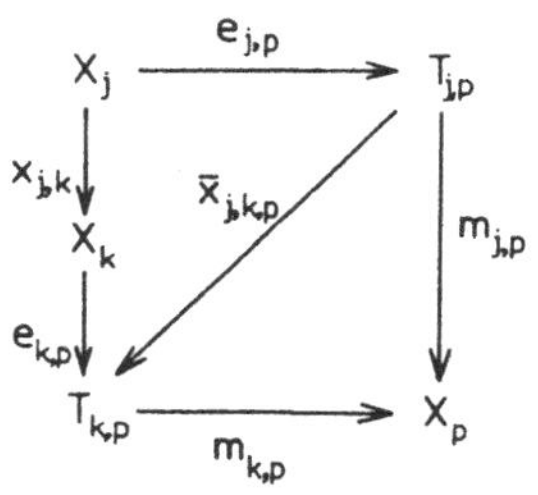

We obtain monos

$$\bar{x}_{j,k,p}: T_{j,p} \twoheadrightarrow T_{k,p} \qquad (j \le k \le p)$$

which form a natural transformation from T^j (restricted to objects $T_{j,p}$ with $p \ge k$) into T^k. Since T^j is directed, the restriction does not change the colimit (Exercise III.1.F) thus, we have a colimit morphism

$$\bar{x}_{j,k} = \operatorname*{colim}_{p \in J} x_{j,k,p}: \bar{X}_j \twoheadrightarrow \bar{X}_k.$$

By (b) above, this is a mono, again.

The following square

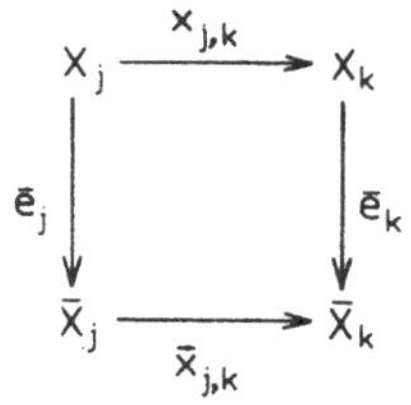

commutes. Indeed, consider the object $T_{j,k}$ of the diagram T^j: we have

$$f_{j,k} \cdot \bar{x}_{j,k} = \bar{x}_{j,k,k} \cdot f_{k,k}.$$

Since $f_{k,k} = \bar{e}_k$ as well as $\bar{x}_{j,k,k} = m_{j,k}$ (because in the square above, $m_{k,k} = 1$), we get

$$\begin{aligned}\bar{e}_j \cdot \bar{x}_{j,k} &= e_{j,k} \cdot f_{j,k} \cdot \bar{x}_{j,k}\\ &= e_{j,k} \cdot \bar{x}_{j,k,k} \cdot f_{k,k}\\ &= e_{j,k} \cdot m_{j,k} \cdot \bar{e}_k\\ &= x_{j,k} \cdot \bar{e}_k.\end{aligned}$$

Hence, the regular epis $\bar{e}_j$ $(j \in J)$ define a natural transformation

$$\bar{e}: X \to \bar{X}.$$

(e) The diagrams X and $\bar{X}$ have the same compatible families. More precisely, to each compatible family of $\bar{X}$,

$$\bar{g}_j: \bar{X}_j \to Z \qquad (j \in J)$$

we can assign a compatible family of X,

$$\bar{e}_j \cdot \bar{g}_j: X_j \to Z \qquad (j \in J),$$

and we claim that this assignment is bijective. It is obviously one-to-one, since $\bar{e}_j$ are epis. Now, consider an arbitrary compatible family of X,

$$g_j: X_j \to Z \qquad (j \in J).$$

Let $g_j = e^* \cdot m^*$ be an image factorization $(j \in J)$, then for each $p \in J$ with $p \geq j$ we have $g_j = x_{j,p} \cdot g_p$, thus

$$e_{j,p} \leq e^*.$$

In fact, consider the diagonal fill-in:

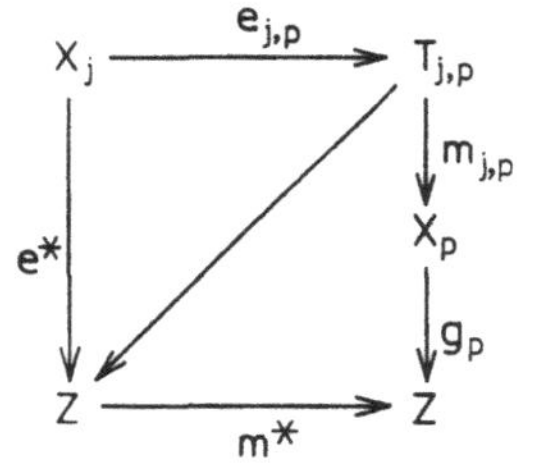

Therefore, $\bar{e}_j \leq e^*$, i.e., $e^* = \bar{e}_j \cdot \tilde{g}_j$. Put $\bar{g}_j = \tilde{g}_j \cdot m^*: \bar{X}_j \to Z$, then

$$g_j = \bar{e}_j \cdot \bar{g}_j$$

and the family $\bar{g}_j$ $(j \in J)$ is compatible since given $j \leq k$, we have $g_j = x_{j,k} \cdot g_k$. Hence,

$$\bar{e}_j \cdot \bar{g}_j = x_{j,k} \cdot \bar{e}_k \cdot \bar{g}_k = \bar{e}_j \cdot (\bar{x}_{j,k} \cdot \bar{g}_k)$$

and $\bar{e}_j$ is an epi.

Since X and $\bar{X}$ have the same compatible families, they have the same colimits, too.

(f) The diagrams $X \cdot F$ and $\bar{X} \cdot F$ are related in the same way as the diagrams X and $\bar{X}$. Indeed, F preserves regular epis [see part (a) above] and monos (since it preserves finite intersections and since monos are characterized by their pullbacks) and hence, F preserves image factorizations. Further, F preserves the colimit of each T^j $(j \in J)$. We conclude that also $X \cdot F$ has the same colimit as $\bar{X} \cdot F$. Therefore, it is sufficient to prove that F preserves the colimit of $\bar{X}$. And $\bar{X}$ is a directed diagram of monos.

Here we use (for the first time) the preservation of directed unions. Let

$$V = \operatorname{colim} \bar{X} \quad \text{and} \quad W = \operatorname{colim} \bar{X} \cdot F$$

with colimit injections

$$v_j : \bar{X}_j \to V \quad \text{and} \quad w_j : \bar{X}_j F \to W \qquad (j \in J).$$

Then each v_j is a mono. Indeed, let $v_j = \tilde{e} \cdot \tilde{m}$ be an image factorization, then for each $k \geq j$ we have

$$e_{j,k} \leq \bar{e}_j \cdot \tilde{e}$$

(because $v_j = x_{j,k} \cdot v_k = e_{j,k} \cdot m_{j,k} \cdot v_k$). Since $\bar{e}_j$ is the cointersection of all $e_{j,k}$ $(k \geq j)$, we conclude

$$\bar{e}_j \leq \bar{e}_j \cdot \tilde{e},$$

in other words, $\tilde{e}$ is an isomorphism. By Remark III.5.4,

$$\bigcup_{j \in J} v_j = 1_Y,$$

and we conclude (since this union is directed) that

$$\bigcup_{j \in J} v_j F = 1_{YF}.$$

Denote by

$$v : W \to VF$$

the morphism defined by

$$w_j \cdot v = v_j F \qquad (j \in J).$$

We are to show that v is an isomorphism. Since $\bigcup v_j F = 1_{YF}$, it is obvious that v is a regular epi (each $v_j F$ factors through the image of v, hence, this must be all of YF).

It remains to prove that v is a mono. Since each $v_j F$ is a mono, we have a directed family of pullbacks

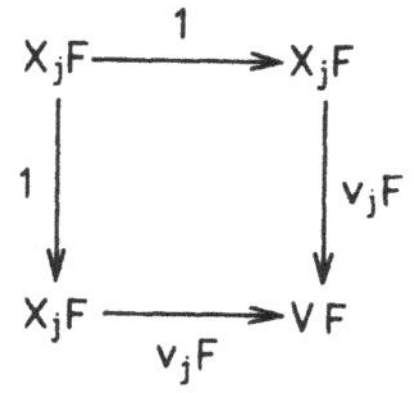

with $j \in J$. The colimit square

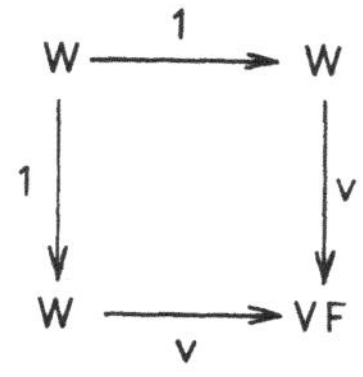

is a pullback, too. Hence, v is a mono (as well as a regular epi) which proves that F preserves the colimit of $\bar{X}$. □

3.10. Examples. We discuss the hypotheses of the Characterization Theorem first for $\mathscr{X}$ and then for F.

(i) **Set** and R-**Mod** : The hypotheses are fulfilled. These categories are complete, cocomplete and well-powered. The amalgamation property has been discussed in V.3.7, and the stability of pullbacks in V.3.6.

(ii) Varieties of finitary algebras: The hypotheses are fulfilled except, conceivably, the amalgamation property which is a non-trivial problem of universal algebra. Some varieties are known to have this property (e.g., groups, R-modules, and lattices) and some are known not to have it (e.g. commutative semigroups).

(iii) **Pos** has the amalgamation property as an (epi, embedding)-category but, unfortunately, not as a (quotient, mono)-category, as required by the Characterization Theorem. For example, let $A = \{0, 1\}$ be the discrete poset and $B = \{0, 1\}$ the chain $0 \leq 1$. Then $m: A \to B$ defined by $(0)m = 0$, $(1)m = 1$ is a mono such that the pushout of m with itself is

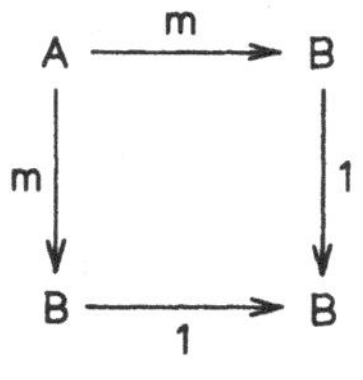

This square is not a pullback.

All other hypotheses are fulfilled by **Pos**. Thus, each right exact, finitary functor on the (quotient, mono)-category **Pos** preserves cointersections. And each functor on the (epi, embedding)-category **Pos** which preserves epis, finite intersections, and cointersections, is finitary. But "finitary" in two factorization systems means two different coinditions, of course.

(iv) **Top** and **Gra**: The situation is the same as in **Pos**, see Exercises V.3. below.

Proposition. Let $\mathcal{K}$ be a category in which

(i) every regular epi splits;

(ii) every equivalence is a kernel equivalence of some morphism.

Then each functor $F: \mathcal{K} \to \mathcal{K}$ is right exact.

Proof. Let $e: X \to Y$ be the coequalizer of an equivalence relation $[r_1, r_2]: X \rightharpoonup X$. Since $[r_1, r_2]$ is the kernel equivalence of some morphism f, it is, in fact, the kernel equivalence of e ($r_1 \cdot f = r_2 \cdot f$ implies that f factors through e, say, $f = e \cdot \bar{f}$; then each pair p_1, p_2 with $p_1 \cdot e = p_2 \cdot e$ fulfils $p_1 \cdot f = p_2 \cdot f$ and hence, $[p_1, p_2] \subset [r_1, r_2]$).

Since e splits, we can choose a morphism $i: Y \to X$

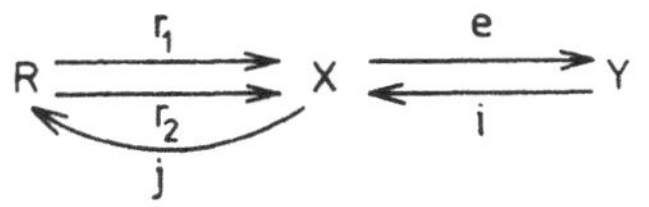

with

$$i \cdot e = 1_Y.$$

The pair 1_X and $e \cdot i$ fulfils

$$1_X \cdot e = e \cdot (i \cdot e) = (e \cdot i) \cdot e,$$

therefore $[1_X, e \cdot i] \subset [r_1, r_2]$. This means that there exists $j: X \to R$ with

$$j \cdot r_1 = 1_X \text{ and } j \cdot r_2 = e \cdot i.$$

All this implies that for each functor $F: \mathcal{K} \to \mathcal{K}$ the coequalizer of r_1F and r_2F is eF. Indeed, let $h: XF \to Z$ be a morphism with $r_1F \cdot h = r_2F \cdot h$. Then the

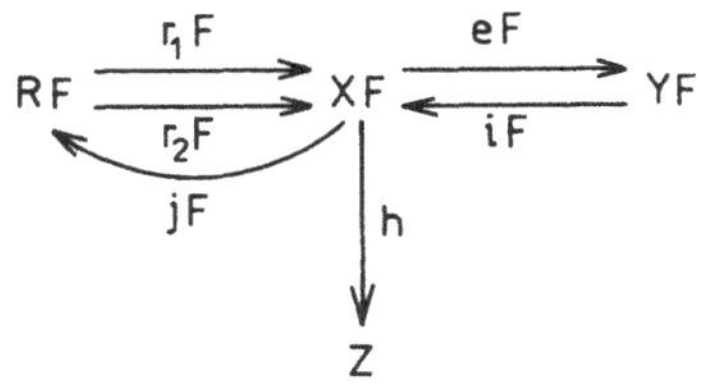

morphism $\bar{h} = iF\cdot h$ fulfils

$$eF\cdot \bar{h} = (e\cdot i)F\cdot h = (j\cdot r_2)F\cdot h = jF\cdot r_1F\cdot h = h.$$

Since eF is a (split) epi, such $\bar{h}$ is unique. □

Corollary. For each set functor the conditions (i)—(vi) of the Characterization Theorem are all equivalent to

(vii) F is a quotient functor of H_Σ for some finitary type Σ.

Indeed, F is right exact by Proposition above. Assuming that F is standard (III.4.5), then it preserves finite intersections, and so all the conditions (i)—(vi) are equivalent; see III.4.3 for the equivalence of (vii). If F is arbitrary then, by III.4.5, there is a functor F' with $F = F'$ on non-empty sets and maps such that F' is naturally isomorphic to a standard functor. Obviously, (i)—(vii) are equivalent for F'. And the empty set does not influence any of these conditions except, possibly, (vi) (because of the initial F-algebra) and (v) (because of difficulties with the empty F-automaton). But, in fact, there are no problems because (iii) and (i) clearly imply (v) and (vi); conversely, (vi) → (v) → (i) has been proved above. □

Remark. We conclude that the only F-automata in **Set** which have minimal realizations are finitary tree automata and their basic varieties (III.3.2).

Corollary. Let R be a commutative field. For each functor F: R-**Vect** → R-**Vect**, the conditions (i)—(vi) of the Characterization Theorem are all equivalent.

Indeed, F is right exact by the Proposition above. And F preserves finite intersections by Ex. III.4.A.

3.11. The category $\mathbf{Set}_\omega$ of countable sets and maps has a special property (shared by R-$\mathbf{Vect}_\omega$, see Exercise D below):

Theorem. Each functor F:$\mathbf{Set}_\omega \to \mathbf{Set}_\omega$ is a finitary varietor with minimal realizations.

Proof. We proved in IV.6.4 that F is a finitary varietor. In that proof we have exhibited a finitary functor G: **Set** → **Set** extending F. Since G has minimal realizations by the Corollary above, so does F. Indeed, each reachable F-automaton is countable, and F coincides with G on countable sets. □

Exercises V.3

A. Stability of pullbacks. (i) Verify that the following categories

Pos, **Gra**, any variety of finitary algebras

have stable pullbacks. (Hint: The functor colim: $\mathcal{K}^{(J,\leq)} \to \mathcal{K}$ preserves finite products and equalizers.)

(ii) The category

$$\mathcal{K} = \sigma\textbf{-Lat}$$

of σ-complete lattices and σ-complete homomorphisms fails to have stable pullbacks. Indeed, consider the ω-chain of pullbacks

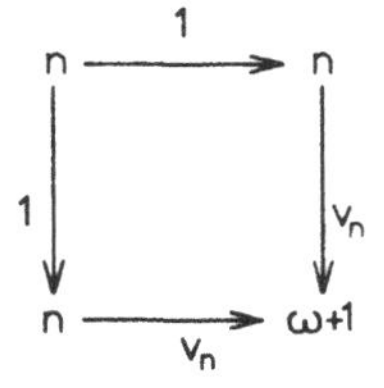

$(n < \omega)$

where $v_n : n \to \omega + 1$ is the inclusion of $n = \{0, 1, \ldots, n - 1\}$ (linearly ordered) to $\omega + 1 = \{0, 1, \ldots\} \cup \{\omega\}$. Prove that the colimit square is no pullback. (Hint: The colimit of the chain $\{n\}_{n<\omega}$ is the free σ-complete lattice X over ω. The canonical map $v_\omega : X \to \omega + 1$ is not one-to-one, hence, the pullback of v_ω and v_ω is not 1_X, 1_X.)

B. Finitary functors. (i) Prove that in **Gra, Pos, Top** with $\mathcal{E}$ = epi, $\mathcal{M}$ = embeddings, a functor F preserving embeddings is finitary iff for every object X and every point $x \in XF$ there exists a finite object Y and a morphism $f: Y \to X$ with $x \in (YF)(fF)$.

(ii) Consider **Top** with regular factorizations. Define a functor F: **Top** $\to$ **Top** as follows: for any space X, let XF be the space on the same set, the open base of which is precisely the set of all closed-and-open subsets of X; on morphisms, $fF = f$. Prove that F is not finitary. (Hint: Let R be the space of real numbers, then XF has only two open sets, $\emptyset$ and R; consider the directed set of all subspaces Y of X, vhich have only a finite number of non-isolated points; then $Y = YF$ and hence $X = \bigcup Y = \bigcup YF \neq XF$).

C. Amalgamation property. (i) Let $\mathcal{K}$ be a concrete $(\mathcal{E}, \mathcal{M})$-category. Assume there is $m \in \mathcal{M}$ which is not an isomorphism though its underlying map is a bijection. Prove that $\mathcal{K}$ fails to have the amalgamation property. [Hint: The pushout of m with itself is no pullback. In fact, if $m \cdot p = m \cdot q$ is this pushout, then for the forgetful functor U, $pU = (mU)^{-1}[(mU)(pU)] = (mU)^{-1}[(mU)(qU)] = qU$ implies $p = q$. If $m \cdot p = m \cdot p$ is a pullback, then $1 \cdot p = 1 \cdot p$ implies that m is a split epi.]

(ii) Deduce that **Gra, Pos, Top** with regular factorizations fail to have the amalgamation property.

(iii) Verify that **Gra, Pos** and **Top** with (epi, regular mono)-factorizations have the amalgamation property. (Hint: The pushout is created on the level of sets and all morphisms in it are regular monos, hence, it is a pullback.)

(iv) Verify that the full subcategory of **Top** formed by all Hausdorff spaces fails to have the amalgamation property with respect to (epi, regular mono)-factorizations as well as (regular epi, mono)-factorizations; but it has the amalgamation property with respect to $(\mathscr{E}, \mathscr{M})$-factorizations with $\mathscr{E}$ = continuous maps onto dense subspaces and $\mathscr{M}$ = embeddings of closed subspaces. (Hint: Consider the pushout of m with itself, where m is the embedding of the space of rational numbers into the space of real numbers—the resulting space is the space of real numbers again.)

D. Minimal realization in R-$\mathbf{Vect}_\omega$. Prove that on the category R-$\mathbf{Vect}_\omega$ of countably-dimensional vector spaces, each functor is a finitary varietor with minimal realizations.

V.4. Consequences of Minimal Reduction

4.1. Each functor with minimal reductions is a finitary varietor. We have proved this (V.3.9) under very restrictive additional hypotheses. In the present section, we prove this more generally. (In V.3.9 we have proved that even the functor itself is finitary which is much stronger—see for example IV.6.7.) We also prove that a functor with minimal reductions preserves epis, thus getting rid of another additional hypothesis.

A pullback in a category $\mathscr{K}$ is said to be *absolute* if each functor from $\mathscr{K}$ to any category preserves it.

Lemma. Each pushout

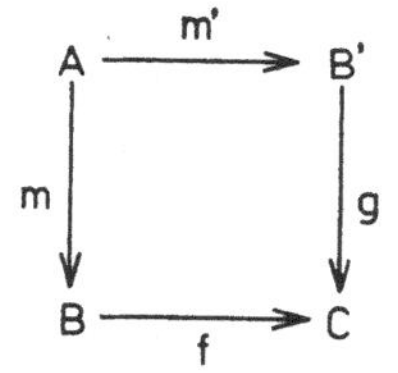

of split monos m and m' is an absolute pullback, and f and g are split monos.

Proof. Choose morphism $\bar{m}: B \to A$ and $\bar{m}': B' \to A$ with

$$m \cdot \bar{m} = m' \cdot \bar{m}' = 1_A.$$

There exists a unique morphism $\bar{f}: C \to B$ such that the following diagram

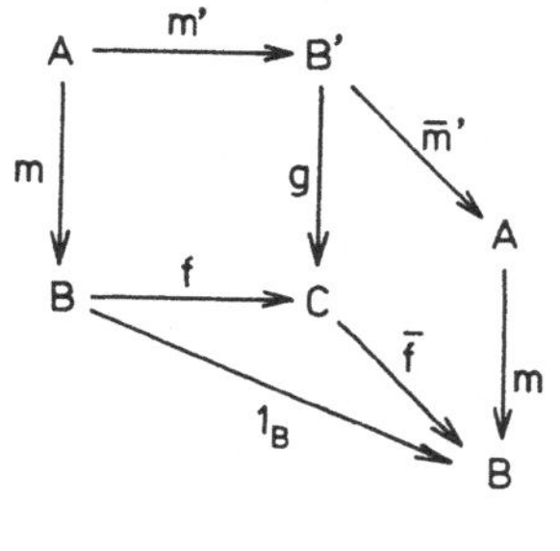

commutes. Analogously define $\bar{g}: C \to B'$. Then we see that f and g are split monos, and we have a diagram

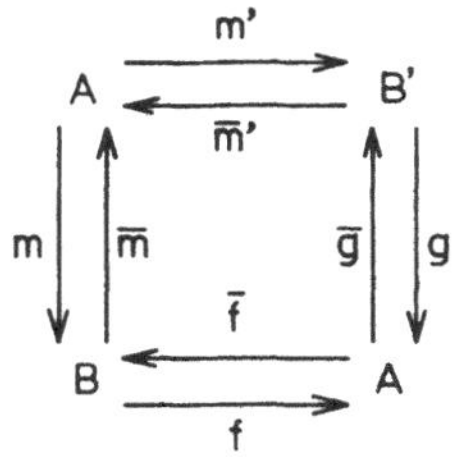

such that

$$(*)\qquad m \cdot \bar{m} = m' \cdot \bar{m}' = 1_A;\quad m \cdot f = m' \cdot g;\quad f \cdot \bar{f} = 1_B;\quad g \cdot \bar{g} = 1_{B'}$$
$$\text{and}\quad g \cdot \bar{f} = \bar{m}' \cdot m;\quad f \cdot \bar{g} = \bar{m} \cdot m'.$$

It is sufficient to prove that $(*)$ implies that the pullback of f and g is formed by m and m': for each functor F the F-image of the diagram above has the property analogous to $(*)$ and hence, fF and gF are then proved to have the pullback mF and $m'F$.

Let p and q be morphisms with

$$p \cdot f = q \cdot g.$$

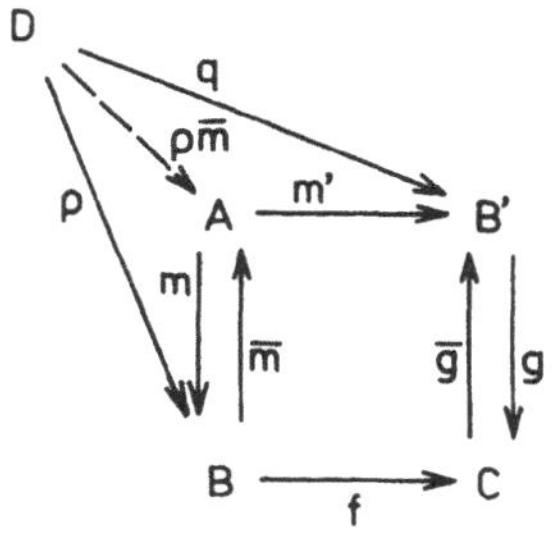

The morphism $p \cdot \bar{m}$ fulfils

$$p = (p \cdot \bar{m}) \cdot m \quad \text{and} \quad q = (p \cdot \bar{m}) \cdot m'.$$

Indeed, the latter follows from $(*)$:

$$q = q \cdot g \cdot \bar{g} = p \cdot f \cdot \bar{g} = p \cdot \bar{m} \cdot m';$$

the former is now a consequence of $(*)$ and $q = p \cdot \bar{m} \cdot m'$:

$$p = p \cdot f \cdot \bar{f} = q \cdot g \cdot \bar{f} = p \cdot \bar{m} \cdot m' \cdot g \cdot \bar{f} = p \cdot \bar{m} \cdot m \cdot f \cdot \bar{f} \cdot =$$
$$p \cdot \bar{m} \cdot m.$$

It is clear that $p \cdot \bar{m}$ is unique with this property since m is a mono. Thus, (*) implies that the pushout above is an (absolute) pullback. □

4.2 Theorem. Let $\mathcal{K}$ be a countably cocomplete $(\mathcal{E}, \mathcal{M})$-category which has cointersections, stable pullbacks and regular finite coproducts. Let $F: \mathcal{K} \to \mathcal{K}$ preserve $\mathcal{E}$-epis.

If F has minimal reductions, then the free-algebra construction stops after ω steps for any object I with $\hom(IF, I) \neq \emptyset$.

Remark. Instead of the stability of pullbacks it would be sufficient to assume the ω-stability of split intersections (i.e., of pullbacks of split monos). Here, ω-stability refers to ω-colimits of pullbacks (whereas stability concerns all directed colimits). The proof of the theorem above uses the same technique as that of V.3.8.

Proof. (i) We prove that in the free-algebra construction, each $w_{n,\omega}: W_n \to W_\omega$ $(n < \omega)$ is a split mono. Let us choose an arbitrary morphism

$$d: IF \to I,$$

and define a chain of morphisms

$$d_{n,m}: W_m \to W_n \quad (n \leq m < \omega)$$

by the following induction:

$$d_{0,1}: I + IF \to I$$

has components 1_I and d;

$$d_{n+1,m+1} = 1_I + d_{n,m} F: I + W_m F \to I + W_n F.$$

For each $n < \omega$ we define a collection of morphisms

$$f_k: W_k \to W_n \quad (k < \omega)$$

by

$$f_k = \begin{cases} w_{k,n} & \text{if } k \leq n; \\ d_{k,n} & \text{if } k > n. \end{cases}$$

It is easy to verify that this collection is compatible with the free-algebra construction. Hence, there exists

$$d_{n,\omega}: W_\omega \to W_n$$

with $f_k = w_{k,\omega} \cdot d_{n,\omega}$ $(k < \omega)$, particularly with $w_{n,\omega} \cdot d_{n,\omega} = f_n = w_{n,n} = 1$.

(ii) To prove that the free-algebra construction stops after ω steps, we are to verify that F preserves the colimit $W_\omega = \operatorname{colim}_{n<\omega} W_n$. We use the following

ω-chain of pushouts

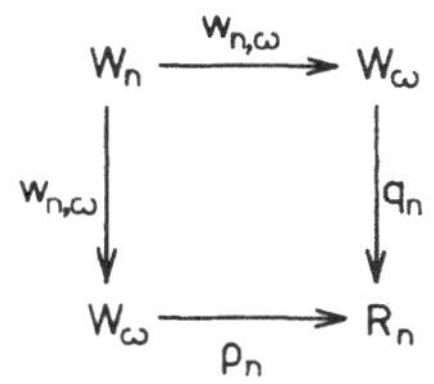

($n < \omega$). More precisely, for each $n < \omega$ we define R_n by the pushout above, and for each $n < m$ we define

$$r_{n,m}: R_n \to R_m$$

by the commutativity of the following diagram

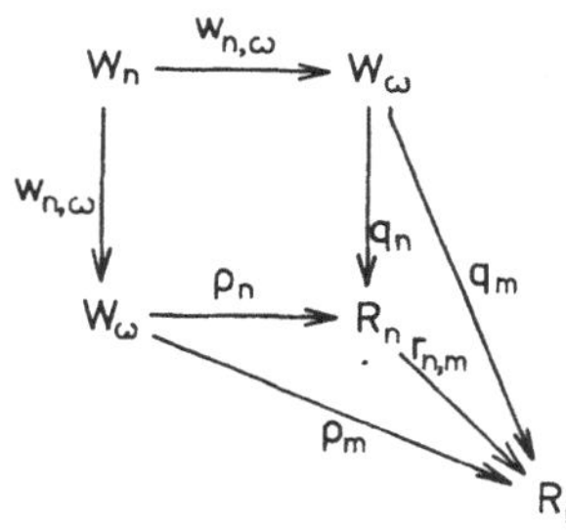

(using the fact that $w_{n,\omega} \cdot p_m = w_{n,m} \cdot w_{m,\omega} \cdot p_m = w_{n,m} \cdot w_{m,\omega} \cdot q_m = w_{n,\omega} \cdot q_m$).

The colimit square:

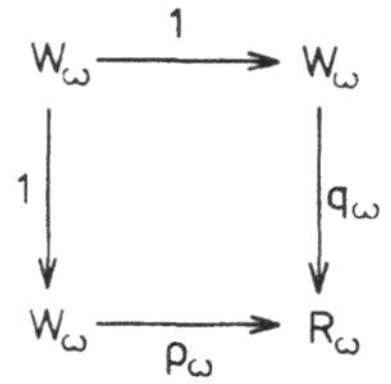

is a pushout (see Remark V.3.6) and hence,

$$R_\omega = W_\omega \quad \text{and} \quad p_\omega = q_\omega = 1_{W_\omega}.$$

Moreover, since $w_{n,\omega}$ are split monos, the following squares

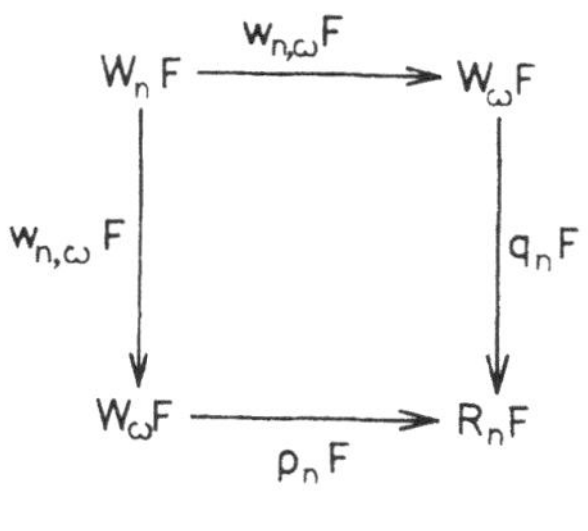

are pullbacks of split monos, by the preceding lemma. Since split intersections are ω-stable, the colimit square

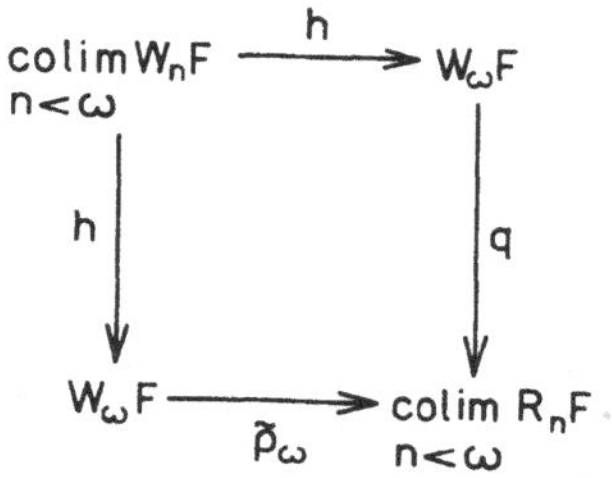

is a pullback, too. We are to prove that h is an isomorphism. By the preceding remark, it is sufficient to prove that F preserves the colimit of R, i.e., that

$$\operatorname{colim} R \cdot F = W_\omega F$$

from which we readily conclude that $\tilde{p}_\omega = \tilde{q}_\omega = 1$: since h, h is the pullback of 1,1, clearly, h is an isomorphism.

(iii) F preserves colim R. Here we use the fact that F weakly preserves cointersections (by Remark V.1.3). First for each $n \leq m$ we have

$$r_{n,m} \in \mathscr{E}.$$

In fact, $r_{n,m}$ is a regular epi because it is the coequalizer of $w_{m,\omega} \cdot p_n$ and $w_{m,\omega} \cdot q_n$. [Proof: Let $f\colon R_n \to S$ be an arbitrary morphism with

$$w_{m,\omega} \cdot p_n \cdot f = w_{m,\omega} \cdot q_n \cdot f.$$

Then, since p_m and q_m form the pushout, there exists $\bar{f}\colon R_m \to S$ with $p_m \cdot \bar{f} = p_n \cdot f$ and $q_m \cdot \bar{f} = q_n \cdot f$. It follows that $f = r_{n,m} \cdot \bar{f}$, because both

$$p_n \cdot f = p_m \cdot \bar{f} = p_n \cdot (r_{n,m} \cdot \bar{f})$$

and

$$q_n \cdot f = q_m \cdot \bar{f} = q_n \cdot (r_{n,m} \cdot \bar{f}).$$

The uniqueness of f is easily seen.]

Thus, the colimit of R is given by the cointersection of $r_{0,m}$ ($m < \omega$). More precisely, if $r_{0,\omega}\colon R_0 \to R_\omega$ denotes this cointersection, then for each $m < \omega$ we have $r_{m,\omega}\colon R_m \to R$ with $r_{0,\omega} = r_{0,m} \cdot r_{m,\omega}$, and these are the colimit injections of $R_\omega = \operatorname{colim}_{n<\omega} R_n$. Analogously, each $r_{n,m}F$ is an $\mathscr{E}$-epi, thus, colim $R \cdot F$ is the cointersection of $r_{0,m}F$ ($m < \omega$). It is sufficient to prove that F preserves the cointersection of $r_{0,m}$ ($m < \omega$). This will be clear when we exhibit morphism

$$\delta_m\colon R_mF \to R_m \quad (m < \omega)$$

such that $r_{0,m}: (R_0, \delta_0) \to (R_m, \delta_m)$ are homomorphisms, i.e.

$$\delta_0 \cdot r_{0,m} = r_{0,m}F \cdot \delta_m \quad (m < \omega).$$

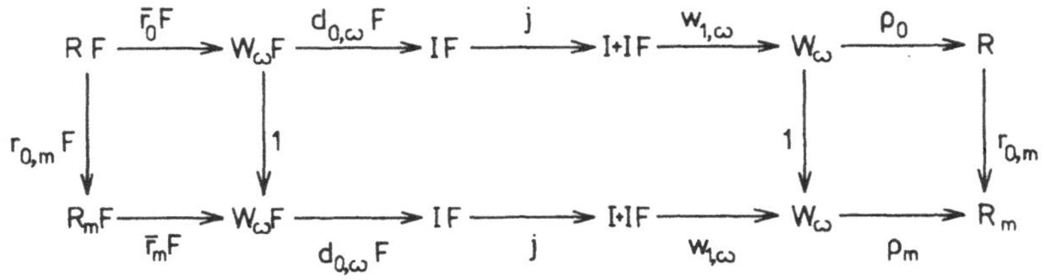

For each $m < \omega$ there exists (by the property of pushouts) a unique morphism $\bar{r}_m: R_m \to W_\omega$ for which the following diagram

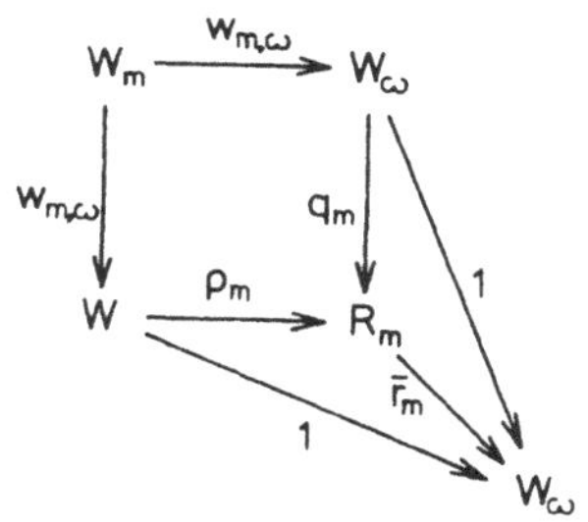

commutes. Clearly,

$$\bar{r}_0 = r_{0,m} \cdot \bar{r}_m \quad (m < \omega).$$

Next, there exists a morphism δ from $W_\omega F$ to W_ω [for example, compose $d_{0,\omega}F$ of part (i) of the present proof with the coproduct injection $j: IF \to W_1$ and $w_{1,\omega}$]. Put

$$\delta_m = \bar{r}_mF \cdot \delta \cdot p_m \quad (m < \omega).$$

These morphisms have the required property because the squares above clearly commute. □

Corollary. Let $\mathscr{K}$ fulfil the hypotheses above and be connected. Let $F: \mathscr{K} \to \mathscr{K}$ preserve $\mathscr{E}$-epis and fulfil $\hom(\perp F, \perp) \neq \emptyset$. If F has minimal reductions, then F is a finitary varietor.

4.3. Theorem. Let $\mathscr{K}$ be a connected $(\mathscr{E}, \mathscr{M})$-category with finite regular coproducts. If a functor $F: \mathscr{K} \to \mathscr{K}$ has minimal reduction then

$$e \in \mathscr{E} \quad \text{implies} \quad eF \text{ is an epi}.$$

Remark. Since $\mathscr{K}$ is connected, the coproduct injections $A \to A + B$ are split (hence, regular) monos whenever $A \neq \perp$. Thus, the regularity of finite coproducts in the theorem above only means that the canonical morphisms $\perp \twoheadrightarrow B$ are $\mathcal{M}$-monos. Equivalently, that $\perp$ is simple, i.e. has no proper $\mathcal{E}$-quotients. In Exercise C below we show that this assumption is essential.

Proof. Let $e: A \twoheadrightarrow B$ be an $\mathcal{E}$-epi. Given $g, h: BF \to C$ with

$$eF \cdot g = eF \cdot h,$$

we are going to prove that $g = h$. This is clear if $A = \perp$ since (by the regularity of finite coproducts) then e is an isomorphism. Assuming $A \neq \perp$, there exists a morphism

$$t: C \to A.$$

Define a non-initial F-automaton:

$$M = (A + C, \delta, B + C, e + 1_C)$$

where

$$\delta: (A + C)F \to A + C$$

is the following morphism. We use the following notation for the coproduct injections

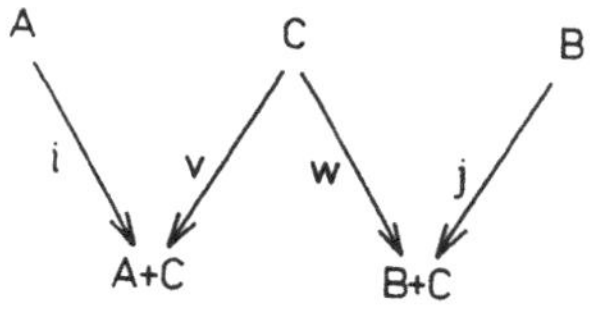

and we define

$$p: A + C \to B \quad \text{and} \quad q: B + C \to B$$

by

$$v \cdot p = w \cdot q = t \cdot e: C \to B;$$
$$v \cdot p = e \quad \text{and} \quad j \cdot q = 1_B.$$

Note that $p = p_0 \cdot e$, where $p_0: A + C \to A$ has the components 1_A and t. Thus, $pF \cdot g = pF \cdot h$. Put

$$\delta = pF \cdot g \cdot v = pF \cdot h \cdot v: (A + C)F \to A + C.$$

To prove that $g = h$, we use the minimal reduction of M:

$$r_0: M \to M_0 = (Q_0, \delta_0, B + C, \gamma_0).$$

We define a reduction of M:

$$e + 1_C: M \to M_g = (B + C, \delta_g, B + C, 1)$$

where

$$\delta_g : (B + C)F \to B + C$$

is defined by

$$\delta_g = qF \cdot g \cdot v.$$

Since $e \in \mathscr{E}$, also $e + 1_C \in \mathscr{E}$ (III.5.5). Moreover, $e + 1_C$ is a morphism of automata, since the following diagram

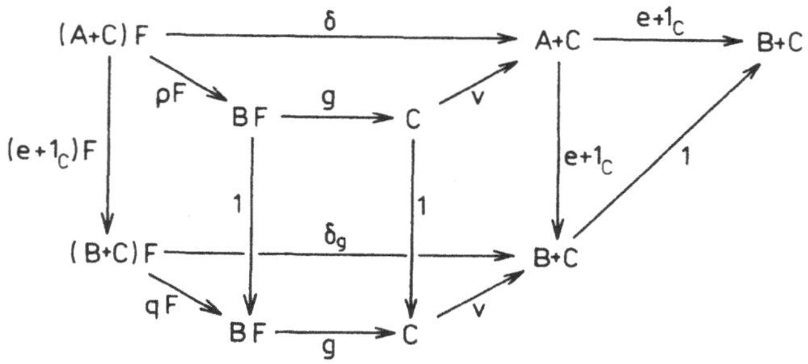

commutes. Thus, M_g is really a reduction of M. Hence, there exists a morphism of automata

$$s_g : M_g \to M_0$$

with

$$r_0 = (e + 1_C) \cdot s_g .$$

Quite analogously, we define a reduction

$$e + 1_C : M \to M_h$$

(always substituting h for g), and we obtain

$$s_h : M \to M_h$$

with

$$r_0 = (e + 1_C) \cdot s_h .$$

Since s_g preserves the outputs, we have

$$1_{B+C} = s_g \cdot \gamma_0 .$$

Thus, s_g is a split mono as well as an $\mathscr{E}$-epi, i.e., s_g is an isomorphism. Analogously,

$$1_{B+C} = s_h \cdot \gamma_0 ,$$

hence,

$$s_g = s_h = \gamma_0^{-1} .$$

Both s_g and s_h are homomorphisms, hence,

$$\begin{aligned}\delta_g &= \delta_g \cdot s_g \cdot s_g^{-1}\\ &= s_g F \cdot \delta_0 \cdot s_g^{-1}\\ &= s_h F \cdot \delta_0 \cdot s_h^{-1}\\ &= \delta_h .\end{aligned}$$

Thus

$$qF \cdot g \cdot v = qF \cdot h \cdot v.$$

By the definition of q we have $j \cdot q = 1_B$ and thus, by multiplying the last equation by jF we get

$$g \cdot v = h \cdot v.$$

And v is a mono (since finite coproducts are regular), thus,

$$g = h. \qquad \square$$

4.4. Recall that (epi, extremal mono)-factorizations exist very often (III.5.6).

Corollary. Let $\mathcal{K}$ be a countably cocomplete, connected (epi, extremal mono)-category with stable pullbacks and cointersections, in which finite coproducts are regular.

For each functor $F: \mathcal{K} \to \mathcal{K}$ with hom $(\perp F, \perp) \neq \emptyset$, equivalent are:

(i) F is a finitary varietor with minimal realizations;
(ii) F has minimal reductions;
(iii) F preserves cointersections.

If fact, (iii) → (ii) is Theorem V.1.3; (ii) → (i) follows from Corollary V.4.2, the hypothesis of which is fulfilled by Theorem V.4.3, and from Proposition V.1.2; (i) → (iii) follows from Theorem V.1.5 (using V.4.3 again).

Example. A functor F: R-**Mod** → R-**Mod** has minimal reductions iff F is finitary. Moreover, F then preserves epis and is a finitary varietor. In fact, R-**Mod** fulfils all hypotheses of the preceding corollary and, moreover, hom$(\perp F, \perp) \neq \emptyset$ is always fulfilled.

Exercises V.4

A. A non-varietor with minimal reductions. Let $\mathcal{K}$ = **Set**^{op} be the dual to the category of sets (III.2.12). The power-set functor (III.3.4) defines a functor

$$P^{op}: \mathbf{Set}^{op} \to \mathbf{Set}^{op}.$$

(i) Verify that P^{op} preserves cointersections by proving that P in **Set** preserves intersections.

(ii) Prove that no object generates a free P^{op}-algebra. (Hint: use IV.2.6 and IV.3.1).

(iii) Explain why this does not contradict Corollary V.4.2.

B. A non-constructive varietor with minimal realizations. Prove that the functor F: **Gra** $\to$ **Gra** of IV.3.A has minimal reductions. Explain why this does not contradict Corollary V.4.2.

C. Preservation of epis is not necessary. We exhibit a finitary varietor on a connected category which has minimal realizations and yet does not preserve epis.

(i) Define a category **Ab*** by enlarging **Ab**, the category of abelian groups and homomorphisms, by a new object, the "empty group" $\emptyset$ and by empty maps $t_A : \emptyset \to A$ for each A in **Ab***. Verify that **Ab*** is an (epi, mono)-category with cointersections and coproducts. Verify that $t_{\{0\}} : \emptyset \to \{0\}$ is an epi in **Ab***; conclude that **Ab*** does not have regular finite coproducts.

(ii) Define a functor

$$F: \mathbf{Ab^*} \to \mathbf{Ab^*}$$

by choosing an arbitrary non-trivial abelian group A and putting

$$XF = A \quad \text{for each } X \text{ in } \mathbf{Ab^*};$$

given $f: X \to Y$ then

$$fF = \begin{cases} 1_A & \text{if } X \neq \emptyset \text{ or } X = Y = \emptyset; \\ \text{zero map} & \text{if } X = \emptyset \neq Y. \end{cases}$$

Verify that F is a finitary varietor with

$$I^\# = I + A \quad \text{if} \quad I \neq \emptyset; \quad \emptyset^\# = \{0\}.$$

(Hint: In the free-algebra construction over $\emptyset$ we have $W_n = A$ and $w_{n,m}$ the zero map ($n \leq m < \omega$) and hence, $W_\omega = \{0\}$.)

(iii) Verify that each behavior $\beta: I^\# \to \Gamma$ has a minimal reduction obtained by the image factorization $\beta = e \cdot m$, $e: I^\# \to Q$ as follows: Q is the state object, m the output morphism and, if $I \neq \emptyset$, the components of $e: I + A \to Q$ are λ and δ.

(iv) Observe that F does not preserve epis: consider $t_{\{0\}}!$.

V.5. Finite Automata

5.1. We are going to prove that each finite automaton has a minimal reduction. The hypotheses under which this holds are quite mild—for example, all functors on **Set** or R-**Vect** are included. Throughout this section, $\mathscr{K}$ is an $(\mathscr{E}, \mathscr{M})$-category.

Definition. An object Q is said to be *finite* if $\hom(Q, -)\colon \mathcal{K} \to \mathbf{Set}$ is a finitary functor. Explicitly, if for each directed collection of $\mathcal{M}$-monos $m_j\colon R_j \to R$ ($j \in J$) with $\bigcup_{j \in J} m_j = m\colon S \to R$ and for each morphism

$$f\colon Q \to S$$

there exists $j_0 \in J$ such that $f \cdot m$ factors through m_{j_0}.

Examples. (i) **Set**: this is the usual concept. Indeed, let Q be a set with finitely many points. Then Q is a finite object since for each

$$f\colon Q \to S = \bigcup_{j \in J} R_j$$

there exists $j_0 \in J$ with $(q)f \in R_{j_0}$ for all $q \in Q$ (because the union is directed).

Conversely, if Q has infinitely many points, the condition above fails even for

$$f = 1_Q\colon Q \to Q = \bigcup_{j \in J} R_j,$$

where R_j ($j \in J$) is the collection of all finite subsets of Q.

(ii) **Pos** with $\mathcal{E}$ = epis and $\mathcal{M}$ = embeddings: finite objects are just finite posets, the proof is as for **Set**. The same is true for $\mathcal{E}$ = quotients and $\mathcal{M}$ = monos, but here we must be more catious. Let us consider a directed union $\bigcup m_j = m$. We can assume that $R_j \subset S$ and m_j is the inclusion map ($j \in J$) but the order of R_j can be weaker than that induced by S. Nevertheless, given $x, y \in R$ with $x \leq y$, then there is $j \in J$ with $x, y \in R_j$ and $x \leq y$ also in R_j. Indeed, since J is directed, the relation $\preceq$ formed by all pairs x, y for which such j exists, is transitive (as well as reflexive and antisymmetric). Hence, $\bigcup m_j$ is the poset S with the order $\preceq$—consequently, the latter coindices with $\leq$.

Now, given a finite poset Q and order-preserving map $f\colon Q \to S = \bigcup_{j \in J} m_j$, for each pair $x, y \in Q$ with $x \leq y$ we have $(x)f \leq (y)f$ in S and we choose $j_{x,y} \in J$ with $(x)f \leq (y)f$ in $R_{j_{x,y}}$. There exists $j_0 \in J$ with $m_{j_{x,y}} \subset m_{j_0}$ for all x, y.

The converse is proved as in (i).

(iii) R-**Mod**: finite means finitely generated. Indeed, let Q be generated by $q_1, \ldots, q_n \in Q$. For each $f\colon Q \to S = \bigcup_{j \in J} R_j$ there exists $j_0 \in J$ with $(q_1)f, \ldots, (q_n)f \in R_{j_0}$. Hence, $(Q)f \subset R_{j_0}$.

Conversely, if Q is not finitely generated, the condition fails even for $f = 1_Q\colon Q \to Q = \bigcup_{j \in J} R_j$, where R_j ($j \in J$) is the collection of all finitely generated submodules.

(iv) More generally, in each variety of finitary algebras, finite means finitely generated.

(v) σ-complete lattices (and σ-complete homomorphisms): no non-empty object is finite. Let R be the free σ-completion of ω and, for each $n < \omega$, let $m_n: n \to R$ be the embedding. Then $\bigcup m_n = 1_R$. Consider any constant map $f: Q \to R - \omega$.

Lemma. Finite objects are closed under finite coproducts and quotients.

Proof. (i) Let Q_1, Q_2 be finite objects, let $f: Q_1 + Q_2 \to S$ be a morphism and let $m = \bigcup_{j \in J} m_j: S \to R$ be a directed union. There exists j_1 such that the first component of $f \cdot m$ factors through m_{j_1}; analogously, the second component factors through some m_{j_2}. If $j \in J$ is larger than j_1 and j_2, then $f \cdot m$ factors through m_j.

(ii) Let Q be a finite object, and let $e: Q \to \bar{Q}$ be a quotient of Q. For each directed union $\bigcup m_j = m: S \to R$ and each morphism $f: Q \to S$ there exists j_0 such that $e \cdot f \cdot m$ factors through m_{j_0}, say, $e \cdot f \cdot m = \bar{f} \cdot m_{j_0}$. Now use the diagonal fill-in:

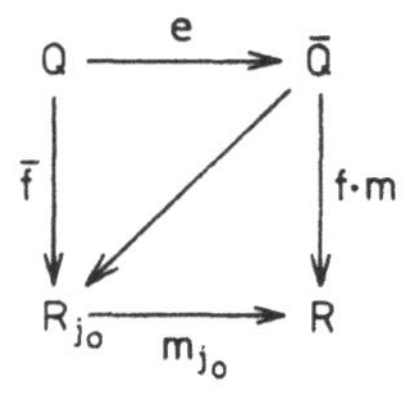

Since $f \cdot m$ factors through m_{j_0}, the object $\bar{Q}$ is finite. □

5.2. Remark. An F-automaton $A = (Q, \delta, \Gamma, \gamma, I, \lambda)$ is said to be *finite* if Q and I are finite objects. Let $F: \mathcal{K} \to \mathcal{K}$ be a finitary varietor preserving finite objects. Then for each finite object I also the steps of the free-algebra construction:

$$I; \quad I + IF; \quad I + (I + IF)F; \ldots$$

are finite. Given an automaton A, its run morphism $\rho: I^{\#} \to Q$ is the colimit morphism of the following "approximations" (see IV.3.1):

$$\rho_0 = \lambda: I \to Q;$$
$$\rho_1: I + IF \to Q$$

has components λ and $\lambda F \cdot \delta$; etc.

We say that an automaton A is *reachable in n steps* if ρ_n is in $\mathcal{E}$.

Each automaton A with I finite, which is reachable in n steps, is finite (because Q is a quotient of W_n) and reachable (because $\rho_n = w_{n,\omega} \cdot \rho$, hence

$\rho \in \mathscr{E}$). Conversely, each finite reachable automaton is reachable in n steps for some $n < \omega$. Indeed,

$$Q = \bigcup_{n < \omega} \operatorname{im} \rho_n$$

and for $f = 1_Q: Q \to Q$ there exists $n_0 < \omega$ and f' such that $1_Q = f' \cdot \operatorname{im} \rho_{n_0}$. This implies that $\operatorname{im} \rho_{n_0} = 1_Q$.

For each functor $F: \mathscr{K} \to \mathscr{K}$ we define its *finitary part* F_ω as follows. For each object X put

$$XF_\omega = \bigcup \operatorname{im}(mF)$$

where the union ranges over all subobjects $m: Y \to X$ of X with Y finite. Let

$$u_X: XF_\omega \to XF$$

be a mono representing the union above. For each morphism

$$f: X \to Y$$

we define fF_ω as the restriction of fF, i.e., as the (unique) morphism for which the following square

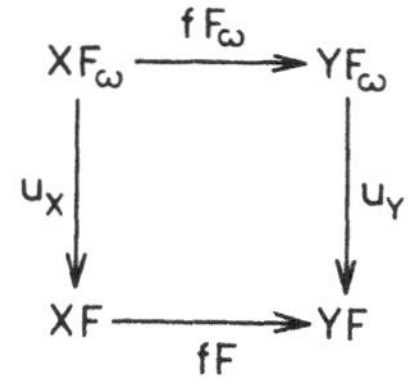

commutes. We must verify that such fF_ω exists—we do this in the following proposition. Then fF_ω is unique because u_Y is a mono. And the preservation of composition and unit morphisms is an easy consequence. Note that

$$u: F_\omega \to F$$

is then a natural transformation. Thus, F_ω is a subfubctor of F—the least subfunctor which coincides with F on the full subcategory of all finite objects.

Proposition. Let $\mathscr{K}$ be a complete and well-powered. For each functor $F: \mathscr{K} \to \mathscr{K}$ the finitary part is well-defined, and it is a finitary subfunctor of F.

Proof. (i) For each morphism $f: X \to Y$ we are to exhibit a morphism $fF_\omega: XF_\omega \to YF_\omega$ with

$$fF_\omega \cdot u_Y = u_X \cdot fF.$$

Let us form the pullback of u_Y and fF:

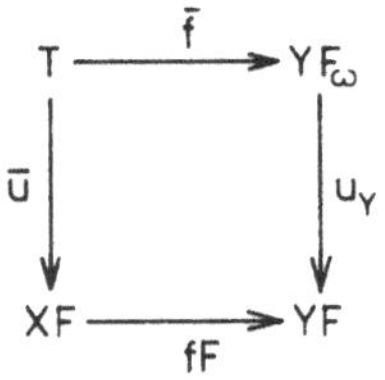

Since u_Y is in $\mathcal{M}$, so is $\bar{u}$ (III.5.1). It suffices to check that for each finite subobject

$$m: Q \to X$$

of X we have

$$\text{im}(mF) \subset \bar{u}.$$

Indeed, then $u_X \subset \bar{u}$, i.e., there exists $t: XF_\omega \to T$ with $u_X = t \cdot \bar{u}$ and we put

$$fF_\omega = t \cdot \bar{f}.$$

Consider the image factorization of $m \cdot f$:

$$m \cdot f = \bar{e} \cdot \bar{m}$$

where $\bar{e}: Q \to \bar{Q}$ is in $\mathcal{E}$, and $\bar{m}: \bar{Q} \to Y$ is in $\mathcal{M}$. Then $\bar{Q}$ is finite, by Lemma V.5.1, and hence,

$$\text{im}(\bar{m}F) \subset u_Y.$$

Thus there exists a morphism $p: \bar{Q}F \to YF_\omega$ with

$$\bar{m}F = p \cdot u_Y.$$

Consequently, there is a unique morphism $\bar{p}$ such that the following diagram

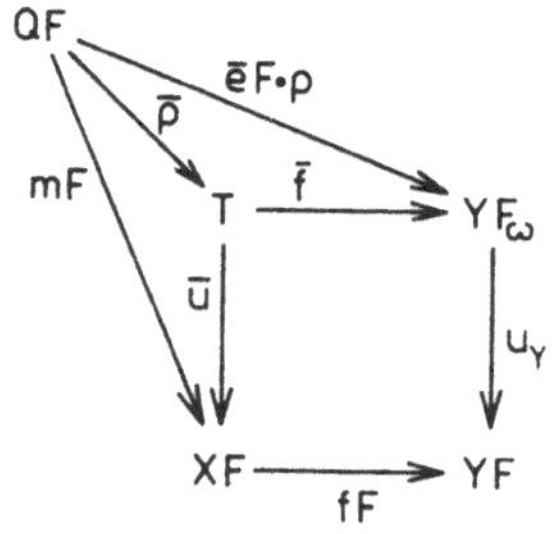

commutes. Thus, $\text{im}(mF) \subset \bar{u}$.

(ii) F_ω is finitary: let $m_j: R_j \to R$ $(j \in J)$ be a directed collection of subobjects with

$$\bigcup_{j \in J} m_j = 1_R$$

(see Lemma V.3.1). For each finite subobject $m\colon Q \to R$ there exist $j_0 \in J$ and $m'\colon Q \to R_{j_0}$ with $m = m' \cdot m_{j_0}$; in other words,

$$m \subset m_{j_0}.$$

Hence,

$$RF_\omega = \bigcup \mathrm{im}(mF) \subset \bigcup_{j \in J} \mathrm{im}(m_j F).$$

Since the opposite inclusion is obvious, we conclude

$$\bigcup_{j \in J} \mathrm{im}(m_j F_\omega) = 1_{RF_\omega}. \qquad \square$$

Corollary. Let $\mathcal{K}$ be a complete, well-powered (regular epi, mono)-category. Let $F\colon \mathcal{K} \to \mathcal{K}$ be a functor, the finitary part of which is right exact. Then each finite F-automaton has a minimal reduction.

Indeed, the finitary part F_ω preserves cointersections by Theorem V.3.5 and hence, each F_ω-automaton has a minimal reduction by Theorem V.1.4 (since $\mathcal{K}$ has cointersections by Observation V.3.5). Further, F coincides with F_ω on finite objects, and hence, each finite F-automaton has a minimal reduction, too. □

Example. For each functor

$$F\colon \mathbf{Set} \to \mathbf{Set}$$

every finite F-automaton has a minimal reduction. For each commutative field R and each functor

$$F\colon R\text{-}\mathbf{Vect} \to R\text{-}\mathbf{Vect},$$

every finite-dimensional F-automaton has a minimal reduction.

Indeed, all these functors have right exact finitary parts (Proposition V.3.10).

Remark. In automata theory, a behavior

$$\beta\colon I^{\#} \to \Gamma$$

is called *recognizable* if there exists a finite automaton the behavior of which is β.

For $\mathcal{K} = \mathbf{Set}$ or $\mathcal{K} = R\text{-}\mathbf{Vect}$, each recognizable behavior has a minimal realization. This is the minimal reduction of (any) finite realization of β.

Exercises V.5

The aim of the following exercises is to present a generalization of F-automata to automata over a monad. We explain first the concept of monad, and then we hint some results on minimal realization.

A. Monoid monad. For each set X, denote by XT the free monoid (of all words) generated by X; for each map $f\colon X \to Y$, let $fT\colon XT \to YT$ be the free extension of f to a homomorphism [defined by $(x_1 \dots x_n)fT = y_1 \dots y_n$ where $y_i = (x_i)f$, and $(\emptyset)fT = \emptyset$].

(i) Verify that

$$T\colon \mathbf{Set} \to \mathbf{Set}$$

is a functor, and the injection of generators defines a natural transformation

$$\eta\colon 1_{\mathbf{Set}} \to T.$$

(ii) For each set X denote by

$$\mu_X\colon (XT)T \to XT$$

the natural map, assigning to each word $w_1 \dots w_n$, where $w_i = x^i_1 \dots x^i_{m_i} \in XT$, the concatenated word

$$(w_1 w_2 \dots w_n)\mu_X = x^1_1 \dots x^1_{m_1} x^2_1 \dots x^2_{m_2} \dots x^n_1 \dots x^n_{m_n} \in XT.$$

Verify that

$$\mu\colon T \cdot T \to T$$

is a natural tranformation such that the following diagram

(1) 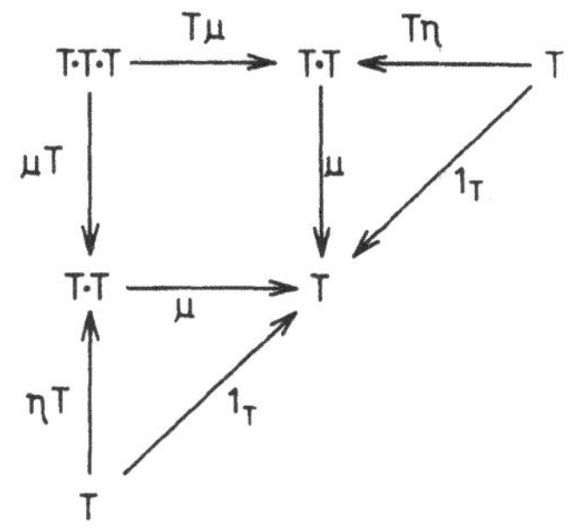

commutes.

(iii) Prove that for each monoid $(Q, \circ, e)$, the map

$$\delta\colon QT \to Q$$

defined by

$$\emptyset\delta = e;$$
$$(q_1 \dots q_n)\delta = q_1 \circ q_2 \circ \dots \circ q_n$$

satisfies the following equations:

(2) $$\eta_Q \cdot \delta = 1_Q \quad \text{and} \quad \delta T \cdot \delta = \mu_Q \cdot \delta.$$

(iv) Prove that each T-algebra satisfying (2) corresponds to a unique monoid on Q. Conclude that the category of monoids and monoid-homomorphism is concretely isomorphic (III.3.8) to the full subcategory of the category T-**Alg** (III.3.1) consisting of all T-algebras satisfying (2).

B. Monads. A *monad* $\mathbb{T} = (T, \eta, \mu)$ on a category $\mathcal{K}$ consists of a functor $T: \mathcal{K} \to \mathcal{K}$ and natural transformations $\eta: 1_{\mathcal{K}} \to T$ and $\mu: T \cdot T \to T$ for which the diagram (1) commutes. The full subcategory of the category T-**Alg** consisting of all T-algebras which satisfy (2) is called the *Eilenberg-Moore category* of $\mathbb{T}$, and it is denoted by $\mathcal{K}^{\mathbb{T}}$.

(i) For each monad $\mathbb{T}$ in **Set**, verify that any set X generates a free object (XT, μ_X) of $\mathbf{Set}^{\mathbb{T}}$ (if the map $\eta_X: X \to XT$ is considered to be the inclusion map).

(ii) Verify that each variety $\mathcal{V}$ of algebras, considered as a concrete category (with homomorphisms as morphisms) is concretely isomorphic to $\mathbf{Set}^{\mathbb{T}_{\mathcal{V}}}$ for the following monad $\mathbb{T}_{\mathcal{V}}$. For each set X, let XT be the (underlying set of) free $\mathcal{V}$-algebra generated by X, $\eta_X: X \to XT$ the injection of generators and $\mu_X: (XT)T \to XT$ the unique homomorphism extending the map 1_{XT}. We call $\mathbb{T}_{\mathcal{V}}$ the *$\mathcal{V}$-free algebra monad.*

(iii) Denote by $\mathbb{P} = (P, \eta, \mu)$ the following monad on **Set**: P is the power-set functor (III.3.4), η is the injection of singleton subsets and $\mu_X: (XP)P \to XP$ assigns to each collection $M \subset XP$ of subsets of X its union $(M)\mu_X = \bigcup\{A; A \in M\}$.

Verify that $\mathbb{P}$ is the $\mathcal{V}$-algebra monad for the variety $\mathcal{V}$ of complete join-semilattices and complete homomorphisms.

(iv) Generalizing (i), verify that for each monad in $\mathcal{K}$ the pair (IT, μ_I) is a $\mathbb{T}$-algebra with the following universal property: for each $\mathbb{T}$-algebra (Q, δ) and each morphism $f: I \to Q$, there exists a unique homomorphism

$$f^{\#}: (IT, \mu_I) \to (Q, \delta)$$

with $f = \eta_I \cdot f^{\#}$. [Hint: $f^{\#} = fT \cdot \delta$.]

C. Monads and varietors. (i) For each varietor $F: \mathcal{K} \to \mathcal{K}$ define a monad $\mathbb{T}_F = (T_F, \eta, \mu)$ on $\mathcal{K}$ as follows. Given I in $\mathcal{K}$, then $IT_F = I^{\#}$ and η_I has the usual meaning; the map $1_{I^{\#}}$ has a unique extension to an F-homomorphism $\mu_I: (I^{\#})^{\#} \to I^{\#}$. Verify that $\mathbb{T}_F$ is a well defined monad, and that the categories

$$F\text{-}\mathbf{Alg} \quad \text{and} \quad \mathcal{K}^{\mathbb{T}_F}$$

are isomorphic.

(ii) Prove that the morphisms

$$\eta_I F \cdot \varphi_I: IF \to I^{\#} = IT_F$$

define a natural transformation $\tau: F \to T_F$ with the following universal property: For each monad $\mathbb{T}' = (T', \eta', \mu')$ and each natural transformation

$\tau': F \to T'$ there exists a unique monad morphism $\sigma: \mathbb{T}_F \to \mathbb{T}'$ (i.e., a natural transformation $\sigma: T_F \to T'$ with $\eta' = \eta \cdot \sigma$ and $\mu \cdot \tau = \sigma T \cdot \mu'$) such that $\tau' = \tau \cdot \sigma$.

D. Monad automata. A T-automaton $(Q, \delta, \Gamma, \gamma, I, \lambda)$ is said to be a $\mathbb{T}$-automaton if (Q, δ) is a $\mathbb{T}$-algebra [i.e., (2) is fulfilled]. The $\mathbb{T}$-run morphism is defined as $\rho = \lambda T \cdot \delta: IT \to Q$, and the behavior is $\beta = \rho \cdot \gamma: IT \to \Gamma$.

If $\mathcal{K}$ is an $(\mathscr{E}, \mathscr{M})$-category, then the $\mathbb{T}$-*minimal realization* of a behavior $\beta: IT \to \Gamma$ is a realization A which is reachable ($\rho \in \mathscr{E}$) and such that any other reachable realization of β can be reduced to A.

(i) Let T preserve $\mathscr{E}$-epis. Prove that each $\mathbb{T}$-automaton has a unique reachable subautomaton.

(ii) Let T preserve $\mathscr{E}$-epis, and let $\mathcal{K}$ have cointersections. Prove that each behavior has a minimal realization iff T preserves cointersections. [Hint: Sufficiency is proved as in V.1.3. For the necessity, let $e = e_i \cdot p_i: Q \to R$ be the cointersection of $\mathscr{E}$-quotients $e_i: Q \to Q_i$ $(i \in I)$. Given morphisms $r_i: Q_iT \to \Gamma$ with $e_iT \cdot r_i = r$ $(i \in I)$, use the $\mathbb{T}$-automaton

$$A = (QT, \mu_Q, \Gamma, r, Q, \eta_Q);$$

each e_i defines a reduction $(Q_iT, \mu_{Q_i}, \Gamma, r_i, Q_i, \eta_{Q_i})$ and using the minimal reduction, it is clear that r factors through eT. Hence, eT is the cointersection of e_iT $(i \in I)$.]

(iii) Prove that in **Set**, monads with minimal realization are precisely the $\mathscr{V}$-free algebra monads of varieties $\mathscr{V}$ of finitary algebras. [Hint: For each $\mathscr{V}$, the functor T of B(i) above is clearly finitary and hence preserves cointersections (V.3.5). Conversely, if T preserves cointersections, then it is finitary (V.3.8) and hence, it is a quotient of H_Σ with Σ finitary (III.4.3). Verify that $\mathbf{Set}^{\mathbb{T}}$ is concretely isomorphic to a variety of Σ-algebras.]

(iv) For each monad $\mathbb{T}$ in **Set**, prove that any behavior with a finite realization has a minimal one. [Hint: The finitary part of a monad is a monad; proceed as in V.5.2.]

Notes to Chapter V

V.1

V. Trnková [1974] proved that a set functor admits minimal realization iff it preserves cointersections, and J. Adámek [1974a] generalized this to an arbitrary category (under rather restrictive additional hypotheses). The general result V.1.5 appeared in J. Adámek [1977a].

V.2

A nice exposition of relations in a (regular epi, mono)-category was presented by P. A. Grillet [1971], where most of the results of sections V.2.1—7 can be found. Relations in an $(\mathscr{E}, \mathscr{M})$-category appear in V. Trnková [1980] which is the source of V.2.8—10.

V.3

The fact that a right exact, finitary functor preserves cointersections was established by M. Barr [1974]. His hypotheses were much stronger; the present form of Theorem V.3.4 is based on the results of J. Adámek [1976b] and J. Adámek and V. Koubek [1981]. The former paper is the source of the converse implication (Theorems V.3.8 and V.3.9).

V.4

All results in this section were proved by J. Adámek [1974a, 1977a].

V.5

The existence of minimal reductions for finite automata was established by J. Adámek [1977a] in case $\mathscr{K}$ is **Set** or R-**Vect**. The general result (Corollary V.5.2) is new.

More information on monads can be found in the monograph of E. G. Manes [1976]. Each varietor defines a "free" monad, i.e., a monad with the universal property of Ex. V.5.C (ii). It was proved by M. Barr [1970] that conversely, if a functor F generates a free monad, then F is a varietor. Thus, F-automata present just the case of monad automata for the free monads.

Minimal realization for monad automata has been investigated by J. Adámek [1976b, 1979b].

Chapter VI: Universal Realization

VI.1. The Concept of Universality

1.1. We say that minimal realization is universal if it has "functorial" nature or, equivalently, if reduced automata form a full, reflective subcategory of the category of F-automata (with minimal reduction as reflection). Whereas sequential automata have this property (II.3.9), tree automata do not: we shall prove that universality of realization implies that the type functor F preserves unions.

Throughout this section we work with an $(\mathscr{E}, \mathscr{M})$-category $\mathscr{K}$. We use a more general concept of automata morphism than before (corresponding to that of II.1.8). A *morphism* from an F-automaton $A = (Q, \delta, \Gamma, \gamma, I, \lambda)$ into an F-automaton $A' = (Q', \delta', \Gamma', \gamma', I', \lambda')$ is a triple

$$(f, f_{\text{in}}, f_{\text{out}}) : A \to A'$$

of morphisms $f: Q \to Q'$, $f_{\text{out}} : \Gamma \to \Gamma'$ and $f_{\text{in}} : I \to I'$ such that the following diagram

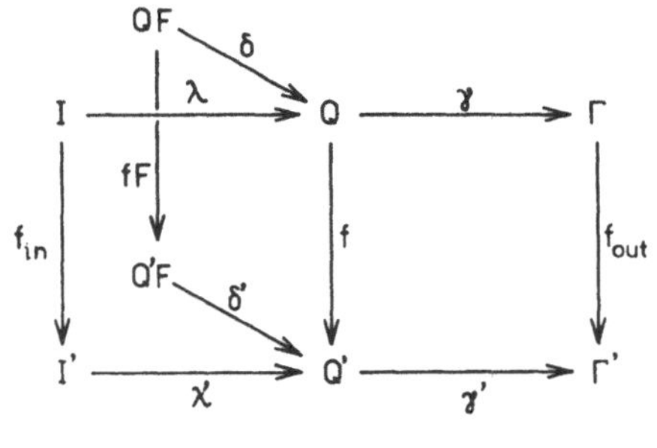

commutes. Analogously, a morphism of non-initial automata is a pair (f, f_{out}).

In case $\Gamma = \Gamma'$ and $f_{\text{out}} = \text{id}_\Gamma$ as well as $I = I'$ and $f_{\text{in}} = \text{id}_I$, we write f instead of $(f, \text{id}, \text{id})$ (as before).

We formulate the concept of universality first for reduction and later for realization.

1.2. Definition. Let $F: \mathscr{K} \to \mathscr{K}$ be a functor such that each non-initial F-automaton A has a minimal reduction

$$e_A : A \to A_r.$$

Then F is said to have *universal reduction* provided that for each morphism of automata $(f, f_{out}): A \to A'$ there exists a morphism of the minimal reductions $(f_r, f_{out}): A_r \to A'_r$ such that the following square

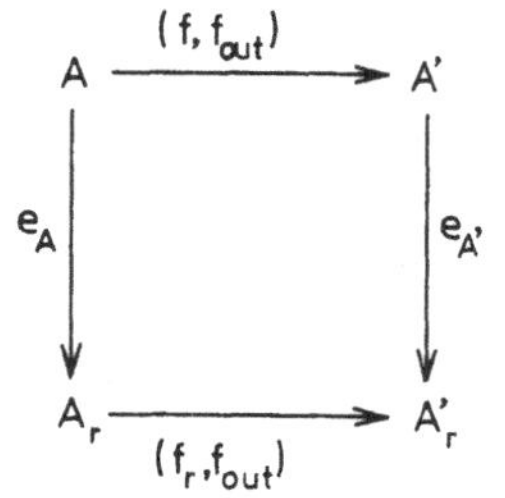

commutes.

Remarks. (a) Since e_A abbreviates $(e_A, \mathrm{id}_\Gamma)$ and $e_{A'}$ abbreviates $(e_{A'}, \mathrm{id}_\Gamma)$, it is clear that the morphism $A_r \to A'_r$ above must have the output-part equal to f_{out}.

(b) A simple categorical formulation of the universality can be stated by means of the category

$$\mathbf{Aut}(F)$$

of non-initial F-automata and their morphisms:

> Reduction is universal iff reduced automata form a full reflective subcategory of $\mathbf{Aut}(F)$, the reflections of which are the minimal reductions.

Examples. (i) Sequential automata: reduction is universal. For each sequential Σ-automaton $A = (Q, \delta, \Gamma, \gamma)$, the minimal reduction is $A_r = A/{\sim_A}$ where $q_1 \sim_A q_2$ holds iff q_1 and q_2 have the same behavior, and e_A is the canonical map (I.2:5).

For each morphism

$$(f, f_{out}): (Q, \delta, \Gamma, \gamma) \to (Q', \delta', \Gamma', \gamma')$$

we know that

$$q_1 \sim_A q_2 \text{ in } A \quad \text{implies} \quad (q_1)f \sim_{A'} (q_2)f \text{ in } A',$$

see I.1.8. Hence, we can define

$$f_r: Q/{\sim_A} \to Q'/{\sim_{A'}}$$

by $e_A \cdot f_r = f \cdot e_{A'}$ (i.e., $[q]f_r = [(q)f]$). This defines $(f_r, f_{out}): A_r \to A'_r$ with $e_A \cdot (f_r, f_{out}) = (f, f_{out})e_{A'}$.

1.3. Recall that a *subautomaton* of a non-initial F-automaton $A = (Q, \delta, \Gamma,$

γ) is an automaton $A' = (Q', \delta', \Gamma, \gamma')$ together with a morphism $m: A' \to A$ such that $m \in \mathcal{M}$.

Proposition. Let $F: \mathcal{K} \to \mathcal{K}$ preserve $\mathcal{E}$-epis and have minimal reductions. Then F has universal reduction iff each subautomaton of any reduced F-automaton is reduced.

Proof. (i) Assume that subautomata of reduced F-automata are reduced. Consider a morphism of automata and the minimal reductions of these automata:

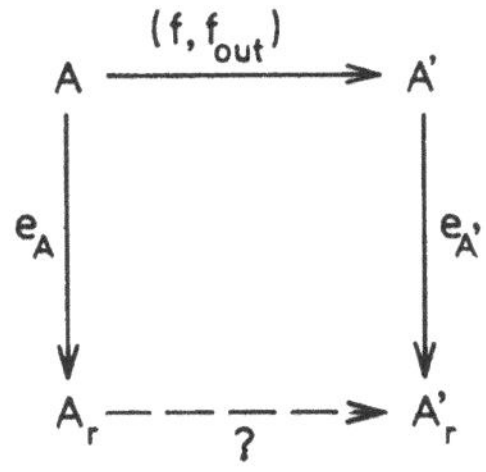

Put $A = (Q, \delta, \Gamma, \gamma)$ and $A' = (Q', \delta', \Gamma', \gamma')$; let the subscript r denote the corresponding data in the minimal reductions.

Since F preserves $\mathcal{E}$-epis, also **F-Alg** is an $(\mathcal{E}, \mathcal{M})$-category, see IV.8.5. The homomorphism

$$f \cdot e_{A'}: (Q, \delta) \to (Q'_r, \delta'_r)$$

can be factored as $f = e^* \cdot m^*$ with

$$e^*: (Q, \delta) \to (Q^*, \delta^*), \quad e^* \in \mathcal{E}$$

and

$$m^*: (Q^*, \delta^*) \to (Q'_r, \delta'_r), \quad m^* \in \mathcal{M}.$$

Define an automaton

$$A^* = (Q^*, \delta^*, \Gamma', m^* \cdot \gamma'_r).$$

Then

$$m^*: A^* \to A'_r$$

is clearly a morphism of automata. Since $m^* \in \mathcal{M}$ and A'_r is reduced, A^* is also reduced. Define another automaton

$$\tilde{A} = (Q, \delta, \Gamma', \gamma \cdot f_{out}).$$

Then

$$e^*: A \to A^*$$

is a morphism because the following diagram

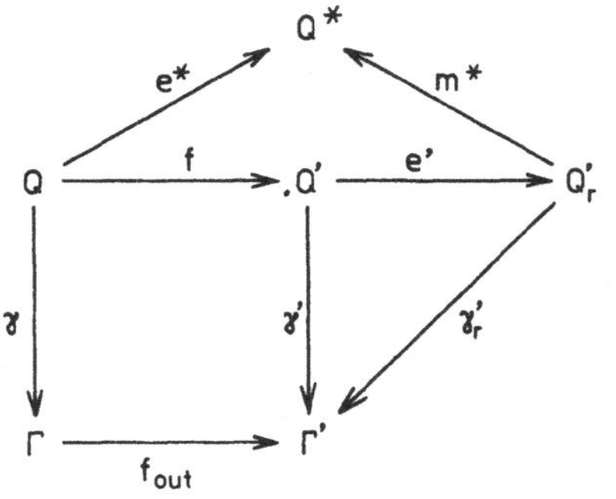

commutes. Since $e^* \in \mathscr{E}$, we see that A^* is a reduction of $\tilde{A}$. This is, in fact, the minimal reduction (because A^* is reduced and hence, it cannot be further reduced to a minimal reduction of $\tilde{A}$).

Redefining the outputs in A_r, we get a new automaton

$$\tilde{A}_r = (Q_r, \delta_r, \Gamma', \gamma_r \cdot f_{out}).$$

Then, obviously,

$$e_A : \tilde{A} \to \tilde{A}_r$$

is a reduction. Therefore, $\tilde{A}_r$ can be further reduced to the minimal reduction, i.e., there exists

$$e : \tilde{A}_r \to A^*$$

with

$$e^* = e_A \cdot e.$$

Put

$$f_r = e \cdot m^* : (Q_r, \delta_r) \to (Q'_r, \delta'_r).$$

This is a homomorphisms such that the following diagram

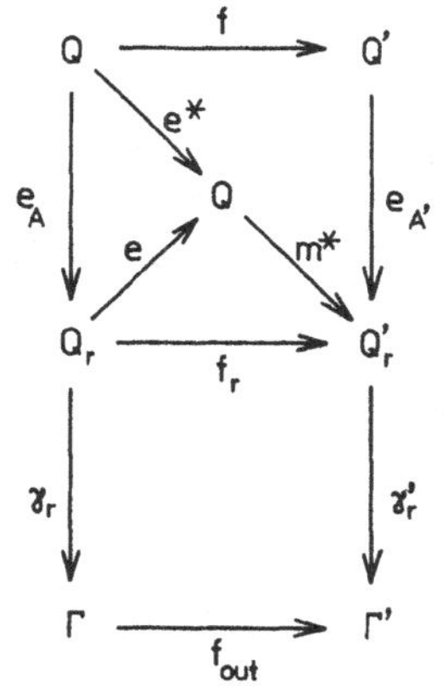

commutes. This implies (since e_A is an epi) that

$$f_r \cdot \gamma'_{r} = \gamma_r \cdot f_{out}.$$

Hence,

$$(f_r, f_{out}): A_r \to A'_r$$

is the desired morphism of reduced automata.

(ii) Let reduction be universal. Let $m: A' \to A$ be a subautomaton of a reduced automaton A. Let $e_{A'}: A' \to A'_r$ be the minimal reduction of A'; note that $1_A: A \to A$ is the minimal reduction of A. Hence, there exists a morphism m_r such that the following square

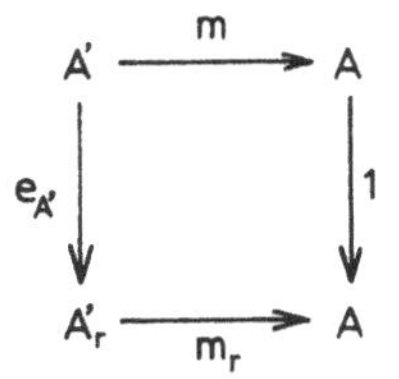

commutes. Since $e_{A'} \cdot m_r = m \in \mathcal{M}$ implies $e_{A'} \in \mathcal{M}$ (III.5.1) and since $e_{A'} \in \mathcal{E}$, we conclude that $e_{A'}$ is an isomorphism. This proves that A' is reduced. □

1.4. Functors with universal reduction can be characterized in a way analogous to that for minimal reduction in Theorem V.1.5.

Recall that a *preimage* of a subobject $m: Q' \to Q$ ($\in \mathcal{M}$) under a morphism $f: P \to Q$ is the subobject $\bar{m}$ of P defined by the following pullback:

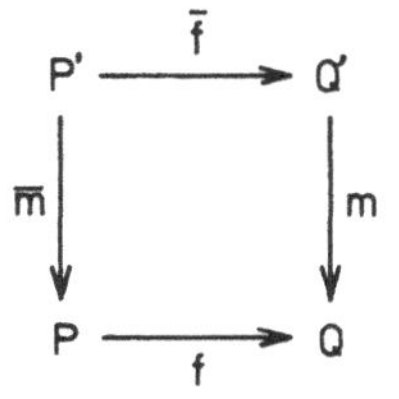

Dually, a *co-preimage* is any pushout

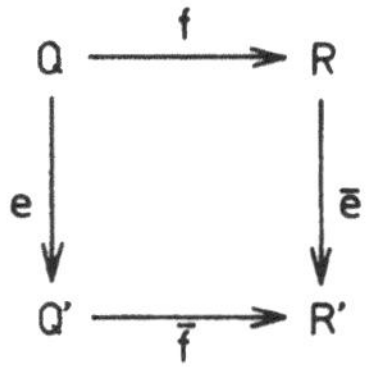

with $e \in \mathcal{E}$. A functor F is said to *preserve co-preimages* if it preserves each pushout of a morphism and an $\mathcal{E}$-epi.

Theorem. Let $\mathcal{K}$ be a connected, finitely cocomplete $(\mathscr{E}, \mathscr{M})$-category with cointersections. Let $F: \mathcal{K} \to \mathcal{K}$ preserve $\mathscr{E}$-epis. Then F has universal reduction iff F preserves both cointersections and co-preimages.

Remark. For the sufficiency, i.e., for the proof that preservation of cointersections and co-preimages implies universal reduction, we do not need the hypothesis that $\mathcal{K}$ be connected. This will be clear from the proof.

Proof. (i) Let F preserve cointersections and co-preimages. Then F has minimal reductions by Theorem V.1.3. By Proposition VI.1.3, it is enough to prove that each subautomaton $m: A \to A'$ of a reduced automaton A' is reduced. Put

$$A = (Q, \delta, \Gamma, \gamma) \quad \text{and} \quad A' = (Q', \delta', \Gamma, \gamma').$$

For each reduction

$$e: A \to A_0 = (Q_0, \delta_0, \Gamma, \gamma_0)$$

of A we are to show that e is an isomorphism. Consider the co-preimage of e under m, i.e., the following pushout

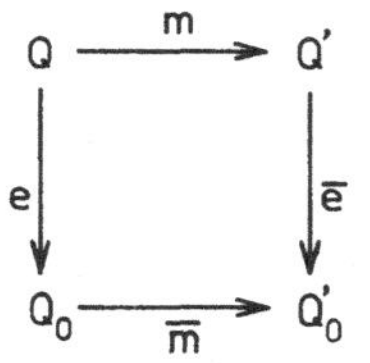

It suffices to find an automaton A'_0 on Q'_0 such that $\bar{e}: A' \to A'_0$ is a morphism. Then $\bar{e}$ is a reduction of the reduced automaton A' and hence, $\bar{e}$ is an isomorphism, and therefore,

$$e \cdot \bar{m} \cdot \bar{e}^{-1} = m \in \mathscr{M} \quad \text{implies} \quad e \in \mathscr{M}.$$

Thus, $e \in \mathscr{E} \cap \mathscr{M}$ is an isomorphism.

The output morphism γ_0 of A'_0 is defined by the universal property of the above pushout:

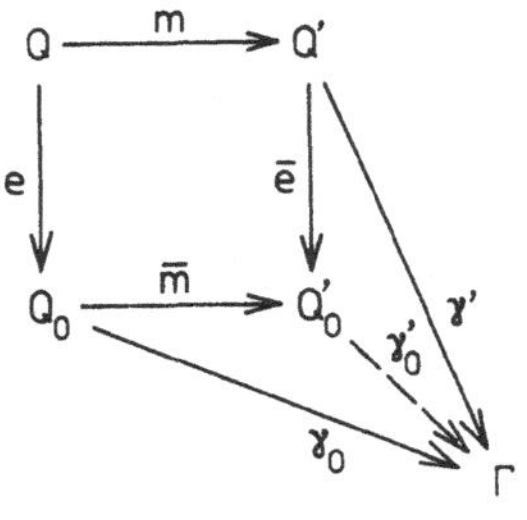

Since $e \cdot \gamma_0 = \gamma = m \cdot \gamma'$, there exists a unique γ'_0 making the above diagram commutative.

The next-state morphism δ'_0 of A' is defined by the universal property of the F-image of the above pushout (which is a pushout, too, since F preserves co-preimages):

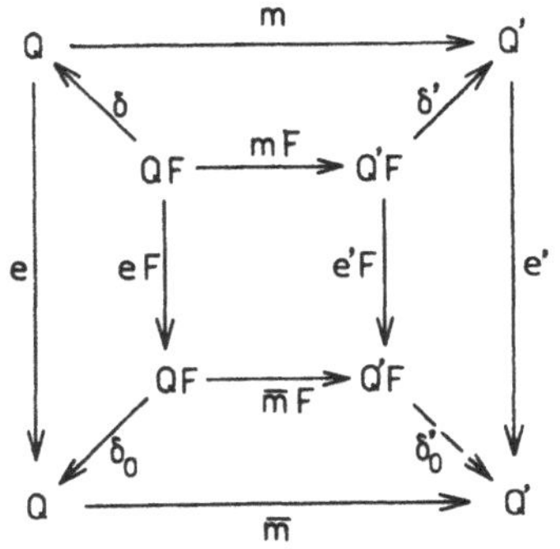

Since both the squares commute, and since m and e are homomorphisms, there exists a unique δ'_0 making the diagram above commutative.

Clearly, $e: A' \to (Q'_0, \delta'_0, \Gamma, \gamma'_0)$ is a morphism, which concludes the proof.

(ii) Let F have universal reduction. Then F preserves cointersections by Theorem V.1.5. Let us prove that F preserves any pushout

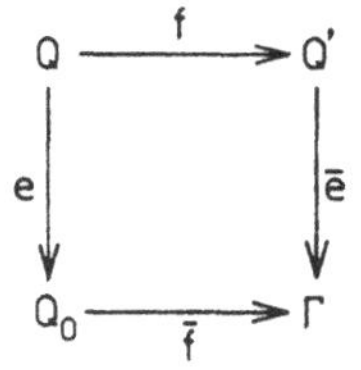

with $e \in \mathscr{E}$.

The category $\mathscr{K}$ has a terminal object T (by Remark V.1.5). For each object X denote the unique morphism into T by

$$t_X: X \to T.$$

(a) Let $Q = \perp$. Then the pushout above is trivially preserved:

(a_1) If Q' is also initial, then f is an isomorphism and the pushout is the following square

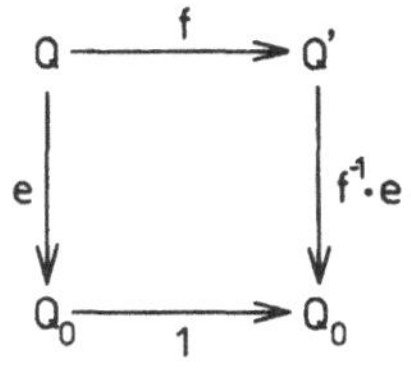

(a$_2$) If Q' is non-initial, then there exists a morphism $j: Q_0 \to Q'$. Since Q is initial, we have $e \cdot j = f$, and the pushout is the following square

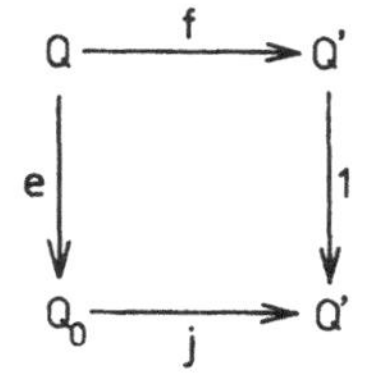

(b) Let $Q \neq \perp$. Since $\mathscr{K}$ is connected, there exists a morphism $r: T \to Q$. We are to prove that F preserves the considered pushout. Thus, let p' and p_0 be arbitrary morphisms with $eF \cdot p_0 = fF \cdot p'$:

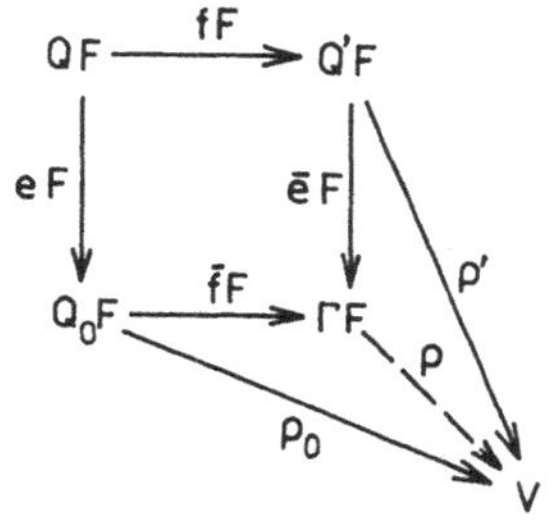

We shall prove that there is a morphism $p: \Gamma F \to V$ with

$$p' = \bar{e}F \cdot p.$$

Then (i) p is unique since $\bar{e}F$ is epi, ($e \in \mathscr{E}$ implies $\bar{e} \in \mathscr{E}$ and hence, $\bar{e}F \in \mathscr{E}$) and (ii) $p_0 = \bar{f}F \cdot p$ because eF is epi and

$$eF \cdot p_0 = fF \cdot p' = fF \cdot \bar{e}F \cdot p = eF \cdot (\bar{f}F \cdot p).$$

To find the morphism p, we define automata A (on the state-object $Q + QF$), A' (on $Q' + Q'F$), and A_0 (on $Q_0 + Q_0F$) in such a way that both $f + fF: A \to A'$ and $e + eF: A \to A_0$ become automata-morphisms. Using the universality of reduction on the morphism $f + fF$, we shall be able to produce p as required. The joint output object of all these automata will be $\Gamma + V$. Denote the various coproduct injections as follows:

$$Q \xrightarrow{i_Q} Q + QF \xleftarrow{j_Q} QF$$

(analogously $i_{Q'}, j_{Q'}$ and i_{Q_0}, j_{Q_0}) and

$$\Gamma \xrightarrow{i} \Gamma + V \xleftarrow{j} V.$$

To define the automaton A, first consider the morphism

$$\bar{\delta}: Q + QF \to Q$$

with components 1_Q and $t_{QF} \cdot r\colon QF \to Q$, and put

$$\delta = \bar{\delta}F \cdot j_Q : (Q + QF)F \to Q + QF.$$

Next, the output map will be

$$\gamma = (f + fF) \cdot (\bar{e} + p') = (e + eF) \cdot (\bar{f} + p_0) : Q + QF \to \Gamma + V.$$

Thus, we obtain an automaton

$$A = (Q + QF, \delta, \Gamma + V, \gamma).$$

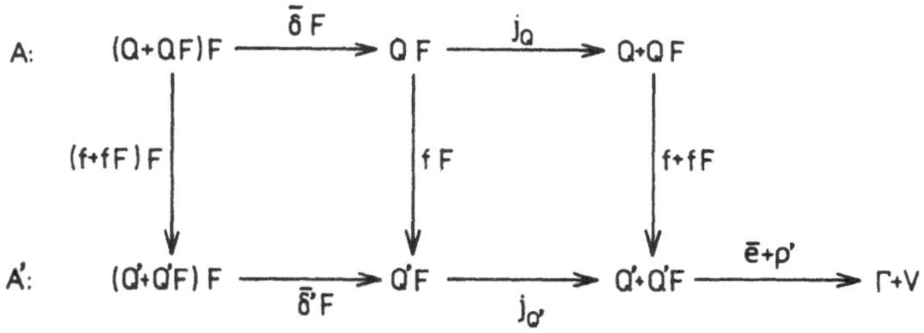

Analogously, define

$$A' = (Q' + Q'F, \delta', \Gamma + V, \bar{e} + p')$$

where $\delta' = \bar{\delta}'F \cdot j_{Q'}$ with $\bar{\delta}' : Q' + Q'\mathrm{F} \to Q'$ having components $1_{Q'}$ and $t_{Q'F} \cdot r \cdot f\colon Q'F \to Q'$. Clearly,

$$f + fF\colon A \to A'$$

is a morphism of automata. Denote the minimal reductions by

$$e_A : A \to A_r = (Q_r, \delta_r, \Gamma + V, \gamma_r)$$

and

$$e_{A'}\colon A' \to A'_r = (Q'_r, \delta'_r, \Gamma + V, \gamma'_r).$$

By hypothesis, reduction is universal. Therefore, there exists a morphism $f_r\colon A_r \to A_{r'}$ such that the following square

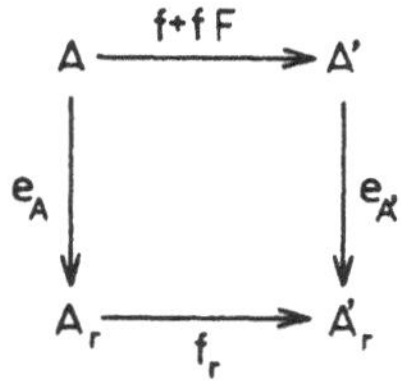

commutes.

Finally, define another automaton analogously to A and A':

$$A_0 = (Q_0 + Q_0F, \delta_0, \Gamma + V, \bar{f} + p_0)$$

where $\delta_0 = \bar{\delta}_0 F \cdot j_Q$ with $\bar{\delta}_0 : Q_0 + Q_0F \to Q_0$ having components 1_{Q_0} and $t_{Q_0F} \cdot r \cdot e$.

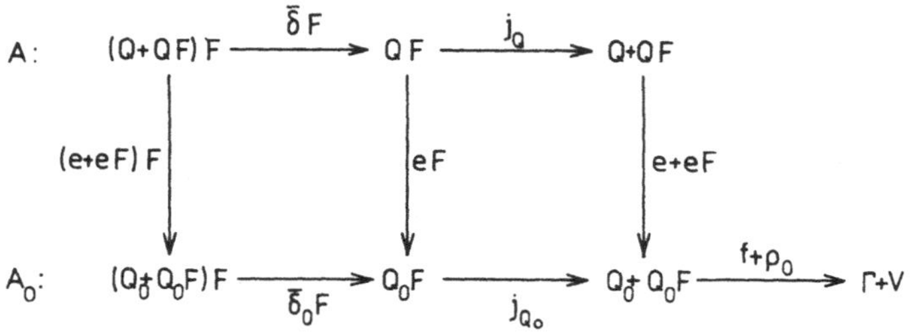

Clearly,

$$e + eF : A \to A_0$$

is a morphism. This is a reduction of A, since $e \in \mathscr{E}$ implies $eF \in \mathscr{E}$ and hence, by III.5.5, $e + eF \in \mathscr{E}$. Since A_r is the minimal reduction of A, there exists a reduction

$$h : A_0 \to A_r$$

with

$$e_A = (e + eF) \cdot h.$$

We have

$$i_Q \cdot (f + fF) = f \cdot i_{Q'} \quad \text{and} \quad i_Q \cdot (e + eF) = e \cdot i_Q$$

and therefore,

$$\begin{aligned} f \cdot i_{Q'} \cdot e_{A'} &= i_Q \cdot (f + fF) \cdot e_{A'} \\ &= i_Q \cdot e_A \cdot f_r \\ &= i_Q \cdot (e + eF) \cdot h \cdot f_r \\ &= e \cdot i_{Q_0} \cdot h \cdot f_r. \end{aligned}$$

We use the universal property of the given pushout:

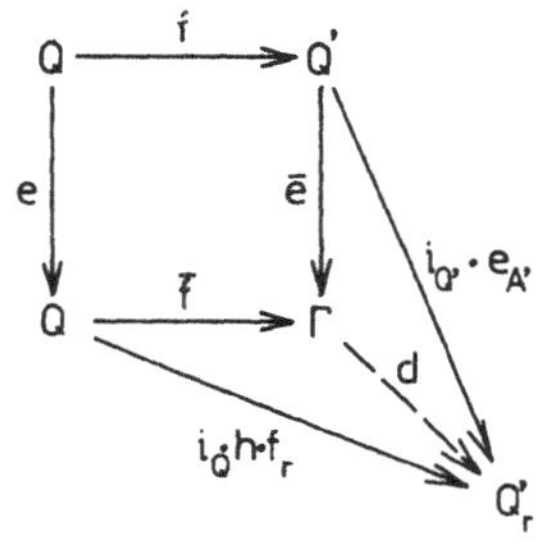

We obtain a unique d for which the above diagram commutes, in particular, for which

$$\bar{e} \cdot d = i_{Q'} \cdot e_{A'} .$$

Also the following diagram

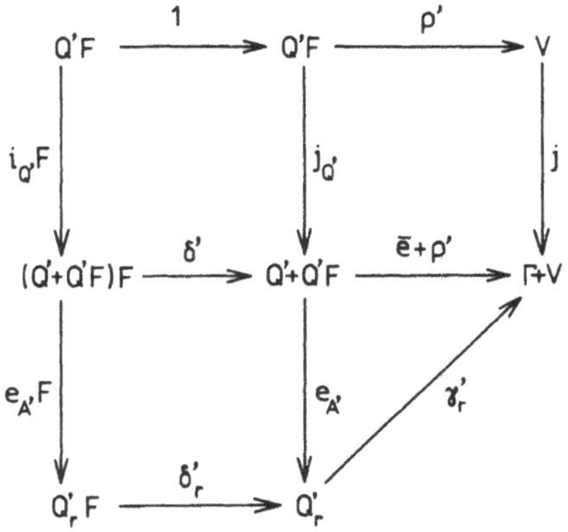

commutes. Indeed, the lower part commutes because $e_{A'}$ is a morphism of automata. In the upper part, we have $j_{Q'} = (\bar{e} + p') = p' \cdot j$, which is trivial, and

$$j_{Q'} = i_{Q'} F \cdot \delta' .$$

The last follows from the definition of $\delta' = \bar{\delta}' F \cdot j_Q$, since $i_{Q'} \cdot \bar{\delta}' = 1_{Q'}$.

We conclude that

$$p' \cdot j = (i_{Q'} \cdot e_{A'})F \cdot \delta'_r \cdot \gamma'_r = \bar{e}F \cdot dF \cdot \delta'_r \cdot \gamma'_r .$$

If $\hom(\Gamma, V) \neq \emptyset$, choose a morphism

$$k \colon \Gamma \to V;$$

we denote by $\bar{k} \colon \Gamma + V \to V$ the morphism with components k and 1_V. Then

$$p' = p' \cdot j \cdot \bar{k} = eF \cdot dF \cdot \delta'_r \cdot \gamma'_r \cdot \bar{k}$$

and to conclude the proof, it suffices to put

$$p = dF \cdot \delta'_r \cdot \gamma'_r \cdot \bar{k}.$$

If

$$\hom(\Gamma, V) = \emptyset,$$

then $V = \perp$ is the initial object; moreover, for each non-initial object X we have $\hom(X, \perp) = \emptyset$ (for, there is some morphism $h \colon \Gamma \to X$ and assuming the existence of $g \colon X \to \perp$, we obtain $f \cdot g \colon \Gamma \to \perp$, a contradiction). Since $p' \cdot fF \in \hom(QF, \perp)$, it follows that $QF = \perp$. By our hypothesis, we have a morphism $rF: TF \to QF = \perp$, thus also $TF = \perp$. It follows immediately that F is the constant functor to $\perp$: for each object X we have the morphism $t_X F: XF \to TF = \perp$, thus, $XF = \perp$. Then F preserves all colimits. The proof is concluded. □

1.5. The above theorem is rather abstract—it is usually hard to decide whether a given functor preserves co-preimages or not. In the next section, we prove a more concrete criterion. But first we formulate the universality in terms of realization.

Let $F: \mathcal{K} \to \mathcal{K}$ be a varietor. Denote by

$$\mathbf{Aut}_{\mathrm{i}}(F)$$

the category of all reachable (!) initial F-automata and their homomorphisms. Denote by

$$\mathbf{Beh}(F)$$

the *category of behaviors*: objects are triples (I, b, Γ), where I and Γ are $\mathcal{K}$-objects and $b: I^\# \to \Gamma$ is a morphism; morphisms are pairs

$$(f_{\mathrm{in}}, f_{\mathrm{out}}): (I, b, \Gamma) \to (I', b', \Gamma')$$

of morphisms in $\mathcal{K}$, $f_{\mathrm{in}}: I \to I'$ and $f_{\mathrm{out}}: \Gamma \to \Gamma'$, such that the following square

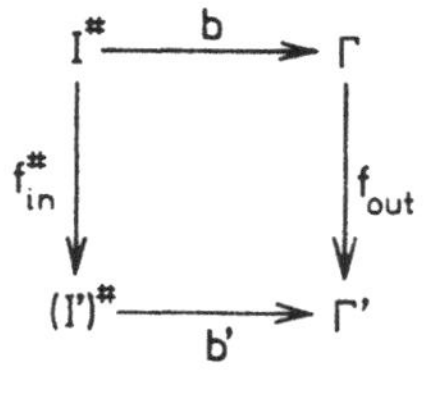

commutes. It is easy to check that this defines a category.

Assigning to each reachable automaton A its behavior $b_A: I^\# \to \Gamma$, we get a functor

$$B: \mathbf{Aut}_{\mathrm{i}}(F) \to \mathbf{Beh}(F).$$

On objects,

$$AB = (I, b_A, \Gamma).$$

Given a morphism

$$(f, f_{\mathrm{in}}, f_{\mathrm{out}}): A \to A',$$

then

$$(f, f_{\mathrm{in}}, f_{\mathrm{out}})B = (f_{\mathrm{in}}, f_{\mathrm{out}}): (I, b_A, \Gamma) \to (I', b_{A'}, \Gamma').$$

We call B the *behavior functor*.

It turns out that, even in case that F has minimal realizations, there need not be any way how to define a functor

$$M\colon \mathbf{Beh}(F) \to \mathbf{Aut}_{\mathrm{i}}(F),$$

assigning to each behavior its minimal realization. In fact, the existence of such a functor is equivalent to the universality of reduction, as we shall prove presently.

1.6. Definition. A varietor F is said to have *universal realization* if there exists a functor

$$M\colon \mathbf{Beh}(F) \to \mathbf{Aut}_{\mathrm{i}}(F)$$

assigning to each behavior its minimal realization and such that $M \cdot B = \mathrm{id}_{\mathbf{Beh}(F)}$.

Explicitly, the condition $M \cdot B = \mathrm{id}$ means the following. For each morphism of behaviors

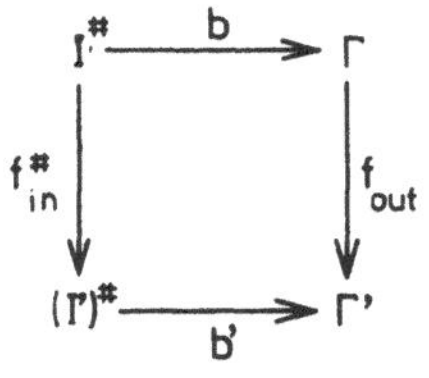

we can find the minimal realization of b (with run morphism $\rho\colon I^{\#} \to Q_b$ and with output $\gamma\colon Q_b \to \Gamma$) and that of b' (with ρ', γ') and we can fill-in the above diagram by an $f\colon Q_b \to Q_{b'}$ in such a way that the following diagram

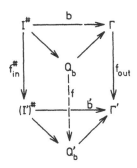

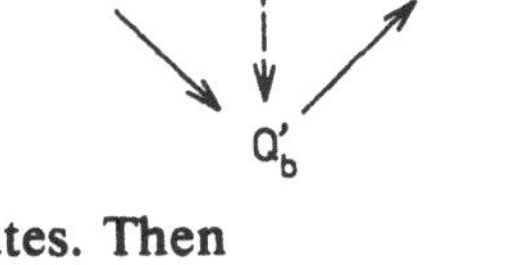

commutes. Then

$$(f_{\mathrm{in}}, f_{\mathrm{out}})M = (f, f_{\mathrm{in}}, f_{\mathrm{out}}).$$

If the functor M exists, we call it the *minimal realization functor.* The reason why we have restricted ourselves to reachable automata when defining $\mathbf{Aut}_{\mathrm{i}}(F)$ is that this fill-in is uniquely determined, hence, canonical.

Observation. The functors M and B are adjoint. I.e., for each automaton A_0 with behavior $b_0: I_0^\# \to \Gamma_0$ and for each behavior $b: I^\# \to \Gamma$ we have a natural bijection between behavior morphisms and automata morphisms:

$$\frac{(I_0, b_0, \Gamma_0) \to (I, b, \Gamma)}{A_0 \to (I, b, \Gamma)M}$$

This bijection assigns to each morphism of behaviors

$$(f_{\text{in}}, f_{\text{out}}): (I_0, b_0, \Gamma_0) \to (I, b, \Gamma)$$

the unique morphism of the form

$$(f, f_{\text{in}}, f_{\text{out}}): A_0 \to (I, b, \Gamma)M.$$

1.7. Proposition. A varietor has universal reduction iff it has universal realization.

Proof. (i) Let F have universal reduction.
Given a morphism of behaviors:

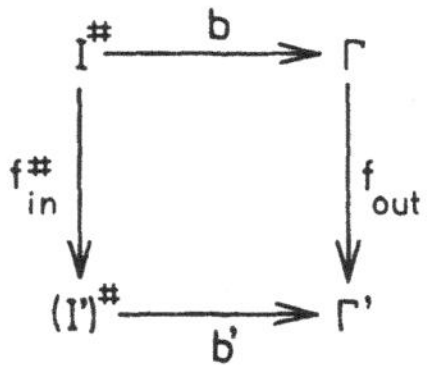

we form non-initial F-automata

$$A = (I^\#, \varphi, \Gamma, b) \quad \text{and} \quad A' = ((I')^\#, \varphi', \Gamma', b').$$

Then

$$(f_{\text{in}}^\#, f_{\text{out}}): A \to A'$$

is clearly a morphism of automata. Let

$$e_A: A \to A_0 = (Q_0, \delta_0, \Gamma, \gamma_0)$$

and

$$e_{A'}: A' \to A_0' = (Q_0', \delta_0', \Gamma', \gamma_0')$$

denote the minimal reductions of A and A', respectively. Since reduction is universal, there exists a morphism

$$(f_r, f_{\text{out}}): A_0 \to A_0'$$

with

$$e_A \cdot (f_r, f_{\text{out}}) = (f_{\text{in}}^\#, f_{\text{out}}) \cdot e_{A'}$$

i.e., with

$$e_A \cdot f_r = f_{in}^{\#} \cdot e_{A'}.$$

The minimal realization of b is

$$A_0^* = (Q_0, \delta_0, \Gamma, \gamma_0, I, \eta \cdot e_A)$$

with the run morphism e_A. This is proved in V.1.2 (the first part of the proof of the proposition). Analogously, the minimal realization of b' is A'^* with the run morphism $e_{A'}$. The morphism of minimal realizations "filling-in" the given morphism of behaviors is $(f_r, f_{in}, f_{out}): A_0^* \to A_0'^*$:

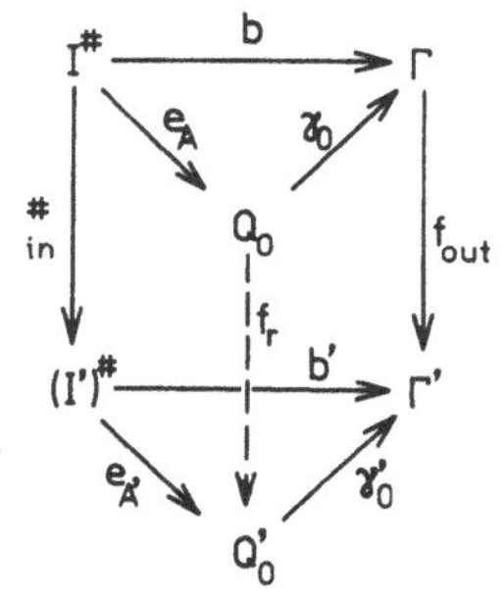

(ii) Let F have universal realization. Let $(f, f_{out}): A \to A'$ be a morphism of non-initial F-automata, where

$$A = (Q, \delta, \Gamma, \gamma) \quad \text{and} \quad A' = (Q', \delta', \Gamma', \gamma').$$

Define corresponding reachable initial automata

$$\bar{A} = (Q, \delta, \Gamma, \gamma, Q, 1_Q) \text{ and } \bar{A}' = (Q', \delta', \Gamma', \gamma', Q', 1_{Q'}).$$

Then put $f_{in} = f$ to obtain a morphism

$$(f, f_{in}, f_{out}): \bar{A} \to \bar{A}'$$

in $\mathbf{Aut}_i(F)$. Denote by $\rho: (Q^{\#}, \varphi) \to (Q, \delta)$ the run morphism of $\bar{A}$. We have $\eta \cdot \rho = 1_Q$ and hence, behavior of $\bar{A}$ is $\rho \cdot \gamma: Q^{\#} \to \Gamma$. Let

$$e: A \to A_0 = (Q_0, \delta_0, \Gamma, \gamma_0)$$

be the minimal reduction of A, then

$$A_0^* = (Q_0, \delta_0, \Gamma, \gamma_0, Q, e)$$

is the minimal realization of $\rho \cdot \gamma$. In fact, the run morphism of A_0^* is $\rho \cdot e$ (since $\eta \cdot \rho \cdot e = e$) and hence, A_0^* is a reachable realization of $\rho \cdot e \cdot \gamma_0 = \rho \cdot \gamma$. Consequently, the minimal realization $\tilde{A}$ of $\rho \cdot \gamma$ is a reduction of A_0^*. But A_0 is reduced and hence, A_0^* is isomorphic to $\tilde{A}$, in other words, A_0^* is minimal realization of $\rho \cdot \gamma$.

Analogously, denote by ρ' the run morphism if $\bar{A}'$ and by

$$e' : A' \to A'_0 = (Q'_0, \delta'_0, \Gamma'_0, \gamma'_0)$$

the minimal reduction of A'. The minimal realization of $\rho' \cdot \gamma'$ is $A'^*_0 = (Q'_0, \delta'_0, \Gamma', \gamma'_0, Q'_0, e')$. The morphism of behaviors

$$(f, f_{\text{out}}) : (Q, \rho \cdot \gamma, \Gamma) \to (Q', \rho' \cdot \gamma', \Gamma')$$

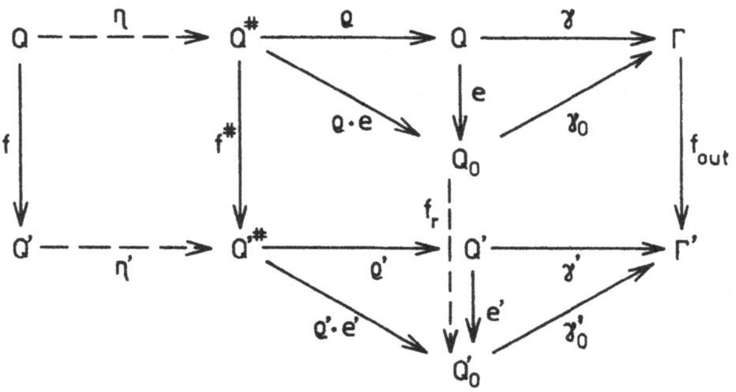

can be "filled-in" to a morphism of the minimal realizations:

$$(f_r, f, f_{\text{out}}) : A_0^* \to A'^*_0.$$

Then

$$(f_r, f_{\text{out}}) : A_0 \to A'_0$$

is a morphism of the minimal reduction with

$$e \cdot f_r = \eta \cdot (\rho \cdot e \cdot f_r) = \eta \cdot (f^\# \cdot \rho \cdot e') = f \cdot \eta' \cdot \rho' \cdot e' = f \cdot e',$$

i.e., with

$$e \cdot (f_r, f_{\text{out}}) \cdot (f, f_{\text{out}}) \cdot e'.$$

□

1.8. Example. Sequential automata with resets. Recall the functors

$$S_{\Sigma_1, \Sigma_0} = S_{\Sigma_1} + C_{\Sigma_0} : \mathbf{Set} \to \mathbf{Set}$$

from III.4.8. An S_{Σ_1, Σ_0}-automaton can be considered as a special case of a sequential Σ-automaton, where

$$\Sigma = \Sigma_1 \cup \Sigma_0$$

(assuming $\Sigma_1 \cap \Sigma_0 = \emptyset$). In fact, the next-state map

$$\delta : Q \times \Sigma_1 + \Sigma_0 \to Q$$

yields a map from $Q \times \Sigma$ to Q which for pairs (q, σ), $\sigma \in \Sigma_0$, is independent of q. Thus S_{Σ_1, Σ_0}-automata are just sequential Σ-automata with resets in Σ_0, see Exercise II.1.E.

The functors S_{Σ_1, Σ_0} have universal realization (II.3.10) and we shall prove in the next section that these are the only set functors with universal realization.

1.9. Example. Coadjoints. Let $\mathscr{K}$ be an $(\mathscr{E}, \mathscr{M})$-category with countable products and coproducts, and let $F: \mathscr{K} \to \mathscr{K}$ be a coadjoint preserving $\mathscr{E}$-epis. By III.2.14, F has minimal realization, obtained by image factorization. We shall prove that F has universal realization.

Let

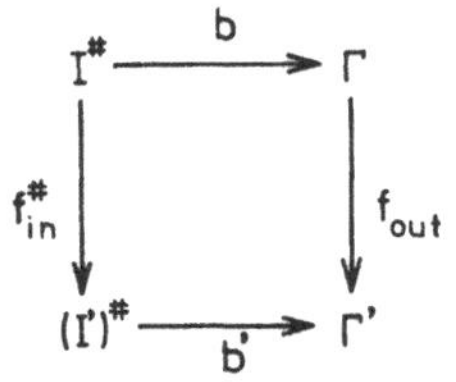

be a behavior morphism. Consider

$$b_{\#} : (I^{*}, \varphi) \to (\Gamma_{\#}, \psi) \text{ and } b'_{\#} : (I'^{*}, \varphi) \to (\Gamma'_{\#}, \psi').$$

Then the following square

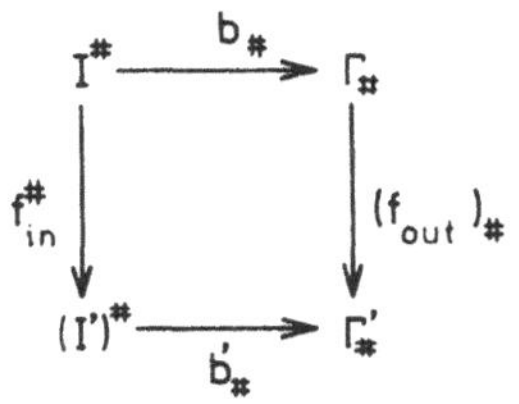

commutes too, where $(f_{out})_{\#}$ denotes the unique homomorphism

$$(f_{out})_{\#} : (\Gamma_{\#}, \psi) \to (\Gamma'_{\#}, \psi')$$

with

$$\pi \cdot f_{out} = (f_{out})_{\#} \cdot \pi'$$

(see III.2.13).

The minimal realizations of b and b' are obtained by the image factorizations of $b_{\#}$ and $b'_{\#}$, respectively:

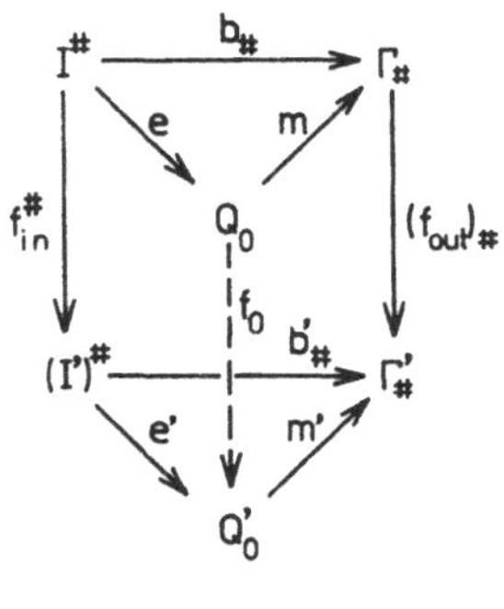

Then the morphism f_0 which "fills-in" the behavior morphism is obtained by the diagonal fill-in property:

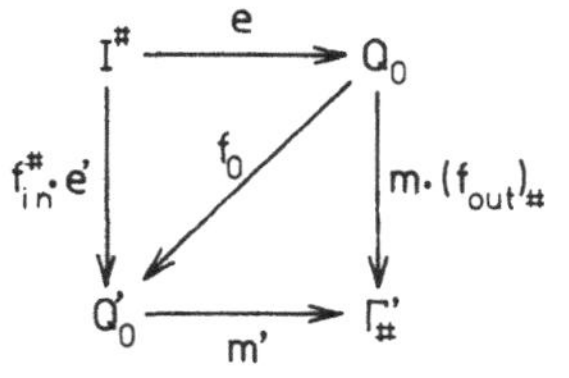

Exercises VI.1

A. Weak preservation of co-preimages. A functor F is said to preserve co-preimages *weakly* iff it preserves each co-preimage

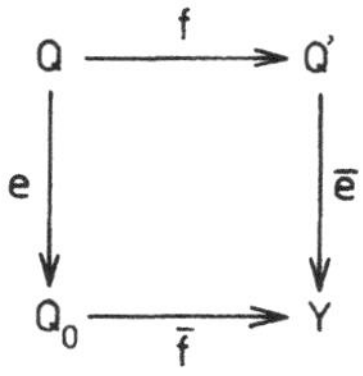

for which there exist morphisms $\delta: QF \to Q$; $\delta': Q'F \to Q'$ and $\delta_0: Q_0F \to Q_0$ turning both f and e into homomorphisms.

Prove the following generalization of Theorem VI.1.4. Let $\mathscr{K}$ be a finitely cocomplete $(\mathscr{E}, \mathscr{M})$-category with cointersections and finite regular coproducts. Then an $\mathscr{E}$-epis preserving functor F has universal reduction iff it weakly preserves both cointersections and co-preimages.

[Hint: The proof is quite analogous to that in VI.1.4, based on Theorem V.1.4, with two exceptions. (i) The operation $\delta: (Q + QF)F \to Q + QF$ is defined by the given operation on Q, say $\bar{\delta}: QF \to Q$, via $\delta = \bar{\bar{\delta}}F \cdot j_Q$, where $\bar{\bar{\delta}}: Q + QF \to Q$ has components 1_Q and $\bar{\delta}$, analogously with δ' and δ_0. (ii) In the end, j is not a split mono but only $j \in \mathscr{M}$; the diagonal fill-in must be used for $j \in \mathscr{M}$ and $\bar{e}F \in \mathscr{E}$.]

B. Sequential automata in a category. (i) A category $\mathscr{K}$ with finite products is said to be *cartesian closed* if for each object Σ the functor $S_\Sigma = - \times \Sigma$ is a coadjoint. This implies that sequential Σ-automata in $\mathscr{K}$ have universal realization.

Verify that **Set**, **Pos** and **Gra** are cartesian closed categories.

(ii) Verify that linear sequential automata have universal realization (though R-**Mod** is not cartesian closed). [Hint: Prove that S_Σ preserves co-preimages.]

(iii) Verify that sequential Σ-automata in **Top** have universal realization whenever Σ is a compact Hasdorff space, both for (regular epi, mono)-factorizations and for (epi, regular mono)-factorizations. [Hint: Prove that S_Σ: **Top** $\to$ **Top** preserves co-preimages and use Exercise V.1.B.]

(iv) Prove that **Top** is not cartesian closed. [Hint: Exercise V.1.B(ii).]

C. Construction of functors. Let $\mathscr{K}$ fulfil the hypotheses of Theorem VI.1.4.

(i) Verify that given functors F_1, F_2 preserving $\mathscr{E}$-epis and having universal reduction, then $F_1 + F_2$ has the same properties.

(ii) Conclude that $F + C_{\Sigma_0}$ (where C_{Σ_0} is the constant functor with value Σ_0) has universal reduction for each $\mathscr{E}$-epis preserving functor F with universal reduction.

(iii) Can (i) be generalized to arbitrary coproducts? If $\hat{S}_{\Sigma_1, \Sigma_0}$ denotes the coproduct $C_{\Sigma_0} + \coprod_{\sigma \in \Sigma} F_\sigma$ where $F_\sigma = 1_{\mathscr{K}}$ is it clear that $\hat{S}_{\Sigma_1, \Sigma_0}$ has universal reduction?

VI.2. Universal Reduction Theorem

2.1. The aim of the present section is to prove that any functor with universal reduction preserves unions. This shows, of course, that universal reduction is quite rare. The proof is rather complicated and it requires additional hypotheses that might look discouraging at first sight. Fortunately, we are able to prove that a concrete category $\mathscr{K}$ fulfils these hypotheses whenever its forgetful functor preserves finite colimits (which includes **Set**, **Pos**, **Gra**, **Top**, etc.), or whenever $\mathscr{K}$ is additive (like R-**Mod**).

Recall that a functor $F: \mathscr{K} \to \mathscr{K}$ on an $(\mathscr{E}, \mathscr{M})$-category $\mathscr{K}$ *preserves unions* if for each collection $m_i: A_i \to B$ $(i \in I)$ of $\mathscr{M}$-monos

$$\bigcup_{i \in I} m_i = m \quad \text{implies} \quad \bigcup_{i \in I} \operatorname{im}(m_i F) = \operatorname{im}(mF).$$

2.2. Conventions. (i) For each object X we denote by X_I (I a set) the coproduct $X_I = \coprod_{i \in I} X_i$ where $X_i = X$ for all $i \in I$.

(ii) Given a coproduct $\coprod_{i \in I} X_i$ where I is a union $I = \bigcup_{t \in T} I_t$, we have a *canonical morphism*

$$\coprod_{t \in T} \left(\coprod_{j \in I_t} X_j \right) F \to \left(\coprod_{i \in I} X_i \right) F$$

with components $v_t F$, where $v_t: \coprod_{j \in I_t} X_j \to \coprod_{i \in I} X_i$ is given by the inclusion $I_t \subset I$ ($t \in T$). We are going to use the term canonical morphism quite freely—in each case it will be clear which union $I = \bigcup I_t$ is meant. For example, the case $I = \bigcup_{i \in I} \{i\}$ leads to a canonical morphism

$$\varepsilon: \coprod_{i \in I} X_i F \to \left(\coprod_{i \in I} X_i\right) F.$$

Proposition. Let $\mathscr{K}$ be a cocomplete $(\mathscr{E}, \mathscr{M})$-category, and let $F: \mathscr{K} \to \mathscr{K}$ preserve $\mathscr{E}$-epis. Then F preserves unions iff for each coproduct $\coprod_{i \in I} X_i$, the canonical morphism

$$\varepsilon: \coprod_{i \in I} X_i F \to \left(\coprod_{i \in I} X_i\right) F$$

is an $\mathscr{E}$-epi.

Proof. (i) Let F preserve unions. For each coproduct $X = \coprod_{i \in I} X_i$ we denote the image factorization of the injections v_i by $v_i = e_i \cdot m_i$ ($i \in I$), then $\bigcup_{i \in I} m_i = 1_X$ by Remark III.5.5 and hence,

$$\bigcup_{i \in I} \operatorname{im}(m_i F) = 1_{XF}.$$

Since $v_i F = e_i F \cdot m_i F$ and $e_i F \in \mathscr{E}$, clearly $\operatorname{im}(v_i F) = \operatorname{im}(m_i F)$. Thus,

$$\bigcup_{i \in I} \operatorname{im}(v_i F) = 1_{XF}.$$

Each $v_i F$ factors through ε, consequently, $\operatorname{im} \varepsilon = 1_{XF}$. This is equivalent to $\varepsilon \in \mathscr{E}$.

(ii) Let $\varepsilon \in \mathscr{E}$ for each coproduct. Given subobjects $m_i: A_i \to B$ ($i \in I$) with the union $m: A \to B$, we denote by $u_i: A_i \to A$ the morphisms with

$$m_i = u_i \cdot m, \quad i \in I.$$

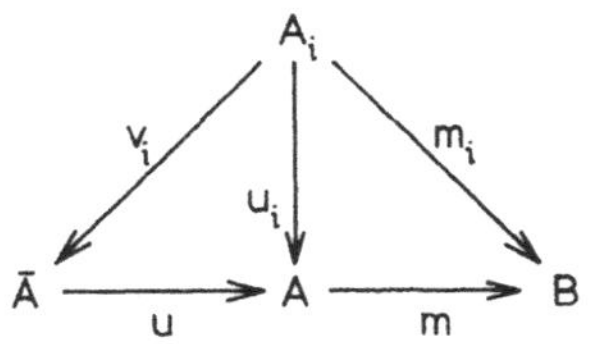

Let $\bar{A} = \coprod_{i \in I} A_i$ be a coproduct with injections $v_i: A_i \to \bar{A}$. Then the morphism $u: \bar{A} \to A$ defined by $v_i \cdot u = u_i \quad (i \in I)$ is in $\mathscr{E}$. In fact, if $u = u_e \cdot u_m$ is an image factorization of u, then $u_m \cdot m$ is a subobject of B containing all m_i and hence, containing m. Then u_m is an isomorphism.

Consider the image factorizations

$$mF = e^* \cdot m^* \quad \text{and} \quad m_iF = e_i^* \cdot m_i^* \qquad (i \in I).$$

For each subobject $d: D \to BF$ such that $m_i^* \subset d$ (i.e., $m_i^* = d_i \cdot d$) for all $i \in I$, we are to show that also $m^* \subset d$.

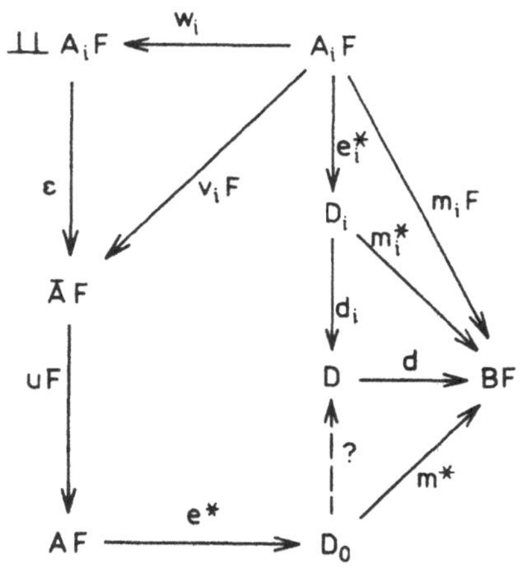

By hypothesis, the canonical morphism

$$\varepsilon: \coprod_{i \in I} A_iF \to \bar{A}F$$

is in $\mathscr{E}$. Hence, so is

$$\varepsilon \cdot uF \cdot e^*: \coprod A_iF \to D_0.$$

For each $i \in I$ we have $v_iF = w_i \cdot \varepsilon$ (where w_i are the injections of $\coprod A_iF$) and hence, $m_i = v_i \cdot u \cdot m$ implies

$$(e_i^* \cdot d_i) \cdot d = m_iF = (v_i \cdot u \cdot m)F = w_i \cdot \varepsilon \cdot uF \cdot e^* \cdot m^*.$$

Define $r: \coprod A_iF \to D$ by

$$w_i \cdot r = e_i^* \cdot d_i \quad (i \in I),$$

then

$$w_i \cdot (r \cdot d) = w_i \cdot (\varepsilon \cdot uF \cdot e^* \cdot m^*) \quad (i \in I)$$

and this implies

$$r \cdot d = \varepsilon \cdot uF \cdot e^* \cdot m^*.$$

We apply the diagonal fill-in

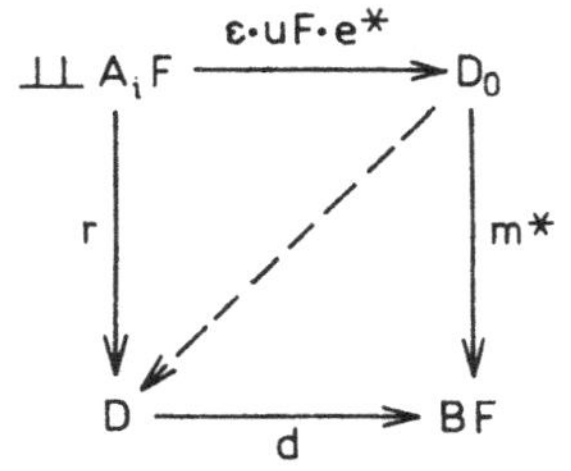

to conclude that $m^* \subset d$. □

2.3. An object S of a category $\mathcal{K}$ is a *generator* if for arbitrary distinct morphisms $f, g\colon C \to B$ there exists morphism $s\colon S \to C$ with $s \cdot f \neq s \cdot g$. In other words, if the functor

$$U = \hom(S, -)\colon \mathcal{K} \to \mathbf{Set}$$

is faithful.

Given a generator S in $\mathcal{K}$, we consider $\mathcal{K}$ as a concrete category (see III.3.8) with the forgetful functor U.

If S is a generator in an $(\mathcal{E}, \mathcal{M})$-category $\mathcal{K}$, then S is said to be *projective* if the following holds for any morphism $f\colon C \to B$:

$f \in \mathcal{E}$ iff for each morphism $s\colon S \to B$ there exists a morphism $\bar{s}\colon S \to C$ with $s = \bar{s} \cdot f$.

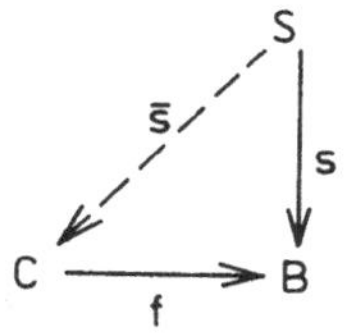

Equivalently, S is a projective generator iff the forgetful functor $U = \hom(S, -)$ fulfils

$$\mathcal{E} = \{f; fU \text{ is onto}\}.$$

Remarks. (i) All current concrete categories have their forgetful functor of the form $\hom(S, -)$ for some generator S. Thus, the assumption "S be a projective generator" does not exclude, essentially, any important category—but we prove that it does pick up the factorization system.

(ii) An *embedding* in a concrete category $\mathcal{K}$ is a mono $m\colon B_0 \to B$ such that for any morphism $p: C \to B$ and any map $p_0 : CU \to B_0U$ with $pU = p_0 \cdot (mU)$ $p_0 : C \to B_0$ is also a morphism.

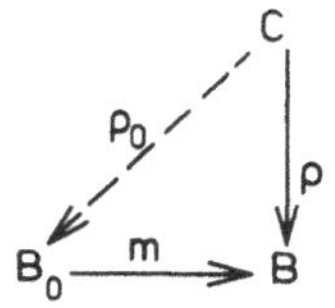

Proposition. Let S be a projective generator of a complete $(\mathscr{E}, \mathscr{M})$-category $\mathscr{K}$. Then

$$\mathscr{E} = \text{onto morphisms} \quad \text{and} \quad \mathscr{M} = \text{embeddings},$$

and each object is a quotient of some coproduct S_I (I a set).

Proof. (i) We start with the last statement. For each object A denote by I the set $\hom(S, A)$ of all morphisms from S to A; let S_I be the coproduct with the injections

$$s_f: S \to S_I \quad (f: S \to A).$$

Define a morphism

$$e: S_I \to A$$

by

$$s_f \cdot e = f \quad \text{for each } f: S \to A.$$

Then $e \in \mathscr{E}$ because S is a projective generator and for each morphism $f: S \to A$ we have $\bar{f} = s_f: S \to S_I$ with $f = \bar{f} \cdot e$. Hence, A is a quotient of S_I.

(ii) The fact that $\mathscr{E}$ = onto morphisms is actually the definition of projectivity.

(iii) To prove that $\mathscr{M}$ contains all embeddings, let

$$m: A \to B$$

be an embedding. Let $m = e \cdot p$ be an image factorization,

$$e: A \to \bar{A} \text{ in } \mathscr{E}; \quad p: \bar{A} \to B \text{ in } \mathscr{M}.$$

We present a map $p_0 : \bar{A}U \to AU$ with $pU = p_0 \cdot (mU)$.

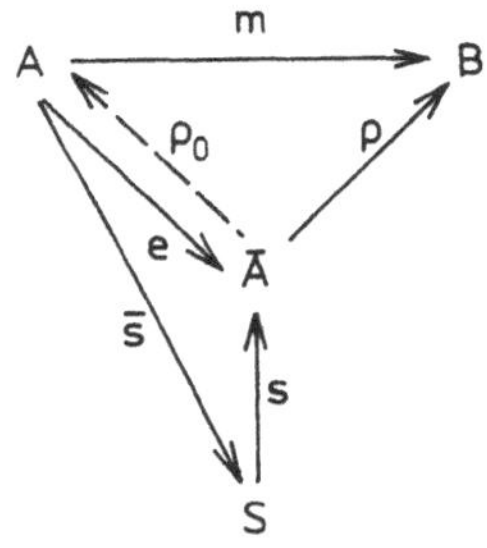

Then $p_0 : \bar{A} \to A$ is a morphism (since m is an embedding) with $p = p_0 \cdot m$ and hence, $m = e \cdot p_0 \cdot m$. Since m is a mono, the last implies $e \cdot p_0 = 1$ and therefore, e is an isomorphism, which proves $m \in \mathcal{M}$. The map p_0 is defined as follows: for each "point"

$$s \in \bar{A}U = \hom(S, \bar{A})$$

there exists $\bar{s}: S \to A$ with

$$s = \bar{s} \cdot e$$

(because $e \in \mathcal{E}$ and S is projective); put

$$(s)p_0 = \bar{s}.$$

Then $pU = p_0 \cdot (mU)$ because for each $s \in \bar{A}U$,

$$(s)pU = s \cdot p = \bar{s} \cdot e \cdot p = \bar{s} \cdot m = [(s)p_0]mU.$$

(iv) Let $m: A \to B$ be an $\mathcal{M}$-mono. To prove that m is an embeding, consider an arbitrary morphism $p: C \to B$. By (i), there exists an $\mathcal{E}$-epi

$$e: S_I \to C$$

for some set I; let $s_i: S \to S_I$ $(i \in I)$ denote the coproduct injections.

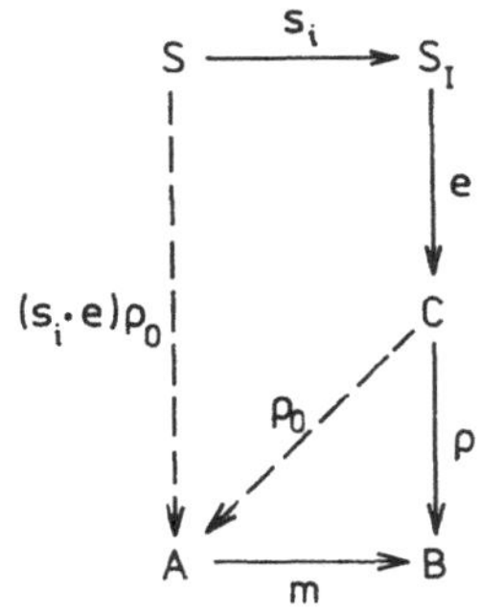

Assuming that $pU = p_0 \cdot (mU)$ for some map $p_0 : \hom(S, C) \to \hom(S, A)$, we define

$$f: S_I \to A$$

to have components $(s_i \cdot e)p_0 : S_I \to A$ $(i \in I)$. Then

$$f \cdot m = e \cdot p : S_I \to B$$

because the i-th component of each of these morphisms is

$((s_i \cdot e)p_0)mU = (s_i \cdot e)pU = s_i \cdot e \cdot p$. The morphism obtained from the diagonal fill-in:

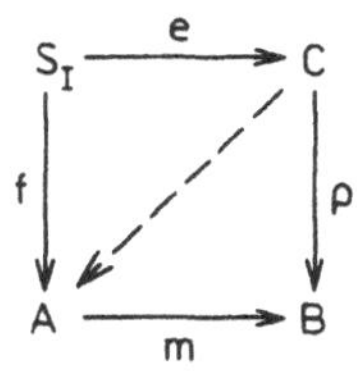

has the underlying map p_0. This concludes the proof. □

Corollary. Let $\mathcal{K}$ be a cocomplete $(\mathcal{E}, \mathcal{M})$-category with a projective generator S. A functor $F: \mathcal{K} \to \mathcal{K}$ preserving $\mathcal{E}$-epis preserves unions iff for each set K, the canonical morphism

$$\varepsilon_K : (SF)_K \to S_K F$$

is an $\mathcal{E}$-epi.

In fact, by Proposition VI.2.2 it suffices to prove that each canonical map

$$\varepsilon: \coprod_{i \in I} X_i F \to \left(\coprod_{i \in I} X_i\right) F$$

is in $\mathcal{E}$, assuming $\varepsilon_K \in \mathcal{E}$ for all sets K. For each $i \in I$ we have an $\mathcal{E}$-epi $e_i: S_{K_i} \to X_i$; put $K = \coprod_{i \in I} K_i$. The morphism

$$e = \coprod_{i \in I} e_i : S_K \to \coprod_{i \in I} X_i$$

is an $\mathcal{E}$-epi (III.5.5). The components of

$$\varepsilon_K \cdot eF: \coprod_{i \in I} (SF)_{K_i} = (SF)_K \to \left(\coprod_{i \in I} X_i\right) F$$

are $\varepsilon_{K_i} \cdot e_i F \cdot v_i F$ (where v_i is the i-th injection of $\coprod X_i$) and hence,

$$\varepsilon_K \cdot eF = \coprod_{i \in I} (\varepsilon_{K_i} \cdot e_i F) \cdot \varepsilon.$$

Since F preserves $\mathcal{E}$-epis, clearly $\varepsilon_K \cdot eF \in \mathcal{E}$, and this proves $\varepsilon \in \mathcal{E}$. □

2.4. Definition. A projective generator S is said to be *perfect* if for each functor $F: \mathcal{K} \to \mathcal{K}$ preserving $\mathcal{E}$-epis, the following holds: if the canonical morphism

$$\bar{\varepsilon}: S_{\{0,1\}} F + S_{\{0,2\}} F \to S_{\{0,1,2\}} F$$

is an $\mathscr{E}$-epi, then also the canonical morphism

$$\varepsilon \colon SF + SF \to (S + S)F$$

is an $\mathscr{E}$-epi.

Remark. We prove below that in quite a number of concrete categories each projective generator is perfect.

Lemma. Let $\mathscr{K}$ be a cocomplete $(\mathscr{E}, \mathscr{M})$-category with a perfect generator S. A functor $F \colon \mathscr{K} \to \mathscr{K}$ preserving $\mathscr{E}$-epis preserves unions iff for each collection $\mathscr{A}$ of sets with $\bigcap_{K \in \mathscr{A}} K \neq \emptyset$, the canonical morphism

$$\bar{\varepsilon} \colon \coprod_{K \in \mathscr{A}} S_K F \to (S_{\bigcup \mathscr{A}})F$$

is an $\mathscr{E}$-epi.

Proof. (i) Let F preserve unions. Then the canonical morphism

$$\varepsilon \colon (SF)_B \to S_B F \qquad (B = \bigcup \mathscr{A})$$

is an $\mathscr{E}$-epi. Denote by L the disjoint union of the sets in $\mathscr{A} = \{K_i;\ i \in I\}$,

$$L = \coprod_{i \in I} K_i.$$

Then $(SF)_L = \coprod_{i \in I} (SF)_{K_i}$, and the morphism

$$f \colon (SF)_L \to (SF)_B$$

with components $(SF)_{K_i} \to (SF)_B$ given by the inclusion $K_i \subset B$ is a split epi. Hence,

$$f \cdot \varepsilon \in \mathscr{E}.$$

Further, for each $i \in I$ we have a canonical morphism

$$\varepsilon_i \colon (SF)_{K_i} \to (S_{K_i})F \quad \text{in } \mathscr{E}$$

and hence, $\coprod_{i \in I} \varepsilon_i \in \mathscr{E}$ by III.5.5. The following square

$$\begin{array}{ccc} (SF)_L & \xrightarrow{\coprod \varepsilon_i} & \coprod (S_{K_i}F) \\ {\scriptstyle f}\big\downarrow & & \big\downarrow{\scriptstyle \bar{\varepsilon}} \\ (SF)_B & \xrightarrow[\varepsilon]{} & S_B F \end{array}$$

commutes. Thus $\bar{\varepsilon} \in \mathscr{E}$.

(ii) Asume that F does not preserves unions. By VI.2.2., there exists a set $B \neq \emptyset$ such that the canonical morphism

$$\varepsilon : (SF)_B \to S_B F$$

is not an $\mathscr{E}$-epi. We choose such a set B with the smallest cardinality possible; B has at least two elements, of course.

(a) Let card $B = 2$. Then the fact that

$$\varepsilon : SF + SF \to (S + S)F$$

is not an $\mathscr{E}$-epi implies that neither is the canonical morphism

$$\bar{\varepsilon} : S_{\{0,1\}}F + S_{\{0,2\}}F \to S_{\{0,1,2\}}F$$

(since S is a perfect generator). For $K_1 = \{0, 1\}$ and $K_2 = \{0, 2\}$, F does not fulfil the condition above.

(b) Let card $B > 2$. Choose an element $k_0 \in B$ and let

$$K_i, \quad i \in I,$$

be the collection of all subsets of B with

$$k_0 \in K_i \quad \text{and} \quad \text{card } K_i < \text{card } B.$$

By the choice of B, each of the canonical morphisms

$$\varepsilon_i : (SF)_{K_i} \to S_{K_i}F \qquad (i \in I)$$

is an $\mathscr{E}$-epi. Hence, in the square above,

$$\coprod_{i \in I} \varepsilon_i \in \mathscr{E}.$$

Since $\varepsilon \notin \mathscr{E}$ implies $f \cdot \varepsilon \notin \mathscr{E}$, we have

$$\left(\coprod_{i \in I} \varepsilon_i\right) \cdot \bar{\varepsilon} \notin \mathscr{E}.$$

Then $\coprod \varepsilon_i \in \mathscr{E}$ implies $\bar{\varepsilon} \notin \mathscr{E}$. Again, F does not fulfil the condition above. □

2.5. Definition. An $(\mathscr{E}, \mathscr{M})$-category is said to have *exact co-preimages* if each pushout

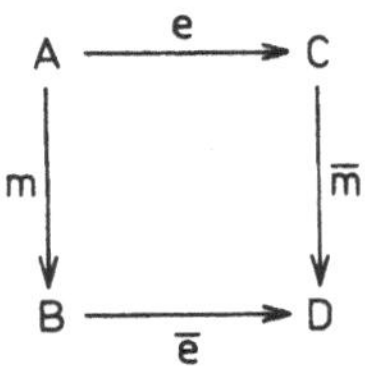

with $e \in \mathscr{E}$ and $m \in \mathscr{M}$ is also a pullback.

Examples. (i) **Set** has exact co-preimages. In fact, assuming that $A \subset B$ (and m is the inclusion map), the pushout above can be described as follows: $D = C + (B - A)$, where $\bar{m}$ is the first injection and $(a)\bar{e} = (a)e$ for $a \in A$, $(b)\bar{e} = b$ for $b \in B - A$. If $(b)\bar{e} = (c)\bar{m}$, then $b \in A$ is the unique element with $b = (b)m$ and $c = (b)e$—thus, the square above is a pullback.

(ii) R-**Mod** has exact co-preimages. In fact, assuming that $A_0 \subset A \subset B$ are submodules with m the inclusion map and $e: A \to C = A/A_0$ the quotient map, the pushout can be described as follows: $D = B/A_0$ with $\bar{e}$ the quotient map and $\bar{m} = m/1_{A_0}: A/A_0 \to B/A_0$. It is easy to see that this square is a pullback, too.

2.6. Universal Reduction Theorem. Let $\mathscr{K}$ be a cocomplete $(\mathscr{E}, \mathscr{M})$-category with a perfect generator S and with exact co-preimages. Let $F: \mathscr{K} \to \mathscr{K}$ preserve $\mathscr{E}$-epis and fulfil $\hom(SF, S) \neq \emptyset$.

If F has universal reduction, then F preserves unions.

Proof. (i) Plan of the proof. We are going to verify the condition of Lemma VI.2.4: for each collection $\mathscr{A}$ of sets with, say,

$$0 \in \bigcap_{K \in \mathscr{A}} K \quad \text{and} \quad B = \bigcup_{K \in \mathscr{A}} K$$

we are going to prove that canonical morphism

$$\bar{\varepsilon}: \coprod_{K \in \mathscr{A}} S_K F \to S_B F$$

is an $\mathscr{E}$-epi.

Denote by

$$v_K: S_K \to S_B \quad (K \in \mathscr{A})$$

the injection given by the inclusion $K \subset B$, and by

$$w_K: S_K F \to \coprod_{L \in \mathscr{A}} S_L F \quad (K \in \mathscr{A})$$

the coproduct injections. Further, the 0-th coproduct injection will be denoted by

$$u_K: S \to S_K \quad (K \in \mathscr{A}).$$

Note that the injection $v_0: S \to S_B$ of the 0-th coproduct injection fulfils

(1) $\qquad v_0 = u_K \cdot v_K \quad$ for each $K \in \mathscr{A}$.

Finally, denote by

$$\nabla_K: S_K \to S \quad (K \in \mathscr{A}) \quad \text{and} \quad \nabla_B: S_B \to S$$

the co-diagonal morphisms (all components of which are 1_S); note that

(2) $\qquad u_K \cdot \nabla_K = 1_S \quad$ for all $K \in \mathscr{A}$

and

(3) $\nabla_K = \nu_K \cdot \nabla_B$ for all $K \in \mathscr{A}$.

Define a morphism

$$t: \coprod_{K \in \mathscr{A}} S_K F \to SF$$

by

(4) $w_K \cdot t = \nabla_K F: S_K F \to SF \quad (K \in \mathscr{A})$.

We can factor the morphism $\bar{\varepsilon}$ as $\bar{\varepsilon} = e \cdot m$ with $e: \coprod_{K \in \mathscr{A}} S_K F \to D$ in $\mathscr{E}$ and $m: D \to S_B F$ in $\mathscr{M}$. To prove $\bar{\varepsilon} \in \mathscr{E}$, we are going to verify that $m \in \mathscr{E}$.

Let us form the pushout of e and t:

(5)

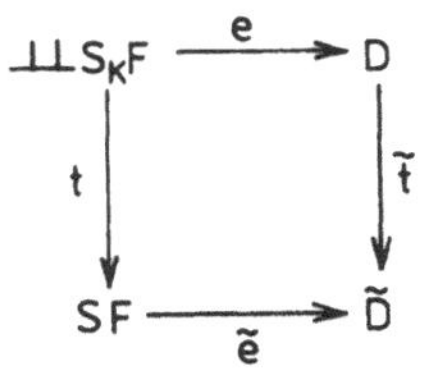

Note that $t \in \mathscr{E}$ (indeed, t is a split epi since $u_K F \cdot w_K \cdot t = 1$), hence, $\tilde{t} \in \mathscr{E}$. Further, let us form the pushout of m and $\tilde{t}$:

(6)

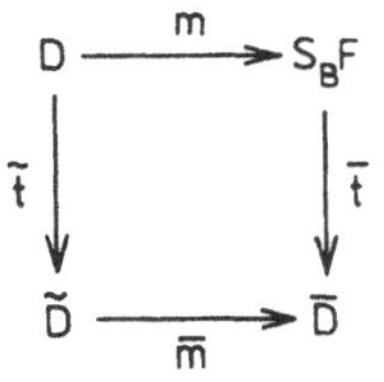

Then $\tilde{t} \in \mathscr{E}$ and $m \in \mathscr{M}$ imply that

(6) is a pullback

(since co-preimages are supposed to be exact).

To prove $m \in \mathscr{E}$, we use the fact that S is projective: it suffices to show that for each $g: S \to S_B F$ there exists $\bar{g}: S \to D$ with $g = \bar{g} \cdot m$:

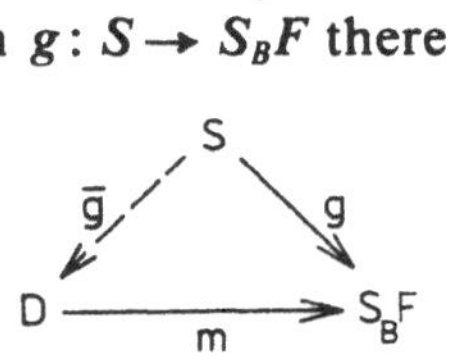

We are going to verify that

(7) $\quad \bar{t} = h \cdot \bar{m} \qquad$ for some $h\colon S_BF \to \tilde{D}$.

Then we use the universal property of the pullback (6):

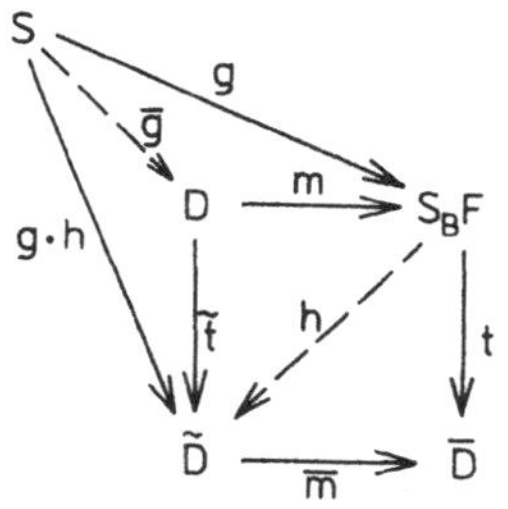

Thus, the proof will be concluded when we exhibit h such that (7) holds. This we do by defining automata A_K on the objects $S_K + S_KF$ $(K \in \mathscr{A})$ and an automaton A_B on the object $S_B + S_BF$ in such a way that

(α) $\quad v_K + v_KF$ become morphisms from A_K to A_B and

(β) $\quad$ the output of A_B is $\nabla_B + \bar{t}\colon S_B + S_BF \to S + \bar{D}$.

Using the universality of reduction, we present then the required morphism h.

(ii) The definition of automata. By hypothesis, there exists a morphism

$$\xi\colon SF \to S.$$

For each $K \in \mathscr{A}$ we define an automaton

$$A_K = (S_K + S_KF, \delta_K, S + \bar{D}, \gamma_K)$$

as follows. Denote by i_K and j_K the coproduct injections of $S_K + S_KF$, and define a morphism $\bar{\delta}_K$ by the commutativity of the following diagram

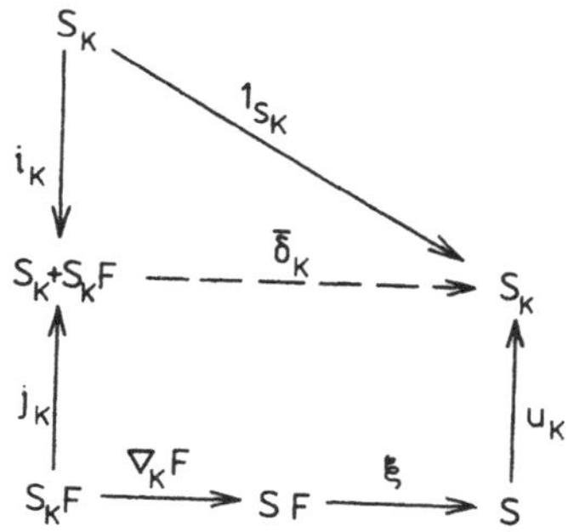

Put

$$\delta_K = \bar{\delta}_KF \cdot j_K\colon (S_K + S_KF)F \to S_K + S_KF$$

and using $\tilde{e} \cdot \bar{m}: SF \to \bar{D}$ [see (5), (6)], put

$$\gamma_K = \nabla_K + (\nabla_K F \cdot \tilde{e} \cdot \bar{m}): S_K + S_K F \to S + \bar{D}.$$

Without loss of generality we can assume

$$\{0\} \in \mathscr{A}.$$

(In fact, Lemma VI.2.4 remains true if only such collections $\mathscr{A}$ are considered.) Then we have an automaton A_0 (we write 0 instead of $\{0\}$ in the indices),

$$A_0 = (S + SF, \delta_0, S + \bar{D}, 1_S + \tilde{e} \cdot \bar{m}).$$

Further, define an automaton

$$A_B = (S_B + S_B F, \delta_B, S + \bar{D}, \nabla_B + \bar{t}),$$

(the outputs of which are defined in a different way than in A_0, but the next-state morphism is analogous):

$$\delta_B = \bar{\delta}_B F \cdot j_B$$

where i_B and j_B are coproduct injections and $\bar{\delta}_B: S_B + S_B F \to S_B$ is defined by

$$i_B \cdot \bar{\delta}_B = 1_{S_B} \quad \text{and} \quad j_B \cdot \bar{\delta}_B = \nabla_B F \cdot \xi \cdot v_0.$$

Note that

(8) $$i_B F \cdot \delta_B = (i_B \cdot \bar{\delta}_B) F \cdot j_B = j_B.$$

We claim that for each $K \in \mathscr{A}$,

$$v_K + v_K F: A_K \to A_B$$

is a morphism of automata, i.e., that the following diagram

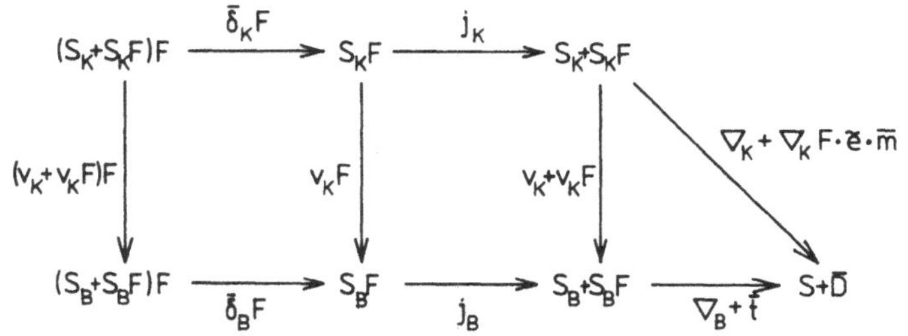

commutes. The verification of the two squares is routine. For the triangle, use (3) on the first summand; for the second one, we compute (using the fact that $w_K \cdot \bar{\varepsilon} = v_K F$, by definition of $\bar{\varepsilon}$):

$$\begin{aligned} v_K F \cdot \bar{t} &= w_K \cdot \bar{\varepsilon} \cdot \bar{t} \\ &= w_K \cdot e \cdot m \cdot \bar{t} \qquad (\bar{\varepsilon} = e \cdot m) \end{aligned}$$

$$= w_K \cdot e \cdot \tilde{t} \cdot \bar{m} \qquad \text{[by (6)]}$$
$$= w_K \cdot t \cdot \tilde{e} \cdot \bar{m} \qquad \text{[by (5)]}$$
$$= \nabla_K F \cdot \tilde{e} \cdot \bar{m} \qquad \text{[by (4)].}$$

Denote the minimal reduction of A_K ($K \in \mathscr{A}$) by

$$e_K : A_K \to A_K^* = (Q_K, \delta_K^*, S + \bar{D}, \gamma_K^*)$$

and the minimal reduction of A_B by

$$e_B : A_B \to A_B^* = (Q_B, \delta_B^*, S + \bar{D}, \gamma_B^*).$$

Since reduction is universal, there exists a unique morphism of automata v_K^*, for each $K \in \mathscr{A}$, such that the following square

(9) 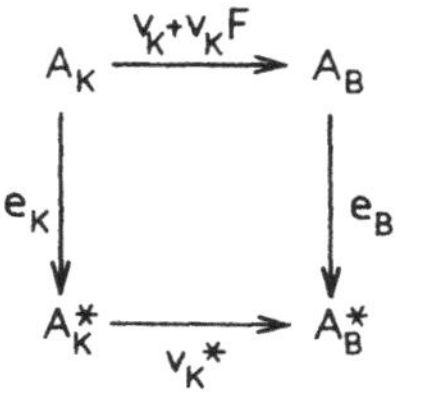

commutes.

It is easy to check that, for each $K \in \mathscr{A}$, another morphism of automata is

$$\nabla_K + \nabla_K F : A_K \to A_0.$$

Since ∇_K is a split epi by (2), so is $\nabla_K + \nabla_K F$; hence, A_0 is a reduction of A_K. Therefore, there exists a unique morphism

$$\bar{e}_K : A_0 \to A_K^*$$

with

(10) $$e_K = (\nabla_K + \nabla_K F) \cdot \bar{e}_K.$$

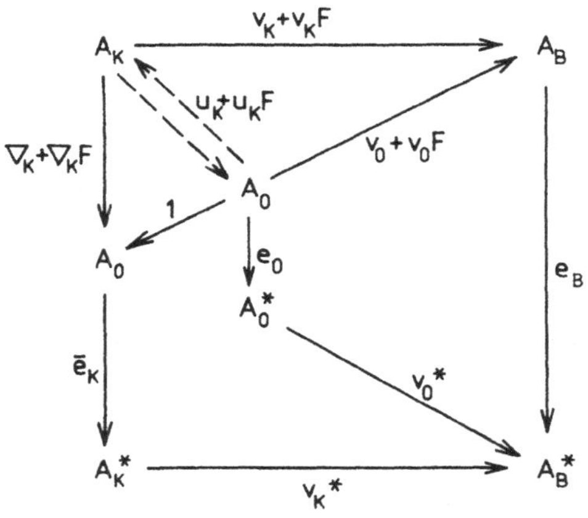

Then $\bar{e}_K \cdot v_K^* : A_0 \to A_B^*$ is independent of K, i.e.,

(11) $\quad \bar{e}_K \cdot v_K^* = e_0 \cdot v_0^*$ for each $K \in \mathscr{A}$.

Indeed, by (2), $(u_K + u_K F) \cdot (\nabla_K + \nabla_K F) = 1$, hence

$$\begin{aligned} \bar{e}_K \cdot v_K^* &= (u_K + u_K F) \cdot (\nabla_K + \nabla_K F) \cdot \bar{e}_K \cdot v_K^* \\ &= (u_K + u_K F) \cdot e_K \cdot v_K^* && \text{[by (10)]} \\ &= (u_K + u_K F) \cdot (v_K + v_K F) \cdot e_B && \text{[by (9)]} \\ &= (v_0 + v_0 F) \cdot e_B && \text{[by (1)]} \\ &= e_0 \cdot v_0^* && \text{[by (9)].} \end{aligned}$$

(iii) The proof of (7). For each $K \in \mathscr{A}$, the following diagram

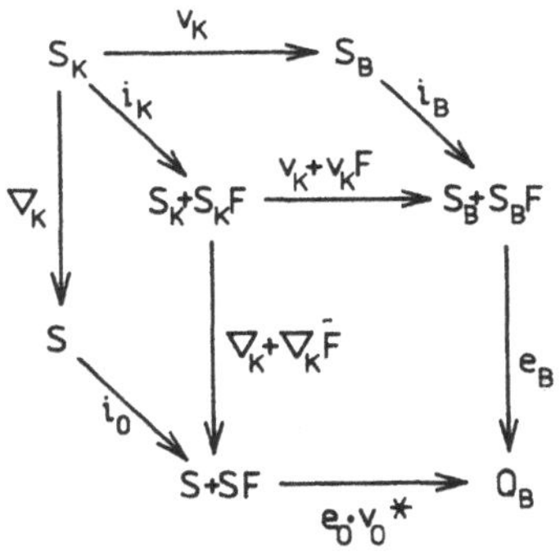

commutes—combine (9) and (11). Hence,

(12) $\quad i_B \cdot e_B = \nabla_B \cdot i_0 \cdot (e_0 \cdot v_0^*): S_B \to Q_B.$

Indeed, since $\bigcup_{K \in \mathscr{A}} K = B$, to prove (12) it suffices to show that the two morphisms are equal when preceeded by any $v_K : S_K \to S_B$. We have by (3),

$$\begin{aligned} v_K \cdot (i_B \cdot e_B) &= \nabla_K \cdot i_0 \cdot (e_0 \cdot v_0^*) \\ &= v_K \cdot (\nabla_B \cdot i_0 \cdot e_0 \cdot v_0^*). \end{aligned}$$

Further, denote by $\bar{j}: \bar{D} \to S + \bar{D}$ the coproduct injection. Since

$e_B : A_B \to A_B^*$

is a morphism of automata, the following diagram

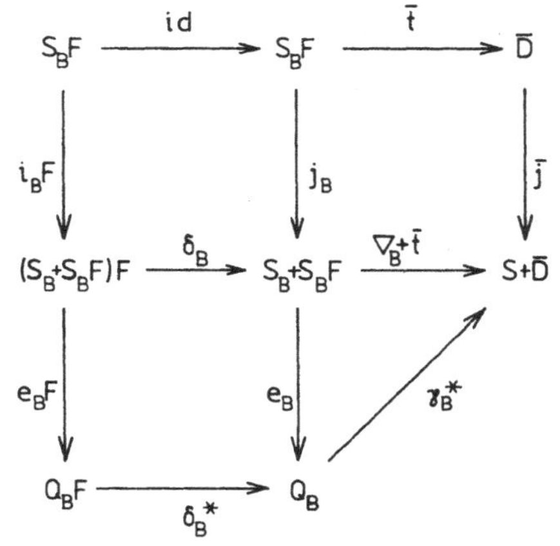

commutes: see (8) for the upper left square, and use (12) to get

$$\begin{aligned}\bar{t} \cdot \bar{j} &= (i_B \cdot e_B)F \cdot \delta_B^* \cdot \gamma_B^* \\ &= \nabla_B F \cdot k,\end{aligned}$$

where

$$k = i_0 F \cdot (e_0 \cdot v_0^*)F \cdot \delta_B^* \cdot \gamma_B^* \colon SF \to S + D.$$

(a) Assume hom$(S, \bar{D}) \neq \emptyset$. Choose a morphism $j^*\colon S + \bar{D} \to \bar{D}$ with the first component arbitrary and the second one $\mathrm{id}_{\bar{D}}$. Then $\bar{j} \cdot j^* = \mathrm{id}$, therefore

(13) $\quad \bar{t} = \bar{t} \cdot \bar{j} \cdot j^* = \nabla_B F \cdot (k \cdot j^*).$

Further by (5),

$$v_0 F = w\bar{e} = w_0 \cdot e \cdot m$$

and since $(v_0 \cdot \nabla_B)F = \mathrm{id}_{SF}$, we conclude

(14) $\quad w_0 \cdot e \cdot m \cdot \nabla_B F = \mathrm{id}_{SF}.$

Thus,

$$\begin{aligned}\bar{t} &= \nabla_B F \cdot (k \cdot j^*) && \text{[by (13)]} \\ &= \nabla_B F \cdot (w_0 \cdot e \cdot m \cdot \nabla_B F) \cdot (k \cdot j^*) && \text{[by (14)]} \\ &= \nabla_B F \cdot w_0 \cdot e \cdot m \cdot \bar{t} && \text{[by (13)]} \\ &= \nabla_B F \cdot w_0 \cdot e \cdot \tilde{t} \cdot \bar{m} && \text{[by (6)].}\end{aligned}$$

In (7), it suffices to put

$$h = \nabla_B F \cdot w_0 \cdot e \cdot \tilde{t}.$$

(b) Assume $\hom(S, \bar{D}) = \emptyset$. Since $\bar{t}: S_B F \to D$ is a morphism, it follows that $\hom(S, S_B F) = \emptyset$. Now, S is a projective generator, hence, the latter implies that any morphism into $S_B F$ is in $\mathscr{E}$. In particular, $\bar{\varepsilon} \in \mathscr{E}$. This concludes the proof. □

2.7. Let us turn to a discussion which categories fulfil the hypotheses of the Universal Reduction Theorem. By VI.2.3 we know that they are cocomplete, concrete (onto, embedding)-categories with the forgetful functor $\hom(S, -)$; we prove now that it is sufficient to assume that the forgetful functor preserves finite colimits. This includes a large number of current categories. But the important case of R-**Mod** is not included. Therefore, we prove that additivity is also sufficient. Let us briefly present some of the basic facts about additive categories we need below.

By definition, an *additive category* is a category $\mathscr{K}$ such that

(i) $\hom(A, B)$ is endowed with the structure of an abelian group for arbitrary objects A and B.

(ii) Composition preserves this structure, i.e., given $f_1, f_2: A \to B$ then

$$h \cdot (f_1 + f_2) = h \cdot f_1 + h \cdot f_2 \qquad \text{for each } h: A' \to A$$

and

$$(f_1 + f_2) \cdot k = f_1 \cdot k + f_2 \cdot k \qquad \text{for each } k: B \to B'.$$

(iii) $\mathscr{K}$ has a zero 0, i.e., an object which is both initial and terminal.

From this definition we can easily derive further properties:

(iv) Each finite coproduct is also a product: let A_1, A_2 be two objects and let $\pi_1: A_1 + A_2 \to A_1$ have components 1_{A_1} and $0: A_2 \to A_1$, analogously $\pi_2: A_1 + A_2 \to A_2$. Then

$$A_1 + A_2 = A_1 \times A_2$$

with the projections π_1 and π_2. [In fact, given $f_i: X \to A_i$, $i = 1, 2$, and denoting by $v_i: A_i \to A_1 + A_2$ the coproduct injections, then $f_1 \cdot v_1 + f_2 \cdot v_2: X \to A_1 + A_2$ is the unique morphism with $(f_1 \cdot v_1 + f_2 \cdot v_2) \cdot \pi_i = f_i$.]

(v) The operation + can be derived from the compositions: given $f_1, f_2: A \to B$, let $f: A + A \to B$ be the morphism with components f_1 and f_2 and let $\Delta_A: A \to A \times A$ $(= A + A)$ be the "diagonal" morphism with the components 1_A and 1_A. Then

$$f_1 + f_2 = \Delta_A \cdot f.$$

(This follows from $\Delta_A = v_1 + v_2$.)

(vi) The zero element of the group $\hom(A, B)$ is obtained as the composition of the unique morphism $A \to 0$ and the unique morphism $0 \to B$.

(vii) Let $m: A \to B$ be a regular mono. If $c: B \to C$ is the *cokernel* of m, i.e., the coequalizer of m and $0: A \to B$, then m is the *kernel* of c, i.e., the equalizer of c and $0: B \to C$.

Dually, if $c\colon B \to C$ is a regular epi and $m\colon A \to B$ is a kernel of c, then c is a cokernel of m.

(viii) For each object S of $\mathscr{K}$ we obtain a functor

$$\hom(S, -)\colon \mathscr{K} \to \mathbf{Ab} \ (= \text{abelian groups})$$

assigning to $A \in \mathscr{K}^{\circ}$ the abelian group $\hom(S, A)$ and to $f\colon A \to B$ the homomorphism $\hom(S, f)\colon p \mapsto p \cdot f$ (for each $p\colon S \to A$). This functor preserves all limits which exist in $\mathscr{K}$.

2.8. Theorem. Let $\mathscr{K}$ be a cocomplete, concrete $(\mathscr{E}, \mathscr{M})$-category with a projective generator S. Assume that

either the forgetful functor preserves finite colimits,
or $\mathscr{K}$ is additive and $\mathscr{M}$ = regular monos.

Then S is a perfect generator, and $\mathscr{K}$ has exacts co-preimages.

Remark. Since S is a projective generator, we know that $\mathscr{M}$ = embeddings. Therefore, the hypothesis (in the additive case) that $\mathscr{M}$ = regular monos is rather weak; it is equivalent to assuming that a composition of two regular monos is a regular mono.

Proof. (i) Denote by U either the hom- functor

$$U = \hom(S, -)\colon \mathscr{K} \to \mathbf{Set}$$

in case it preserves finite colimits, or

$$U = \hom(S, -)\colon \mathscr{K} \to \mathbf{Ab},$$

see (viii) above, if $\mathscr{K}$ is additive and $\mathscr{M}$ = regular monos. We prove first that also in the additive case, U preserves finite colimits, i.e., finite coproducts and coequalizers. Note first that since S is a projective generator, and since in **Ab** the epis (and the regular epis) are precisely the homomorphisms onto, the functor U preserves epis.

(a) U preserves finite coproducts. Since U preserves limits [see (viii) above], it preserves the terminal object, i.e.,

$$0U = 0.$$

And U preserves binary coproducts because they coincide with products [(iv) above]:

$$(A + B)U = (A \times B)U = AU \times BU = AU + BU.$$

(b) U preserves coequalizers. First, U is additive, i.e.,

$$(f_1 + f_2)U = f_1U + f_2U$$

for arbitrary $f_1, f_2\colon A \to B$. In fact, we have

$$f_1 + f_2 = \Delta_A \cdot f$$

[see (vi) above]. Since U preserves finite products, $\Delta_A U = \Delta_{AU}$. Since U preserves finite coproducts, the components of $fU: AU + AU \to BU$ are $f_1 U$ and $f_2 U$. Hence,

$$f_1 U + f_2 U = \Delta_{AU} \cdot fU = (\Delta_A \cdot f)U = (f_1 + f_2)U.$$

Let

$$f_1, f_2 : A \to B$$

be morphisms with a coequalizer

$$c: B \to C.$$

Since $f_1 \cdot c = f_2 \cdot c$ is equivalent to $(f_1 - f_2) \cdot c = 0 \cdot c = 0$, we see that c is the cokernel of $f_1 - f_2$. Let

$$f_1 - f_2 = e \cdot m$$

be an image factorization ($e: A \to A_0$ an epi, and $m: A_0 \to B$ a regular mono). Then c is a cokernel of m hence by (vii), m is a kernel of c. Since U preserves limits by (viii), also mU is a kernel of cU. Since U preserves epis, and each epi in **Ab** is regular, we conclude from (vii) that cU is a cokernel of mU. Since also eU is an epi, cU is also a cokernel of

$$eU \cdot mU = (f_1 - f_2)U = f_1 U - f_2 U.$$

Equivalently, cU is the coequalizer of $f_1 U$ and $f_2 U$.

(ii) $\mathscr{K}$ has exact co-preimages. This follows easily from the fact that both **Set** and **Ab**(= Z-**Mod**) have this property—see Examples VI.2.5. Let

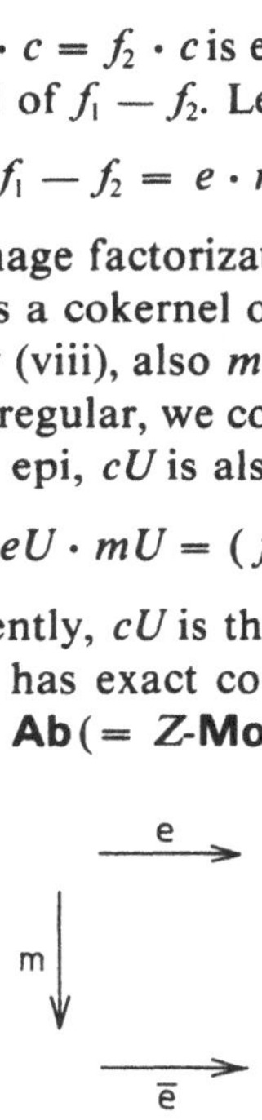

be a pushout with $e \in \mathscr{E}$ and $m \in \mathscr{M}$. Then eU is epi and mU is mono (in **Set** or **Ab**) because $\mathscr{E}$ = onto morphisms and $\mathscr{M}$ = embeddings, by Proposition VI.2.3. Hence, the following square

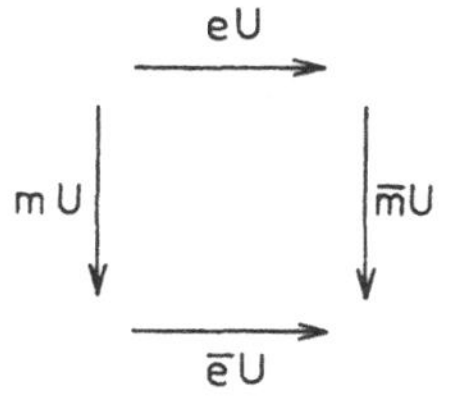

is a pullback: we know that it is a pushout, since U preserves pushouts and hence, it is also a pullback (in **Set** or **Ab**). To verify that the original square is a pullback, let p, q be morphisms in $\mathscr{K}$ with

$$p \cdot \bar{e} = q \cdot \bar{m}.$$

In **Set** or **Ab** there exists a unique morphism p_0 for which the following diagram

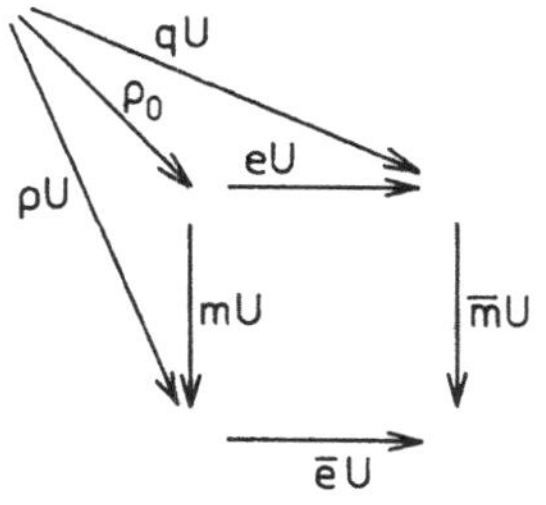

commutes. Since m is an embedding, $pU = p_0 \cdot (mU)$ implies that p_0 carries a morphism in $\mathscr{K}$. Then $p = p_0 \cdot m$ and $q = p_0 \cdot e$ (because U is faithful) and p_0 is unique because m is mono. This proves that $\mathscr{K}$ has exact co-preimages.

(iii) S is perfect. Consider the following square of coproduct injections:

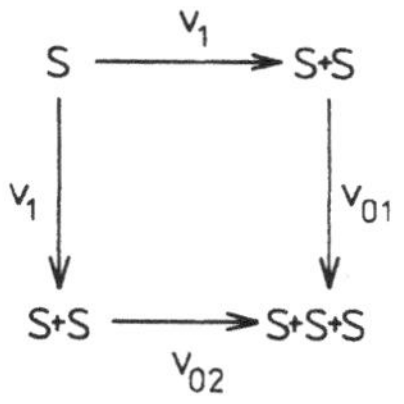

where the indices show which injections are meant. It is easy to see that (i) this square is a pushout and (ii) v_1 and v_2 are split monos. By Lemma V.4.1., this square is an absolute pullback. We use this three times, forming the following diagram of coproduct injections:

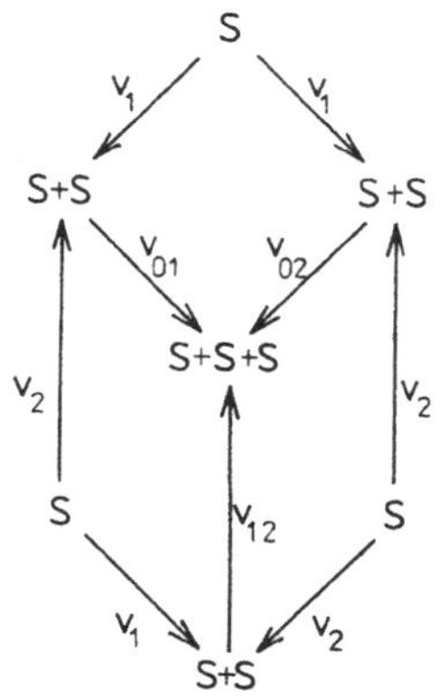

Suppose that $F: \mathcal{K} \to \mathcal{K}$ is an $\mathcal{E}$-epis-preserving functor such that the canonical morphism

$$\bar{\varepsilon}: (S + S)F + (S + S)F \to (S + S + S)F,$$

with components

$$v_{01}F \quad \text{and} \quad v_{02}F,$$

is an $\mathcal{E}$-epi. We are going to prove that then the canonical morphism

$$\varepsilon: SF + SF \to (S + S)F,$$

with components

$$v_1F \quad \text{and} \quad v_2F,$$

is an $\mathcal{E}$-epi, too. Since U preserves finite coproducts, the components of the onto map $\bar{\varepsilon}U$ are $v_{01}F \cdot U$ and $v_{02}F \cdot U$. It is sufficient to prove that εU is onto (then $\varepsilon \in \mathcal{E}$); its components are $v_1F \cdot U$ and $v_2F \cdot U$.

(a) $U: \mathcal{K} \to$ **Set**. The fact that $\bar{\varepsilon}U$ is onto means that

$$(S + S + S)F \cdot U = \operatorname{im}(v_{01}F \cdot U) \cup \operatorname{im}(v_{02}F \cdot U).$$

To prove that εU is onto, consider any point

$$x \in (S + S)F \cdot U$$

and put

$$y = (x)v_{12}F \cdot U \in (S + S + S)F \cdot U.$$

Then for $i = 1$ or $i = 2$ we have $y \in \operatorname{im}(v_{0i}F \cdot U)$—say, $i = 1$. Choose any

$$y_1 \in (S + S)F \cdot U$$

with

(1) $\quad (y_1)v_{01}F \cdot U = y = (x)v_{12}F \cdot U.$

Since the diagram above consists of absolute pullbacks, the following square

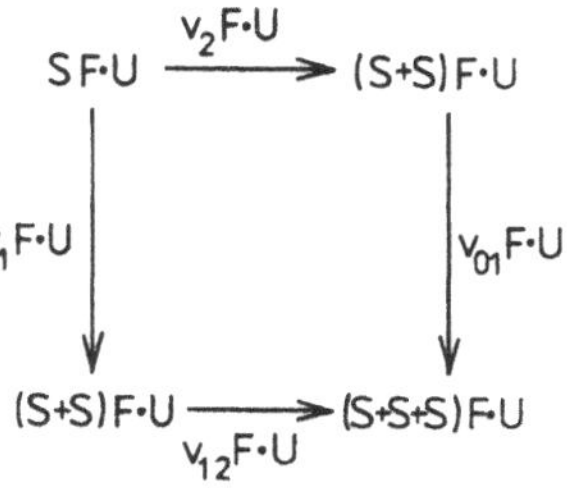

is a pullback in **Set**. Hence, (1) implies that there exists $z_1 \in SF \cdot \mathrm{U}$ with

$$x = (z_1)v_1F \cdot U \quad \text{and} \quad y_1 = (z_1)v_2F \cdot U.$$

Since $v_1F \cdot U$ is the first component of εU, this proves that $x \in \operatorname{im} \varepsilon U$. Hence, εU is onto.

(b) U: $\mathscr{K} \to$ **Ab**. We define a "dual" diagram, using the zero morphisms [see VI.2.7 (iv)]:

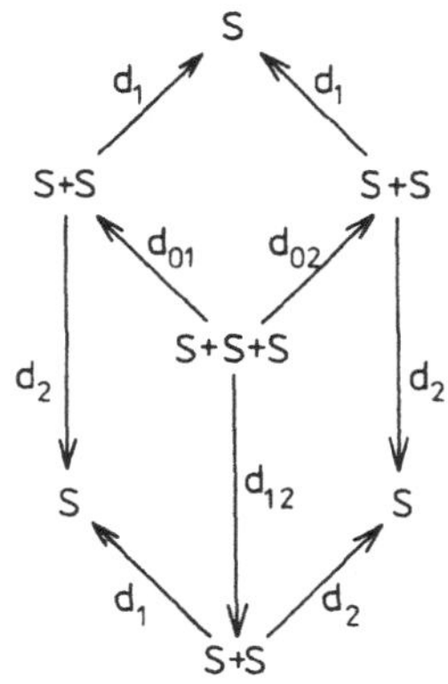

Here,

$$d_1, d_2 : S + S \to S$$

are the projections of the biproduct, i.e., d_1 has the components 1_S and 0 and d_2 has the components 0 and 1_S. Analogously,

$$d_{ij}: S + S + S \to S + S$$

has the i-th and j-th components equal to 1_S while the remaining component is 0. It is easy to check that

(2) $$v_{ij} \cdot d_{ij} = 1_{S+S} \qquad (i, j = 0, 1, 2;\ i \neq j)$$

and

$$v_{01} \cdot d_{12} = d_2 \cdot v_1 \quad \text{and} \quad v_{02} \cdot d_{12} = d_2 \cdot v_2.$$

Since $\bar{\varepsilon}U$ is onto, the group $(S + S + S)F \cdot U$ is generated by $\operatorname{im}(v_{01}F \cdot U) \cup \operatorname{im}(v_{02}F \cdot U)$, i.e., each element $y \in (S + S + S)F \cdot U$ has the form

(4) $$y = (y_1)v_{01}F \cdot U + (y_2)v_{02}F \cdot U$$

for some $y_1, y_2 \in (S + S)F \cdot U$.

To prove that εU is onto, consider any point

$$x \in (S + S)F \cdot U$$

and put

$$y = (x)v_{12}F \cdot U.$$

There exist y_1, y_2 with the property (4). Then

$$\begin{aligned} x &= (x)(v_{12} \cdot d_{12})F \cdot U && \text{[see (2)]} \\ &= (y)\, d_{12}F \cdot U \\ &= (y_1)(v_{01} \cdot d_{12})F \cdot U + (y_2)(v_{02} \cdot d_{12})F \cdot U && \text{[see (4)]} \\ &= (y_1)(d_2 \cdot v_1)F \cdot U + (y_2)(d_2 \cdot v_2)F \cdot U && \text{[see (3)].} \end{aligned}$$

This proves that each element $x \in (S + S)F$ lies in the subgroup generated by $\mathrm{im}(v_1F \cdot U) \cup \mathrm{im}(v_2F \cdot U)$. Hence, the homomorphism εU, which has components $v_1F \cdot U$ and $v_2F \cdot U$, is onto.

The proof is concluded. □

Examples. Each of the following (onto, embedding)-categories $\mathscr{K}$ fulfils the hypotheses of the Universal Reduction Theorem.

(i) **Set, Pos, Gra, Top**. The forgetful functor of each of these categories is $U = \mathrm{hom}(S, -)$ for the object S on a singleton set. Since U preserves finite (in fact, all) colimits, S is a prefect generator and $\mathscr{K}$ has exact co-preimages.

(ii) The category of unary Σ-algebras. Here the forgetful functor is $\mathrm{hom}(\Sigma^*, -)$, where Σ^* is the free algebra on one generator (I.1.6). This functor preserves colimits. Thus, Σ^* is a perfect generator of S_Σ-**Alg** and S_Σ-**Alg** has exact co-preimages.

(iii) R-**Mod** is a cocomplete additive category in which each mono is regular. The ring R, considered as a module over itself, is a projective (and hence, perfect) generator.

(iv) **TAb**, the category of topological abelian groups and continuous homomorphisms. This category is cocomplete: colimits are constructed as in **Ab**, and then endowed with the finest possible topology. Further, **TAb** is additive and each embedding $m: A \to B$ (i.e., a mono such that the topology of A is induced by that of B) is regular. This is again proved as in **Ab** : if m is an equalizer in **Ab** of $f_1, f_2: B \to C$, then we endow C with the indiscrete topology to get $f_1, f_2: B \to C$ in **TAb**. Finally, the discrete group $(Z, +)$ of integers is a projective generator. Hence, $(Z, +)$ is a perfect generator, and **TAb** has exact co-preimages.

Corollary. In the category of sets, reduction (or realization) is universal precisely for the sequential automata with resets (VI.1.8).

Indeed, the sequential automata with resets have universal reduction, see II.3.10. On the other hand, if F-automata have universal reduction for some

$$F: \textbf{Set} \to \textbf{Set},$$

then F preserves unions. [This follows from the Universal Reduction Theorem because F preserves epis and $\mathrm{hom}(SF, S) \neq \emptyset$.] Then F is naturally equivalent to some S_{Σ_1, Σ_0} (see III.4.8) and, by convention III.3.1, we identify F-algebras with S_{Σ_1, Σ_0}-algebras. Therefore, F-automata are then S_{Σ_1, Σ_0}-automata, i.e. the sequential $(\Sigma_1 \cup \Sigma_0)$-automata with resets in Σ_0.

Remark. There are, of course, important concrete categories which are not additive and the forgetful functor of which does not preserve finite colimits. These often fail to have exact co-preimages. See Exercise D below.

Exercises VI.2

A. Universality without union preservation. Define the following functor F: **Gra** $\to$ **Gra** : for each discrete graph $(X, \emptyset)$ let $(X, \emptyset)F$ be the complete graph on the set $X \times X$; all other graphs are mapped by F to the singleton loop $(\{0\}, \{(0, 0)\})$. Analogously on morphisms.

(i) Verify that F is a finitary varietor with universal realization.

(ii) Prove that F preserves epis (= onto morphismms) but does not preserve unions. Why does it not contradict to the Universal Reduction Theorem? (Hint: F does not preserve unions because the set functor H_2 does not.)

B. Universality for linear functors. Prove that a linear (Exercise. III.4.C) functor F: R-**Vect** $\to$ **Vect** has universal realization iff F is naturally isomorphic to some S_Σ (Hint: Each linear functor preserves finite colimits; if F preserves unions, it preserves all colimits and hence, it has an adjoint.)

C. Universality for non-linear functors. Prove that for arbitrary modules Σ_0, Σ_1 the functor

$$S_{\Sigma_1, \Sigma_0} = V_{\Sigma_1} + C_{\Sigma_0} : R\text{-}\mathbf{Mod} \to R\text{-}\mathbf{Mod}$$

has universal reduction. Is it linear?

D. Non-exact co-preimages. (i) Let B be a simple group (i.e., a group without nontrivial quotients) and let $A \neq 0$ be its proper subgroup. Verify that the following square

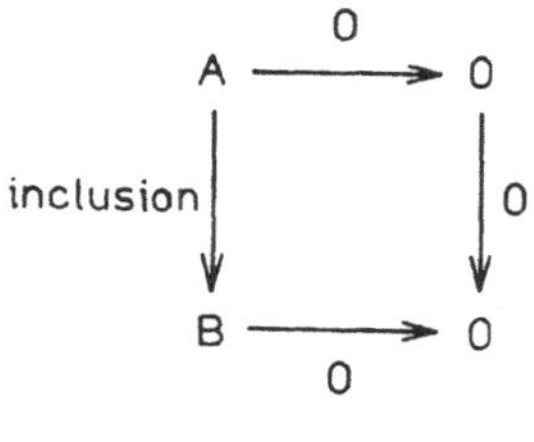

is a pushout but not a pullback.

Conclude that the category of groups does not have exact co-preimages.

(ii) Generalize the above to any category $\mathcal{K}$ which has a simple object B with a proper non-singleton subobject: then $\mathcal{K}$ does not have exact co-preimages.

This includes semigroups, lattices, rings, etc. And also various categories of metric spaces, and the category of Hausdorff topological spaces (consider B = real line and A = the rationals).

VI.3. Nerode Equivalences

3.1. Since universal reduction implies the preservation of unions, we are interested in the converse: does each unions-preserving functor have universal reduction? We are going to present a construction of minimal realizations which has the property that (i) if it works, realization is universal and (ii) if the functor preserves unions, the construction works. Hence, this gives an affirmative answer to our question.

There is a basic catch, however, analogous to that of V.3.10. In the preceding section we worked with "weak" epis: the class $\mathscr{E}$ consisted of all surjective morphisms of a concrete category. Here, we shall need "strong" epis: the class $\mathscr{E}$ will be that of all regular epis. Thus, a necessary and sufficient condition will be obtained, e.g. in the categories **Set**, R-**Mod** and unary algebras, (having just one factorization system) but not in **Top**, **Pos**, etc., where the two factorization systems differ. Throughout this section we assume that $\mathscr{K}$ is a finitely complete category with regular factorizations.

3.2. Recall (I.2.6) that for sequential Σ-automata, the minimal realization of a behavior

$$b: \Sigma^* \to \Gamma$$

has been obtained by factoring the free algebra Σ^* through the *Nerode equivalence E.* This is the equivalence E defined by $u_1 E u_2$ iff $(u_1 w)b = (u_2 w)b$ for each word $w \in \Sigma^*$, i.e., iff

(i) $(u_1)b = (u_2)b$,

(ii) $(u_1\sigma)b = (u_2\,\sigma)b$ for all $\sigma \in \Sigma$,

(iii) $(u_1\sigma\sigma')b = (u_2\sigma\sigma')b$ for all $\sigma, \sigma' \in \Sigma$, etc.

These conditions (i), (ii), (iii), . . . can be expressed by arrows, considering the equivalence E as a relation $[e_1, e_2]: \Sigma^* \rightharpoonup \Sigma^*$ where we denote the projections of $E \subset \Sigma^* \times \Sigma^*$ by

$$e_1, e_2: E \to \Sigma^*.$$

Then (i) says

$$e_1 \cdot b = e_2 \cdot b: E \to Y,$$

i.e., for each $x \in E$ with $x = (u_1, u_2)$ we have

$$(u_1)b = (x)e_1 \cdot b = (x)e_2 \cdot b = (u_2)b.$$

For the condition (ii), consider the concatenation

$$\varphi\colon \Sigma^* \times \Sigma \to \Sigma^* \quad (u, \sigma)\varphi = u\sigma.$$

Then (ii) says for $e_i \times 1_\Sigma = e_i S_\Sigma$ (III.2.3) that

$$e_1 S_\Sigma \cdot \varphi \cdot b = e_2 S_\Sigma \cdot \varphi \cdot b\colon E \times \Sigma \to \Gamma,$$

i.e., for each $(x, \sigma) \in E \times \Sigma$ with $x = (u_1, u_2)$ we have

$$(u_1\sigma)b = (u_1, \sigma)\varphi \cdot b = (x, \sigma)e_1 S_\Sigma \cdot \varphi \cdot b$$
$$= (x, \sigma)e_2 \cdot S_\Sigma \cdot \varphi \cdot b = (u_2\sigma)b.$$

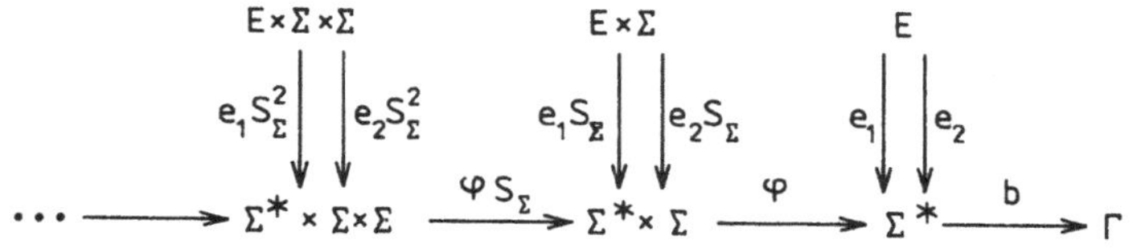

Analogously, (iii) says that

$$e_1 S_\Sigma^2 \cdot (\varphi S_\sigma \cdot \varphi \cdot b) = e_2 S_\Sigma^2 \cdot (\varphi S_\Sigma \cdot \varphi \cdot b),$$

etc.

The Nerode equivalence is clearly the largest relation satisfying these conditions (i), (ii), (iii), ... This desription can be used for a generalization of the Nerode equivalence. We shall work with relations in a category, see V.2.

3.3. Definition. Let $\mathcal{K}$ be a finitely complete category with regular factorizations, and let F be a varietor preserving regular epis. Let $b\colon I^\# \to \Gamma$ be a behavior. A relation $[e_1, e_2]\colon I^\# \rightharpoonup I^\#$ is said to be *externally b-equivalent* if for each $n < \omega$ it fulfils

$$e_1 F^{n-1} \cdot (\varphi F^{n-2} \cdot \varphi F^{n-3} \cdot \ldots \cdot \varphi F \cdot \varphi) \cdot b$$

(n)

$$= e_2 F^{n-1} \cdot (\varphi F^{n-2} \cdot \varphi F^{n-3} \cdot \ldots \cdot \varphi F \cdot \varphi) \cdot b.$$

The largest externally b-equivalent relation is called the *external Nerode equivalence* of the behavior b.

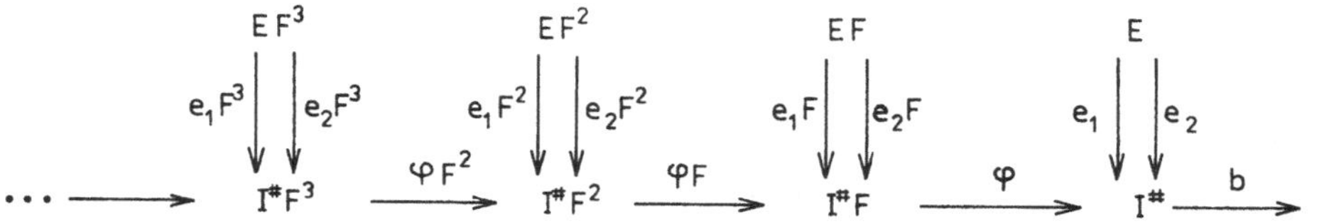

Remarks. (a) The external b-equivalence of $[e_1, e_2]$ means that
(i) $e_1 \cdot b = e_2 \cdot b$

(ii) $e_1 F \cdot \varphi \cdot b = e_2 F \cdot \varphi \cdot b$
(iii) $e_1 F^2 \cdot (\varphi F \cdot \varphi) \cdot b = e_2 F^2 \cdot (\varphi F \cdot \varphi) \cdot b$

etc. It is easy to check that these conditions are independent of the concrete representation of the relation: given a regular epi $c: E' \to E$ then $[e_1, e_2]$ fulfils (i), (ii), (iii), etc. iff $[c \cdot e_1, c \cdot e_2]$ does. Here we use the hypothesis that F (hence F^2, F^3, etc.) preserves regular epis.

(b) In the sequential case, the state object of the minimal realization is Σ^*/E, i.e., the coequalizer of $e_1, e_2: E \to \Sigma^*$. We generalize this:

3.4. Construction. Let F be a varietor preserving regular epis. Let

$$b: I^\# \to \Gamma$$

be a behavior. Assume that b has an external Nerode equivalence

$$[e_1, e_2]: I^\# \to I^\#.$$

Assume, moreover, that the coequalizer of e_1 and e_2,

$$c: I^\# \to Q$$

is a *congruence*, i.e., there is a $\delta: QF \to Q$ with

$$c: (I^\#, \varphi) \to (Q, \delta)$$

a homomorphism.

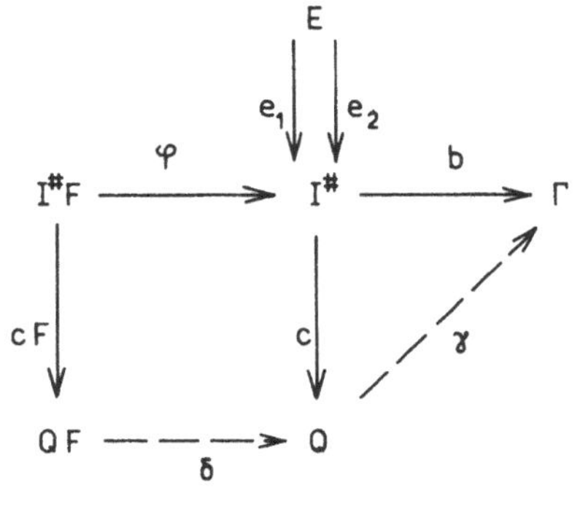

Since by the condition (i), $e_1 \cdot b = e_2 \cdot b$, there exists a unique morphism $\gamma: Q \to \Gamma$ with

$$b = c \cdot \gamma.$$

Then the F-automaton

$$A = (Q, \delta, \Gamma, \gamma, I, \eta \cdot c)$$

is the minimal realization of b.

Proof. First, the homomorphism $c:(I^{\#},\rho)\rightarrow(Q,\delta)$ extends $\eta\cdot c$ and hence, c is the run morphism of A. This implies both that A is reachable (since c is a coequalizer) and that it realizes $b = c\cdot\gamma$.

Next, let

$$A' = (Q', \delta', \Gamma, \gamma', I, \lambda')$$

be a reachable realization of b, i.e.,

$$b = c'\cdot\gamma'$$

for the run morphism $c'\colon I^{\#}\rightarrow Q'$ of A' which is a regular epi. Let

$$e_1', e_2'\colon E'\rightarrow I^{\#}$$

be the kernel equivalence of c'. Then c' is a coequalizer of e_1', e_2', see V.3.2. The relation $[e_1', e_2']$ is externally b-equivalent because the following diagram

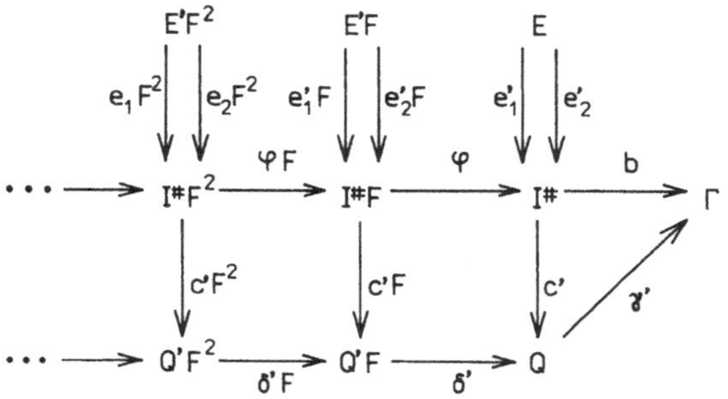

commutes.

By the definition of the external Nerode equivalence, we get

$$[e_1', e_2'] \subset [e_1, e_2].$$

Then $e_1\cdot c = e_2\cdot c$ implies

$$e_1'\cdot c = e_2'\cdot c$$

and, since c' is a coequalizer of e_1' and e_2', there exists

$$r\colon Q'\rightarrow Q$$

with

$$c = c'\cdot r.$$

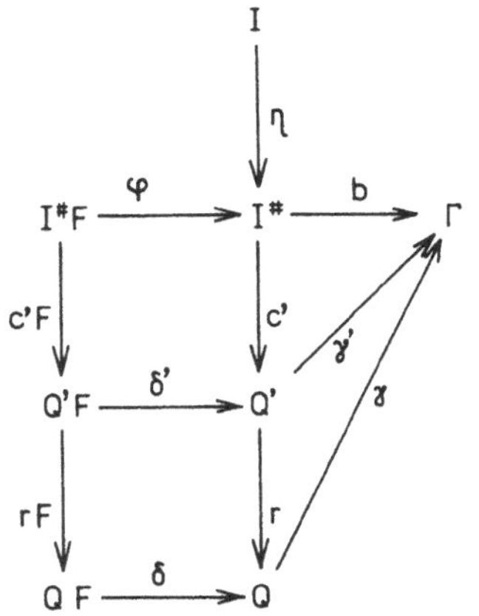

Then r is a regular epi, since $c' \cdot r$ is a regular epi, and we claim that

$$r: A' \to A$$

is a reduction. To prove

$$\delta' \cdot r = rF \cdot \delta,$$

we use the fact that $c'F$ is a regular epi (since F preserves regular epis), and

$$c'F \cdot (\delta' \cdot r) = \varphi \cdot c' \cdot r = \varphi \cdot c = cF \cdot \delta = c'F \cdot (rF \cdot \delta).$$

To prove

$$r \cdot \gamma = \gamma',$$

we use the fact that c' is a regular epi and the behavior $c' \cdot \gamma'$ of A' is b:

$$c' \cdot (r \cdot \gamma) = c \cdot \gamma = b = c' \cdot \gamma'.$$

Finally, to prove

$$\lambda' \cdot r = \eta \cdot c,$$

we use the fact that $\lambda' = \eta \cdot c'$ and thus, $\lambda' \cdot r = \eta \cdot c' \cdot r = \eta \cdot c$. This proves that A is the minimal realization of b. □

Remark. We say that *F has external Nerode realization* if (a) each beahvior has an external Nerode equivalence and (b) its coequalizer is a congruence. It will be shown that (b) holds quite generally: whenever F has minimal realizations, then (a) implies (b). Thus, it is the *existence* of Nerode equivalences which plays the basic role.

We prove that the existence of Nerode equivalences is trivial if F preserves unions, i.e., given a union of monos

$$m = \bigcup_{j \in J} m_j$$

then

$$\operatorname{im}(mF) = \bigcup_{j \in J} \operatorname{im}(m_j F).$$

Recall from IV.7.11 that the preservation of unions implies that F is a varietor (in reasonable categories).

3.5. Observation. Let $\mathscr{K}$ have unions and let F be a varietor preserving regular epis. If F preserves unions, then each behavior $b\colon I^\# \to \Gamma$ has an external Nerode equivalence. Namely, the union of all b-equivalent relations on $I^\#$.

In fact, let ε_t, $t \in T$, be the collection of all b-equivalent relations on $I^\#$. It clearly suffices to verify that the relation

$$\varepsilon^* = \bigcup_{t \in T} \varepsilon_t$$

is also b-equivalent. Put $\varepsilon_t = [\varepsilon_{t1}, \varepsilon_{t2}]$.

(i) For each t we have $\varepsilon_{t1} \cdot b = \varepsilon_{t2} \cdot b$, i.e., ε_t is contained in the kernel equivalence of b. Therefore, also ε^* is contained in this equivalence, which proves that

$$\varepsilon_1^* \cdot b = \varepsilon_2^* \cdot b.$$

(ii) For each t we have $\varepsilon_{t1}F \cdot (\varphi \cdot b) = \varepsilon_{t2}F \cdot (\varphi \cdot b)$. Since F preserves regular epis, $[\varepsilon_{t1}F, \varepsilon_{t2}F] = \varepsilon_t F$, see V.2.8. Thus, the relation $\operatorname{im}(\varepsilon_t F)$ is contained in the kernel equivalence of $\varphi \cdot b$. Therefore, also the relation

$$\bigcup_{t \in T} \operatorname{im}(\varepsilon_t F) = \operatorname{im}\left(\bigcup_{t \in T} \varepsilon_t\right) F = \operatorname{im} \varepsilon^* F$$

is contained in this equivalence. This proves that

$$\varepsilon_1^* F \cdot (\varphi \cdot b) = \varepsilon_2^* F \cdot (\varphi \cdot b).$$

Etc.

3.6. Lemma. Each external Nerode equivalence $[e_1, e_2]\colon E \rightharpoonup I^\#$ is indeed an equivalence relation on $I^\#$.

Proof. Let $\varepsilon\colon I^\# \rightharpoonup I^\#$ be the Nerode equivalence of a behavior $b\colon I^\# \to \Gamma$, put $e_1 = \varepsilon_{(1)}$ and $e_2 = \varepsilon_{(2)}$.

I. Reflexivity. The diagonal relation

$$\Delta = [1_{I^\#}, 1_{I^\#}]\colon I^\# \rightharpoonup I^\#$$

is evidently b-equivalent and hence,

$$\Delta \subset \varepsilon.$$

II. Symmetry. The inverse relation

$$\varepsilon^{-1} = [e_2, e_1]: I^{\#} \to I^{\#}$$

is evidently b-equivalent and hence,

$$\varepsilon^{-1} \subset \varepsilon.$$

III. Transitivity. The relation $\varepsilon \circ \varepsilon$ is defined by the following pullback

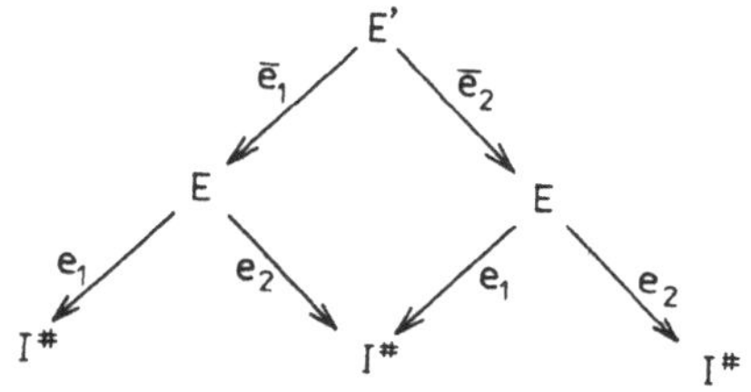

as the relation

$$\varepsilon \circ \varepsilon = [\bar{e}_1 \cdot e_1, \bar{e}_2 \cdot e_2].$$

It suffices to prove that $\varepsilon \circ \varepsilon$ is externally b-equivalent.

(i) $e_1 \cdot b = e_2 \cdot b$ implies

$$(\bar{e}_1 \cdot e_1) \cdot b = \bar{e}_1 \cdot e_2 \cdot b = \bar{e}_2 \cdot e_1 \cdot b = (\bar{e}_2 \cdot e_2) \cdot b;$$

(ii) $e_1 F \cdot (\varphi \cdot b) = e_2 F \cdot (\varphi \cdot b)$ implies

$$(\bar{e}_1 \cdot e_1)F \cdot (\varphi \cdot b) = \bar{e}_1 F \cdot e_2 F \cdot (\varphi \cdot b) = \bar{e}_2 F \cdot e_1 F \cdot (\varphi \cdot b) = (\bar{e}_2 \cdot e_2)F \cdot \varphi \cdot b,$$

etc. Hence, ε is an equivalence relation. □

3.7. Theorem. Let $\mathcal{K}$ be a finitely complete category with regular factorizations, which has cointersections and regular finite coproducts. Let F be a varietor preserving regular epis.
If F has minimal realizations, then the coequalizer of any external Nerode equivalence is a congruence.

Corollary. If F has minimal realizations and each behavior has an external Nerode equivalence, then F has external Nerode realization.

Proof. Let $b: I^{\#} \to \Gamma$ be a behavior which has a Nerode equivalence, represented by a pair $e_1, e_2: E \to I^{\#}$.

I. There exists $\psi: EF \to E$ such that both e_1 and e_2 are homomorphisms,

$$e_1, e_2: (E, \psi) \to (I^{\#}, \varphi).$$

To prove this, it is sufficient to observe that the relation

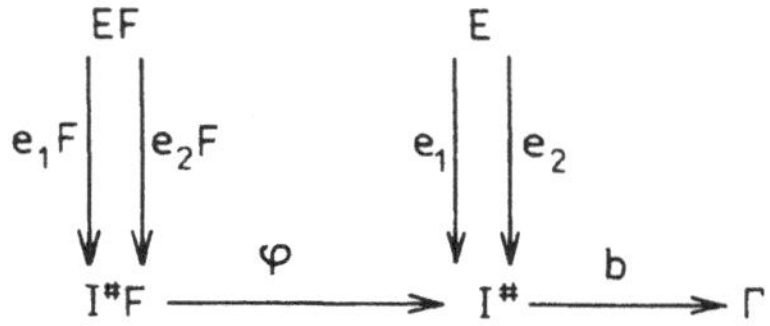

$[e_1F \cdot \varphi,\ e_2F \cdot \varphi]$ is externally b-equivalent. Indeed, the condition (ii) for $[e_1, e_2]$ is just the condition

(i) $\qquad (e_1F \cdot \varphi) \cdot b = (e_2F \cdot \varphi) \cdot b$

for $[e_1F \cdot \varphi, e_2F \cdot \varphi]$. Analogously, (iii) for $[e_1, e_2]$ is

(ii) $\qquad (e_1F \cdot \varphi)F \cdot (\varphi \cdot b) = (e_2F \cdot \varphi)F \cdot (\varphi \cdot b),$

etc.

Therefore, $[e_1F \cdot \varphi, e_2F \cdot \varphi] \subset [e_1, e_2]$, which means that there exists ψ with

$$e_1F \cdot \varphi = \psi \cdot e_1 \quad \text{and} \quad e_2F \cdot \varphi = \psi \cdot e_2.$$

II. F preserves the pushout of e_1 and e_2:

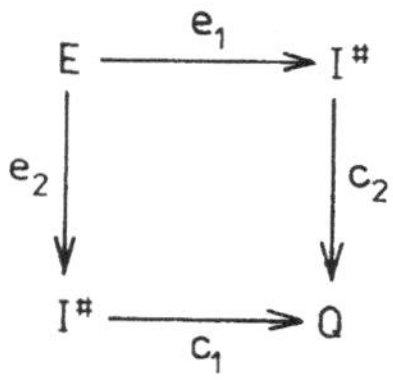

First, note that e_1 and e_2 are regular epis—indeed, split epis, since $\Delta \subset [e_1, e_2]$ by Lemma VI.3.6. Thus, their pushout is a cointersection.

By Theorem V.1.6, F weakly preserves cointersections. Since e_1 and e_2 are homomorphisms, it follows that F preserves their cointersection.

III. F preserves the coequalizer of e_1 and e_2. Indeed, we prove that $c_1 = c_2$ is the coequalizer c of e_1 and e_2; analogously, $c_1F = c_2F$ is then the coequalizer of e_1F and e_2F. The relation Δ is contained in $[e_1, e_2]$ and hence, there exists a morphism $j\colon I^{\#} \to E$ with

$$j \cdot e_1 = j \cdot e_2 = 1_{I^{\#}}.$$

This implies

$$c_1 = j \cdot e_2 \cdot c_1 = j \cdot e_1 \cdot c_2 = c_2.$$

Put $c = c_1 = c_2$. Then $e_1 \cdot c = e_2 \cdot c$. If c' is another morphism with $e_1 \cdot c' = e_2 \cdot c'$ then c' factors through c because of the universal property of pushouts:

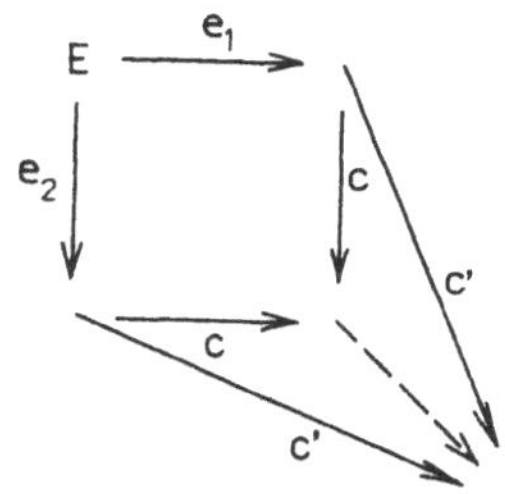

IV. The coequalizer c is a congruence. By III above,

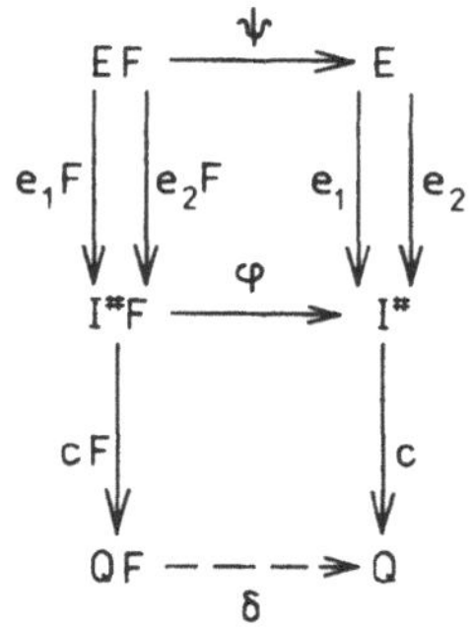

cF is the coequalizer of e_1F and e_2F. By I, we have

$$e_1F \cdot (\varphi \cdot c) = \psi \cdot e_1 \cdot c = \psi \cdot e_2 \cdot c = e_2\mathrm{F} \cdot (\varphi \cdot c).$$

Therefore, $\varphi \cdot c$ factors through cF. □

Remark. In the preceding proof, the hypothesis that F have minimal realizations was needed only to prove that F preserves the coequalizer of the Nerode equivalence. This can be concluded also whenever F is right exact, i.e., preserves the coequalizers of equivalences (V.3.4).

Thus, if $\mathscr{K}$ is a finitely complete category with regular factorizations and coequalizers, and if F is a right exact varietor, then F has external Nerode realization iff each behavior has an external Nerode equivalence.

3.8. Theorem. Each varietor with the external Nerode realization has universal realization.

Proof. For each morphism of behaviors $(f_{\mathrm{in}}, f_{\mathrm{out}})$: $(I, b, \Gamma) \to (I', b', \Gamma')$ we are to present the corresponding morphism of minimal realizations.

Let $[e_1, e_2]$ be the external Nerode equivalence of b. By hypothesis, b has a minimal realization

$$A = (Q, \delta, \Gamma, \gamma, I, \eta \cdot c),$$

where $c: I^{\#} \to Q$ is the coequliazer of e_1 and e_2 (which is also the run morphism of A). Analogously, let $[e_1', e_2']$ be the external Nerode equivalence of b' and $A' = (Q', \delta', \Gamma', \gamma', I', \eta' \cdot c')$ the corresponding minimal realization.

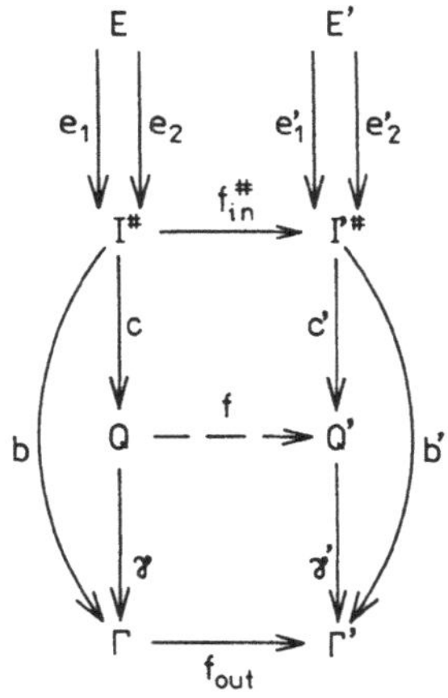

The relation $[e_1 \cdot f_{in}^{\#}, e_2 \cdot f_{in}^{\#}]$ is externally b'-equivalent:

(i) $$\begin{aligned} e_1 \cdot f_{in}^{\#} \cdot b' &= e_1 \cdot b \cdot f_{out} \\ &= e_2 \cdot b \cdot f_{out} \\ &= e_2 \cdot f_{in}^{\#} \cdot b'; \end{aligned}$$

(ii) $$\begin{aligned} (e_1 \cdot f_{in}^{\#})F \cdot (\varphi \cdot b') &= e_1 F \cdot (\varphi \cdot f_{in}^{\#}) \cdot b' \\ &= e_1 F \cdot (\varphi \cdot b) \cdot f_{out} \\ &= e_2 F \cdot (\varphi \cdot b) \cdot f_{out} \\ &= e_2 F \cdot (\varphi \cdot f_{in}^{\#}) \cdot b' \\ &= (e_2 \cdot f_{in}^{\#})F \cdot (\varphi \cdot b'); \end{aligned}$$

etc.

Therefore, this relation is contained in $[e_1', e_2']$. Consequently,

$$e_1 \cdot f_{in}^{\#} \cdot c' = e_2 \cdot f_{in}^{\#} \cdot c'.$$

Since c is the coequalizer of e_1 and e_2, this implies that there exists $f: Q \to Q'$ with

$$c \cdot f = f_{in}^{\#} \cdot c'.$$

To conclude the proof, it is sufficient to show that

$$(f, f_{\text{in}}, f_{\text{out}}): A \to A'$$

is a morphism of automata.

First, it commutes with the initializations:

$$(\eta \cdot c) \cdot f = \eta \cdot f_{\text{in}}^{\#} \cdot c' = f_{\text{in}} \cdot (\eta' \cdot c').$$

For the outputs, we use the fact that c is an epi:

$$\begin{aligned} c \cdot (f \cdot \gamma') &= f_{\text{in}}^{\#} \cdot c' \cdot \gamma' \\ &= f_{\text{in}}^{\#} \cdot b' \\ &= b \cdot f_{\text{out}} \\ &= c \cdot (\gamma \cdot f_{\text{out}}), \end{aligned}$$

and this implies

$$f \cdot \gamma' = \gamma \cdot f_{\text{out}}.$$

Finally, f is a homomorphism, since cF is an epi and

$$\begin{aligned} cF \cdot (fF \cdot \delta') &= f_{\text{in}}^{\#} F \cdot c'F \cdot \delta' \\ &= f_{\text{in}}^{\#} F \cdot \varphi' \cdot c' \\ &= \varphi \cdot f_{\text{in}}^{\#} \cdot c' \\ &= \varphi \cdot c \cdot f \\ &= cF \cdot (\delta \cdot f). \end{aligned}$$

This concludes the proof that $(f, f_{\text{in}}, f_{\text{out}})$, is a morphism. □

Corollary. Let $\mathscr{K}$ be a finitely complete category with regular factorizations, coequalizers and unions.

Each right exact varietor preserving unions has universal realizations.

(Every behavior has an external Nerode equivalence by Observation VI.3.5 and thus, F has external Nerode realization by Remark VI.3.7.)

3.9. Theorem. The following are equivalent for any functor

F: **Set** → **Set**,

and any right exact functor

F: R-**Mod** → R-**Mod** (R commutative ring):

(i) F has universal reduction;
(ii) F has external Nerode realization;
(iii) F is a varietor with universal realization;
(iv) F preserves unions.

Proof. This is just a combination of the preceding results with Theorem VI.2.6. The categories **Set** and R-**Mod** fulfil the hypotheses of both. Moreover, each set functor is right exact (V.3.10). □

Remark. In the theorem above, we have restricted ourselves to **Set** and R-**Mod** because in these categories we have a unique factorization system and hence, Theorem VI.2.6 can be combined with the preceding Corollary. (Also the category of unary algebras fulfils the hypotheses of both results.) The difference between onto morphisms, used in VI.2.6, and regular epis makes it impossible to combine these two results in categories like **Pos, Top**, etc.

3.10. We conclude this section by an observation, not related to the universal realization. Since the external Nerode equivalences exist only rarely, we ask whether there exists a general construction of minimal realizations by means of equivalences? The answer is affirmative for all finitary varietors (which is a small restriction only, in view of V.4.2).

We start with the observation that the conditions for E in the sequential case can be reformulated as follows:

(i′) $(u_1)b = (u_2)b$;

(ii′) $(u_1)b = (u_2)b$ and $(u_1\sigma)b = (u_2\sigma)b$ for all $\sigma \in \Sigma$;

(iii′) $(u_1 w)b = (u_2 w)b$ for all words $w \in \Sigma^*$ of length ≤ 2;

etc.

These conditions differ only in considering all words of length $\leq n$ rather than of length $= n$. And they correspond to the free-algebra construction over E: $W_0 = E$; $W_1 = E + E \times \Sigma$; $W_2 = E + W_1 S_\Sigma, \ldots$. Indeed, (i′) states that $e_1 \cdot b = e_2 \cdot b: W_0 \to \Gamma$; (ii′) states that $(e_1 + e_1 S_\Sigma) \cdot b = (e_2 + e_2 S_\Sigma) \cdot b$: $W_1 \to \Gamma$, etc.

In the following definition, we use the symbols W_0, W_1, $W_2, \ldots$ as functors from $\mathcal{K}$ to $\mathcal{K}$: on objects X

$$XW_0 = X \quad \text{and} \quad XW_{n+1} = X + XW_n \cdot F;$$

on morphisms $f: X \to Y$

$$fW_0 = f \quad \text{and} \quad fW_{n+1} = f + fW_n \cdot F.$$

In other words, we define functors $W_n: \mathcal{K} \to \mathcal{K}$ as follows:

$$W_0 = 1_{\mathcal{K}}; \quad W_1 = 1_{\mathcal{K}} + F; \quad W_2 = 1_{\mathcal{K}} + (1_{\mathcal{K}} + F) \cdot F, \ldots.$$

3.11. Definition. Let $\mathcal{K}$ be a finitely complete, finitely cocomplete category with regular factorizations, and let F be a varietor, preserving regular epis.

Given a behavior $b: I^\# \to \Gamma$, a relation $[e_1, e_2]: I^\# \rightharpoonup I^\#$ is said to be *inner b-equivalent* if it satisfies the following conditions (with $\bar{\varphi}: I^\# + I^\# F \to I^\#$ having components $1_{I^\#}$ and φ):

(i′) $e_1 \cdot b = e_2 \cdot b$;

(ii′) $e_1 W_1 \cdot (\bar{\varphi} \cdot b) = e_2 W_1 \cdot (\bar{\varphi} \cdot b)$;

(iii′) $e_1 W_2 \cdot (\bar{\varphi} W_1 \cdot \bar{\varphi} \cdot b) = e_2 W_2 \cdot (\bar{\varphi} W_1 \cdot \bar{\varphi} \cdot b)$;

in general:

$$\text{(n}'\text{)}\quad e_1 W_{n-1} \cdot (\bar{\varphi} W_{n-2} \cdot \bar{\varphi} W_{n-3} \cdot \ldots \cdot \bar{\varphi}) \cdot b = e_2 W_{n-1} \cdot (\bar{\varphi} W_{n-2} \cdot \bar{\varphi} W_{n-3} \cdot \ldots \cdot \bar{\varphi}) \cdot b.$$

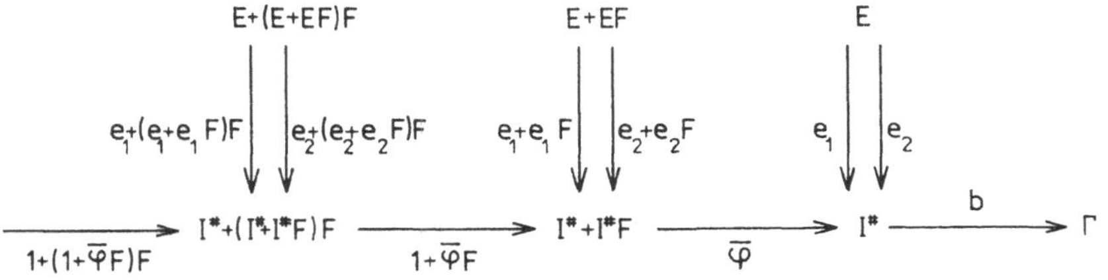

The largest reflexive, inner b-equivalent relation on $I^{\#}$ is the *inner Nerode equivalence* of b.

Remark. The conditions (i′), (ii′), (iii′), ... are again independent of the representing pair, since F preserves regular epis. Note that the definition of the inner equivalence differs from that of the external one not only in these conditions but also in considering only reflexive relations.

We say that F has *inner Nerode realization* if each behavior has an inner Nerode equivalence and its coequalizer is a congruence. (This is completely analogous to VI.3.4.) And if this is the case, than the minimal realizations are again easily seen to be constructed as quotients under the inner Nerode equivalence.

We shall prove now that, roughly speaking, minimal realizations are always inner Nerode realizations.

3.12. Theorem. Let $\mathscr{K}$ be a finitely complete and finitely cocomplete category with regular factorizations, cointersections and finite regular coproducts. Let F be a finitary varietor, preserving regular epis. Then F has minimal realizations iff F has inner Nerode realization.

Proof. Let F have minimal realizations. For each behavior

$$b: I^{\#} \to \Gamma$$

we have a minimal realization

$$A_0 = (Q_0, \delta_0, \Gamma, \gamma_0, I, \lambda_0)$$

with a run map $c_0: I^{\#} \to Q_0$. We shall prove that the kernel equivalence

$$[e_1^0, e_2^0]: I^{\#} \rightharpoonup I^{\#}$$

of c_0 is the inner Nerode equivalence of b. This will clearly prove that F has inner Nerode realizations.

(a) $[e_1^0, e_2^0]$ is a reflexive relation (indeed, an equivalence) and is inner b-equivalent:

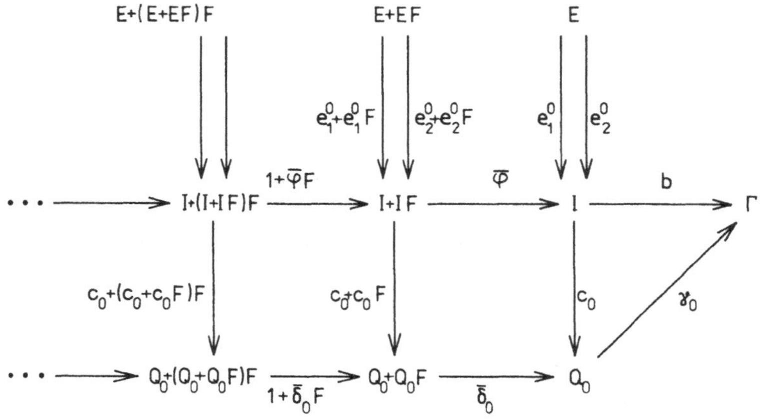

(i) $e_1^0 \cdot b = e_1^0 \cdot c_0 \cdot \gamma_0 = e_2^0 \cdot c_0 \cdot \gamma_0 = e_2^0 \cdot b$.

(ii) Let $\bar{\delta}_0 : Q_0 + Q_0F \to Q_0$ have components 1_{Q_0} and δ_0, then clearly $\varphi \cdot c_0 = c_0F \cdot \delta_0$ implies $\bar{\varphi} \cdot c_0 = c_0W_1 \cdot \bar{\delta}_0$. Thus

$$\begin{aligned} e_1^0W_1 \cdot (\bar{\varphi} \cdot b) &= e_1^0W_1 \cdot \bar{\varphi} \cdot c_0 \cdot b \\ &= e_1^0W_1 \cdot c_0W_1 \cdot \bar{\delta}_0 \\ &= e_2^0W_1 \cdot c_0W_1 \cdot \bar{\delta}_0 \\ &= e_2^0W_1 \cdot (\bar{\varphi} \cdot b) \end{aligned}$$

etc.

(b) Let $[e_1, e_2]$ be a reflexive, inner b-equivalent relation of $I^\#$. To prove that it is contained in $[e_1^0, e_2^0]$, it is sufficient to show that

$$e_1 \cdot c_0 = e_2 \cdot c_0.$$

Extending e_1, e_2 to homomorphisms

$$e_1^\#, e_2^\# : (E^\#, \varphi_E) \to (I^\#, \varphi_I),$$

we have

$$e_1^\# \cdot b = e_2^\# \cdot b.$$

In fact F is a finitary varietor, hence,

$$E^\# = \operatorname*{colim}_{n<\omega} EW_n,$$

therefore, it suffices to prove that $e_1^\# \cdot b$ and $e_2^\# \cdot b$ coincide on each of EW_n, $n < \omega$. For $n = 0$ this is (i′) $e_1 \cdot b = e_2 \cdot b$; for $n = 1$ this is (ii′) $(e_1 + e_1F \cdot \varphi) \cdot b = e_1W_1 \cdot \bar{\varphi} \cdot b = e_2W_1 \cdot \bar{\varphi} \cdot b = (e_2 + e_2F \cdot \varphi) \cdot b$; etc.

Moreover, the relation $[e_1^\#, e_2^\#]$ is reflexive. Indeed, since $[e_1, e_2]$ is reflexive, there exists $d: E \to I^\#$ with

$$d \cdot e_1 = d \cdot e_2 = 1_E.$$

Then $d^\# : (E^\#, \varphi) \to (I^\#, \varphi)$ fulfils

$$d^\# \cdot e_1^\# = d^\# \cdot e_2^\# = 1_{E^\#}.$$

Thus, the pushout of $e_1^\#$ and $e_2^\#$

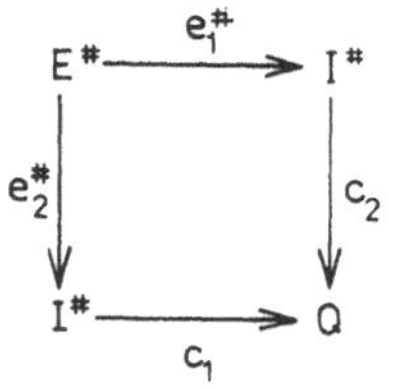

has the property that $c_1 = c_2$ $(= c)$ is the coequalizer of $e_1^\#$ and $e_2^\#$ (because $c_1 = d^\# \cdot e_2^\# \cdot c_1 = d^\# \cdot e_1^\# \cdot c_2 = c_2$).

Since F has minimal realizations, F weakly preserves cointersections by Theorem V.1.4. Thus, F preserves the pushout above (since both $e_1^\#$ and $e_2^\#$ are split epis) and consequently, cF is coequalizer of $e_1^\# F$ and $e_2^\# F$.

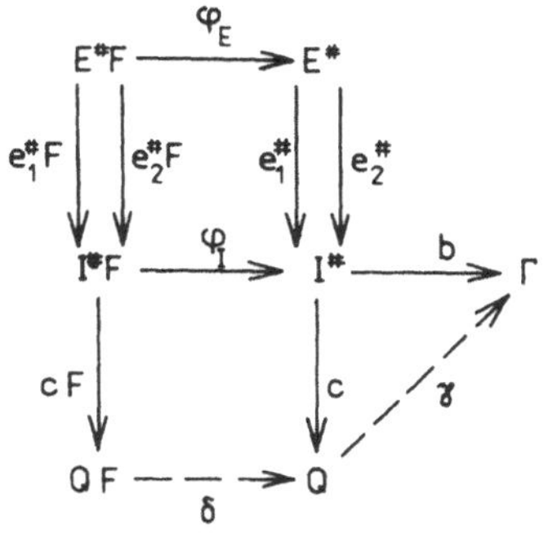

Since

$$e_1^\# F \cdot (\varphi \cdot c) = \varphi_E \cdot e_1^\# \cdot c = \varphi_E \cdot e_2^\# \cdot c = e_2^\# F \cdot (\varphi \cdot c),$$

there exists δ: $QF \to Q$ such that

$$\varphi \cdot c = cF \cdot \delta,$$

i.e., such that $c: (I^\#, \varphi) \to (Q, \delta)$ is a homomorphism. Further, since $e_1^\# \cdot b = e_2^\# \cdot b$, there exists γ: $Q \to \Gamma$ with

$$b = c \cdot \gamma.$$

Then we get an automaton

$$A = (Q, \delta, \Gamma, \gamma, I, \eta \cdot c)$$

with the run morphism c. Since c is a coequalizer and the behavior of A is $c \cdot \gamma = b$, this automaton is a reachable realization of b.

Therefore, A_0 is a reduction of A: we have a morphism $r: A \twoheadrightarrow A_0$, satisfying

$$c_0 = c \cdot r.$$

Then

$$e_1 \cdot c_0 = e_1 \cdot c \cdot r = \eta \cdot e_1^{\#} \cdot c \cdot r = \eta \cdot e_2^{\#} \cdot c \cdot r = e_2 \cdot c \cdot r = e_2 \cdot c_0.$$

This concludes the proof that $[e_1, e_2]$ is contained in the kernel equivalence of c_0. □

Corollary. Let $\mathcal{K}$ be as above. Each finitary, right exact functor has inner Nerode realization.

Exercises VI.3

A. Each external equivalence is inner equivalence. Prove that every externally b-equivalent relation is inner b-equivalent. Conclude that the external Nerode equivalence of any behavior is its inner Nerode equivalence.

B. The non-existence of the external equivalence. Prove that for each behavior b with the external Nerode equivalence the minimal realization A_b is "hereditarily reduced", i.e., each subautomaton of A_b is reduced.

Use this to find a behavior for $F = H_2$: **Set** $\to$ **Set** with no external Nerode equivalence. Find the inner Nerode equivalence. (Hint: The kernel equivalence of the run morphism of A_b.)

Notes to Chapter VI

VI.1—2

We have been inspired to study the universality by the paper of J. A. Goguen [1973]. Results of the first two sections were annouced by V. Trnková [1975] (who proved that for set functors, the characterizing condition is preservation of unions) and V. Trnková and J. Adámek [1977]. The proofs appear in the present book for the first time.

A generalization of these results was presented by J. Adámek, H. Ehrig and V. Trnková [1980], where an abstract category $\mathcal{D}$ (for example of automata) is considered together with a faithful functor from $\mathcal{D}$ to an $(\mathcal{E}, \mathcal{M})$-category $\mathcal{K}$. The concept of minimal reduction can be presented in this generality; it is universal iff the faithful functor preserves cointersections and copreimages.

VI.3

External Nerode equivalence was introduced by M. A. Arbib and E. G. Manes [1974a] who proved that in case this equivalence exists and its coequalizer is a congruence, then it constructs the minimal realization (VI.3.4). They found out later that external Nerode equivalence need not exist for tree automata, and they introduced several related concepts of Nerode equivalence, see P. O. Anderson, M. A. Arbib and E. G. Manes [1976]. The inner Nerode equivalence was defined by J. Adámek [1979a], who proved Theorems VI.3.7, VI.3.8 and VI.3.12.

Chapter VII: Nondeterministic Automata and Kleene Theorem

VII.1. Nondeterministic Behavior

1.1. In the present chapter we investigate the behavior of finite F-automata in the category of sets, and nondeterministic F-automata in a general category. Recall from Chapters I and II that nondeterministic automata are introduced because some operations on them are easier to perform than on the deterministic ones and, fortunately, finite nondeterministic and deterministic automata have the same behaviors. It turns out that in the category **Set**, the functors F for which finite nondeterministic and deterministic F-automata have the same behaviors are rather special. (This will be proved in VII.2.) And these special functors are the only ones for which an analogy of Kleene Theorem holds. (This will be proved in VII.3.)

The first section is devoted to the introduction of nondeterministic F-automata in a general category. We investigate the concept of behavior which is natural, though not entirely obvious.

1.2. Standing hypothesis: Throughout the present section, $\mathcal{K}$ denotes a finitely complete $(\mathcal{E}, \mathcal{M})$-category.

We work with relations in $\mathcal{K}$ in the sense of V.2.

Definition. For each functor $F: \mathcal{K} \to \mathcal{K}$, a *nondeterministic F-automaton* is a sixtuple

$$A = (Q, \delta, \Gamma, \gamma, I, \lambda)$$

consisting of objects Q, Γ and I, and of relations

$$\delta: QF \rightharpoonup Q;$$
$$\gamma: Q \rightharpoonup \Gamma;$$
$$\lambda: I \rightharpoonup Q.$$

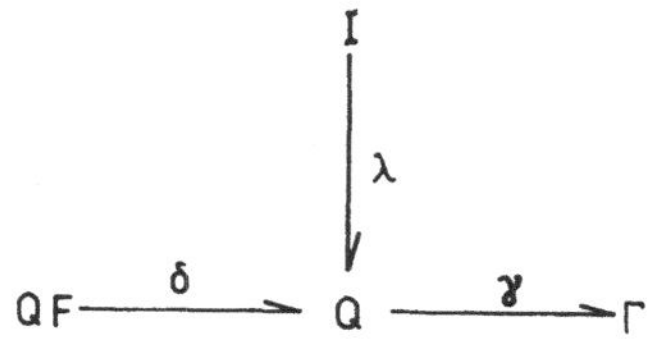

If δ, λ and γ are all partial morphism, then A is called a *partial automaton.*

Remark. Assuming that (i) $\mathcal{K}$ is well-powered and fulfils the pullback axiom (V.2.7) and (ii) F covers pullbacks (V.2.10), then we can extend F to

$$F\colon \mathbf{Rel}\,\mathcal{K} \to \mathbf{Rel}\,\mathcal{K},$$

see V.2.10. Then a nondeterministic F-automaton is simply an F-automaton in the category $\mathbf{Rel}\,\mathcal{K}$. We shall *not* impose these (rather severe) restrictions on $\mathcal{K}$ and F. As a consequence, we have to present a new definition of the run morphism and behavior for nondeterministic automata. We first recall the example of the first two chapters.

Examples. (i) Nondeterministic sequential automata. Here $\mathcal{K}$ = **Set** and $F = S_\Sigma$. A relation

$$\delta\colon Q \times \Sigma \rightharpoonup Q$$

assigns to each state q and each input σ the set

$$(q, \sigma)\delta \subset Q$$

of all possible next states. For a singleton set $I = \{i\}$, the relation

$$\lambda\colon I \rightharpoonup Q$$

assigns to i the set $(i)\lambda \subset Q$ of all initial states. And the output relation

$$\gamma\colon Q \rightharpoonup \Gamma$$

generalizes the output map considered in Chapter I.

The run relation

$$\rho\colon \Sigma^* \rightharpoonup Q$$

assigns to each sequence of inputs $\sigma_1 \dots \sigma_n$ the set $(\sigma_1 \dots \sigma_n)\rho \subset Q$ of all states that can be possibly reached from one of the initial states when inputs $\sigma_1, \sigma_2, \dots, \sigma_n$ are successively applied.

The *behavior* of this automaton is the relation

$$\beta\colon \Sigma^* \rightharpoonup \Gamma$$

assigning to each string $\sigma_1 \dots \sigma_n$ the set $(\sigma_1 \dots \sigma_n)\beta \subset \Gamma$ of all possible outputs resulting from an application of $\sigma_1, \dots, \sigma_n$ to any of the initial states. Thus,

$$(\sigma_1 \dots \sigma_n)\beta = \bigcup_{q \in (\sigma_1 \dots \sigma_n)\rho} (q)\gamma \quad \text{for each } \sigma_1 \dots \sigma_n \in \Sigma^*.$$

In other words,

$$\beta = \rho \circ \gamma\colon \Sigma^* \rightharpoonup \Gamma.$$

(ii) Nondeterministic tree automata. Consider, for simplicity, the type $\Sigma = \{\sigma\}$ with $|\sigma| = 2$. A nondeterministic tree Σ-automaton consists of a set Q, a nondeterministic binary operation

$$\square : Q \times Q \rightharpoonup Q,$$

an output relation

$$\gamma : Q \rightharpoonup \Gamma,$$

and an initialization relation

$$\lambda : I \rightharpoonup Q.$$

For each variable $x \in I$, we have a set $(x)\lambda \subset Q$ of possible interpretations. Thus, given a binary tree

$$t \in I^{\#},$$

we have several ways how to interpret the labels on the leaves and then several ways how to compute the tree using all the possible values of $q_1 \square q_2$ to compute the subtree

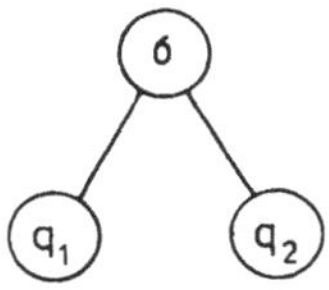

The run relation

$$\rho : I^{\#} \rightharpoonup Q$$

assigns to each tree $t \in I^{\#}$ the set $(t)\rho$ of all possible results of computation of t (with all possible interpretations of the leaves). The behavior relation is again

$$\beta = \rho \circ \gamma : I^{\#} \rightharpoonup \Gamma.$$

1.3. We want to define behavior of nondeterministic automata for each varietor F. The problem is to generalize the concept of the free extension $\lambda^{\#} : (I^{\#}, \varphi) \to (Q, \delta)$ to the case that both λ and δ are relations. We cannot define $\lambda^{\#}$ by the condition that the following diagram

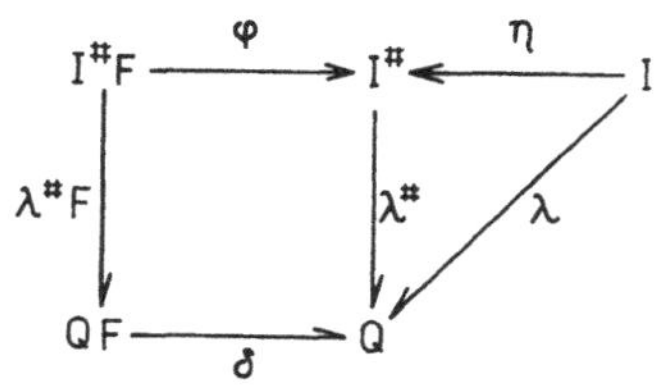

commutes because

(i) such $\lambda^{\#}$ need not exist (even in the case of linear sequential automata, see Exercise VII.1.B below) and
(ii) such $\lambda^{\#}$ need not be unique (even for a finitary varietor in **Set**, see Exercise VII.1.A below).

The concept of $\lambda^{\#}$ we define now has the following important features which are proved below:
(a) If δ and λ are morphisms, $\lambda^{\#}$ is the previously studied concept,
(b) $\lambda^{\#}$ exists and is unique for arbitrary relations λ and δ;
(c) if F is a constructive varietor, $\lambda^{\#}$ is obtained by a construction which naturally generalizes the case of morphisms;
(d) for sequential automata and tree automata, the concept agrees with the examples above.

Convention. Let (Q, δ) and (Q', δ') be relational F-algebras (i.e., $\delta: QF \rightharpoonup Q$ and $\delta': Q'F \rightharpoonup Q'$ are relations). A relation $f: Q \rightharpoonup Q'$ is called a *state relation* if

$$fF \circ \delta' \subset \delta \circ f.$$

This is the opposite inclusion to homomorphism (V.2.8).

Definition. Let F be a varietor and let (Q, δ) be a relational F-algebra. For each relation $\lambda: I \rightharpoonup Q$ we define the *free extension* as the least state relation $\lambda^{\#}: (I^{\#}, \varphi) \rightharpoonup (Q, \delta)$ extending λ. Explicitly, $\lambda^{\#}: I^{\#} \rightharpoonup \mathrm{Q}$ is the least relation with

$$\lambda \subset \eta \circ \lambda^{\#} \quad \text{and} \quad \lambda^{\#}F \circ \delta \subset \varphi \circ \lambda^{\#}.$$

Remark. For each nondeterministic F-automaton $A = (Q, \delta, \Gamma, \gamma, I, \lambda)$, we call $\rho = \lambda^{\#}$ the *run relation* of A, and

$$\beta = \rho \circ \gamma: I^{\#} \rightharpoonup \Gamma$$

the *behavior* of A.

Example: Sequential linear automata. Consider the functor

$$S_{\Sigma}: R\text{-}\mathbf{Mod} \to R\text{-}\mathbf{Mod}$$

and put $I = \{0\}$. For each nondeterministic sequential linear Σ-automaton

$$A = (Q, \delta, \Gamma, \gamma, \{0\}, \lambda)$$

we describe the run relation

$$\rho: \Sigma[z] \rightharpoonup \mathrm{Q}$$

(where $I^{\#} = \Sigma[z]$ is the module of polynomials, see III.2.4). First, let us consider the polynomial

$$0 = 0 + 0z + 0z^2 + \dots .$$

We have a subspace

$$(0)\lambda \subset Q$$

and for each state $q_0 \in (0)\lambda$, on receiving the input 0 we can transfer to any state $q_1 \in (q_0, 0)\delta$, and from q_1 to $q_2 \in (q_1, 0)\delta$, etc. Therefore,

$$(1) \qquad (0)\rho = Q_I = \bigcup_{n=0} Q_I^{(n)}$$

where

$$Q_I^{(0)} = (0)\lambda$$

and for each $n < \omega$,

$$Q_I^{(n+1)} = (Q_I^{(n)}, 0)\delta,$$

here the union means the union of subobjects in the category R-**Mod**, i.e., the linear envelope of the set-theoretical union. Given a polynomial

$$(z)w = \sigma_0 + \sigma_1 z + \ldots + \sigma_n z^n, \quad \sigma_n \neq 0,$$

then

$$(2) \qquad (\sigma + z \cdot w)\rho = (0 + \sigma + z \cdot w)\rho = ((w)\rho, \sigma)\delta \cup Q_I.$$

To prove that (1) and (2) define the run relation of A, we must first check that ρ is linear (i.e., a subspace of $\Sigma[z] \times Q$) and that it satisfies

$$\lambda \subset \eta \circ \rho \quad \text{and} \quad \rho S_\Sigma \circ \delta \subset \varphi \circ \rho.$$

Both statements are easy. Let $\rho'\colon \Sigma[z] \longrightarrow Q$ also satisfy

$$\lambda \subset \eta \circ \rho', \quad \text{i.e., } (0)\lambda \subset (0)\rho',$$

and

$$\rho' S_\Sigma \circ \delta \subset \varphi \circ \rho'.$$

Then we prove that

$$(w)\rho \subset (w)\rho' \quad \text{for each } w \in \Sigma[z]$$

by induction in the degree of w. For $w = 0$, we have $Q_I^{(0)} = (0)\lambda \subset (0)\rho'$ and by $\rho' S_\Sigma \circ \delta \subset \varphi \circ \rho'$, we prove that $Q_I^{(n)} \subset (0)\rho'$ implies $Q_I^{(n+1)} \subset (0)\rho'$. Hence,

$$(0)\rho = \bigcup_{n=0}^{\infty} Q_I^{(n)} \subset (0)\rho'.$$

Let $(w)\rho \subset (w)\rho'$, then

$$\begin{aligned}(\sigma + z \cdot w)\rho &= ((w)\rho, \sigma)\delta \cup Q_I \subset ((w)\rho', \sigma)\delta \cup Q_I \\ &= (w, \sigma)(\rho' S_\Sigma \circ \delta) \cup Q_I \subset (w, \sigma)(\varphi \circ \rho') \cup Q_I \\ &= (\sigma + z \cdot w)\rho' \cup Q_I.\end{aligned}$$

It remains to verify that

$$Q_I \subset (\bar{w})\rho' \quad \text{for each } \bar{w} \in \Sigma[z].$$

In fact, $0 + \bar{w} = \bar{w}$ implies $(0)\rho' + (\bar{w})\rho' \subset (\bar{w})\rho'$, and $Q_I \subset (0)\rho' \subset (0)\rho' + (\bar{w})\rho'$.

1.4. In order to prove the basic properties of λ^*, we need the following results concerning composition of relations.

Proposition. Let $f\colon X \to Y$ be a morphism, and let $g\colon Y \rightharpoonup Z$ be a relation represented by an $\mathcal{M}$-mono $m\colon R \to Y \times Z$. Let us form the pullback of $f \times 1_Z$ and m:

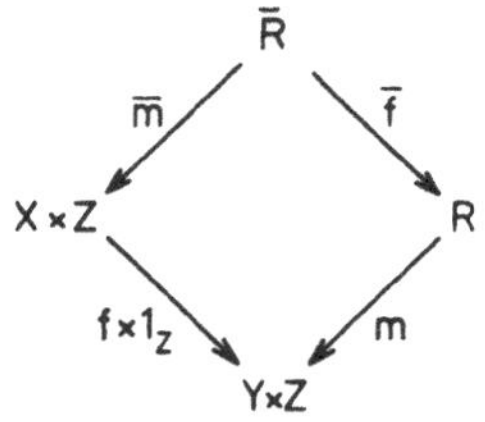

Then $\bar{m}\colon \bar{R} \to X \times Z$ is an $\mathcal{M}$-mono representing $f \circ g$.

Proof. It is easy to verify that the following square

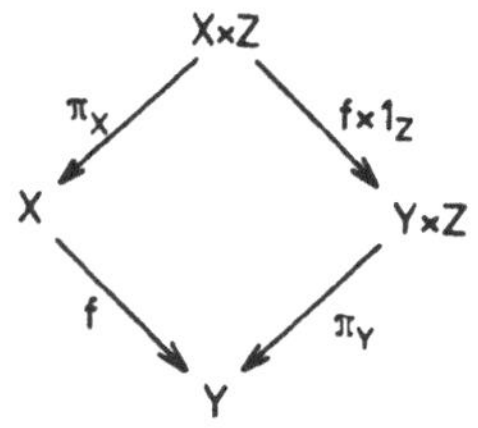

is a pullback (where π_X and π_Y are the projections). Consequently, in the following diagram

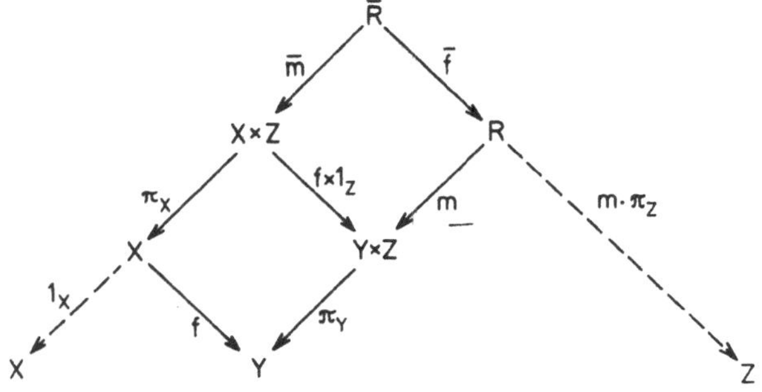

the outward square is the pullback of f and $m \cdot \pi_Y$. The relation g is represented by m and hence,

$$g_{(1)} = m \cdot \pi_Y \quad \text{and} \quad g_{(2)} = m \cdot \pi_Z,$$

where

$$\pi'_Z: X \times Z \to Z \quad \text{and} \quad \pi_Z: Y \times Z \to Z$$

are the projections. By the definition of composition, we have

$$\begin{aligned} f \circ g &= [\bar{m} \cdot \pi_X \cdot 1_X, \bar{f} \cdot m \cdot \pi_Z] \\ &= [\bar{m} \cdot \pi_X, \bar{m} \cdot (f \times 1_Z) \cdot \pi_Z] \\ &= [\bar{m} \cdot \pi_X, \bar{m} \cdot \pi'_Z]. \end{aligned}$$

Since, moreover $\bar{m}$ is an $\mathcal{M}$-mono [see III.5.1(iii)], it follows that $f \circ g$ is represented by $\bar{m}$. □

Corollary. For arbitrary morphisms $f: X \to Y$ and $g: Y \to Z$, and for each relation $h: Z \rightharpoonup T$, we have

$$(f \cdot g) \circ h = f \circ (g \circ h).$$

In fact, in the following diagram

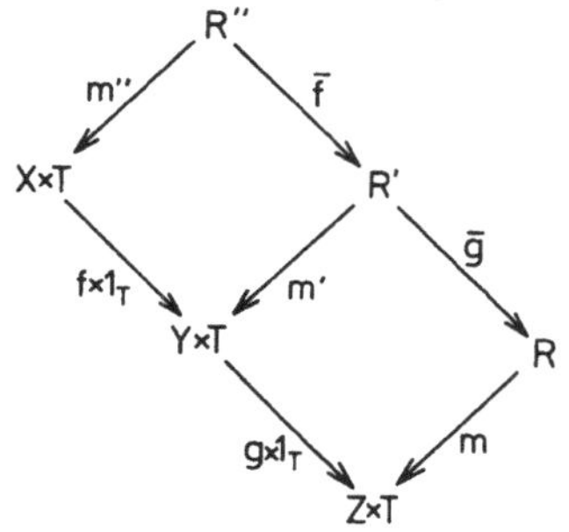

with pullbacks as both of the inner squares, also the outward square is a pullback. Let m be an $\mathcal{M}$-mono representing h. Then m'' represents $(f \cdot g) \circ h$, and since m' represents $g \circ h$, it follows that m'' also represents $f \circ (g \circ h)$.

Remark. The proposition above can be formulated in terms of the pullback of f and $g_{(1)}$:

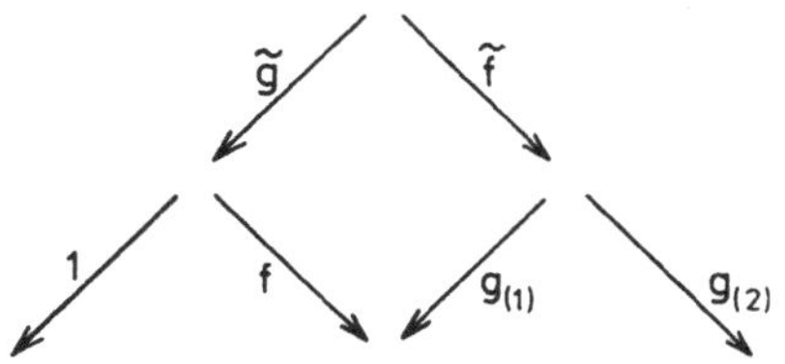

It states that for each morphism $f: X \to Y$ and each relation $g: Y \rightharpoonup Z$ we have

$$(f \circ g)_{(1)} = \bar{g} \quad \text{and} \quad (f \circ g)_{(2)} = \bar{f} \cdot g_{(2)}.$$

1.5. Lemma. Let $f: X \to Y$ be a morphism. For arbitrary relations $g: Y \rightharpoonup Z$ and $h: Z \rightharpoonup T$, we have

$$(f \circ g) \circ h \subset f \circ (g \circ h).$$

Proof. Let $m: R \to Y \times Z$ be an $\mathcal{M}$-mono representing g, and $n: S \to Z \times T$ and $\mathcal{M}$-mono representing h. In the following pullback

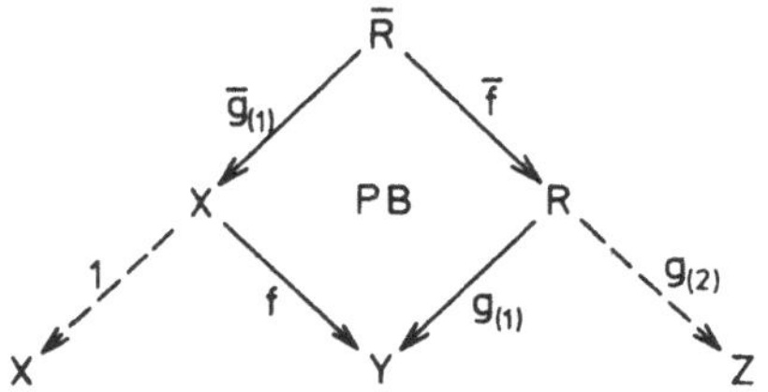

we have, by the preceding Remark,

$$(f \circ g)_{(1)} = \bar{g}_{(1)} \quad \text{and} \quad (f \circ g)_{(2)} = \bar{f} \cdot g_{(2)}.$$

The relation $(f \circ g) \circ h$ is given by the pullback of $\bar{f} \cdot g_{(2)}$ with $h_{(1)}$; this is obtained by "joining" the pullbacks of $g_{(2)}$ with $h_{(1)}$ and then of $\bar{f}$ with $\bar{h}$:

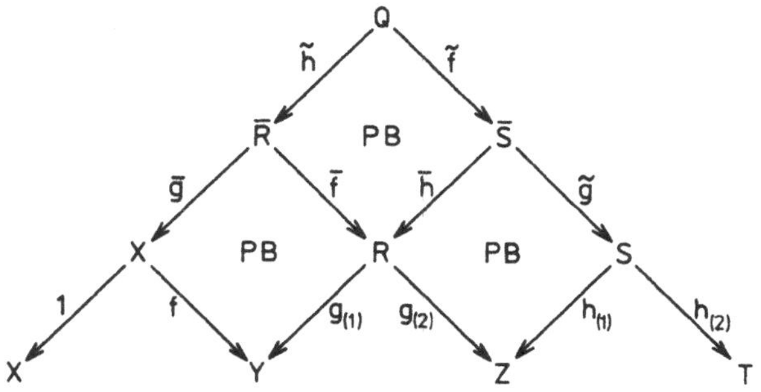

Thus

$$(f \circ g) \circ h = [\tilde{h} \cdot \bar{g}, \tilde{f} \cdot \tilde{g} \cdot h_{(2)}].$$

On the other hand,

$$g \circ h = [\bar{h} \cdot g_{(1)}, \tilde{g} \cdot h_{(2)}],$$

and this means that there exists an $\mathscr{E}$-epi $e: \bar{S} \to \tilde{S}$ such that the canonical representing pair $(g \circ h)_{(1)}: \tilde{S} \to Y$ and $(g \circ h)_{(2)}: S \to T$ fulfils

$$\bar{h} \cdot g_{(1)} = e \cdot (g \circ h)_{(1)} \quad \text{and} \quad g \cdot h_{(2)} = e \cdot (g \circ h)_{(2)}.$$

Let us form the pullback of f with $(g \circ h)_{(1)}$:

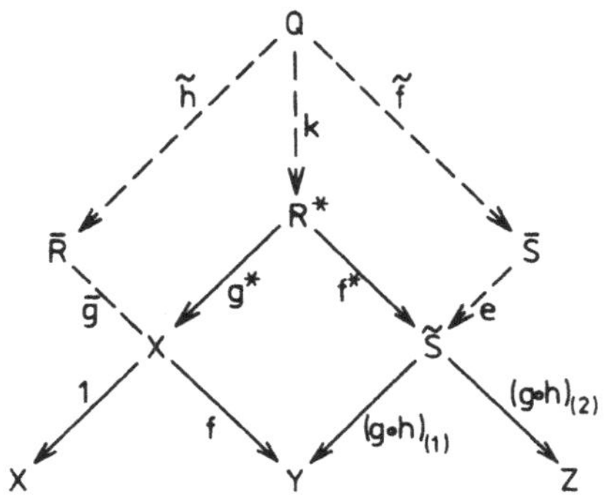

Then

$$f \circ (g \circ h) = [g^*, f^* \cdot (g \circ h)_{(2)}].$$

We have $(\tilde{h} \cdot \bar{g}) \cdot f = (\tilde{f} \cdot e) \cdot (g \circ h)_{(1)}$ because

$$\begin{aligned} \tilde{h} \cdot \bar{g} \cdot f &= \tilde{h} \cdot \bar{f} \cdot g_{(1)} \\ &= \tilde{f} \cdot \bar{h} \cdot g_{(1)} \\ &= \tilde{f} \cdot e \cdot (g \circ h)_{(1)}. \end{aligned}$$

Thus, there exists a unique $k: Q \to R^*$ for which the diagram above commutes. Hence,

$$\begin{aligned} (f \circ g) \circ h &= [\tilde{h} \cdot \bar{g}, \tilde{f} \cdot \tilde{g} \cdot h_{(2)}] \\ &= [\tilde{h} \cdot \bar{g}, \tilde{f} \cdot e \cdot (g \circ h)_{(2)}] \\ &= [k \cdot g^*, k \cdot f^* \cdot (g \circ h)_{(2)}] \\ &\subset [g^*, f^* \cdot (g \circ h)_{(2)}] \\ &= f \circ (g \circ h). \end{aligned}$$ □

Remark. Composition of relations with a morphism preserves intersections. More precisely, for each morphism $f: X \to Y$ and arbitrary relations $g_i: Y \rightharpoonup Z$ $(i \in I)$, we have

$$\bigcap_{i \in I} f \circ g_i = f \circ \bigcap_{i \in I} g_i : X \rightharpoonup Z$$

(the intersection of subobjects of $X \times Z$ on the left-hand side, and of subobjects of $Y \times Z$ on the right-hand side). This follows easily from the Proposition VII.1.4. Let m_i be an $\mathcal{M}$-mono representing g_i $(i \in I)$. Then $f \circ g_i$ is represented by the pullback of $f \times 1_Z$ with m_i. Pullbacks commute with limits, in particular with intersections (which are multiple pullbacks, see III.5.2). Thus, the pullback of $f \times 1_Z$ with $\bigcap_{i \in I} g_i$ is obtained by the intersection of the pullbacks with g_i, $i \in I$.

The formula above holds also for empty collections: here $\bigcap_{i \in \emptyset} g_i$ is the largest relation $1_{Y \times Z}$ and it is easy to verify that $f \circ 1_{Y \times Z} = 1_{X \times Z} = \bigcap_{i \in \emptyset} f \circ g_i$.

We can prove now that $\lambda^{\#}$ exists quite generally.

Proposition. If $\mathscr{K}$ has intersections, and F is a varietor, then for each relational algebra (Q, δ) and each relation $\lambda: I \rightharpoonup Q$, the relation $\lambda^{\#}$ exists.

Proof. Let $\lambda^{\#} = \bigcap h$ where the intersection is taken over all state relations $h: (I^{\#}, \varphi) \rightharpoonup (Q, \delta)$ extending λ. It is sufficient to prove that also $\lambda^{\#}$ is a state relation extending λ. For each h we have $\lambda \subset \eta \circ h$ and hence, by the preceding remark,

$$\lambda \subset \bigcap \eta \circ h = \eta \circ \bigcap h = \eta \circ \lambda^{\#}.$$

And each h is a state relation, i.e., $hF \circ \delta \subset \varphi \circ h$, and hence, by the same remark, and since $\lambda^{\#} F \circ \delta \circ \subset hF \circ \delta$ for each h, we get

$$\lambda^{\#} F \circ \delta \subset \bigcap hF \circ \delta \subset \bigcap \varphi \circ h = \varphi \circ \bigcap h = \varphi \circ \lambda^{\#}. \qquad \square$$

1.6. Next we prove that the present concept of $\lambda^{\#}$ coincides with that studied previously in case both δ and λ are morphisms.

Proposition. Let F be a varietor, and let (Q, δ) be an F-algebra. For each morphism $\lambda: I \to Q$, the homomorphism $\lambda^{\#}: (I, \varphi) \to (Q, \delta)$ is the least state relation extending λ.

Proof. The only statement to be proved is that $\lambda^{\#}$ is the least one. Thus, for each state relation $h: I^{\#} \rightharpoonup Q$ extending λ, we shall prove that

$$\lambda^{\#} \subset h.$$

Let $m: R \to I^{\#} \times Q$ be an $\mathscr{M}$-mono representing h, i.e., with

$$h_{(1)} = m \cdot \pi_1 \quad \text{and} \quad h_{(2)} = m \cdot \pi_2,$$

where π_1 and π_2 are the projections of $I^{\#} \times Q$. Since $\lambda^{\#}$ is represented as $[1_{I^{\#}}, \lambda^{\#}]$, we have to find a morphism $p: I^{\#} \to R$ with

$$p \cdot h_{(1)} = 1_{I^{\#}} \quad \text{and} \quad p \cdot h_{(2)} = \lambda^{\#}.$$

I. We present a morphism

$$\tilde{\delta}: RF \to R$$

for which both $h_{(1)}: (R, \tilde{\delta}) \to (I^{\#}, \varphi)$ and $h_{(2)}: (R, \tilde{\delta}) \to (Q, \delta)$ become homomorphisms. We use the fact that h is a state relation, i.e.,

$$hF \circ \delta \subset \varphi \circ h.$$

Since $hF = [h_{(1)}F, h_{(2)}F]$ (by definition), and since $\delta = [1_{QF}, \delta]$, clearly

$$hF \circ \delta = [h_{(1)}F, h_{(2)}F \cdot \delta].$$

Furthermore, by Proposition VII.1.4, in the following pullback

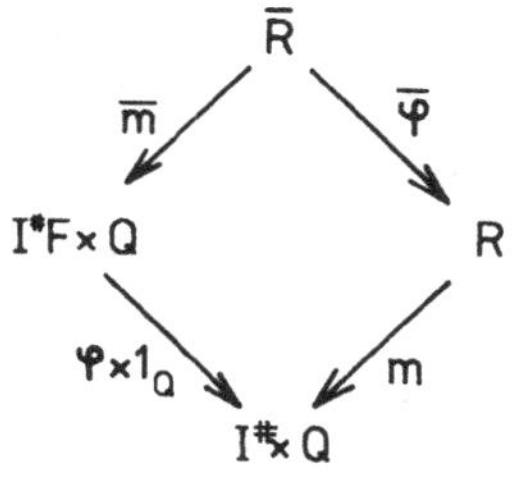

the $\mathcal{M}$-mono $\bar{m}:\bar{R} \to I^{\#}F \times Q$ represents $\varphi \circ h$. Thus, denoting by π_1' and π_2' the projections of $I^{\#}F \times Q$, we have $(\varphi \circ h)_{(1)} = \bar{m} \cdot \pi_1'$ and $(\varphi \circ h)_{(2)} = \bar{m} \cdot \pi_2'$. Since $hF \circ \delta \subset \varphi \circ h$, there exists a mophism $d: RF \to \bar{R}$ with

$$h_{(1)}F = d \cdot \bar{m} \cdot \pi_1' \quad \text{and} \quad h_{(2)}F \cdot \delta = d \cdot \bar{m} \cdot \pi_2'.$$

Put

$$\bar{\delta} = d \cdot \bar{\varphi}.$$

Then $h_{(1)}$ becomes a homomorphism because the following diagram

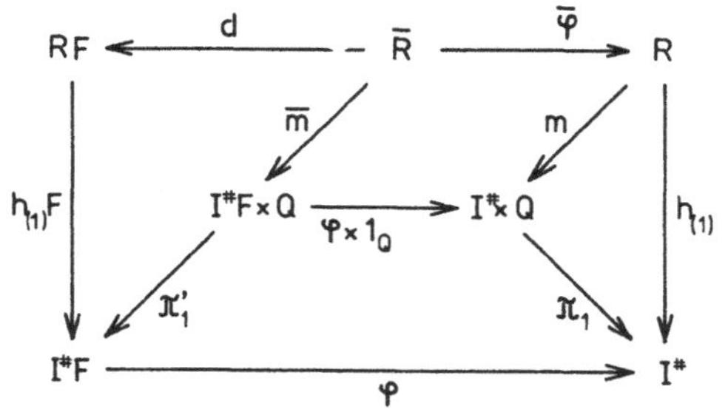

commutes. And $h_{(2)}$ becomes a homomorphism because the folowing one

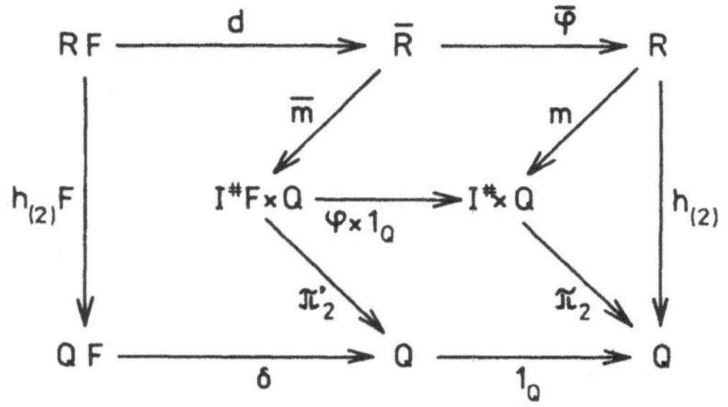

commutes too.

II. We use the condition

$$\lambda \subset \eta \circ h.$$

The composition $\eta \circ h$ is represented in the following pullback

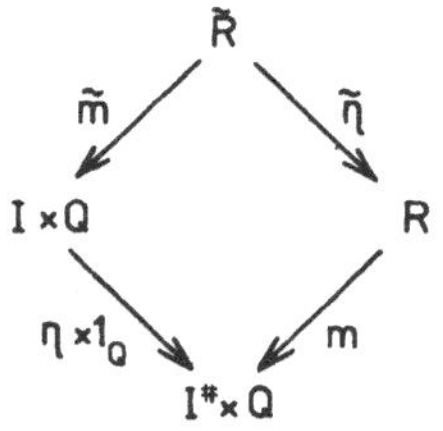

by $\tilde{m}$ (see VII.1.4). Thus, denoting by π_1'' and π_2'' the projections of $I \times Q$, then $(\eta \circ h)_{(1)} = \tilde{m} \cdot \pi_1''$ and $(\eta \circ h)_{(2)} = \tilde{m} \cdot \pi_2''$. Since $\eta \circ h$ contains λ, there exists a morphism

$$q : I \to \tilde{R}$$

with

$$1_I = q \cdot \tilde{m} \cdot \pi_1'' \quad \text{and} \quad \lambda = q \cdot \tilde{m} \cdot \pi_2''.$$

We can extend $q \cdot \tilde{\eta} : I \to R$ to a homomorphism $(q \cdot \tilde{\eta})^{\#} : (I^{\#}, \varphi) \to (R, \bar{\delta})$. Then

$$(q \cdot \tilde{\eta})^{\#} \cdot h_{(1)} = 1_{I^{\#}}$$

because $(I^{\#}, \varphi)$ is the free algebra and $(q \cdot \tilde{\eta})^{\#} \cdot h_{(1)}$ is its endomorphism with

$$\begin{aligned}\eta \cdot (q \cdot \tilde{\eta})^{\#} \cdot h_{(1)} &= q \cdot \tilde{\eta} \cdot h_{(1)} \\ &= q \cdot \tilde{\eta} \cdot m \cdot \pi_1 \\ &= q \cdot \tilde{m} \cdot [(\eta \times 1_Q) \cdot \pi_1]\end{aligned}$$

$$= q \cdot \tilde{m} \cdot [\pi_1'' \cdot \eta]$$
$$= \eta.$$

Furthermore,

$$(q \cdot \tilde{\eta})^{\#} \cdot h_{(2)} = \lambda^{\#}$$

because $(q \cdot \tilde{\eta})^{\#} \cdot h_{(2)}$ is a homomorphism extending λ:

$$\begin{aligned} \eta \cdot (q \cdot \tilde{\eta})^{\#} \cdot h_{(2)} &= q \cdot \tilde{\eta} \cdot m \cdot \pi_2 \\ &= q \cdot \tilde{m} \cdot (\eta \times 1_Q) \cdot \pi_2 \\ &= q \cdot \tilde{m} \cdot \pi_2'' \\ &= \lambda. \end{aligned}$$

Thus, $p = (q \cdot \tilde{\eta})^{\#}$ is the morphism we needed. □

1.7. For constructive varietors (IV.3.2), the free extension $\lambda^{\#}$ can be obtained constructively. Recall first that in case of morphisms, we have

$$I^{\#} = W_k \quad \text{and} \quad \lambda^{\#} = \lambda^{(k)}$$

for a sufficiently large ordinal, where $\lambda^{(0)} = \lambda$, the components of

$$\lambda^{(n+1)}: I + W_n F \to Q$$

are λ and $\lambda^{(n)} F \cdot \delta$, and for each limit ordinal i, the components of

$$\lambda^{(i)}: \operatorname*{colim}_{n<i} W_n \to Q$$

are $\lambda^{(n)}$ for $n < i$. In case $\lambda: I \rightharpoonup Q$ is a relation, we expect that $\lambda^{(n)}: W_n \rightharpoonup Q$ will be relations defined by an analogous induction. For this, we have to specify what it means for a relation $r: A' + A'' \rightharpoonup \mathrm{B}$ to have components $r': A' \rightharpoonup \mathrm{B}$ and $r'': A'' \rightharpoonup B$. It turns out that the folowing condition

$$v' \circ r = r' \quad \text{and} \quad v'' \circ r = r'' \ (v', v'' \text{ injections})$$

is not satisfactory: Such r need not exist, and it need not be unique (see Exercise VII.1 C below).

Definition. Let $A' + A''$ be a coproduct with injections v' and v''. Given relations $r': A' \rightharpoonup \mathrm{B}$ and $r'': A'' \rightharpoonup \mathrm{B}$, we say that a relation $r: A' + A'' \rightharpoonup \mathrm{B}$ *has components* r' and r'' provided that

$$(*) \qquad r' \subset v' \circ r \quad \text{and} \quad r'' \subset v'' \circ r,$$

and r is the least relation satisfying $(*)$.

The relations r' and r'' are represented by pairs of morphisms:

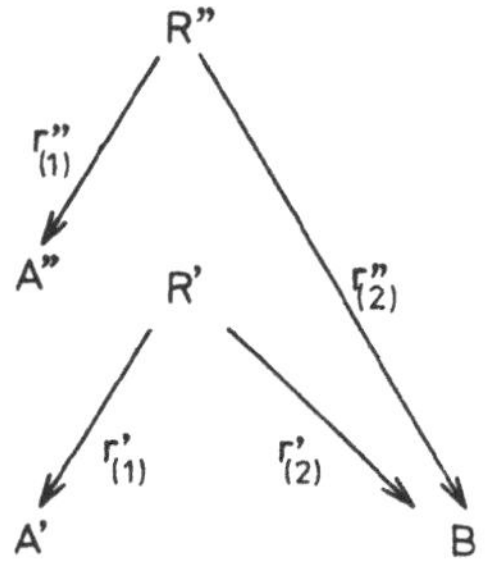

and these form a pair consisting of $r'_{(1)} + r''_{(1)}$, and the morphism $\tilde{r}$ with components $r'_{(2)}$ and $r''_{(2)}$:

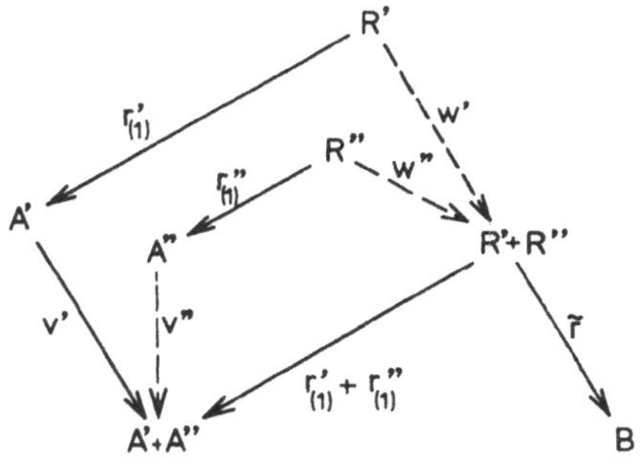

Lemma. The relation $r = [r'_{(1)} + r''_{(1)}, \tilde{r}]: A' + A'' \to B$ has components r' and r''.

Proof. Denote by v', v'', w' and w'' the coproduct injections as in the diagram above. We prove first that

$$r' \subset v' \circ r.$$

Let us form the pullback of v' and $r'_{(1)} + r''_{(1)}$:

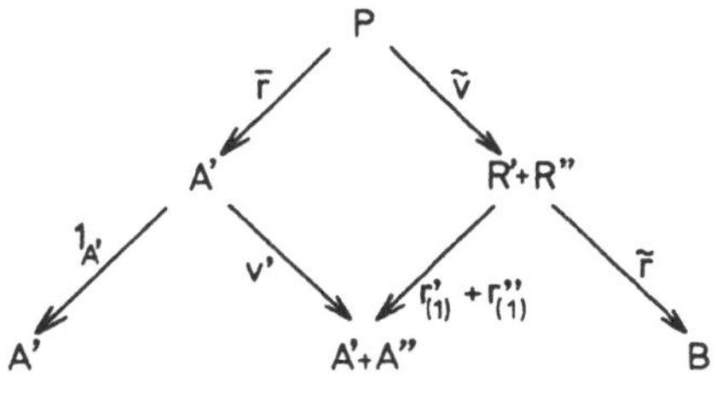

Then $(v' \circ r)_{(1)} = \bar{r}$ and $(v' \circ r)_{(2)} = \tilde{v} \cdot \bar{r}$ by Remark VII.1.4. Since $r'_{(1)} \cdot v' = w' \cdot (r'_{(1)} + r''_{(1)})$, there exists a unique morphism $k \colon R' \to P$ for which the following diagram

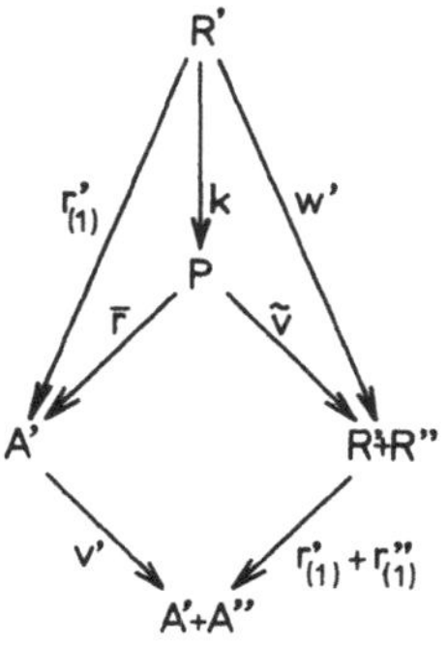

commutes. Then $r'_{(2)} = w' \cdot \bar{r}$ implies

$$r' = [r'_{(1)}, r'_{(2)}] = [k \cdot \bar{r}, k \cdot (\tilde{v} \cdot \bar{r})] \subset [\bar{r}, \tilde{v} \cdot \bar{r}] = v' \circ r.$$

Analogously,

$$r'' \subset v'' \circ r.$$

Next, we prove that r contains any relation $s \colon A' + A'' \rightharpoonup B$ with

$$r' \subset v' \circ s \quad \text{and} \quad r'' \subset v'' \circ s.$$

Let us form the pullback of v' and $s_{(1)}$:

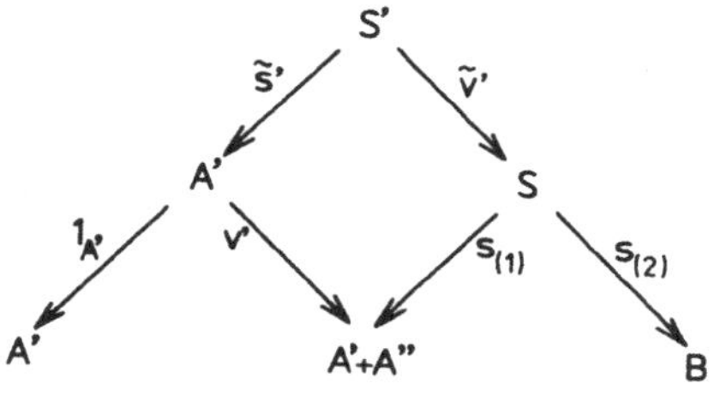

By Remark VII.1.4, the condition $r' \subset v' \circ s$ means that there exists

$$p' \colon R' \to S'$$

with

$$r'_{(1)} = p' \cdot \tilde{s}' \quad \text{and} \quad r'_{(2)} = p' \cdot \tilde{v}' \cdot s_{(2)}.$$

Analogously, the condition $r'' \subset v'' \circ s$ means that there is

$$p'' \colon R'' \to S''$$

with

$$r''_{(1)} = p'' \cdot \tilde{s}'' \quad \text{and} \quad r''_{(2)} = p'' \cdot \tilde{v}'' \cdot s_{(2)}.$$

The morphism

$$p: R' + R'' \to S$$

with components $p' \cdot \tilde{v}'$ and $p'' \cdot \tilde{v}''$ fulfils

$$r'_{(1)} + r''_{(1)} = p \cdot s_{(1)} \quad \text{and} \quad \tilde{r} = p \cdot s_{(2)}.$$

In fact

$$w' \cdot (p \cdot s_{(1)}) = (p' \cdot v') \cdot s_{(1)} = p' \cdot \tilde{s}' \cdot v' = r'_{(1)} \cdot v'$$

and analogously $w'' \cdot (p \cdot s_{(1)}) = r''_{(1)} \cdot v''$—this proves the first equation. Further,

$$w' \cdot (p \cdot s_{(2)}) = p' \cdot \tilde{v}' \cdot s_{(2)} = r'_{(2)}$$

and analogously, $w'' \cdot (p \cdot s_{(2)}) = r''_{(2)}$—this proves the latter. Consequently,

$$r = [r'_{(1)} + r''_{(1)}, \tilde{r}] = [p \cdot s_{(1)}, p \cdot s_{(2)}] \subset [s_{(1)}, s_{(2)}] = s. \qquad \square$$

Remark. The concept of components can be generalized to other types of colimits. For example let

$$W: \alpha \to \mathscr{K}$$

be an α-chain with a colimit $w_i: W_i \to A$ $(i \in \alpha)$. A collection of relations

$$r^i: W_i \rightharpoonup B \quad (i \in \alpha)$$

is said to be *compatible* if

$$i \leq j \quad \text{implies} \quad r^i \subset w_{i,j} \circ r^j \quad (i, j \in \alpha).$$

We say that $r: A \rightharpoonup B$ *has components* r^i if r is the least relation with

$$r^i \subset w_i \circ r \quad (i \in \alpha).$$

We can express r as

$$r = [\text{colim}\ r^i_{(1)}, \tilde{r}]$$

where $\tilde{r}$ is the morphism with components $r^i_{(2)}$:

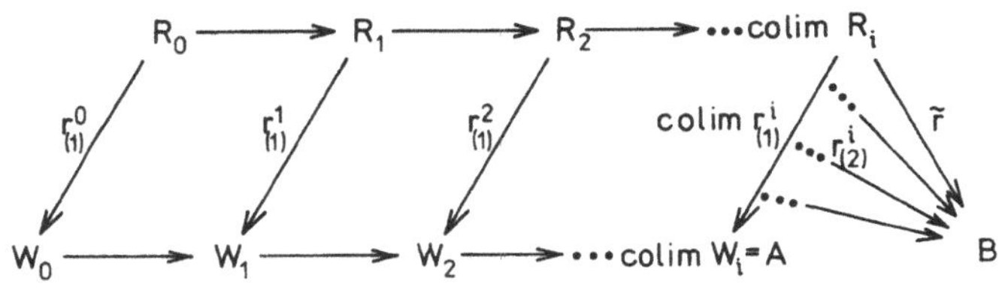

The proof is analogous to the preceding one.

Construction. Let F be a constructive varietor in a cocomplete, finitely complete and $\mathscr{M}$-well-powered $(\mathscr{E}, \mathscr{M})$-category.

For each relational algebra (Q, δ) and each relation $\lambda: I \rightharpoonup Q$, we define a compatible collection of relations

$$\lambda^{(n)}: W_n \rightharpoonup Q \quad (n \in \mathbf{Ord})$$

by the following induction.

(a) $\lambda^{(0)} = \lambda: I \rightharpoonup Q$;

(b) given $\lambda^{(n)}$, then $\lambda^{(n+1)}: I + W_n F \rightharpoonup Q$ has components λ and $\lambda^{(n)}F \circ \delta$;

(c) given a limit ordinal i and $\lambda^{(n)}$ for each $n < i$, then $\lambda^{(i)}$: colim $W_n \rightharpoonup Q$ has components $\lambda^{(n)}$ for $n < i$.

We must verify that the relations $\lambda^{(n)}$ $(n \in \mathbf{Ord})$ form a compatible family and hence, $\lambda^{(i)}$ is well-defined: to prove

$$\lambda^{(n)} \subset w_{n,m} \circ \lambda^{(m)} \quad (n < m)$$

we proceed by induction as in Remark IV.2.4:

(a) $\lambda^{(0)} \subset w_{0,1} \circ \lambda^{(1)}$ because $\lambda^{(0)} = \lambda$ is the first component of $\lambda^{(1)}$:

(b_1) if $\lambda^{(n)} \subset w_{n,m} \circ \lambda^{(m)}$, then $\lambda^{(n)}F \subset w_{n,m}F \circ \lambda^{(m)}F$; hence $\lambda^{(n)}F \circ \delta \subset w_{n,m}F \circ \lambda^{(m)}F \circ \delta$ and thus, $\lambda^{(n+1)}$ (with components λ and $\lambda^{(n)}F \circ \delta$) is contained in $w_{n+1,m+1} \circ \lambda^{(m+1)}$ (with components λ and $w_{n,m}F \circ \lambda^{(m)}F \circ \delta$);

(b_2) if $\lambda^{(n)} \subset w_{n,m} \circ \lambda^{(m)}$ for all $n < n_0$, where n_0 is a limit ordinal, then $w_{n_0,m} \circ \lambda^{(m)}$ is a relation with $\lambda^{(n)} \subset w_{n,n_0} \circ [w_{n_0,m} \circ \lambda^{(m)}]$ (see Corollary VII.1.4) and hence, $w_{n_0,m} \circ \lambda^{(m)}$ contains $\lambda^{(n_0)}$.

(c) if $\lambda^{(n)} \subset w_{n,m} \circ \lambda^{(m)}$ for all $n < m < i$, where i is a limit ordinal, then we have

$$\lambda^{(n)} \subset w_{n,i} \circ \lambda^{(i)} \quad (n < i)$$

by definition of $\lambda^{(i)}$.

1.8. Proposition. There exists and ordinal k with $\lambda^{\#} = \lambda^{(k)}$.

Remark. Since F is a constructive varietor, there exists an ordinal n with $I^{\#} = W_n$. It follows that for each *morphism* λ we have $\lambda^{\#} = \lambda^{(n)}$. Nevertheless, given a relation λ, the ordinal k with $\lambda^{\#} = \lambda^{(k)}$ can be much larger than n. In Exercise VII.1.E below we show a case with $n = 1$ and k arbitrarily large.

Proof. Let n be an ordinal for which the free-algebra construction stops, i.e., such that $w_{n,k}$ is an isomorphism for each $k \geq n$ (see IV.2.5). The relations

$$w_{n,k} \circ \lambda^{(k)}: W_n \rightharpoonup Q \quad (k \geq n)$$

form an increasing chain of subobjects of $W_n \times Q$. Since $\mathscr{K}$ is $\mathscr{M}$-well-powered, there exists an ordinal $k \geq n$ with

$$w_{n,k} \circ \lambda^{(k)} = w_{n,k+1} \circ \lambda^{(k+1)}.$$

We shall prove that then

$$\lambda^{\#} = \lambda^{(k)}.$$

Here, $I^{\#} = W_k$ (because $k \geq n$) with $\varphi = \varphi_k \cdot w_{k,k+1}^{-1}: W_kF \to W_k$, where $\varphi_k: W_kF \to I + W_kF$ is the coproduct injection. Since $\lambda^{(k)}F \circ \delta$ is the second component of $\lambda^{(k+1)}$, we have

$$\lambda^{(k)}F \circ \delta \subset \varphi_k \circ \lambda^{(k+1)}.$$

Further, by Corollary VII.1.4,

$$w_{n,k} \circ \lambda^{(k)} = w_{n,k+1} \circ \lambda^{(k+1)} = w_{n,k} \circ (w_{k,k+1} \circ \lambda^{(k+1)}),$$

which implies

$$\lambda^{(k)} = w_{k,k+1} \circ \lambda^{(k+1)}$$

because $w_{n,k}$ is an isomorphism. This proves that $\lambda^{(k)}: (W_k, \varphi) \to (Q, \delta)$ is a state relation:

$$\lambda^{(k)}F \circ \delta \subset \varphi_k \circ \lambda^{(k+1)} = \varphi_k \circ w_{k,k+1}^{-1} \circ \lambda^{(k)} = \varphi \circ \lambda^{(k)}.$$

Furthermore, $\lambda^{(k)}$ extends λ because

$$\lambda = \lambda^{(0)} \subset w_{0,k} \circ \lambda^{(k)} = \eta \circ \lambda^{(k)}.$$

Finally, let $f: (W_k, \varphi) \to (Q, \delta)$ be a state relation extending λ; we prove that $\lambda^{(k)} \subset f$. We verify by induction on n that

$$\lambda^{(n)} \subset w_{n,k} \circ f.$$

(i) $n = 0$: since f extends λ, we have $\lambda \subset w_{0,k} \circ f$.

(ii) Assume $\lambda^{(n)} \subset w_{n,k} \circ f$. Recall that $\eta_n = w_{0,n+1}$ and φ_n are the coproduct injections of $W_{n+1} = I + W_nF$ [Remark IV.3.2 (ii)]. We prove that $w_{n+1,k} \circ f$ has the following property:

$$\lambda \subset \eta_n \circ (w_{n+1,k} \circ f) \quad \text{and} \quad \lambda^{(n)}F \circ \delta \subset \varphi_n \circ (w_{n+1,k} \circ f).$$

Since $\lambda^{(n+1)}$ is the least relation with this property, this will prove that $\lambda^{(n+1)} \subset w_{n+1,k} \circ f$. We use Corollary VII.1.4 for the first inequality:

$$\lambda \subset \eta \circ f = \eta_k \circ f = (\eta_{n+1} \cdot w_{n+1,k}) \circ f.$$

For the latter, we apply the induction hypothesis and the fact f is a state relation, and we use Lemma VI.1.4:

$$\begin{aligned}\lambda^{(n)}F \circ \delta &\subset (w_{n,k} \circ f)F \circ \delta \\ &\subset (w_{n,k}F \circ fF) \circ \delta \\ &\subset w_{n,k}F \circ (fF \circ \delta) \\ &\subset w_{n,k}F \circ (\varphi \circ f).\end{aligned}$$

Next, $\varphi = \varphi_k \cdot w_{k,k+1}^{-1}$, and since $w_{n+1,k+1} = 1_I + w_{n,k}F$, clearly

$$w_{n,k}F \cdot \varphi_k = \varphi_n \cdot w_{n+1,k+1}.$$

Hence, applying Corollary VII.1.4, we get

$$\begin{aligned} w_{n,k}F \circ (\varphi \circ f) &= (w_{n,k}F \cdot \varphi) \circ f \\ &= (\varphi_n \cdot w_{n+1,k+1} \cdot w_{k,k+1}^{-1}) \circ f \\ &= (\varphi_n \cdot w_{n+1,k}) \circ f \\ &= \varphi_n \circ (w_{n+1,k} \circ f). \end{aligned}$$

(iii) Let i be a limit ordinal with $\lambda^{(n)} \subset w_{n,k} \circ f$ for each $n < i$. Then using Corollary VII.1.4, we have

$$\lambda^{(n)} \subset w_{n,k} \circ f = w_{n,i} \circ (w_{i,k}f) \quad (n < i).$$

Since $\lambda^{(i)}$ is the least relation with $\lambda^{(n)} \subset w_{n,i} \circ \lambda^{(i)}$, it follows that $\lambda^{(i)} \subset w_{i,k} \circ f$. This concludes the proof. □

1.9. For partial automata, it is important that behavior is also a partial morphism. This we are able to prove for constructive varietors preserving monos of a constructive class (see IV.4.2):

Proposition. Let $\mathscr{K}$ be a cocomplete, finitely complete and $\mathscr{M}$-well-powered $(\mathscr{E}, \mathscr{M})$-category with $\mathscr{M}$ a constructive class. Let F be a varietor preserving $\mathscr{M}$-monos. Then the run relation of each partial F-automaton is a partial morphism.

Proof. By IV.4.2, the varietor F is constructive. Hence, the run relation ρ is $\lambda^{(k)}$ for some ordinal k (by the preceding proposition). We prove that $\lambda^{(n)}$ is a partial morphism by induction in n.

(i) $\lambda = \lambda^{(0)}$ is a partial morphism in each partial automaton.

(ii) Let $\lambda^{(n)}$ be a partial morphism. Since F preserves $\mathscr{M}$-monos, also $\lambda^{(n)}F$ is a partial morphism (V.2.9), and since δ is a partial morphism, we conclude that $\lambda^{(n)}F \circ \delta$ is a partial morphism too (V.2.2). By Lemma VII.1.7, we have

$$\lambda^{(n+1)} = [1_I + (\lambda^{(n)}F \circ \delta)_{(1)}, \bar{r}].$$

We know that $(\lambda^{(n)}F \circ \delta)_{(1)} \in \mathscr{M}$ and since $\mathscr{M}$ is constructive, we conclude that $1_I + (\lambda^{(n)}H \circ \delta)_{(1)} \in \mathscr{M}$.

(iii) Let i be a limit ordinal such that each $\lambda^{(n)}$, $n < i$, is a partial morphism. Then $(\lambda^{i})_{(1)} = \operatorname{colim}(\lambda^{(n)})_{(1)}$, by Remark VII.1.7. Since all the morphisms $(\lambda^{(n)})_{(1)}$ are in $\mathscr{M}$ and $\mathscr{M}$ is constructive, $(\lambda^{(i)})_{(1)}$ is also in $\mathscr{M}$, i.e., $\lambda^{(i)}$ is a partial morphism. □

1.10. Proposition. In the category **Set**, each varietor has the property that given a relational algebra (Q, δ) and a relation $\lambda: I \rightharpoonup Q$, then

$$(*) \qquad \lambda = \eta \circ \lambda^{\#} \quad \text{and} \quad \lambda^{\#}F \circ \delta = \varphi \circ \lambda^{\#}.$$

Proof. It is easy to verify that for arbitrary relations $r_1 : A_1 \rightharpoonup B$ and $r_2 : A_2 \rightharpoonup B$ in **Set**, the least relation $r: A_1 + A_2 \rightharpoonup B$ with $r_1 \subset v_1 \circ r$ and $r_2 \subset v_2 \circ r$ actually fulfils

$$r_1 = v_1 \circ r \quad \text{and} \quad r_2 = v_2 \circ r.$$

Since each varietor in **Set** is constructive (IV.4.3), we have

$$\lambda^{\#} = \lambda^{(k)}$$

for some ordinal k with

$$\lambda^{(k+1)} = w_{k,k+1}^{-1} \circ \lambda^{(k)}.$$

It is easy to verify by induction that since η_n is a coproduct injection, we have

$$\eta_n \circ \lambda^{(n)} = \lambda \quad \text{for each ordinal } n.$$

Hence,

$$\eta \circ \lambda^{\#} = \lambda.$$

And since φ_k is the other coproduct injection, we get

$$\begin{aligned} \varphi \circ \lambda^{\#} &= (\varphi_k \cdot w_{k,k+1}^{-1}) \circ \lambda^{(k)} \\ &= \varphi_k \circ \lambda^{(k+1)} \\ &= \lambda^{(k)}F \circ \delta. \end{aligned}$$

□

Remark. Even in **Set** we cannot *define* $\lambda^{\#}$ by the equations (∗) because such $\lambda^{\#}$ need not be unique (even if λ and δ are partial maps and F is a finitary varietor, see Exercise VII.1.A below). For partial maps λ and δ, we prove that (∗) defines $\lambda^{\#}$ at least among partial maps:

1.11. Proposition. Let F be a standard varietor in **Set**. For each partial F-algebra (Q, δ) and each partial map $\lambda: I \rightharpoonup Q$ there exists a unique partial map $\lambda^{\#}$ satisfying (∗) above.

Proof. I. F preserves the compositions of a mono $m: X \to Y$ followed by a partial map $f: Y \rightharpoonup Z$. In fact, the composition is given by the following pullback

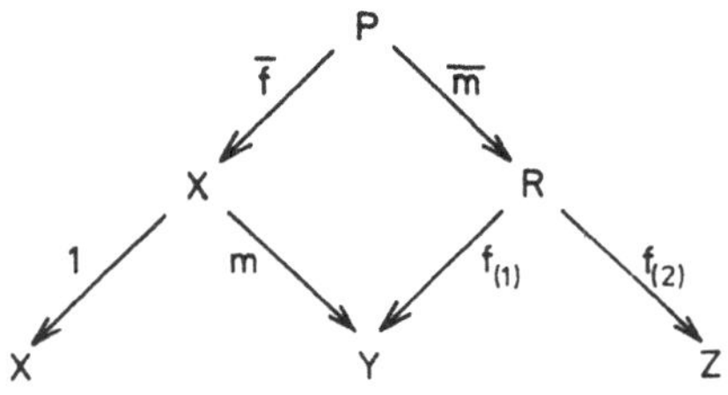

as

$$m \circ f = [\bar{f}, \bar{m} \cdot f_{(2)}].$$

Since both m nad $f_{(1)}$ are monos, this pullback is an intersection and hence, F preserves the pullback by III.4.6. It follows that

$$(m \circ f)F = [\bar{f}F, \bar{m}F \cdot f_{(2)}F] = mF \circ fF.$$

II. Let (Q, δ) be a partial F-algebra, and let $\lambda : I \rightharpoonup Q$ be a partial map. For each partial map $h : I^{\#} \rightharpoonup Q$ satisfying

$$\lambda = \eta \circ h \quad \text{and} \quad \varphi \circ h = hF \circ \delta,$$

we prove that $h = \lambda^{\#}$. Let k be an ordinal with

$$\lambda^{\#} = \lambda^{(k)}$$

(see VII.1.8). We prove that

$$\lambda^{(n)} = w_{n,k} \circ h \quad \text{for each } n \leq k$$

by induction in n.

(a) $\lambda^{(0)} = \lambda = \eta \circ h = w_{0,k} \circ h$.
(b) Let $\lambda^{(n)} = w_{n,k} \circ h$. By I. above,

$$\lambda^{(n)}F = w_{n,k}F \circ hF.$$

We prove that the components of $w_{n+1,k} \circ h$ are the same as those of $\lambda^{(n+1)}$, viz. λ and $\lambda^{(n)}F \circ \delta$.

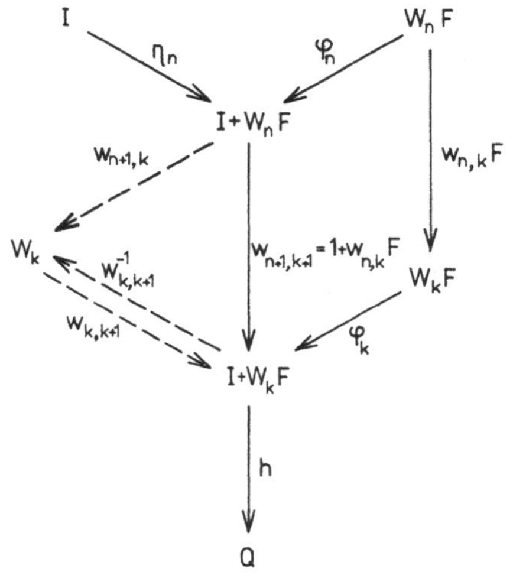

First, since $\eta_n = w_{0,n+1}$ and $\eta = \eta_k = w_{0,k+1}$, we have

$$\eta_n \circ (w_{n+1,k} \circ h) = \eta \circ h = \lambda.$$

Second, since $\varphi \circ h = hF \circ \delta$ and $\varphi = \varphi_k \cdot w_{k,k+1}^{-1}$, we have

$$\begin{aligned}\varphi_n \circ (w_{n+1,k} \circ h) &= (\varphi_n \cdot w_{n+1,k+1} \cdot w_{k,k+1}^{-1}) \circ h \\ &= (w_{n,k}F \cdot \varphi_k \cdot w_{k,k+1}^{-1}) \circ h \\ &= (w_{n,k}F \cdot \varphi) \circ h \\ &= (w_{n,k}F \circ hF) \circ \delta.\end{aligned}$$

By I. again,

$$\varphi_n \circ (w_{n+1,k} \circ h) = (w_{n,k} \circ h)F \circ \delta = \lambda^{(n)}F \circ \delta.$$

(c) The limit step is clear: $\lambda^{(i)} = w_{i,k} \circ h$ because the components are $\lambda^{(n)} = w_{n,i} \circ (w_{i,k} \circ h)$ by induction hypothesis. □

Remark. For relations λ and δ, the solution of (∗) is unique provided that the varietor F preserves preimages (V.2.10). The proof is the same: in part I we prove that F preserves the composition of a mono m and a relation f, since the pullback defining $m \circ f$ is a preimage.

Exercises VII.1

A. Free extensions in Set. Define a set functor F as follows:

$$XF = \{Z \subset X;\ \text{card } Z = 2 \text{ or } Z = \emptyset\};$$
$$fF: Z \mapsto (Z)f \quad \text{if card}(Z)f = 2, \quad \text{else, } Z \mapsto \emptyset.$$

(i) Verify that F is a finitary varietor.

(ii) Define a (deterministic!) automaton

$$A = (\{1, 2\}, \delta, \{1, 2\}, \gamma, \{1\}, \lambda)$$

as follows:

$$(\{1, 2\})\delta = 2 \quad \text{and} \quad (\emptyset)\delta = 1;$$
$$(i)\gamma = 1;$$
$$(1)\lambda = 1.$$

Verify that the run map $\rho: \{1\}^{\#} \to \{1, 2\}$ is the constant map to $\{1\}$.

(iii) Prove that the relation $r: \{1\}^{\#} \rightharpoonup \{1, 2\}$ defined by $(1)r = 1$, $(x)r = \{1, 2\}$ if $x \neq 1$ also fulfils

$$\varphi \circ r = fF \circ \delta \quad \text{and} \quad \eta \circ \rho = \lambda,$$

although $r \neq \rho$ by (ii).

[Hint: The relation r is represented by the projections of $R = \{(1, 1)\} \cup \{(x, i);\ x \in \{1\}^{\#} - \{1\} \text{ and } i = 1, 2\}$. To check $(\emptyset)\varphi \circ r = (\emptyset)rF \circ \delta$, use $z = \{(x, 1), (x, 2)\} \in RF$ to get $\{1, 2\} = (z)r_{(2)}F \in (\emptyset)rF$ and hence, $(\emptyset)rF \circ \delta = \{\emptyset, \{1, 2\}\}\delta = (\emptyset)\varphi \circ r$, etc.]

B. Free extensions in R-Vect. Let F be the identity functor of R-**Vect**.

(i) Define a relational algebra (R, δ) by $\delta = R \times R$. Prove that for $\lambda = 1_R: R \to R$, we have

$$\lambda \neq \eta \circ \lambda^{\#}.$$

(ii) Conclude that no relation $r: R^{\#} \rightharpoonup$ R fulfils

$$\lambda = \eta \circ r \quad \text{and} \quad \varphi \circ r = rF \circ \delta.$$

(iii) Define a partial algebra $(R \times R, \delta)$ as follows:

$$(x, 0)\delta = (x, 0) \quad \text{and else undefined.}$$

Prove that for $\lambda = \delta: R^2 \rightharpoonup R^2$ there exist two distinct partial morphisms r_1, $r_2: (R^2)^{\#} \rightharpoonup R^2$ with

$$\lambda = \eta \circ r_i \quad \text{and} \quad \varphi \circ r_i = r_i F \circ \delta \quad (i = 1, 2).$$

[Hint: $(R^2)^{\#} = R^2[z]$ and we put for each $(z)w = (a_0, b_0) + (a_1, b_1)z + \dots + (a_n, b_n)z^n$:

$$(w)r_1 = \left(\sum_{k=0}^{n} a_k, 0\right) \quad \text{if } b_0 = \dots = b_n = 0, \quad \text{else undefined;}$$

$$(w)r_2 = \left(\sum_{k=0}^{n} a_k, 0\right) \quad \text{if } 0 = b_0 + \dots + b_n, \quad \text{else undefined.]}$$

C. Categories of relations. Let $\mathscr{K}$ be an $(\mathscr{E}, \mathscr{M})$-category for which relations form a category **Rel** $\mathscr{K}$ (see Remark VII.1.2). Then $\mathscr{K}$ is a subcategory of **Rel** $\mathscr{K}$.

(i) Prove that **Set** is closed in **Rel Set** under colimits (Hint: See Lemma VII.1.7. Let $D: \mathscr{D} \to$ **Set** be a diagram. For each compatible family of relations $r_d: dD \rightharpoonup X$, consider the corresponding family of maps $\tilde{r}_d: dD \to \exp X$.)

(ii) Prove that R-**Vect** is not closed in **Rel** R-**Vect** under finite coproducts: there even exists a partial map $p: A \rightharpoonup B$ and two partial maps r_1, r_2: $A + A \rightharpoonup B$ with $p = v_i \circ r_j$ for $i, j = 1, 2$ (where v_1, v_2 are the injections). (Hint: Define $p: R^2 \rightharpoonup R^2$ by $(x, y)p = (x, y)$ if $y = 0$, and else undefined; define r_1, r_2: $R^4 \rightharpoonup R^2$ by

$$(x, 0, y, 0)r_1 = (x, y) \quad \text{and} \quad (x, y, z, -y)r_2 = (x, z),$$

else undefined.)

(iii) Let $\mathscr{K}$ be a concrete category with limits and colimits preserved by the forgetful functor U. Prove that $\mathscr{K}$ is closed under colimits in **Rel** $\mathscr{K}$.

(iv) Prove that if $\mathscr{K}$ is closed under colimits in **Rel** $\mathscr{K}$, then for every constructive varietor $F: \mathscr{K} \to \mathscr{K}$, any relational algebra (Q, δ) and any relation $\lambda: I \rightharpoonup Q$, the free extension $\lambda^{\#}: (I^{\#}, \varphi) \rightharpoonup (Q, \delta)$ fulfils

(*) $\quad \eta \circ \lambda^{\#} = \lambda$ and $\varphi \circ \lambda^{\#} = \lambda^{\#} F \circ \delta$.

If, moreover, $\mathcal{M}$ is constructive and F preserves $\mathcal{M}$-monos and preimages, then $\lambda^{\#}$ is the unique relation which fulfils the equations (*). (Hint: See the proof of Proposition VII.1.10 and Remark VII.1.11.)

D. Partial algebras. (i) For each varietor F: R-**Vect** → R-**Vect** and each partial F-algebra (Q, δ) verify that the free extension of each partial map $\lambda: I \rightharpoonup Q$ fulfils (*) above. (Hint: Use the fact that F is a constructive varietor.)

(ii) Can (*) be used to define $\lambda^{\#}$? (Hint: Exercise VII.1.B.)

(iii) Under what conditions on a category $\mathcal{K}$ is it true that for each constructive varietor $F: \mathcal{K} \to \mathcal{K}$ the equations (*) hold for all partial morphisms δ and λ?

E. Steps required by $\lambda^{\#}$. (i) Denote by G the set functor defined on sets X by

$$XG = \{M \subset X;\ M \text{ is uncountable}\} \cup \{\emptyset\}$$

and on maps $f: X \to Y$ by

$$(M)fG = (M)f \quad \text{if} \quad \text{card } M = \text{card}(M)f; \quad (M)fG = \emptyset \text{ else.}$$

Verify that G is a constructive varietor, and for each countable set I, the free-algebra construction stops after 1 step with

$$I^{\#} = I + \{\emptyset\}.$$

(ii) Describe the functor G acting on relations. In particular, prove that for each relation $f: A \rightharpoonup B$, the set $(\emptyset)fG$ contains any uncountable set $M \subset B$ with $M = (a)f$ for some $a \in A$. (Hint: Representing f by projections of a set $R_f \subset A \times B$, consider $\{a\} \times M \in R_f G$.)

(iii) Define a relational G-algebra (Q, δ) as follows:

$$Q = \{x\} \cup (M \times (\omega + 1)),$$

where M is an uncountable set and $\omega + 1 = \{0, 1, 2, \ldots\} \cup \{\omega\}$, and δ is defined by

$$\begin{aligned} (\emptyset)\delta &= M \times \{0\}, \\ (M \times \{0\})\delta &= M \times \{0, 1\}, \\ (M \times \{0, 1\})\delta &= M \times \{0, 1, 2\}, \\ &\ldots \\ (M \times \omega)\delta &= M \times (\omega + 1) \end{aligned}$$

and $(Z)\delta = \emptyset$ for each remaining $Z \in QG$.

Prove that for the inclusion map $\lambda: \{x\} \to Q$, we have

$$\lambda^{\#} = \lambda^{(\omega+1)} \quad \text{but} \quad \lambda^{\#} \neq \lambda^{(\omega)}.$$

[Hint: For $* = (\emptyset)\varphi \in \{x\}^{\#}$, we have $\lambda^{(1)} = (\emptyset)\lambda^{(0)}F \circ \delta = M \times \{0\}$ and by (ii), $(*)\lambda^{(2)} = (\emptyset)\lambda^{(1)}F \circ \delta = M \times \{0, 1\}$, etc., thus $(*)\lambda^{(\omega)} = M \times \omega$ but, by (ii) again $(*)\lambda^{(\omega+1)} = M \times (\omega + 1)$.]

(iv) For each ordinal k find a relational G-algebra and a map λ with

$$\lambda^{\#} = \lambda^{(k+1)} \quad \text{but} \quad \lambda^{\#} \neq \lambda^{(k)}.$$

VII.2. Nondeterministic Languages in Set

2.1. We are going to characterize varietors in the category of sets for which languages recognized by nondeterministic acceptors coincide with those recognized by (deterministic) acceptors. As in case of tree automata (see II.4.2) we define an *F-acceptor* as a quadruple

$$A = (Q, \delta, T, I)$$

where (Q, δ) is a finite F-algebra and T and I are subsets of Q (of terminal and initial states, respectively). The inclusion map $\lambda: I \to Q$ is extended to the run map $\rho: (I^{\#}, \varphi) \to (Q, \delta)$. Then the *language recognized* by A is the set

$$L_A = \{w \in I^{\#};\, (w)\rho \in T\} = (T)\rho^{-1}.$$

In general, sets

$$L \subset I^{\#} \quad (I \text{ a finite set})$$

are called *languages*; a language is *recognizable* if it is recognized by an F-acceptor.

Again in an analogy to II.4.5, a *nondeterministic F-acceptor* is a quintuple

$$A = (Q, \delta, T, I, \lambda)$$

where (Q, δ) is a finite relational F-algebra, $T \subset Q$ is the set of terminal states and $\lambda: I \rightharpoonup Q$ is a relation. We extend λ freely to the run relation $\rho: (I^{\#}, \varphi) \rightharpoonup (Q, \delta)$ (VII.1.3) and the *language recognized* by A is the set

$$L_A = \{w \in I^{\#};\, (w)\rho \cap T \neq \emptyset\}.$$

If both δ and λ are partial maps, we call A a *partial F-acceptor*.

We say that a language is

N-recognizable

if it is recognized by a nondeterministic F-acceptor; and

P-recognizable

if it is recognized by a partial F-acceptor.

For tree automata we know that the classes of recognizable, P-recognizable and N-recognizable languages coincide (II.4.6). This, as we prove below, is caused by the fact that the functors H_Σ cover pullbacks.

Remark. A deterministic F-acceptor is a special case of a deterministic F-automaton with $\Gamma = \{0, 1\}$ as far as (i) the set $T \subset Q$ is expressed by the characteristic function $\gamma: Q \to \{0, 1\}$ [with $(q)\gamma = 1$ iff $q \in T$] and (ii) $\lambda: I \to Q$ is the inclusion map of $I \subset Q$.

For the nondeterministic acceptors, we have λ arbitrary (as in the case of tree automata in Chapter II) in order to allow a nondeterministic interpretation of variables. The special choice of λ for deterministic automata does not loose generality because the following proposition can be proved for each varietor F: **Set** $\to$ **Set** analogously as II.4.4.:

Proposition. A behavior

$$\beta: I^\# \to \Gamma \quad (I \text{ finite})$$

has a realization by a finite F-automaton iff

(1) the language $(y)\beta^{-1}$ is recognizable for each $y \in \Gamma$;

(2) the set $(I^\#)\beta \subset \Gamma$ is finite.

2.2. The main result of the present section is that for a certain class of "sufficiently small" set functors F

(i) each N-recognizable language is recognizable iff F covers pullbacks (V.2.10);

(ii) each P-recognizable language is recognizable iff F preserves preimages (V.2.10).

The sufficiency of these conditions is easy, and we present the proof now (without restrictions on F, in fact). The necessity is quite difficult to prove, and we devote the rest of the present section to this proof.

Proposition. Let F be a varietor in **Set** which covers pullbacks. Then each N-recognizable language is recognizable.

Proof. For each set X denote by

$$\varepsilon: \exp X \rightharpoonup X$$

the "membership relation", i.e., the set $\varepsilon \subset [\exp X] \times X$ of all pairs (A, a), where $A \subset X$ and $a \in A$. Then for each relation $f: P \rightharpoonup X$ we have

$$f = \tilde{f} \circ \varepsilon,$$

where $\tilde{f}: P \to \exp X$ denotes the corresponding map (I.3.4). Since F preserves the composition of relations (V.2.10), it follows that

$$fF = \tilde{f}F \circ \varepsilon F.$$

Let $A = (Q, \delta, T, I, \lambda)$ be a nondeterministic F-acceptor. Define a deterministic F-automaton

$$\tilde{A} = (\exp Q, \bar{\delta}, \{0, 1\}, \gamma, I, \tilde{\lambda})$$

as follows:

$$\bar{\delta} = \bar{\Delta} : (\exp Q)F \to \exp Q, \text{ where } \Delta = \varepsilon F \cdot \delta$$
$$\gamma : \exp Q \to \{0, 1\}; \quad (M)\gamma = 1 \quad \text{iff} \quad M \cap T \neq \emptyset$$

and $\tilde{\lambda}: I \to \exp Q$ corresponds to the relation λ. We prove that A and $\tilde{A}$ recognize the same languages (which will conclude the proof by Remark VII.2.1).

If $\rho: I^{\#} \rightharpoonup Q$ is the run relation of A, then we prove that $\tilde{\rho}: I^{\#} \to \exp Q$ is the run map of $\tilde{A}$. The following diagram

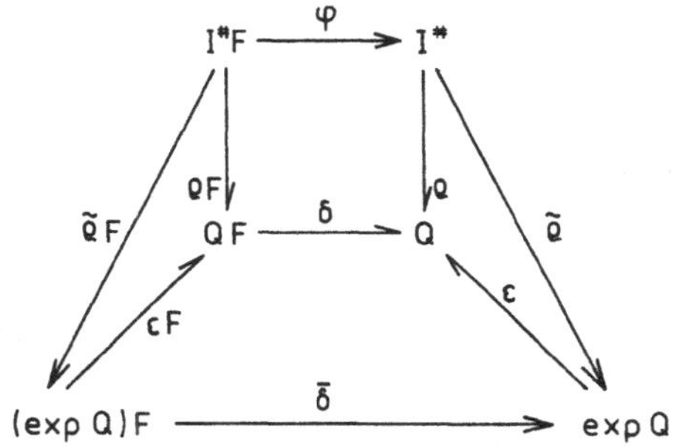

commutes (VII.1.10). Therefore,

$$\tilde{\rho}: (I^{\#}, \varphi) \to (\exp Q, \bar{\delta})$$

is a homomorphism. Furthermore,

$$\tilde{\lambda} = \eta \circ \tilde{\rho}.$$

The language recognized by $\tilde{A}$ is

$$\{w \in I^{\#}; (w)\tilde{\rho} \cdot \gamma = 1\} \quad = \quad \{w \in I^{\#}; (w)\tilde{\rho} \cap T \neq \emptyset\},$$

and this is just the language recognized by A. □

Proposition. Let F be a varietor in **Set** which preserves preimages. Then each P-recognizable language is recognizable.

Remark. For partial maps, we follow the usual conventions: we write $(x)f = y$ [not $(x)f = \{y\}$] and we say that $(x)f$ is undefined if $(x)f = \emptyset$.

Proof. For each set X denote by

$$\varepsilon: X \cup \{a\} \rightharpoonup X$$

the partial map with $(a)\varepsilon$ undefined (a is any element not in X) and $(x)\varepsilon = x$ for each $x \in X$. Then for each partial map $f\colon P \rightharpoonup X$ we have

$$f = \bar{f} \circ \varepsilon$$

where $\bar{f}\colon P \to X \cup \{a\}$ is the extension of f with $(p)\bar{f} = a$ iff $(p)f$ is undefined. Since F preserves the composition of partial maps (V.2.10), it follows that

$$fF = \bar{f}F \circ \varepsilon F.$$

The rest of the proof is completely analogous to that of the preceding proposition. □

2.3. Example: A non-recognizable language which is *P*-recognizable. Let D_2 be the functor defined on sets X by $XD_2 = \{(x, y);\ x, y \in X,\ x \neq y\} \cup \{*_X\}$ and on maps $f\colon X \to X'$ by $(x, y)fD_2 = ((x)f, (y)f)$ if $(x)f \neq (y)f$ and $(x, y)fD_2 = *_{X'} = (*_X)fD_2$ if $(x)f = (y)f$ (see V.2.10). Consider the free D_2-algebra on one generator $\{x\}^{\#}$. Denoting by $*$ the element $(*_{\{x\}^{\#}})\varphi \in \{x\}^{\#}$, we can describe the algebra $\{x\}^{\#}$ by the following trees:

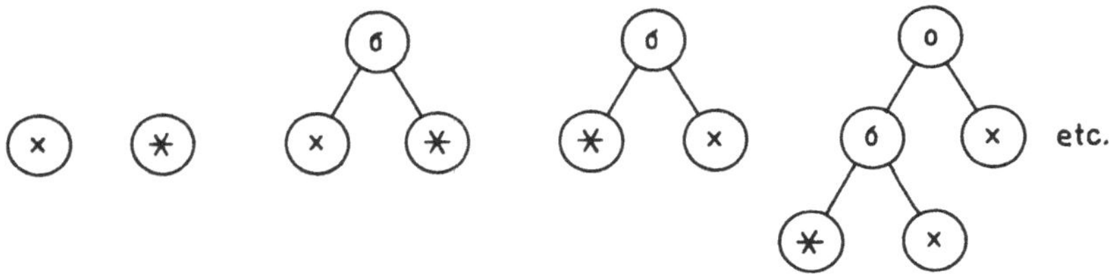

The singleton language

$$\{*\} \subset \{x\}^{\#}$$

is *P*-recognizable. For example, the following partial D_2-acceptor recognizes it:

$$A_0 = (Q, \delta_0, \{y\}, \{x\}, \lambda_0)$$

where $Q = \{x, y\}$ $(x \neq y)$, $(*_Q)\delta_0 = y$ and $(x, y)\delta_0$, $(y, x)\delta_0$ are undefined, and $(x)\lambda_0 = x$. In fact, the run relation is the following partial map $\rho\colon \{x\}^{\#} \rightharpoonup \{x, y\}$:

$$(x)\rho = x, \quad (*) = y, \quad \text{else undefined.}$$

This language $\{*\}$ is not recognizable, however. For each deterministic D_2-acceptor

$$A = (Q, \delta, T, \{x\})$$

with $* \in L_A$, we prove that L_A is infinite. Since Q is finite, there exists an infinite set $\tilde{L} \subset \{x\}^{\#}$ on which the run map ρ is constant. Since $* \in L_A$, we have

$$(*)\rho = (*_{\{x\}^{\#}})\varphi \cdot \rho = (*_{\{x\}^{\#}})\rho D_2 \cdot \delta = (*_Q)\delta \in T.$$

Then for arbitrary distinct $v, w \in \tilde{L}$, the following tree

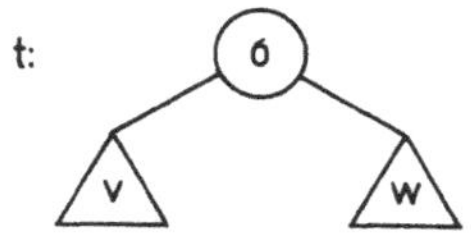

is in L_A because

$$(t)\rho = (v, w)\varphi \cdot \rho = (v, w)\rho D_2 \cdot \delta = (*_Q)\delta \in T.$$

2.4. For the rest of the present section we restrict ourselves to super-finitary set functors. Recall the concept of standard functor (III.4.5).

Definition. A standard set functor F is said to be *super-finitary* if it preserves finite sets, and there exists a natural number n with

$$XF = \bigcup_{T \subset X, \text{ card } T \leq n} TF$$

for each set X.

Examples. (i) The functor H_Σ is super-finitary iff the type Σ is super-finitary, i.e., consists of finitely many finitary operations. In particular, S_Σ is super-finitary iff Σ is finite.

(ii) The functor P_f (III.3.3) assigning to each set X the set XP_f of all finite subsets of X is not super-finitary (though it is finitary and preserves finite sets). For each natural number n and each set X of power $n + 1$ clearly

$$X \in XP_f - \bigcup_{T \subset X, \text{ card } T \leq n} TP_f.$$

Convention. For each natural number n we put

$$[n] = \{1, 2, \ldots, n\}$$

(in contrast to n considered as an ordinal, i.e., $n = \{0, 1, \ldots, n - 1\}$). This will simplify notation below. Elements of X^n can be identified with maps from $[n]$ to X. Given a type Σ, an element $\sigma \in \Sigma_n$ and a set X, we write

$$f \in X^n = XH_\sigma$$

to indicate that $f \in XH_\Sigma$ is an element in the σ-summand X^n of XH_Σ.

Remark. For each super-finitary functor F there exists a super-finitary type

Σ and an epitransformation $\varepsilon\colon H_\Sigma \to F$. The proof is analogous to that of Proposition III.4.3. We call ε a *presentation* of F.

A presentation is often given in the form of equations

$$(x_1, \ldots, x_n)\sigma = (y_1, \ldots, y_m)\tau$$

where $\sigma \in \Sigma_n$, $\tau \in \Sigma_m$ and x_i, y_i are elements of some set X (of variables). It is then understood that ε is the least congruence on H_Σ such that the element $(x_1, \ldots, x_n) \in X^n = XH_\sigma$ is congruent with the element $(y_1, \ldots, y_m) \in X^m = XH_\tau$. For example, the functor D_2 can be presented as a quotient of $H_2 + H_0$ by the following equation

$$(x, x)\sigma = 0$$

(where σ denotes the binary operation and 0 the nullary operation).

2.5. Each presentation

$$\varepsilon\colon H_\Sigma \to F$$

of a super finitary functor can be "minimized" as follows. First, choose a set $\bar{\Sigma} \subset \Sigma$ such that

(i) for each $\sigma \in \Sigma$ there exists $\tau \in \bar{\Sigma}$ with $(H_\sigma)\varepsilon \subset (H_\tau)\varepsilon$ and

(ii) if $\sigma, \tau \in \bar{\Sigma}$, $\sigma \neq \tau$, then neither $(H_\sigma)\varepsilon \subset (H_\tau)\varepsilon$ nor $(H_\tau)\varepsilon \subset (H_\sigma)\varepsilon$. [The inclusion $(H_\sigma)\varepsilon \subset (H_\tau)\varepsilon$ means that for each set X we have $(XH_\sigma)\varepsilon_X \subset (XH_\tau)\varepsilon_X$.] Since Σ is finite, such a choice of $\bar{\Sigma}$ is clearly possible. Next, define the arity $(\sigma)\overline{\mathrm{ar}}$ for each $\sigma \in \bar{\Sigma}$ as the smallest number n for which there is a natural transformation $\bar{\varepsilon}\colon H_n \to F$ with

$$(H_\sigma)\varepsilon = (H_n)\bar{\varepsilon}.$$

Then $\bar{\Sigma}$ and $\overline{\mathrm{ar}}$ is a new type for which there clearly exists an epitransformation

$$\bar{\varepsilon}\colon H_{\bar{\Sigma}} \to F.$$

We call $\bar{\varepsilon}$ a *minimal presentation* of F. Thus, a presentation $\bar{\varepsilon}$ is minimal iff

(i) $\sigma \neq \tau$ implies $(H_\sigma)\bar{\varepsilon} \not\subset (H_\tau)\bar{\varepsilon}$

and

(ii) if $(H_\sigma)\bar{\varepsilon}$ is a quotient of H_n, then $|\sigma|$ (the arity of σ) is at most n.

Another characterization:

Proposition. A presentation

$$\varepsilon\colon H_\Sigma \to F$$

is minimal iff the following two conditions hold:

(1) For each set X, each one-to-one $f \in XH_\sigma$ $(\sigma \in \Sigma_n, n > 0)$ and each $g \in XH_\tau (\tau \in \Sigma)$ with

$$(f)\varepsilon_X = (g)\varepsilon_X,$$

we have $\sigma = \tau$ and

$$g = p \cdot f$$

for some permutation $p: [n] \to [n]$ such that

$$(p)\varepsilon_{[n]} = (1_{[n]})\varepsilon_{[n]}.$$

(2) For distinct $\sigma, \tau \in \Sigma_0$ we have $(H_\sigma)\varepsilon \neq (H_\tau)\varepsilon$.

Remarks. (i) In the language of equations in the variety F-**Alg** (see III.3.2), the condition above can be reformulated as follows:

(1) Let an equation

$$(x_1, \ldots, x_n)\sigma = (y_1, \ldots, y_m)\tau$$

hold in F-**Alg**, and let $n > 0$ and $x_1, \ldots, x_n$ be pairwise distinct. Then $\sigma = \tau$ and there is a permutation p with $y_1 = x_{(1)p}, \ldots, y_n = x_{(n)p}$, for which the following equation

$$(1, 2, \ldots, n)\sigma = ((1)p, (2)p, \ldots, (n)p)\sigma$$

holds in F-**Alg**.

(2) For no pair of distinct nullary symbols $\sigma, t \in \Sigma$ does

$$\sigma = \tau$$

hold in F-**Alg**.

(ii) For unary operations σ, condition (1) states that

$$(f)\varepsilon_X = (g)\varepsilon_X \quad \text{implies} \quad f = g,$$

because f is one-to-one and $p = \mathrm{id}$.

Proof. I. Let ε be a minimal presentation. Then (2) is obvious; let us prove (1). Since f is one-to-one, there exists a map

$$\bar{f}: X \to [n]$$

with $f \cdot \bar{f} = 1_{[n]}$. Then

$$(1_{[n]})\varepsilon_{[n]} = (g \cdot \bar{f})\varepsilon_{[n]}$$

because

$$\begin{aligned}(1_{[n]})\varepsilon_{[n]} &= (f \cdot \bar{f})\varepsilon_{[n]} \\ &= (f)\bar{f}H_\Sigma \cdot \varepsilon_{[n]} \\ &= (f)\varepsilon_X \cdot \bar{f}F\end{aligned}$$

$$= (g)\varepsilon_X \cdot \bar{f}F$$
$$= (g)\bar{f}H_\Sigma \cdot \varepsilon_{[n]}$$
$$= (g \cdot \bar{f})\varepsilon_{[n]}.$$

This implies that $(H_\sigma)\varepsilon \subset (H_\tau)\varepsilon$ because for each $h \in YH_\sigma$ we have

$$\begin{aligned}(h)\varepsilon_Y &= (1_{[n]} \cdot h)\varepsilon_Y \\ &= (1_{[n]})hH_\Sigma \cdot \varepsilon_Y \\ &= (1_{[n]})\varepsilon_Y \cdot hF \\ &= (g \cdot \bar{f})\varepsilon_{[n]} \cdot hF \\ &= (g \cdot \bar{f})hH_\Sigma \cdot \varepsilon_Y \\ &= (g \cdot \bar{f} \cdot h)\varepsilon_Y \in (YH_\tau)\varepsilon_Y.\end{aligned}$$

By the minimality, $\sigma = \tau$.

Next, we prove that if

$$(1_{[n]})\varepsilon_{[n]} = (p)\varepsilon_{[n]} \quad (p \in [n]H_\sigma),$$

then p is a permutation. Indeed, if not, then p is not onto, say, $n \notin ([n])p$. Let $j: [n-1] \to [n]$ be the inclusion map, then

$$p = p_1 \cdot j \quad \text{for some } p_1: [n] \to [n-1].$$

Define a natural transformation $\tilde{\varepsilon}: H_{n-1} \to H_n$ by

$$(h)\tilde{\varepsilon}_X = p_1 \cdot h \quad \text{for each } h \in X^{n-1}.$$

Then for each $h \in X^n$ we have

$$\begin{aligned}(h)\varepsilon_X &= (1_{[n]}) \cdot hH_\Sigma \cdot \varepsilon_X \\ &= (1_{[n]}) \cdot \varepsilon_{[n]} \cdot hF \\ &= (p_1 \cdot j)\varepsilon_{[n]} \cdot hF \\ &= (p_1 \cdot j)hH_\Sigma \cdot \varepsilon_X \\ &= (p_1 \cdot j \cdot h)\varepsilon_X \\ &= (j \cdot h) \cdot (\tilde{\varepsilon} \cdot \varepsilon)_X.\end{aligned}$$

Hence the natural transformation $\tilde{\varepsilon} \cdot \varepsilon: H_{n-1} \to F$ has the same image as $\varepsilon: H_n \to F$—a contradiction with the minimality of ε.

Returning to $p = g \cdot f$, we see that for *any* map f with $f \cdot f = 1$, the map $g \cdot \bar{f}$ is a permutation. This clearly implies that g is one-to-one and has the same image as f. Then there exists a permutation p with $f = p \cdot g$. And $(1_{[n]})\varepsilon = (p)\varepsilon$ follows from the fact that gF is one-to-one:

$$((1_{[n]})\varepsilon_{[n]})gF = (1_{[n]})gH_\Sigma \cdot \varepsilon_X = (g)\varepsilon_X$$

and

$$((p)\varepsilon_{[n]})gF = (p)gH_\Sigma \cdot \varepsilon_X = (p \cdot g)\varepsilon_X = (f)\varepsilon_X,$$

and hence, $(g)\varepsilon_X = (f)\varepsilon_X$ implies $(1_{[n]})\varepsilon_{[n]} = (p)\varepsilon_{[n]}$.

II. Let ε fulfil the condition above.

(a) For $\sigma \in \Sigma_n$ and $\tau \in \Sigma_m$, we prove that

$$(H_\sigma)\varepsilon \subset (H_\tau)\varepsilon \quad \text{implies} \quad \sigma = \tau.$$

If $n = 0$, then $\emptyset H_\sigma = \emptyset$ and hence, $(\emptyset H_\sigma)\varepsilon_\emptyset \subset (\emptyset H_\tau)\varepsilon_\emptyset$ implies $\emptyset H_\sigma \neq \emptyset$, i.e., $m = 0$. Thus, $\sigma = \tau$ by (2).

Assume $n > 0$. For $f = 1_{[n]}$ we get

$$(f)\varepsilon_{[n]} \in ([n]H_\sigma)\varepsilon_{[n]} \subset ([n]H_\tau)\varepsilon_{[n]}$$

and hence, there exists $g \in [n]H_\tau$ with

$$(f)\varepsilon_{[n]} = (g)\varepsilon_{[n]}.$$

This implies $\sigma = \tau$ by (1).

(b) Let $\tilde{\varepsilon}\colon H_k \to F$ be a natural transformation with

$$(H_\sigma)\varepsilon = (H_k)\tilde{\varepsilon} \quad (\sigma \in \Sigma_n),$$

then we prove $n \leq k$. We can suppose $n > 0$ (since $0 \leq k$ anyhow).

For $1_{[n]} \in [n]H_\sigma$, we can choose $h \in [n]H_k$ with

$$(1_{[n]})\varepsilon_{[n]} = (h)\tilde{\varepsilon}_{[n]}.$$

Let $g\colon [n] \to [n]$ be a map with

$$([n])g = ([k])h \quad \text{and} \quad h \cdot g = h.$$

Then

$$\begin{aligned}(1_{[n]})\varepsilon_{[n]} &= (h \cdot g)\tilde{\varepsilon}_{[n]}\\ &= (h)gH_k \cdot \tilde{\varepsilon}_{[n]}\\ &= (h)\tilde{\varepsilon}_{[n]} \cdot gF\\ &= (1_{[n]})\tilde{\varepsilon}_{[n]} \cdot gF\\ &= (g)\varepsilon_{[n]}.\end{aligned}$$

Thus, there exists a permutation p with $g = p \cdot 1_{[n]} = p$. Hence, g is a permutation, and $([k])h = ([n])g = [n]$. This implies $n \leq k$. □

2.6. Convention. Let

$$\varepsilon\colon H_\Sigma \to F$$

be a minimal presentation. For each $\sigma \in \Sigma$ of arity $n > 0$ put

$$P_\sigma = \{p \in [n]^n;\ (p)\varepsilon_{[n]} = (1_{[n]})\varepsilon_{[n]}\}.$$

Then P_σ is a group of permutations.

Indeed, by the preceding Proposition each $p \in P_\sigma$ is a permutation. Since $1_{[n]} \in P_\sigma$ and $[n]$ is a finite set, it suffices to show that P_σ is closed under composition. If $p_1, p_2 \in P_\sigma$ then

$$\begin{aligned}(p_1 \cdot p_2)\varepsilon_{[n]} &= (p_1)p_2H_\Sigma \cdot \varepsilon_{[n]} \\ &= (p_1)\varepsilon_{[n]} \cdot p_2F \\ &= (1_{[n]})\varepsilon_{[n]} \cdot p_2F \\ &= (1_{[n]})p_2H_\Sigma \cdot \varepsilon_{[n]} \\ &= (p_2)\varepsilon_{[n]} \\ &= (1_{[n]})\varepsilon_{[n]}.\end{aligned}$$

Hence, $p_1 \cdot p_2 \in P_\sigma$.

Remarks. (i) For each natural isomorphism

$$\alpha: H_n \to H_m \quad (n, m \text{ finite})$$

we have

$$n = m$$

and there exists a permutation p on $[n]$ such that

$$(f)\alpha_X = p \cdot f \quad \text{for each set } X \text{ and each } f \in X^n.$$

In fact, put $p = 1_{[n]} \in [n]H_n$. Since $fH_n \cdot \alpha_X = \alpha_{[n]} \cdot fH_m$, we have $(f)\alpha_X = p \cdot f$. Proceeding analogously with α^{-1}, we find out that p is invertible.

(ii) More in general, let Σ and $\bar{\Sigma}$ be two super-finitary types for which a natural isomorphism

$$\alpha: H_\Sigma \to H_{\bar{\Sigma}}$$

exists. Then for each $\sigma \in \Sigma_n$ there exists $\bar{\sigma} \in \bar{\Sigma}_n$ with

$$(H_\sigma)\alpha = H_{\bar{\sigma}}$$

[i.e., with $(XH_\sigma)\alpha_X = XH_{\bar{\sigma}}$ for each set X]. This establishes a bijection $\sigma \mapsto \bar{\sigma}$ from Σ_n to $\bar{\Sigma}_n$ (for each n). Hence, Σ is essentially the same type as $\bar{\Sigma}$.

Thus, α is determined by permutations p_σ on $[|\sigma|]$ (for $\sigma \in \Sigma$) in the sense that

$$(f)\alpha_X = p_\sigma \cdot f \in XH_{\bar{\sigma}} \quad \text{for each } f \in XH_\sigma.$$

(iii) We prove now that minimal presentation is unique up to a (non-essential) permutation of variables.

Proposition. (Uniqueness of minimal presentation.) For arbitrary two minimal presentations $\varepsilon: H_\Sigma \to F$ and $\bar{\varepsilon}: H_{\bar{\Sigma}} \to F$ there exists a natural isomorphism $\alpha: H_\Sigma \to H_{\bar{\Sigma}}$ for which the following triangle

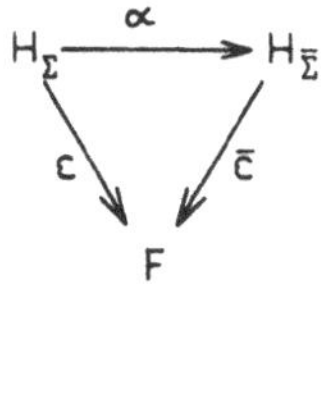

commutes.

Proof. For each $\sigma \in \Sigma_n$ consider $1_{[n]} \in [n]H_\sigma$. Then $(1_{[n]})\varepsilon_{[n]} \in [n]F$, and there exists $\bar{\sigma} \in \Sigma_m$ and $p_\sigma \in [n]H_{\bar{\sigma}}$ with

$$(1_{[n]})\varepsilon_{[n]} = (p_\sigma)\bar{\varepsilon}_{[n]}.$$

It follows that

$$(H_\sigma)\varepsilon \subset (H_{\bar{\sigma}})\bar{\varepsilon}$$

because for each $f \in X^n = XH_\sigma$ we have $fH_\Sigma \cdot \varepsilon_X = \varepsilon_{[n]} \cdot fF$ and $fH_{\bar{\Sigma}} \cdot \bar{\varepsilon}_X = \bar{\varepsilon}_{[n]} \cdot fF$: applying this to $1_{[n]}$, we get

$$\begin{aligned}(f)\varepsilon_X &= ((1_{[n]})\varepsilon_{[n]})fF \\ &= ((p_\sigma)\bar{\varepsilon}_{[n]})fF \\ &= (p_\sigma \cdot f)\bar{\varepsilon}_X.\end{aligned}$$

Since ε is minimal, we conclude that $n \leq m$.

Analogously, for $1_{[m]} \in [m]H_\sigma$ we can find $\sigma' \in \Sigma_{n'}$ and p'_σ with $(1_{[m]})\bar{\varepsilon}_{[m]} = (p'_\sigma)\varepsilon_{[n]}$ and hence,

$$(H_{\bar{\sigma}})\bar{\varepsilon} \subset (H_{\sigma'})\varepsilon.$$

Since ε is minimal,

$$(H_\sigma)\varepsilon \subset (H_{\sigma'})\varepsilon \quad \text{implies} \quad \sigma = \sigma',$$

and since $\bar{\varepsilon}$ is minimal, we conclude that $m \leq n' = n$. Thus,

$$n = m.$$

Moreover $p'_\sigma \cdot p_\sigma$ is a permutation, because ε is minimal and

$$\begin{aligned}(p'_\sigma \cdot p_\sigma)\varepsilon_{[n]} &= (p'_\sigma)p_\sigma H_\Sigma \cdot \varepsilon_{[n]} \\ &= (p'_\sigma)\varepsilon_{[n]} \cdot p_\sigma F \\ &= (1_{[n]})\bar{\varepsilon}_{[n]} \cdot p_\sigma F \\ &= (1_{[n]})p_\sigma H_\Sigma \cdot \bar{\varepsilon}_{[n]} \\ &= (p_\sigma)\bar{\varepsilon}_{[n]} \\ &= (1_{[n]})\varepsilon_{[n]}.\end{aligned}$$

Consequently, $p_\sigma = (p'_\sigma)^{-1}$ is a permutation.

Define a natural isomorphism

$$\alpha: H_\Sigma \to H_{\bar{\Sigma}}$$

by

$$(f)\alpha_X = p_\sigma \cdot f$$

for each set X, each $\sigma \in \Sigma$ and each $f \in XH_\sigma$. Then $(1_{[n]})\varepsilon_{[n]} = (p_\sigma)\bar{\sigma}_{[n]}$ clearly implies that

$$((f)\alpha_X)\bar{\varepsilon}_X = (f)\varepsilon_X \quad \text{for each } f \in XH_\Sigma$$

i.e.,

$$\alpha \cdot \bar{\varepsilon} = \varepsilon. \qquad \square$$

2.7. Definition. Let F be a super-finitary functor with a minimal presentation $\varepsilon\colon H_\Sigma \twoheadrightarrow F$.

We say that F is *perfect* if for each $\sigma, \tau \in \Sigma$, each $f \in XH_\sigma$ (not necessarily one-to-one!) and each $g \in HX_\tau$,

$$(f)\varepsilon_X = (g)\varepsilon_X \quad \text{implies } \sigma = \tau \text{ and } g = p \cdot f \text{ for some } p \in P_\sigma.$$

We say that F is *regular* if for each $\sigma \in \Sigma_n$, $\tau \in \Sigma_m$, each $f \in XH_\sigma$ and $g \in XH_\tau$,

$$(f)\varepsilon_X = (g)\varepsilon_X \quad \text{implies} \quad ([n])f = ([m])g;$$

in other words, if $(f)\varepsilon_X = (g)\varepsilon_X$, then there exist maps p and p' with $g = p \cdot f$ and $f = p' \cdot g$.

(This definition is obviously independent of the concrete choice of minimal presentation.)

Examples. (i) P_2 is perfect. A minimal presentation is given by the transformation $\varepsilon\colon H_2 \twoheadrightarrow P_2$ with

$$(x, y)\varepsilon_X = \{x, y\},$$

i.e., by the equation

$$(x, y)\sigma = (y, x)\sigma.$$

If $(f)\varepsilon = (g)\varepsilon$, then either $f = g$ or $f = (x, y)$ and $g = (y, x)$ in which case $g = t \cdot f$ for the transposition $t\colon [2] \twoheadrightarrow [2]$. Here

$$P_\sigma = \{1_{[2]}, t\}.$$

(ii) P_3 is regular but not perfect. A minimal presentation is given by $\varepsilon\colon H_3 \twoheadrightarrow P_3$ with

$$(x, y, z)\varepsilon_X = \{x, y, z\}.$$

The regularity is obvious. But for $f = (x, x, y)$ and $g = (x, y, y)$ which fulfil $(f)\varepsilon = (g)\varepsilon$ there is no permutation p with $g = p \cdot f$.

(iii) The functor D_2 is not regular. A minimal presentation is given by $\varepsilon\colon H_2 + H_0 \twoheadrightarrow D_2$ with

$$(0)\varepsilon_X = *_X,$$

$$(x, y)\varepsilon_X = \begin{cases} (x, y) & \text{if } x \neq y \\ *_X & \text{if } x = y \end{cases}$$

i.e., by the equation

$$(x, x)\sigma = 0.$$

If $f = (x, x)$ and $g = (y, y)$ with $x \neq y$, then $(f)\varepsilon = (g)\varepsilon$ but the images of f and g are disjoint.

(iv) Let P be a permutation group on the set $[n]$. Denote by

$$H_{n,P}$$

the quotient of H_n given by the equations

$$(x_1, x_2, \ldots, x_n)\,\sigma = (x_{(1)p}, x_{(2)p}, \ldots, x_{(n)p}) \quad (p \in P).$$

That is, $XH_{n,P} = X^n/\sim$ where

$$f \sim g \quad \text{iff} \quad g = p \cdot f \quad \text{for some } p \in P \quad (f, g \in X^n).$$

It is clear that $H_{n,P}$ is perfect.

Proposition. A super-finitary functor F is perfect iff

$$F = \coprod_{i=1}^{k} H_{n_i, P_i}$$

for some permutation groups $P_1, \ldots, P_k$.

Proof. Each $H_{n,P}$ is perfect, and a finite coproduct of perfect functors is obviously perfect. Thus, we only have to prove that each perfect functor F has the form above. Let $\varepsilon\colon H_\Sigma \to F$ be a minimal presentation. Since F is perfect,

$$\sigma \neq \tau \quad \text{implies} \quad (H_\sigma)\varepsilon \cap (H_\tau)\varepsilon = \emptyset$$

and hence

$$F = \coprod_{\sigma \in \Sigma} (H_\sigma)\varepsilon.$$

And, clearly, $(H_\sigma)\varepsilon$ with $\sigma \in \Sigma_n$ is naturally isomorphic to

$$H_{n, P_\sigma}.$$

Lemma. A functor F with a minimal presentation $\varepsilon\colon H_\Sigma \to F$ is regular iff for each $\sigma \in \Sigma_n$ and $\tau \in \Sigma_m$ with $n \geq m \geq 2$ we have

$$(f)\varepsilon_{[n]} = (g)\varepsilon_{[n]} \quad \text{implies} \quad ([n])f = ([m])g$$

for any $f \in [n]H_\sigma$ and $g \in [n]H_\tau$.

Proof. If F fulfils the condition above, we prove that F is regular, i.e., for each $\sigma \in \Sigma_n$, $\tau \in \Sigma_m$ and arbitrary $\bar{f} \in XH_\sigma$ and $\bar{g} \in XH_\tau$,

$$(\bar{f})\varepsilon_X = (\bar{g})\varepsilon_X \quad \textit{implies} \quad ([n])\bar{f} = ([m])\bar{g}.$$

Without loss of generality, assume $n \geq m$.

(A) Let $m = 0$. We prove that $n \leq 1$. Assuming the contrary, we can choose maps $p, q: X \to [n]$ which have disjoint images, and we apply the condition above to σ and σ, with

$$f = \bar{f} \cdot p \quad \text{and} \quad g = \bar{f} \cdot q.$$

Since $\bar{g}(: \emptyset \to X)$ is the empty map, we have

$$\bar{g} \cdot p = \bar{g} \cdot q$$

and hence,

$$(f)\varepsilon_{[n]} = (g)\varepsilon_{[n]}.$$

[In fact,

$$(\bar{f} \cdot p)\varepsilon_{[n]} = (\bar{f})pH_\Sigma \cdot \varepsilon_{[n]} = (\bar{f})\varepsilon_X \cdot pF = (\bar{g})\varepsilon_X \cdot pF = (\bar{g} \cdot p)\varepsilon_{[n]}$$

and analogously, $(\bar{f} \cdot q)\varepsilon_{[n]} = (\bar{g} \cdot q)\varepsilon_{[n]}$.] This is a contradiction, since $([n])f \cap ([n])g = \emptyset$.

If $n = 1$, then $\bar{f}$ is one-to-one and hence $\sigma = \tau$ and $\bar{g} = p \cdot \bar{f}$ for some permutation p [see (1) in VII.2.5]. Hence, $\bar{f}$ and $\bar{g}$ have the same image.

If $n = 0$, then $\bar{f} = \bar{g}$ [see (2) in VII.2.5].

(B) Let $m = 1$. Then g is one-to-one and hence $\tau = \sigma$ and $\bar{f} = p \cdot \bar{g}$ for some permutation p [see (1) in VII.2.5], hence, $\bar{f}$ and $\bar{g}$ have the same image.

(C) Let $m \geq 2$. Assuming $([n])\bar{f} \neq ([m])\bar{g}$, there exists a map $p: X \to [n]$ such that the maps

$$f = \bar{f} \cdot p \quad \text{and} \quad g = \bar{g} \cdot p$$

also have distinct images. This contradicts to the condition above, since

$$(f)\varepsilon_{[n]} = (\bar{f})pH_\Sigma \cdot \varepsilon_{[n]} = (\bar{f})\varepsilon_X \cdot pF$$

and analogously, $(g)\varepsilon_{[n]} = (\bar{g})\varepsilon_X \cdot pF$, and this proves that $(f)\varepsilon_{[n]} = (g)\varepsilon_{[n]}$. □

Remarks. (i) There is no descriptive characterization of regular functors, in contrast to the perfect ones.

(ii) An analogous lemma holds for perfectness: F is perfect iff for $\sigma \in \Sigma_n$, $\tau \in \Sigma_m$ with $n \geq m \geq 2$ and for arbitrary $f \in [n]H_\sigma$ and $g \in [n]H_\tau$,

$(f)\,\varepsilon_{[n]} = (g)\varepsilon_{[n]}$ implies $\sigma = \tau$ and $g = p \cdot f$ for $p \in P_\sigma$.

(iii) The lemma above can be formulated more symmetrically if no restrictions on the size of the sets X are required: F is regular iff for each $\sigma, \tau \in \Sigma$ of arity ≥ 2 and any $f \in XH_\sigma$ and $g \in XH_\tau$,

$$(f)\varepsilon_X = (g)\varepsilon_X \quad \text{implies} \quad \text{image } f = \text{image } g.$$

2.8. Proposition. Each regular functor preserves preimages, and each perfect functor covers pullbacks.

Proof. Let $\varepsilon: H \to F$ be a minimal presentation of a regular functor F and let

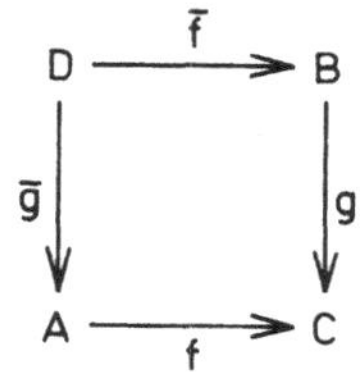

be a pullback. If either g is one-to-one or F is perfect, we shall prove that F covers this pullback. That is, for arbitrary $a \in AF$ and $b \in BF$ with

$$(a)fF = (b)gF,$$

we shall present a $d \in DF$ such that

$$a = (d)\bar{g}F \quad \text{and} \quad b = (d)\bar{f}F.$$

(If g is one-to-one, then so are $\bar{g}$ and $\bar{g}F$, see Remark III.4.5, and hence, such d is then unique. Thus, F actually preserves the pullback.) There exist $a_0 \in AH_\Sigma$ with $(a_0)\varepsilon_A = a$ and $b_0 \in BH_\Sigma$ with $(b_0)\varepsilon_B = b$. Say $a_0 \in AH_\sigma$ where $\sigma \in \Sigma_n$ and $b_0 \in BH_\tau$ where $\tau \in \Sigma_m$. We have

$$(a_0 \cdot f)\varepsilon_C = (a_0)fH_\Sigma \cdot \varepsilon_C = (a_0)\varepsilon_A \cdot fF = (a)fF,$$

analogously with $b_0 \cdot g$. Therefore,

$$(a_0 \cdot f)\varepsilon_C = (b_0 \cdot g)\varepsilon_C.$$

Since ε is regular, the images of $a_0 \cdot f$ and $b_0 \cdot g$ are equal. Hence, there exists a map $p: [n] \to [m]$ with

$$a_0 \cdot f = p \cdot b_0 \cdot g.$$

Let d_0 be the unique map, for which the following diagram

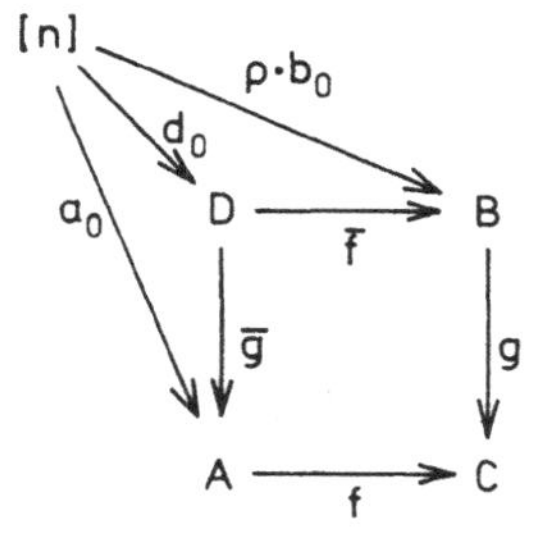

commutes. We claim that

$$d = (d_0)\varepsilon_D \in DF$$

is the element we are looking for. First,

$$a = (a_0)\varepsilon_A = (d_0 \cdot \bar{g})\varepsilon_A = (d_0)\bar{g}H_\Sigma \cdot \varepsilon_A = (d_0)\varepsilon_D \cdot \bar{g}F.$$

For b, we must consider the two cases separately.

(i) If ε is perfect, then $(a_0 \cdot f)\varepsilon_C = (b_0 \cdot g)\varepsilon_C$ implies $\sigma = \tau$ and $p \in P_\sigma$, thus,

$$b = (b_0)\varepsilon_B = (p \cdot b_0)\varepsilon_B = (d_0 \cdot \bar{f})\varepsilon_B = (d_0)\varepsilon_D \cdot \bar{f}F.$$

(ii) If g is one-to-one, then so is gF and we have

$$(b)gF = (a)fF = (d_0)\varepsilon_D \cdot \bar{g}F \cdot fF = (d_0)\varepsilon_D \cdot \bar{f}F \cdot gF,$$

thus, again

$$b = (d_0)\varepsilon_D \cdot \bar{f}F. \qquad \square$$

Remark. We see that for each regular functor, the P-recognizable and recognizable languages coincide. For perfect functors, the N-recognizable and recognizable languages coincide. We are going to prove that the converse holds, too. First, we prove a technical result concerning epitransformations [which actually holds in general $(\mathscr{E}, \mathscr{M})$-categories as well].

2.9. Lemma. Let $\varepsilon: H \twoheadrightarrow F$ be an epitransformation. For each relation $r: A \rightharpoonup B$ in the domain category of F we have

$$rF = \varepsilon_A^{-1} \circ rH \circ \varepsilon_B$$

and therefore,

$$rH \circ \varepsilon_B \subset \varepsilon_A \circ rF.$$

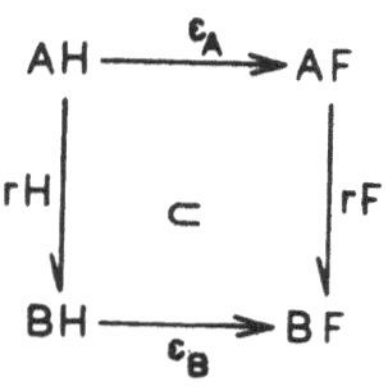

Proof. Put $r = [r_1, r_2]$ with $r_1: R \twoheadrightarrow A$ and $r_2: R \twoheadrightarrow B$. Since $\varepsilon_A^{-1} = [\varepsilon_A, 1_{AH}]$, we have

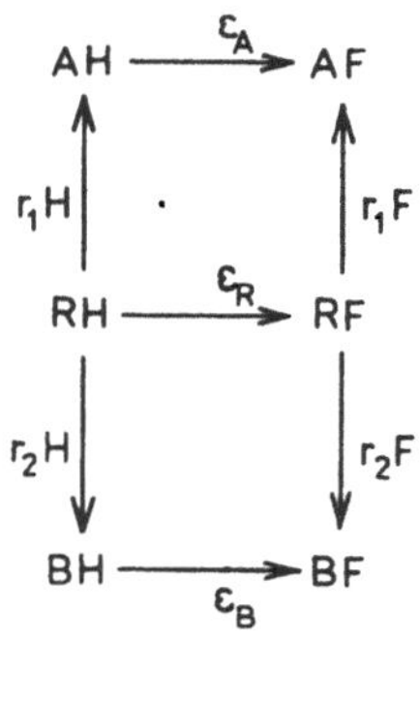

$$\varepsilon_A^{-1} \circ rH = [\varepsilon_A, 1_{AH}] \circ [r_1H, r_2H] = [r_1H \cdot \varepsilon_A, r_2H].$$

Therefore,

$$\varepsilon_A^{-1} \circ rH \circ \varepsilon_B = [r_1H \cdot \varepsilon_A, r_2H \cdot \varepsilon_B] \\ = [\varepsilon_R \cdot r_1F, \varepsilon_R \cdot r_2F].$$

Since ε_R is an epi, the last pair represents the same relation as $[r_1F, r_2F]$. Thus,

$$\varepsilon_A^{-1} \circ rH \circ \varepsilon_B = rF.$$

Since $1_{AH} \subset \varepsilon_A \circ \varepsilon_A^{-1}$, this implies

$$rH \circ \varepsilon_B \subset \varepsilon_A \circ \varepsilon_A^{-1} \circ rH \circ \varepsilon_B = \varepsilon_A \circ rF. \qquad \square$$

Remark. The equation

$$rH \circ \varepsilon_B = \varepsilon_A \circ rF$$

does not hold for relations r in general, see Exercise VII.2.C below.

2.10. Theorem. For each super-finitary functor F, the following are equivalent:

(i) Every P-recognizable language is recognizable:
(ii) F preserves preimages;
(iii) F is regular.

Proof. It suffices to prove that a functor F which fulfils (i) is regular [since (iii) → (ii) is proved in VII.2.8 and (ii) → (i) in VII.2.2]. We first construct P-recognizable languages with special properties for an arbitrary super-finitary functor F. Then we prove that the fact that these languages are recognizable implies that F is regular.

I. Let $\varepsilon: H_\Sigma \to F$ be a minimal presentation of an arbitrary super-finitary functor F. For each $\sigma \in \Sigma_n$, $n \geq 2$, we shall find a partial F-acceptor A_σ recognizing a language

$$L_\sigma \subset I^\#$$

with the following properties:

(a) For each one-to-one n-tuple $w \in I^\# H_\sigma$, the element $(w)\varepsilon_{I^\#} \cdot \varphi \in I^\#$ fulfils

$$(w)\varepsilon_{I^\#} \cdot \varphi \in L_\sigma \quad \text{iff} \quad w_1, \ldots, w_n \in L_\sigma.$$

(b) For each constant m-tuple $w \in I^\# H_\tau$, $t \in \Sigma_m$, $m > 0$,

$$(w)\varepsilon_{I^\#} \cdot \varphi \in L_\sigma \quad \text{provided that } w_1 \in L_\sigma.$$

(c) Both L_σ and $I^\# - L_\sigma$ are infinite sets.

The partial acceptor $A_\sigma = (Q, \delta, T, I, \lambda)$ has only two states 1, 2 with 1 terminal. The partial map

$$\delta\colon [2]F \rightharpoonup [2]$$

is defined in $q \in [2]F$ iff there exists $\tau \in \Sigma_m$, $m > 0$, and a constant m-tuple $a \in [2]H_\tau$ with

$$q = (a)\varepsilon_{[2]};$$

in this case,

$$(q)\delta = 1$$

[and otherwise $(q)\delta$ is undefined]. The object I is $[2n + 4]$, and the partial map

$$\lambda\colon [2n + 4] \rightharpoonup [2]$$

is defined by $(1)\lambda = (2)\lambda = \ldots = (n + 2)\lambda = 1$, else undefined.

Since both δ and λ have only one value 1, also the run relation $\rho\colon I^\# \rightharpoonup Q$ is a partial map which has only value 1. It follows that the language L_σ of A_σ consists of those $u \in I^\#$ for which $(u)\rho$ is defined.

(a) Let w be a one-to-one tuple in $(I^\#)^n = I^\# H_\sigma$. Then $(w)\varepsilon_{I^\#} \cdot \varphi$ is in L_σ iff

$$1 = (w)\varepsilon_{I^\#} \circ \varphi \circ \rho = ((w)_{I^\#} \circ \rho F) \circ \delta,$$

and this holds iff there is a constant m-tuple $a \in [2]H_\tau$, $\tau \in \Sigma_m$, $m > 0$, for which

$$(w)\varepsilon_{I^\#} \circ \rho F = (a)\varepsilon_{[2]}.$$

By VII.2.9,

$$(w)\varepsilon_{I^\#} \circ \rho F = (w)\varepsilon_{I^\#} \circ \varepsilon_{I^\#}^{-1} \circ \rho H_\Sigma \circ \varepsilon_{[2]}.$$

Since ε is minimal and $w = (w_1, \ldots, w_n)$ is one-to-one, we have

$$(w)\varepsilon_{I^\#} \circ \varepsilon_{I^\#}^{-1} = \{p \cdot w;\quad p \in P_\sigma\},$$

by Proposition VII.2.5.

Thus,

$$(w)\varepsilon_{I^\#} \circ \rho F = \{p \cdot w; p \in P_\sigma\}\rho H_\Sigma \circ \varepsilon_{[2]}.$$

Hence, $(w)\varepsilon_{I^\#} \circ \rho F$ is equal to $\{((w_{p_1})\rho, \ldots, (w_{p_n})\rho);\ p \in P_\sigma\}\varepsilon_{[2]}$ provided that all $(w_j)\rho$ are defined ($j = 1, \ldots, n$) and it is undefined else. Therefore, $(w)\varepsilon_{I^\#} \circ \varphi \in L_\sigma$ iff

> $(w_j)\rho$ are defined for all $j = 1, \ldots, n$ and there exists a constant tuple $a \in [2]H_\tau$ and $p \in P_\sigma$ with $(p \circ w \circ \rho)\varepsilon_{[2]} = (a)\varepsilon_{[2]}$.

If this condition is fulfilled, then $w_j \in L_\sigma$ for $j = 1, \ldots, n$ [because $(w_j)\rho$

are defined]. Conversely, if $w_j \in L_\sigma$ for $j = 1, \ldots, n$, then $(w_j)\rho = 1$, $w \circ \rho = (1, \ldots, 1) \in [2]H_\sigma$ and we simply put $p = 1_{[n]} \in P_\sigma$ and $a = w \circ \rho$ to conclude that the condition is fulfilled.

(b) Let $w = (w_1, \ldots, w_1) \in (I^\#)^m = I^\# H_\tau$, $t \in \Sigma_m$, $m > 0$, be a constant tuple with $w_1 \in L_\sigma$. Then $(w)\rho H_\Sigma = (1, \ldots, 1) \in [2]H_\tau$, and therefore

$$(w)\rho H_\Sigma \circ \varepsilon_{[2]} \circ \delta = 1.$$

By VII.2.9,

$$(w)\rho H_\Sigma \circ \varepsilon_{[2]} \circ \delta \subset (w)\varepsilon_{I^\#} \circ \rho F \circ \delta \subset (w)\varepsilon_{I^\#} \circ \varphi \circ \rho.$$

Hence, ρ is defined in $(w)\varepsilon_{I^\#} \circ \varphi$, i.e., $(w)\varepsilon_{I^\#} \circ \varphi \in L_\sigma$.

(c) To prove that L_σ is infinite, we observe first that it contains $1, 2, \ldots, n + 2$ because

$$(i)\sigma = (i)\lambda = 1 \quad \text{for} \quad i = 1, 2, \ldots, n + 2.$$

Since $n \geq 2$, it is sufficient to exhibit a one-to-one map from the set $\exp_n L_\sigma$ of all n-point subsets of L_σ into L_σ: we know that L_σ has at least $n + 2$ elements, and therefore it cannot be a finite set.

For each set $\{w_1, \ldots, w_n\} \in \exp_n L_\sigma$, we choose an arbitrary ordering and we obtain a one-to-one n-tuple $w \in L_\sigma H_\sigma$. By (a), $(w)\varepsilon_{I^\#} \cdot \varphi \in L_\sigma$, and the map assigning $(w)\varepsilon_{I^\#} \cdot \varphi$ to $\{w_1, \ldots, w_n\}$ is one-to-one [because $\varepsilon_{I^\#}$ is "essentially" one-to-one by the minimality of ε, and also φ is one-to-one, see IV.4.2, Remark (i)].

Analogously with the complement $I^\# - L_\sigma$. It has at least $n + 2$ elements: $n + 3, n + 4, \ldots, 2n + 4$. And for each set $\{w_1, \ldots, w_n\}$ of n elements in $I^\# - L_\sigma$ we have, by (a), $(w)\varepsilon_{I^\#} \cdot \varphi \in I^\# - L_\sigma$.

II. Let F be a super-finitary functor for which the above languages L_σ are recognizable. We prove that then F is regular. Let X be a set and let $f \in XH_\sigma$ and $g \in XH_\tau$ be elements with

$$(f)\varepsilon_X = (g)\varepsilon_X.$$

It suffices to prove that im f (the image of f) is a subset of im g, i.e.,

$$\text{if } b \in \text{im } f, \quad \text{then } b \in \text{im } g$$

(by symmetry, then im $f =$ im g). Let n be the arity of σ and m be the arity of τ. We can suppose that n, m are at least 2 [see VII.2.7, Remark (iii)]; we use the language L_σ. Let

$$(Q, \delta, T, I)$$

be F-acceptor recognizing L_σ. Since Q is finite and L_σ is infinite, there exists an infinite set $L \subset L_\sigma$ on which the run map ρ is constant, say, with value q_0:

$$(u)\rho = q_0 \quad \text{for each } u \in L;$$

analogously, there exists an infinite set $L' \subset I^{\#} - L_\sigma$ and $q_0' \in Q$ with

$$(u)\rho = q_0' \qquad \text{for each} \qquad u \in L'.$$

Let $w \in I^{\#} H_\sigma$ be a one-to-one n-tuple such that

$$w_i \in L \quad \text{if} \quad (i)f \neq b; \; w_i \in L' \quad \text{if} \quad (i)f = b.$$

Then $(w_i)\rho = q_0'$ if $(i)f = b$ and $(w_i)\rho = q_0$ if $(i)f \neq b$. Therefore, the map $c: X \to Q$ defined by

$$(b)c = q_0' \quad \text{and} \quad (x)c = q_0 \quad \text{for all } x \in X - \{b\}$$

clearly fulfils

$$f \cdot c = w \cdot \rho : [n] \to Q.$$

By (a)

$$(w)\varepsilon_{I^{\#}} \cdot \varphi \notin L_\sigma,$$

since $w_i \notin L_\sigma$ for at least one $i \in [n]$. Thus,

$$(w)\varepsilon_{I^{\#}} \cdot \varphi \cdot \rho \notin T.$$

And we have

$$\begin{aligned}(w)\varepsilon_{I^{\#}} \cdot \varphi \cdot \rho &= (w)\varepsilon_{I^{\#}} \cdot \rho F \cdot \delta \\ &= (w)\rho H_\Sigma \cdot \varepsilon_Q \cdot \delta \\ &= (f \cdot c)\varepsilon_Q \cdot \delta \\ &= (f)\, c\, H_\Sigma \cdot \varepsilon_Q \cdot \delta \\ &= (f)\varepsilon_X \cdot cF \cdot \delta \\ &= (g)\varepsilon_X \cdot cF \cdot \delta \\ &= (g \cdot c)\varepsilon_Q \cdot \delta.\end{aligned}$$

Hence

$$(g \cdot c)\varepsilon_Q \cdot \delta \notin T.$$

This implies that $b \in \operatorname{im} g$: otherwise $g \cdot c$ is the constant m-tuple $(q_0, q_0, \ldots, q_0) \in QH_\tau$. But

$$(q_0, q_0, \ldots, q_0)\, \varepsilon_Q \cdot \delta \in T$$

by (b) in I: choose a constant m-tuple $\bar{w} \in I^{\#} H_\tau$ with $\bar{w}_i \in L$, then $(\bar{w})\varepsilon_{I^{\#}} \cdot \varphi \in L_\sigma$, hence,

$$(\bar{w})\varepsilon_{I^{\#}} \cdot \varphi \cdot \rho = (\bar{w})\rho H_\Sigma \cdot \varepsilon_Q \cdot \delta = (q_0, q_0, \ldots, q_0)\varepsilon_Q \cdot \delta \in T.$$

This concludes the proof. □

2.11. Lemma. Let F be a super-finitary functor with a minimal presentation

$\varepsilon: H_\Sigma \to F$. Let $\sigma \in \Sigma_n$, $n \geq 2$, and let γ be an equivalence relation of $[n]$. Then there exists an N-recognizable language

$$L \subset I^\#, \ I \text{ a finite set}$$

and subsets $L_1, \ldots, L_n$ of $I^\#$ with the following properties.

(a) Let $w \in I^\# H_\sigma$ be a one-to-one n-tuple with $w_1 \in L_{r_1}, \ldots, w_n \in L_{r_n}$ ($r_1, \ldots, r_n \in [n]$). Then $(w)\varepsilon_{I^\#} \cdot \varphi \in L$ iff there exist permutations $p, p' \in P_\sigma$ with

$$r_{(i)p}\ \gamma\ (i)p' \quad \text{for } i = 1, \ldots, n.$$

(b) For each $\tau \in \Sigma$ of arity $m \leq n$, $\tau \neq \sigma$, the restriction $\bar{w} = (w_1, \ldots, w_m) \in I^\# H_\tau$ of w fulfils $(\bar{w})\varepsilon_{I^\#} \cdot \varphi \notin L$.

(c) The sets $L_1, \ldots, L_n$ are infinite and $L_i = L_j$ iff $i \gamma j$ ($i, j \in [n]$).

Proof. I. First suppose that the given equivalence relation γ is non-trivial, i.e., $\gamma \neq [n] \times [n]$. We define a nondeterministic F-acceptor on the state set

$$Q = [n] \times [n] \cup \{z\},$$

where z is an arbitrary element outside of $[n] \times [n]$. Put

$$A = (Q, \delta, \{z\}, I, \lambda),$$

where

$$I = [n] \times [n+2]$$

and the relations δ and λ are defined as follows. Put

$$Q_i = \{(k, j) \in [n] \times [n];\ k \gamma i\} \quad \text{for } i = 1, \ldots, n$$

and define

$$(i, j)\lambda = Q_i \quad \text{for all } (i, j) \in [n] \times [n+2].$$

Define a map $t \in QH_\Sigma$ (i.e., $t: [n] \to Q$) by

$$(i)t = (i, 1) \quad (i = 1, \ldots, n),$$

and for each $q \in QF$ put

$$(q)\delta = \begin{cases} Q_i \text{ if } q = (a)\varepsilon_Q \text{ for a one-to-one } a \in QH_\sigma \\ \qquad\qquad \text{with } a_1, \ldots, a_n \in Q_i \ (i \in [n]); \\ \{z\} \text{ if } q = (t)\varepsilon_Q; \\ \emptyset \ \text{ else.} \end{cases}$$

This relation is well-defined because ε is minimal:

(i) $(t)\varepsilon_Q \neq (a)\varepsilon_Q$ for any $a \in Q_i^n$ because there exists $j \in [n]$ not equivalent to i under γ, and then $a \neq p \cdot t$ for any permutation p [because $(j, 1)$ is in the image of t but not of a];

(ii) $(a)\varepsilon_Q \neq (b)\varepsilon_Q$ for any one-to-one $a \in Q_i^n \subset QH_\sigma$ and $b \in Q_j^n \subset QH_\sigma$ with $Q_i \neq Q_j$ because, again, $b \neq p \cdot a$ for any permutation p (here a and b have disjoint images).

Let L be the language recognized by A, and let

$$L_i = \{u \in I^\#; (u)\rho = Q_i\}, \quad i = 1, \ldots, n,$$

where $\rho: I^\# \rightharpoonup Q$ is the run relation of A.

(a) Since z is the only terminal state, we have $(w)\varepsilon_{I^\#} \cdot \varphi \in L$ iff

$$z \in (w)\varepsilon_{I^\#} \circ \varphi \circ \rho = (w)\varepsilon_{I^\#} \circ \rho F \circ \delta$$

(see VII.1.12) and this holds iff

$$\begin{aligned}(t)\varepsilon_Q &\in (w)\varepsilon_{I^\#} \circ \rho F \\ &= (w)\varepsilon_{I^\#} \circ \varepsilon_{I^\#}^{-1} \circ \rho H_\Sigma \circ \varepsilon_Q\end{aligned}$$

by VII.2.9. Since ε is minimal and w is one-to-one, we have

$$(w)\varepsilon_{I^\#} \circ \varepsilon_{I^\#}^{-1} = \{(w_{(1)p}, \ldots, w_{(n)p}); p \in P_\sigma\}.$$

Therefore, $(w)\varepsilon_{I^\#} \cdot \varphi \in L$ iff

$$\begin{aligned}(t)\varepsilon_Q &\in \bigcup_{p \in P_\sigma} ((w_{(1)p})\rho \times \ldots \times (w_{(n)p})\rho)\varepsilon_Q \\ &= \bigcup_{p \in P_\sigma} (Q_{r_{(i)p}} \times \ldots \times Q_{r_{(n)p}})\varepsilon_Q\end{aligned}$$

(because $w_i \in L_{r_i}$) or equivalently,

$$(t)\varepsilon_Q = (\bar{t})\varepsilon_Q$$

for some $\bar{t} \in Q_{r_{(1)p}} \times \ldots \times Q_{r_{(n)p}}$. By Proposition VII.2.5, this equation holds iff there exists $p' \in P_\sigma$ with

$$\bar{t} = p' \cdot t.$$

We have $\bar{t}_i = t_{(i)p'} \in Q_{r_{(i)p}}$, in other words,

$$(i)p'\gamma r_{(i)p} \quad \text{for } i = 1, \ldots, n.$$

(b) Assume that $\tau \in \Sigma_m$ $(m \leq n)$ is such that the restriction $\bar{w} \in I^\# H_\tau$ of w fulfils $(\bar{w})\varepsilon_{I^\#} \cdot \varphi \in L$. By an argument similar to (a) above, we verify that

$$(t)\varepsilon_Q \in \bigcup_{p \in P_\tau} ((w_{(1)p})\rho \times \ldots \times (w_{(m)p})\rho)\varepsilon_Q.$$

Since t is one-to-one and ε is minimal, this implies $\sigma = \tau$.

(c) To prove that L_i is infinite, we observe first that $(i, 1), \ldots, (i, n+2) \in L_i$ because $(i, j)\rho = (i, j)\lambda = Q_i$. Since $n \geq 2$, it is sufficient to exhibit a one-to-one map from the set $\exp_n L_i$ of all n-point subsets of L_i into L_i. For each set $\{w_1, \ldots, w_n\} \in \exp_n L_i$ we choose an arbitrary ordering and we obtain a one-to-one n-tuple $w \in L_i^n \subset I^* H_\sigma$; we prove that $(w)\varepsilon_{I^\#} \cdot \varphi \in L_i$. By Lemma VII.2.9,

$$\begin{aligned}(w)\varepsilon_{I^\#} \circ \varphi \circ \rho &= (w)\varepsilon_{I^\#} \circ \rho F \circ \delta \\ &= (w)\varepsilon_{I^\#} \circ \varepsilon_{I^\#}^{-1} \circ \rho H_\Sigma \circ \varepsilon_Q \circ \delta \\ &= \bigcup_{p \in P_\sigma} ((w_{(1)p})\rho \times \ldots \times (w_{(n)p})\rho)\varepsilon_Q \\ &= (Q_i^n)\varepsilon_Q \circ \delta.\end{aligned}$$

By the definition of δ, it is clear that the last set is Q_i. Since φ is one-to-one [IV.4.2, Remark (i)] and ε is minimal, the passage from $\{w_1, \ldots, w_n\}$ to $(w)\varepsilon_{I^\#} \cdot \varphi$ is clearly one-to-one. Hence, L_i is infinite.

Clearly $L_i = L_j$ iff $i\gamma j$.

II. Let $\gamma = [n] \times [n]$. Then we are to present an N-recognizable language $L \subset I^*$ and set $L_1 \subset I^*$ such that

(a) for each one-to-one n-tuple $w \in L_1^n \subset I^* H_\sigma$,

$$(w)\varepsilon_{I^\#} \cdot \varphi \in L;$$

(b) given $\tau \neq \sigma$ of arity $m \leq n$ and denoting by $\bar{w} \in I^* H_\tau$ the restriction of w, then

$$(\bar{w})\varepsilon_{I^\#} \cdot \varphi \in L;$$

(c) L_i is infinite.

Define a nondeterministic acceptor

$$A = (Q, \delta, [n], [n+2], \lambda)$$

as follows: $Q = [n]$, for $1_{[n]} \in [n]H_\sigma$ put $((1_{[n]})\varepsilon_{[n]})\delta = [n]$ and else $(q)\delta = \emptyset$, and $(i)\lambda = [n]$ for all $i = 1, \ldots, n+2$. Let $L = L_1$ be the language recognized by A.

(a) Each state in A is terminal and hence,

$$(w)\varepsilon_{I^\#} \cdot \varphi \in L \quad \text{iff} \quad (1_{[n]})\varepsilon_Q \in \bigcup_{p \in P_\sigma} ((w_{(1)p})\rho \times \ldots \times (w_{(n)p})\rho)\varepsilon_Q.$$

The argument is as in I above. Since $w_i \in L_1 = L$ implies $(w_i)\rho = Q$ for $i = 1, \ldots, n$, we get

$$1_{[n]} \in (w_1)\rho \times \ldots \times (w_n)\rho = Q^n.$$

The proofs of (b) and (c) are analogous to I above. □

2.12. Theorem. For each super-finitary functor F, the following are equivalent:

(i) Every N-recognizable language is recognizable;
(ii) F covers pullbacks;
(iii) F is perfect.

Proof. It is sufficient to prove that each functor with (i) is perfect, since (iii) → (ii) is proved in VII.2.8 and (ii) → (i) in VII.2.2. Let $\varepsilon: H_\Sigma \to F$ be a minimal presentation. We use the fact that the language L in the preceding lemma is recognizable.

We first prove that for $\sigma \in \Sigma_n$ and $\tau \in \Sigma_m$,

$$(f)\varepsilon_X = (g)\varepsilon_X \quad \text{implies} \quad \sigma = \tau \quad (f \in XH_\sigma \text{ and } g \in XH_\tau).$$

We can suppose $n \geq m \geq 2$ (see VII.2.7). We use Lemma VII.2.11 for the trivial equivalence $\gamma = [n] \times [n]$. Let (Q, δ, T, I) be an F-acceptor, recognizing L. Since $L_1 (= L_2 = \ldots = L_n)$ is an infinite set and Q is finite, there exists an infinite subset $\bar{L}_1 \subset L_1$ on which the run map ρ is constant, say, with value $q_0 \in Q$. Choose distinct $w_1, \ldots, w_n \in \bar{L}_1$. This yields $w \in I^\# H_\sigma$ such that, by (a), $(w)\varepsilon_{I^\#} \cdot \varphi \in L$. Thus, the following state

$$\begin{aligned} q_1 = (w)\varepsilon_{I^\#} \cdot \varphi \cdot \rho &= (w)\varepsilon_{I^\#} \cdot \rho F \cdot \delta \\ &= (w)\rho H_\Sigma \cdot \varepsilon_Q \cdot \delta \\ &= (q_0, q_0, \ldots, q_0)\varepsilon_Q \cdot \delta \end{aligned}$$

is terminal.

Denote by $c: X \to Q$ the constant map with value q_0, then $(q_0, q_0, \ldots, q_0) = f \cdot c$ and thus,

$$\begin{aligned} q_1 &= (f \cdot c)\varepsilon_Q \cdot \delta \\ &= (f)cH_\Sigma \cdot \varepsilon_Q \cdot \delta \\ &= (f)\varepsilon_X \cdot cF \cdot \delta \\ &= (g)\varepsilon_X \cdot cF \cdot \delta \\ &= (g \cdot c)\varepsilon_Q \cdot \delta. \end{aligned}$$

Let $\bar{w} = (w_1, \ldots, w_m) \in I^\# H_\tau$ be the restriction of w, then clearly $(\bar{w})\rho H_\Sigma = (q_0, q_0, \ldots, q_0) = g \cdot c$ and hence

$$\begin{aligned} q_1 &= (\bar{w})\rho H_\Sigma \cdot \varepsilon_Q \cdot \delta \\ &= (\bar{w})\varepsilon_{I^\#} \cdot \varphi \cdot \rho. \end{aligned}$$

This implies $(\bar{w})\varepsilon_{I^\#} \cdot \varphi \in L$. By (b), this proves $\tau = \sigma$.

II. We prove that given $\sigma \in \Sigma_n$ with $n \geq 2$ and $f, g \in XH_\sigma$, then

$$(f)\varepsilon_X = (g)\varepsilon_X \quad \text{implies} \quad g = p \cdot f \quad \text{for some } p \in P_\sigma.$$

We apply Lemma VII.2.11 to the kernel equivalence γ of f [i.e., $i\gamma j$ iff $(i)f = (j)f$]. Let $A = (Q, \delta, T, I)$ be an F-acceptor recognizing the language L. For each $i \in [n]$, the set $L_i \in I^*$ is infinite and hence, it has an infinite subset $\bar{L}_i$ on which the run map ρ is constant. By (c) in VII.2.11, we can choose these sets $\bar{L}_i$ so that $\bar{L}_i = \bar{L}_j$ whenever $(i)f = (j)f$. Denote by $q_i \in Q$ the value of ρ on $\bar{L}_i$, $i \in [n]$.

Let $w \in I^\# H_\sigma$ be a one-to-one n-tuple with $w_i \in \bar{L}_i$. By VII.2.11 (a) (with $r = p = p' = 1$)

$$(w)\varepsilon_{I^\#} \cdot \varphi \in L.$$

Thus, the state

$$\begin{aligned} q_w &= (w)\varepsilon_{I^\#} \cdot \varphi \cdot \rho \\ &= (w)\varepsilon_{I^\#} \cdot \rho F \cdot \delta \\ &= (w)\rho H_\Sigma \cdot \varepsilon_Q \cdot \delta \end{aligned}$$

is terminal. Since $(w_i)\rho = q_i$, $i \in [n]$, and since $(i)f = (j)f$ implies $q_i = q_j$, we can find a map

$$c: X \to Q$$

such that

$$((i)f)c = q_i, \quad i = 1, \ldots, n,$$

i.e.

$$f \cdot c = (w)\rho H_\Sigma.$$

Thus,

$$\begin{aligned} q_w &= (f \cdot c)\varepsilon_Q \cdot \delta \\ &= (f)cH_\Sigma \cdot \varepsilon_Q \cdot \delta \\ &= (f)\varepsilon_X \cdot cF \cdot \delta \\ &= (g)\varepsilon_X \cdot cF \cdot \delta \\ &= (g \cdot c)\varepsilon_Q \cdot \delta. \end{aligned}$$

Since F is regular (by Theorem VII.2.10) and $(f)\varepsilon_X = (g)\varepsilon_X$, the images of f and g are the same and hence,

$$g = r \cdot f \quad \text{for some } r \in [n]^n.$$

Choose a one-to-one n-tuple $\bar{w} \in I^\# H_\sigma$ with

$$\bar{w}_1 \in \bar{L}_{r_1}, \ldots, \bar{w}_n \in \bar{L}_{r_n}.$$

Then clearly

$$(\bar{w})\rho H_\Sigma = (q_{r_1}, \ldots, q_{r_n}) = r \cdot f \cdot c = g \cdot c,$$

therefore

$$\begin{aligned} q_w &= (g \cdot c)\varepsilon_Q \cdot \delta \\ &= (\bar{w})\rho H_\Sigma \cdot \varepsilon_Q \cdot \delta \\ &= (\bar{w})\varepsilon_{I^\#} \cdot \rho F \cdot \delta \\ &= ((\bar{w})\varepsilon_{I^\#} \cdot \varphi) \cdot \rho. \end{aligned}$$

Hence, $q_w \in T$ implies that

$$(\bar{w})\varepsilon_{I^\#} \cdot \varphi \in L,$$

which, by VII.2.11(a), means that there exist $p, p' \in P_\sigma$ with

$$r_{(i)p}\gamma(i)p' \quad (i \in [n])$$

i.e., with

$$p \cdot r \cdot f = p' \cdot f.$$

Then

$$q = r \cdot f = (p^{-1} \cdot p) \cdot f$$

and, since

$$p^{-1} \cdot p' \in P_\sigma,$$

this concludes the proof. □

Exercises VII.2

A. Minimal presentations of super-finitary functors. (i) Let $\varepsilon\colon H_\Sigma \to F$ be a minimal presentation. Prove that for each presentation $\varepsilon'\colon H_{\Sigma'} \to F$ there exists a monotransformation $\mu\colon H_\Sigma \to H_{\Sigma'}$ with

$$\varepsilon = \mu \cdot \varepsilon'.$$

(ii) Why is the epitransformation $\varepsilon\colon H_2 \to F$, given by

$$(x, x)\varepsilon = (y, y)\varepsilon$$

not a minimal presentation of a superfinitary functor? (Hint: F is not standard.)

(iii) Change $\emptyset F$ of the above functor F to obtain a standard functor and find its minimal presentation. Compare with Example (iii) in VII.2.7.

B. Functors P_n (see III.4.1).

(i) Prove that for each natural number n, the functor P_n is regular; for which n is P_n perfect?

(ii) Write down a system of equations for a minimal presentation of P_3.

(iii) Define a nondeterministic P_3-acceptor

$$A = ([3]), \delta, \{1\}, \{x, y\}, \lambda)$$

by $([3])\delta = [3]$ and $(M)\delta = \{3\}$ if $M \neq [3]$; $(x)\lambda = [3]$ and $(y)\lambda = \{3\}$. Prove that the language L_A recognized by A is not recognizable. [Hint: Verify that L_A and $I^\# - L_A$ are infinite sets and for $t_1, t_2, t_3 \in I^\#$ pairwise distinct, $(\{t_1, t_2, t_3\})\varepsilon_{I^\#} \cdot \varphi \in L_A$ iff at least two of the elements t_1, t_2, t_3 are in L_A.]

C. Relations and epitransformations. (i) Prove that the minimal presentation $\varepsilon\colon H_\Sigma \to F$ of a perfect functor F has the property that for each relation $r\colon A \rightharpoonup B$ we have

$$\varepsilon_A \circ rF = rH_\Sigma \circ \varepsilon_B.$$

(ii) Verify that the minimal presentation of D_2 [VII.2.7 (iii)] does not have this property. [Hint: Consider $r: \{x, y\} \rightharpoonup \{x\}$ with $(x)r = x$ and $(y)r$ undefined.]

D. Recognizability. Let $\Sigma = \Sigma_0 \cup \Sigma_2$ with $\Sigma_0 = \{0\}$ and $\Sigma_2 = \{\sigma, \tau\}$. The quotient F of H_Σ, given by the equations

$$(x, x)\sigma = 0 \quad \text{and} \quad (x, x)\tau = 0$$

has the property that the recognizable, P-recognizable and N-recognizable languages form three distinct classes:

(i) Proceeding analogously as in Example VII.2.2, find a P-recognizable language which is not recognizable.

(ii) Define a nondeterministic acceptor

$$A = ([2], \delta, [2], \{x, y\}, \lambda)$$

by $(x)\lambda = (y)\lambda = [2]$; for $1_{[2]} \in [2]H_\sigma$ put $((1_{[2]})\ \varepsilon_{[2]})\delta = [2]$ and else δ undefined. Verify that the language L_A recognized by A is not P-recognizable by proving the following properties: L_A is infinite and for arbitrary $t_1, t_2 \in L_A$ with $t_1 \neq t_2$

$$((t_1, t_2)\sigma)\varepsilon_{I^\#} \cdot \varphi \in L_A \quad \text{and} \quad ((t_1, t_2)\tau)\varepsilon_{I^\#} \cdot \varphi \notin L_A.$$

(Hint: To prove that a partial acceptor cannot recognize such a language, consider an infinite subset of L_A on which the run relation is constant.)

VII.3. Kleene Theorem

3.1. For sequential and tree automata, Kleene theorem characterizes the languages recognizable by finite automata as the rational languages (i.e., those obtained from finite languages by union, concatenation and iteration). In the present section, we define concatenation and iteration for each super-finitary varietor F in **Set**. We prove that Kleene theorem holds iff F is perfect.

This shows that there is a deep interrelationship of the concepts of recognizability and nondeterminism: rational languages coincide with recognizable languages iff these coincide with N-recognizable ones.

3.2. Definition of concatenation. Recall from II.4.8 that for two Σ-tree languages $L, K \subset I^\#$ (where I is a finite set of variables) and for each $x \in I$, the x-concatenation is the language

$$L \cdot_x K \subset I^\#$$

of all trees obtained from K-trees by substituting some L-tree for each leaf labelled by x. This can be described algebraically as follows. Let us form the free algebra over the disjoint union of $I - \{x\}$ and L:

$$(I - \{x\} + L)^{\#}.$$

Its elements are finite trees with leaves labelled by variables other than x, or by elements of L. (If L contains a variable $y \in L$, we distinguish it from the element in I by writing y'.) Example: $L = \{x, t\}$ with

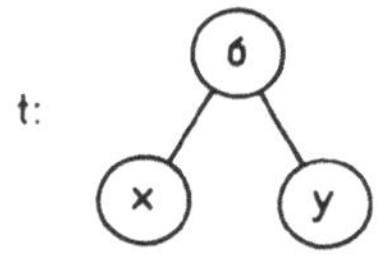

then $s_1, s_2 \in (I - \{x\} + L)^{\#}$ for the following trees

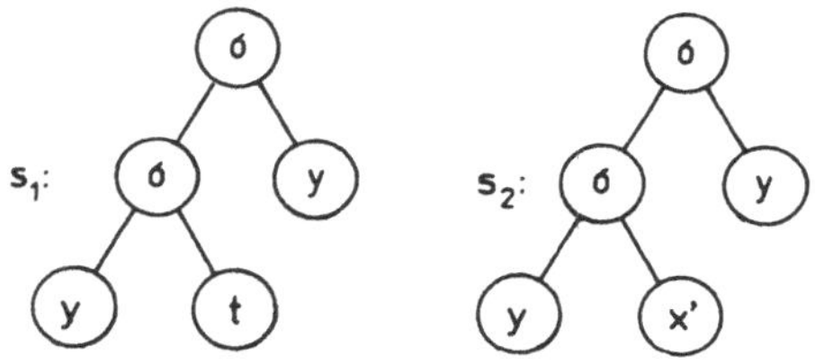

For each $s \in (I - \{x\} + L)^{\#}$ denote by

$$(s)a \in I^{\#}$$

the tree obtained from s by substituting each leaf $t \in L$ by the actual subtree t:

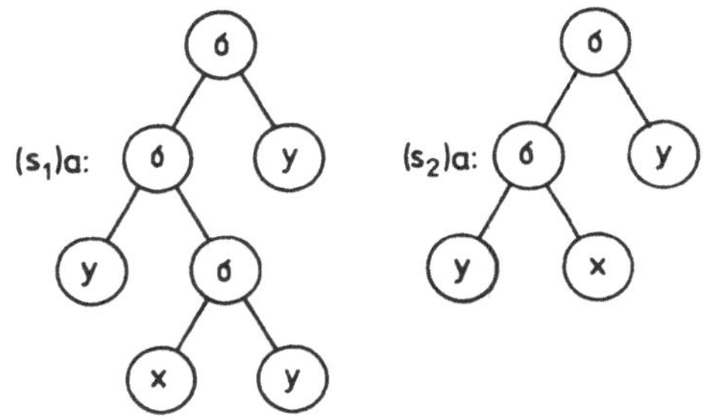

Thus, we obtain a map

$$a : (I - \{x\} + L)^{\#} \rightarrow I^{\#}$$

which is the homomorphism freely extending the map

$$a_0 : (I - \{x\}) + L \rightarrow I^{\#}$$

both components of which are the inclusion maps.

Each tree in $L \cdot_x K$ has the form $(s)a$ for some $s \in (I - \{x\} + L)^{\#}$. More in detail, for every tree $s \in (I - \{x\} + L)^{\#}$ denote by

$$(s)b \in I^{\#}$$

the tree obtained from s by substituting each leaf $t \in L$ by x.

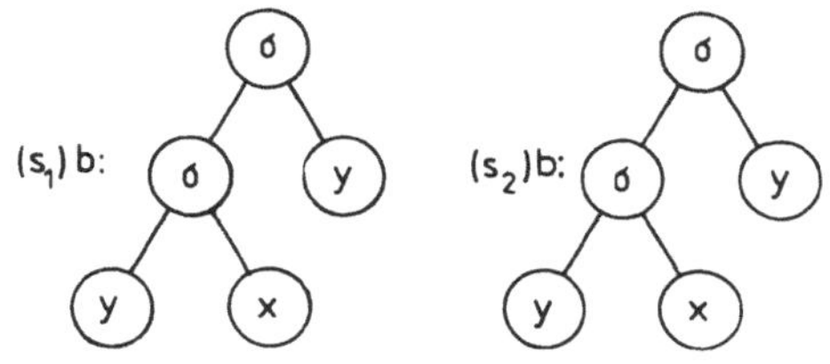

Thus, we obtain a map

$$b: (I - \{x\} + L)^{\#} \to I^{\#}$$

which is the homomorphism freely extending the map

$$b_0: (I - \{x\} + L)^{\#} \to I^{\#}$$

the first component of which is the inclusion map and the second is the constant map with value x. Then

$$L \cdot_x K = \{(s)a;\ s \in (I - \{x\} + L)^{\#} \text{ fulfils } (s)b \in K\}$$

or, shortly,

$$L \cdot_x K = \{(s)a;\ s \in (K)b^{-1}\}.$$

3.3. More generally, let F be a varietor in **Set**, let I be a finite set and let K, $L \subset I^{\#}$ be languages. For each $x \in I$ we denote by

$$a_0,\ b_0: (I - \{x\}) + L \to I^{\#}$$

the following maps: the first component of a_0 and b_0 is the restriction of $\eta: I \to I^{\#}$; the second component of a_0 is the inclusion map, and that of b_0 is the constant map with value $(x)\eta$. These maps can be extended freely to F-homomorphisms

$$a,\ b: (I - \{x\} + L)^{\#} \to I^{\#}.$$

Definition. The language

$$L \cdot_x K = \{(s)a;\ s \in (K)b^{-1}\} \subset I^{\#}$$

is called the *x-concatenation* of L and K in I.

This definition turns out to agree well with the concatenation of tree languages: if $\varepsilon: H_\Sigma \to F$ is a presentation of F-algebras as Σ-algebras (III.3.3), then we prove that concatenation can be performed on the corresponding tree languages with the same result.

Convention. For each presentation $\varepsilon\colon H_\Sigma \to F$ we denote by

$$(I^\#, \varphi) \quad \text{and} \quad \eta\colon I \to I^\#$$

the free F-algebra, and by

$$(I^{\#@}, \varphi^@) \quad \text{and} \quad \eta^@\colon I \to I^{\#@}$$

the free Σ-algebra. Recall from III.3.1 that each F-algebra (Q, δ) defines naturally the H_Σ-algebra $(Q, \varepsilon_Q \cdot \delta)$. We denote by

$$\bar{\varepsilon}_I\colon (I^{\#@}, \varphi^@) \to (I^\#, \varepsilon_{I^\#} \cdot \varphi)$$

the unique Σ-homomorphism extending $\eta\colon I \to I^\#$.

For each $L \subset I^\#$ put

$$L^@ = (L)\bar{\varepsilon}_I^{-1} \subset I^{\#@}.$$

For each tree $t \in I^{\#@}$ put

$$\langle t \rangle = \{s \in I^{\#@};\, (s)\bar{\varepsilon}_I = (t)\bar{\varepsilon}_I\},$$

and given $L \subset I^{\#@}$, then

$$\langle L \rangle = \bigcup_{t \in L} \langle t \rangle.$$

Remark. For each $x \in I$ we have

$$\langle x \rangle = \{x\}.$$

In fact, $I^\# = I + I^\# F$ and $I^{\#@} = I + I^{\#@} H_\Sigma$ (see IV.3.1) and $\bar{\varepsilon}_I = 1_I + (\bar{\varepsilon}_I H_\Sigma) \cdot \varepsilon_{I^\#}$. Hence, $((x)\eta)\bar{\varepsilon}_I^{-1} = \{(x)\eta^@\}$ and since $\eta^@\colon I \to I^{\#@}$ is the inclusion map, $\langle x \rangle = \{x\}$.

3.4. Proposition. For arbitrary languages $K, L \subset I^\#$ and each $x \in I$, we have

$$L \cdot_x K = (L^@ \cdot_x K^@)\bar{\varepsilon}_I.$$

Proof. Denote by

$$a, b\colon (I - \{x\} + L)^\# \to I^\#$$

the homomorphisms from the definition of concatenation of F-languages, and let

$$a^@, b^@\colon (I - \{x\} + L)^{\#@} \to I^{\#@}$$

by the corresponding H_Σ-homomorphisms. Denote by

$$e_0\colon (I - \{x\}) + L^@ \to (I - \{x\}) + L$$

the coproduct of $1_{I-\{x\}}$ and the restriction of $\bar{\varepsilon}_I$ to $L^{@}$. Then clearly

$$e_0 \cdot a_0 = a_0^{@} \cdot \bar{\varepsilon}_I \quad \text{and} \quad e_0 \cdot b_0 = b_0^{@} \cdot \bar{\varepsilon}_I.$$

We have a unique H_Σ-homomorphism

$$e: (I - \{x\} + L^{@})^{\#@} \to (I - \{x\} + L)^{\#}$$

with $\eta^{@} \cdot e = e_0 \cdot \eta$. Then e is easily seen to be onto. We have

$$e \cdot a = a^{@} \cdot \bar{\varepsilon}_I: ((I - \{x\} + L)^{\#@}, \varphi^{@}) \to (I^{\#}, \varepsilon_{I^{\#}} \cdot \varphi)$$

because $e \cdot a$ and $a^{@} \cdot \bar{\varepsilon}_I$ are Σ-homomorphisms with

$$\begin{aligned} \eta^{@} \cdot (e \cdot a) &= e_0 \cdot \eta \cdot a \\ &= e_0 \cdot a_0 \\ &= a_0^{@} \cdot \bar{\varepsilon}_I \\ &= \eta^{@} \cdot (a^{@} \cdot \bar{\varepsilon}_I). \end{aligned}$$

Analogously,

$$e \cdot b = b^{@} \cdot \bar{\varepsilon}_I.$$

Therefore

$$\begin{aligned} (L^{@} \cdot_x K^{@})\bar{\varepsilon}_I &= \{(s)a^{@} \cdot \bar{\varepsilon}_I;\ s \in (K^{@})(b^{@})^{-1}\} \\ &= \{(s)e \cdot a;\ (s)b^{@} \cdot \bar{\varepsilon}_I \in K\} \\ &= \{((s)e)a;\ ((s)e)b \in K\}. \end{aligned}$$

Since e is surjective, we get

$$(L^{@} \cdot_x K^{@})\bar{\varepsilon}_I = \{(t)a;\ (t)b \in K\} = L \cdot_x K.$$ □

Remarks. (i) For each standard finitary varietor F we can assume that

$$I \subset J \quad \text{implies} \quad I^{\#} \subset J^{\#}$$

and for the inclusion map $v: I \to J$, $v^{\#}: I^{\#} \to J^{\#}$ is also the inclusion map. In fact, recall that the free algebra is determined only up to an isomorphism, and the same is true about coproducts. Assuming a standard choice of coproducts in **Set**, we can guarantee that

$$X \subset X' \quad \text{and} \quad Y \subset Y' \quad \text{imply} \quad X + Y \subset X' + Y'.$$

Then $I^{\#} \subset J^{\#}$ follows from the fact that $I^{\#} = \bigcup_{n<\omega} W_n^I$ with

$$W_0^I = I \quad \text{and} \quad W_{n+1}^I = I + W_n F,$$

analogously with $J^{\#}$. We have

$$W_0^I = I \subset J = W_0^J$$

and if $W_n^I \subset W_n^J$, then $W_n^I F \subset W_n^J F$ (see III.4.5) and hence,

$$W_{n+1}^I = I + W_n^I F \subset J + W_n^J = W_{n+1}^J.$$

It is easy to verify that $\nu^\#$ is the inclusion map.

It follows that for $f: I \to Q$ and for $g: J \to Q$ extending f, and for each F-algebra (Q, δ), also $f^\#: (I^\#, \varphi) \to (Q, \delta)$ is extended by $g^\#: (J^\#, \varphi) \to (Q, \delta)$.

(ii) Concatenation can depend on the set of variables: given $I \subset J$ and $x \in I$, then for $K, L \subset I^\#$ ($\subset J^\#$) the languages

$$L \cdot_x K \text{ (in } I) \quad \text{and} \quad L \cdot_x K \text{ (in } J)$$

can be distinct. We prove below that this cannot happen for a regular functor F.

3.5. Definition. The *x-iteration* of a language $L \subset I^\#$ in I is defined for each $x \in I$ with $(x)\eta \in L$ as the following language

$$L^{*x} = L \cup (L \cdot_x L) \cup (L \cdot_x L) \cdot_x L) \cup (((L \cdot_x L) \cdot_x L) \cdot_x L) \ldots$$

Analogously to the concatenation, the iteration agrees well with that for tree languages:

Proposition: For each language $L \subset I^\#$ with $(x)\eta \subset L$, we have

$$L^{*x} = ((L^{@ *x})\bar{\varepsilon}_I.$$

Proof. I. We prove first that given languages L, $K \subset I^\#$ and a tree language $\tilde{L} \subset I^{\#@}$ with $(\tilde{L})\bar{\varepsilon}_I = L$, then

$$L \cdot_x K = (\tilde{L} \cdot_x K^@)\bar{\varepsilon}_I.$$

For this, it is sufficient to prove that

$$L^@ \cdot_x K^@ \subset \langle \tilde{L} \cdot_x K^@ \rangle$$

because then by Proposition VII.3.4 we get $L \cdot_x K \subset (\tilde{L} \cdot_x K^@)\bar{\varepsilon}_I$, and the reverse inclusion is clear.

Let $r \in L^@ \cdot_x K^@$. Then we have a tree $t \in K^@$ and for each x-labelled leaf

$$c \in (x)t^{-1}$$

a tree $s_c \in L^@$ such that r is obtained from t by substituting each $c \in (x)t^{-1}$ by s_c. Since

$$s_c \in L^@ = ((\tilde{L})\bar{\varepsilon}_I)\bar{\varepsilon}_I^{-1},$$

we can choose $s_c' \in \tilde{L}$ with $\langle s_c \rangle = \langle s_c' \rangle$. Let r' be the tree obtained from t by substituing each $c \in (x)t^{-1}$ by s_c'. Since the equivalence $\bar{\varepsilon}_I \cdot \bar{\varepsilon}_I^{-1}$ is a congruence on the free algebra $I^{\#@}$, clearly

$$\langle r \rangle = \langle r' \rangle.$$

Further,

$$r' \in \tilde{L} \cdot_x K^{@},$$

which proves that

$$r \in \langle \tilde{L} \cdot_x K^{@} \rangle.$$

II. By Proposition VII.3.4, we have

$$L \cdot_x L = (L^{@} \cdot_x L^{@})\bar{\varepsilon}_I$$

and using I above, we conclude that

$$(L \cdot_x L) \cdot_x L = ((L^{@} \cdot_x L^{@}) \cdot_x L^{@})\bar{\varepsilon}_I$$

and we use I again, etc. Thus,

$$\begin{aligned} L^{*x} &= L \cup (L \cdot_x L) \cup ((L \cdot_x L) \cdot_x L) \cup \dots \\ &= (L^{@})\bar{\varepsilon}_I \cup (L^{@} \cdot_x L^{@})\bar{\varepsilon}_I \cup ((L^{@} \cdot_x L) \cdot_x L^{@})\bar{\varepsilon}_I \cup \dots \\ &= (L^{@} \cup (L^{@} \cdot_x L^{@}) \cup ((L^{@} \cdot_x L^{@}) \cdot_x L^{@}) \cup \dots)\bar{\varepsilon}_I = ((L^{@})^{*x})\bar{\varepsilon}_I. \quad \square \end{aligned}$$

3.6. By VII.2.5, if $\varepsilon : H_\Sigma \to F$ is a minimal presentation, then for each basic tree $s = (x_1, \dots, x_n)\sigma$ with x_i pairwise distinct, we have

$$\langle s \rangle = \{(x_{(1)p}, \dots, x_{(n)p})\sigma;\ p \in P_\sigma\}.$$

We generalize this to more complex trees, and then we use this generalization in order to prove that Kleene theorem does not hold for functors which are not perfect.

Lemma. Let $\varepsilon: H_\Sigma \to F$ be a minimal presentation of a super-finitary functor, and let

$$s = (s_1, \dots, s_n)\sigma \in I^{\#@} \quad (\sigma \in \Sigma_n,\ n > 0)$$

be a Σ-tree with $\langle s_1 \rangle, \dots, \langle s_n \rangle$ pairwise disjoint. Then

$$\langle s \rangle = \{(t_1, \dots, t_n)\sigma;\ t_i \in \langle s_{(i)p} \rangle\ (i = 1, \dots, n) \text{ for some } p \in P_\sigma\}.$$

Proof. For each tree $t \in I^{\#@}$ we have

$$t \in \langle s \rangle \quad \text{iff} \quad (t)\bar{\varepsilon}_I = (s)\bar{\varepsilon}_I,$$

and if this is the case, then $t \notin I$ by Remark VII.3.3. Thus $t = (t_1, \dots, t_m)\tau$ for some $\tau \in \Sigma_n$.

Since $\bar{\varepsilon}_I$ is a homomorphism, we have

$$(s)\bar{\varepsilon}_I = (s_1, \dots, s_n)\varphi^{@} \cdot \bar{\varepsilon}_I = ((s_1)\bar{\varepsilon}_I, \dots, (s_n)\bar{\varepsilon}_I)\bar{\varepsilon}_{I^\#} \cdot \varphi$$

and analogously,

$$(t)\bar{\varepsilon}_I = ((t_1)\bar{\varepsilon}_I, \dots, (t_m)\bar{\varepsilon}_I)\bar{\varepsilon}_{I^\#} \cdot \varphi.$$

The map φ is one-to-one (Remark IV.4.2) and hence, $t \in \langle s \rangle$ iff

$$((s_1)\bar{\varepsilon}_I, \ldots, (s_n)\bar{\varepsilon}_I)\varepsilon_{I\#} = ((t_1)\bar{\varepsilon}_I, \ldots, (t_m)\bar{\varepsilon}_I)\varepsilon_{I\#}.$$

By hypothesis, $(s_i)\bar{\varepsilon}_I$ are pairwise distinct elements of $I^{\#}$ and hence by VII.2.5, the last equation is equivalent to the existence of $p \in P_\sigma$ with

$$\tau = \sigma \quad \text{and} \quad (t_i)\bar{\varepsilon}_I = (s_{(i)p})\bar{\varepsilon}_I \quad \text{for } i = 1, \ldots, n.$$

This was to be proved. □

Basic Example. Let $\varepsilon\colon H_\Sigma \to F$ be a minimal presentation, and let $\sigma \in \Sigma_n$ with $n > 0$. Consider the following trees in variables $1, 2, \ldots, n$:

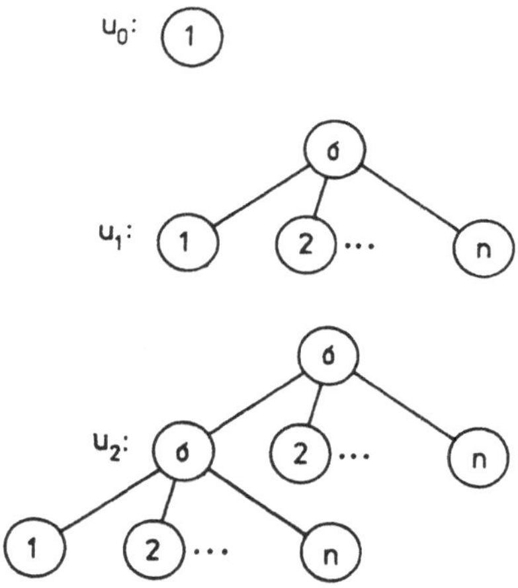

etc. Thus, $u_0 = 1$ and $u_1 = (1, 2, \ldots, n)\sigma$ and

$$u_{k+1} = \{u_k\} \cdot_1 u_1.$$

By the preceding lemma and Remark VII.3.3, $\langle u_1 \rangle$ is the set of all trees

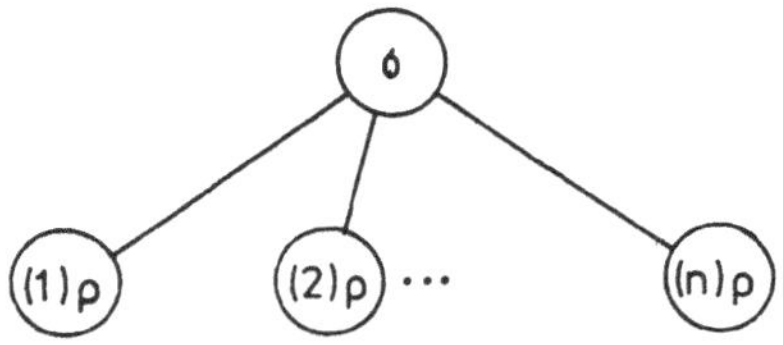

with $p \in P_\sigma$. None of these trees belongs to $\langle 2\rangle, \ldots, \langle n\rangle$ and hence, by the same lemma, $\langle u_2\rangle$ is the set of all trees

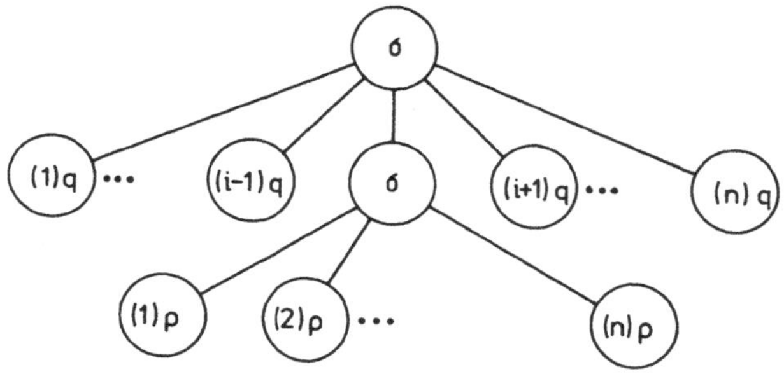

with $p, q \in P_\sigma$ and $i = (1)q^{-1}$.

Analogously, u_3 is the set of all trees

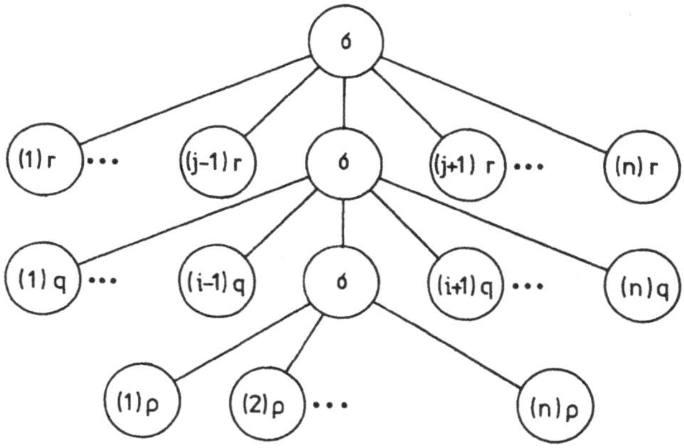

with $p, q, r \in P_\sigma$ and $i = (1)q^{-1}$ and $j = (1)r^{-1}$, etc. Note that

$$\langle u_2\rangle = \langle u_1\rangle \cdot_1 \langle u_1\rangle$$

and $$\langle u_3\rangle = \langle u_1\rangle \cdot_1 \langle u_1\rangle \cdot_1 \langle u_1\rangle.$$

In general

(a) $$\langle u_k\rangle = \langle u_1\rangle \cdot_1 \langle u_1\rangle \cdot_1 \ldots \cdot_1 \langle u_1\rangle \quad (k\text{-times})$$

and hence, for $U_k = \{1, u_k\}$,

(b) $$U_k^{*1} = \langle U_k\rangle \cup \langle U_{2k}\rangle \cup \langle U_{3k}\rangle \cup \ldots .$$

Further, since u_k has just one leaf labelled by 1, clearly

(c) $$\{u_k\} = \{u_1\} \cdot_1 \{u_1\} \cdot_1 \ldots \cdot_1 \{u_1\} \quad (k\text{-times})$$

and hence,

(d) $$U_k^{*1} = \bigcup_{n=0}^{\infty} U_{nk}.$$

Further, for

$$V_k = (U_k)\bar{\varepsilon}_{[n]} \subset [n]^{\#}$$

we have

(e) $$\langle U_k \rangle^{*1} = (V_k^{*1})^{@}.$$

In fact, for each k, $k' \in \omega$ we have by Proposition VII.3.4,

$$\begin{aligned} V_k \cdot_1 V_{k'} &= (U_k)\bar{\varepsilon}_{[n]} \cdot_1 (U_{k'})\bar{\varepsilon}_{[n]} \\ &= (\langle U_k \rangle \cdot_1 \langle U_{k'} \rangle)\bar{\varepsilon}_{[n]} \\ &= (\langle U_k \rangle \cup \langle U_{k'} \rangle \cup \langle U_{k+k'} \rangle)\bar{\varepsilon}_{[n]} \\ &= (U_k \cup U_{k'} \cup U_{k+k'})\bar{\varepsilon}_{[n]} \\ &= V_k \cup V_{k'} \cup V_{k+k'}. \end{aligned}$$

Therefore,

$$\begin{aligned} (\langle U_k \rangle^{*1})\bar{\varepsilon}_{[n]} &= V_k \cup V_{2k} \cup V_{3k} \cup \dots \\ &= V_k \cup (V_k \cdot_1 V_k) \cup \dots \\ &= V_k^{*1}; \end{aligned}$$

since by (b) clearly

$$\langle U_k \rangle^{*1} = \langle \langle U_k \rangle^{*1} \rangle,$$

this proves (e).

Finally,

(f) $$V_k^{*1} \cdot_1 V_k^{*1} = V_k^{*1}$$

because by Proposition VII.3.4

$$\begin{aligned} V_k^{*1} \cdot_1 V_k^{*1} &= [(V_k^{*1})^{@} \cdot_1 (V_k^{*1})^{@}]\bar{\varepsilon}_{[n]} \\ &= (\langle U_k \rangle^{*1} \cdot_1 \langle U_k \rangle^{*1})\bar{\varepsilon}_{[n]}. \end{aligned}$$

Since clearly $\langle U_k \rangle^{*1} \cdot_1 \langle U_k \rangle^{*1} = \langle U_k \rangle^{*1}$, this proves (f).

For each map

$$f\colon [n] \to \omega$$

we put

(g) $$s_f = (u_{(1)f}, u_{(2)f}, \dots, u_{(n)f})\sigma \in [n]^{\#@}.$$

If f is one-to-one, then $\langle u_{(i)f} \rangle$ are pairwise disjoint and hence,

(h) $\langle s_f \rangle = \{(s_1, \ldots, s_n)\sigma;$ there exists $p \in P_\sigma$ with $s_i \in \langle u_{(i)p \cdot f} \rangle$ for $i = 1, \ldots, n\}$.

It is clear that the depth $|t|$ (II.1.5) of each tree $t \in \langle u_k \rangle$ is equal to the depth of u_k which is k. Thus,

(k) $t \in \langle u_k \rangle$ implies $|t| = k$; $t \in \langle U_k \rangle$ implies $|t| \in \{jk\}_{j=0}^{\infty}$.

3.7. Theorem. Let F be a super finitary functor such that the languages

$$\{v, (x)\eta\}^{*x} \cdot_x \{w\} \subset I^\#$$

are recognizable (for each finite set I, each $x \in I$ and $v, w \in I^\#$). Then F is perfect.

Proof. By VII.2.7, it is sufficient to prove that given $f \in [n]H_\sigma$ and $g \in [n]H_\tau$, where

$$\sigma \in \Sigma_n, \tau \in \Sigma_m \quad \text{and} \quad n \geq m \geq 2,$$

then $(f)\varepsilon_{[n]} = (g)\varepsilon_{[n]}$ implies

(*) $\sigma = \tau$ and $g = p \cdot f$ for some $p \in P_\sigma$.

Define a one-to -one map $f': [n] \to \omega$ by

$$(i)f' = (i)f + (i - 1) \cdot n \quad \text{for } i = 1, \ldots, n.$$

To prove (*), we use the recognizability of the language

$$L_f = V_n^{*1} \cdot_1 \{w\}$$

with

$$w = (s_{f'})\bar{\varepsilon}_{[n]},$$

in the notation of the Basic Example above. Note that by (e) above and Proposition VII.3.4 we have

$$L_f^@ = (V_n^{*1})^@ \cdot_1 \{w\}^@ = \langle\langle U_n \rangle^{*1} \cdot_1 \langle s_{f'} \rangle\rangle.$$

(A) For each one-to-one map $r: [n] \to \omega$ with

$$s_r \in L_f^@$$

we prove that there exists a permutation $p \in P_\sigma$ such that

$$(i)r \equiv (i)p \cdot f (\mathrm{mod}\; n) \quad \text{for } i = 1, \ldots, n,$$

i.e., that $(i)r - (i)p \cdot f$ is a multiple of n. Since

$$s_r \in \langle\langle U_n \rangle^{*1} \cdot_1 \langle s_{f'} \rangle\rangle,$$

there exists $t \in \langle s_r \rangle \cap (\langle U_n \rangle^{*1} \cdot_1 \langle s_{f'} \rangle)$. By (h) above, there exists $p \in P_\sigma$ and

$t_i \in \langle u_{(i)p \cdot r} \rangle$ with

$$t = (t_1, \ldots, t_n)\sigma.$$

Since

$$t \in \langle U_n \rangle^{*1} \cdot_1 \langle s_{f'} \rangle,$$

the tree t is obtained from a tree

$$s \in \langle s_{f'} \rangle$$

by substitution of trees in $\langle U_n \rangle^{*1}$ for leaves labelled by 1. Applying (h) to f', we see that there exist $q \in P_\sigma$ and $s_i \in \langle u_{(i)q = f'} \rangle$ with

$$s = (s_1, \ldots, s_n)\sigma.$$

Each tree in $\langle U_n \rangle^{*1}$ has depth $\equiv 0 \pmod n$, see (k) above. Thus, the branches in t and s have the same depths modulo n:

$$|t_i| \equiv |s_i| \pmod n \quad \text{for } i = 1, \ldots, n.$$

By (k) applied to $\langle u_{(i)p \cdot r} \rangle$ and $\langle u_{(i)q \cdot f'} \rangle$, we conclude that

$$(i)p \cdot r \equiv (i)q \cdot f' \pmod n \quad \text{for } i = 1, \ldots, n.$$

Therefore, the permutation

$$p^{-1} \cdot q \in P_\sigma$$

fulfils

$$(i)r \equiv (i)(p^{-1} \cdot q) \cdot f' \pmod n \quad \text{for } i = 1, \ldots, n.$$

(B) For each one-to-one map $r: [m] \to \omega$ and each $\tau \in \Sigma_m$, $m > 0$,

we prove that

$$(u_{(1)r}, \ldots, u_{(m)r})\tau \in L_f^{@} \quad \text{implies} \quad \tau = \sigma.$$

In fact, (k) implies that $\langle u_{(1)r} \rangle, \ldots, \langle u_{(m)r} \rangle$ are all disjoint, hence by Lemma VII.3.6 there exist $p \in P_\sigma$ and $t_i \in \langle u_{(i)p \cdot r} \rangle$ with

$$(t_1, \ldots, t_m)\tau \in \langle U_n \rangle^{*1} \cdot_1 \langle s_{f'} \rangle.$$

Each tree in $\langle U_n \rangle^{*1} \cdot_1 \langle s_{f'} \rangle$ has its root labelled by σ. Therefore, $\tau = \sigma$.

(C) For each $r: [n] \to \omega$ such that

$$(i)f' < (i)r \equiv (i)f' \pmod n \quad \text{for } i = 1, \ldots, n$$

we have

$$s_r = (u_{(1)r}, \ldots, u_{(n)r})\sigma \in L_f^{@}.$$

To prove this, put $(i)d = (i)r - (i)f'$. Since $(i)d$ is a positive multiple of n, clearly

$$u_{(i)d} \in \langle U_n \rangle^{*1} \quad \text{for } i = 1, \ldots, n.$$

The tree

$$\begin{aligned} s_r &= (u_{(1)r}, \ldots, u_{(n)r})\sigma \\ &= (u_{(1)d+(1)f'}, \ldots, u_{(n)d+(n)f'})\sigma \\ &= (u_{(1)d} \cdot_1 u_{(1)f'}, \ldots, u_{(n)d} \cdot_1 u_{(n)f'})\sigma \end{aligned}$$

is obtained from $s_{f'}$ by substituing $u_{(1)d}$ for the 1-labelled leaf in $u_{(1)f'}$, $u_{(2)d}$ for that in $u_{(2)f'}$, etc.:

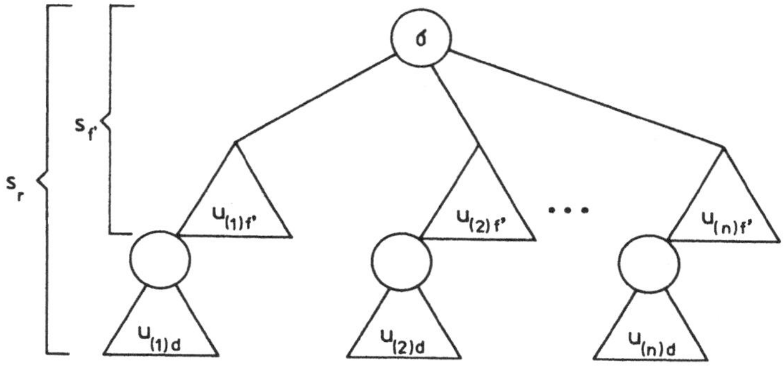

Thus, $s_r \in L_f^{@}$.

(D) The proof of (∗). We use the fact that L_f is recognized by an F-acceptor $A = (Q, \delta, T, [n])$. Clearly, $\bar{\varepsilon}_{[n]}$ is one-to-one on the set $\{u_0, u_1, u_2, \ldots\}$, and hence, the set

$$\{(u_i)\bar{\varepsilon}_{[n]}, (u_{i+n})\bar{\varepsilon}_{[n]}, (u_{i+2n})\bar{\varepsilon}_{[n]}, \ldots\} \subset [n]^{\#}$$

is infinite for each $i \in [n]$. On the other hand, Q is a finite set. Thus, there exists an infinite set

$$B_i \subset \{i, i+n, i+2n, \ldots\}$$

such that the run map $\rho: [n]^{\#} \to Q$ is constant on the set of all $(u_k)\bar{\varepsilon}_{[n]}$ with $k \in B_i$. More precisely, there are states

$$(1)q, \ldots, (n)q \in Q$$

such that

$$k \in B_i \quad \text{implies} \quad (u_k)\bar{\varepsilon}_{[n]} \cdot \rho = (i)q \quad (i = 1, \ldots, n).$$

This defines a map $q: [n] \to Q$.

In the final part of the present proof we shall verify that the map q fulfils

$$(f)(\varepsilon_{[n]} \cdot qF \cdot \delta) \in T.$$

Then we conclude the proof as follows. Since

$$(f)\varepsilon_{[n]} = (g)\varepsilon_{[n]},$$

we have

$$(g \cdot q)\varepsilon_Q \cdot \delta = (g)qH_\Sigma \cdot \varepsilon_Q \cdot \delta = (g)\varepsilon_{[n]} \cdot qF \cdot \delta \in T.$$

Choose a one-to-one map $r: [m] \to \omega$ such that

$$(1)r \in B_{(1)g}, \ldots, (m)r \in B_{(m)g},$$

and put

$$t = (u_{(1)r}, \ldots, u_{(m)r})\tau.$$

Then $t \in L_f^{@}$, i.e. $(t)\bar{\varepsilon}_{[n]} \cdot \rho \in T$ because ρ is a homomorphism and hence,

$$\begin{aligned}(t)\bar{\varepsilon}_{[n]} \cdot \rho &= (u_{(1)r}, \ldots, u_{(m)r})\varphi^{@} \cdot \bar{\varepsilon}_{[n]} \cdot \rho \\ &= (u_{(1)r}, \ldots, u_{(m)r})\bar{\varepsilon}_{[n]}H_\Sigma \cdot \rho H_\Sigma \cdot \varepsilon_Q \cdot \delta \\ &= ((u_{(1)r})\bar{\varepsilon}_{[n]} \cdot \rho, \ldots, (u_{(m)r})\bar{\varepsilon}_{[n]} \cdot \rho)\varepsilon_Q \cdot \delta \\ &= ((1)g \cdot q, \ldots (m)g \cdot q)\varepsilon_Q \cdot \delta \\ &= (g \cdot q)\varepsilon_Q \cdot \delta \in T.\end{aligned}$$

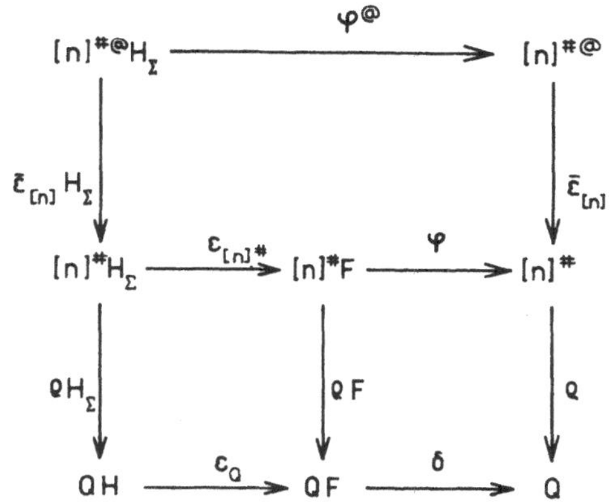

Then by (B), $\tau = \sigma$; consequently, $t = s_r$. By (A), there exists a permutation $p \in P_\sigma$ with $r \equiv p \cdot f \pmod{n}$. Since $(i)r \in B_{(i)g}$ implies $(i)r \equiv (i)g \pmod{n}$, we conclude that

$$(i)g \equiv (i)p \cdot f \pmod{n}, \quad i = 1, \ldots, n.$$

Since g and f have just the values $1, \ldots, n$, this implies

$$g = p \cdot f.$$

(E) The proof of $(f)(\varepsilon_{[n]} \cdot qF \cdot \delta) \in T$. Choose a map $r: [n] \to \omega$ with $(i)r \in B_{(i)f}$ and $(i)r > (i)f'$ for $i = 1, \ldots, n$. We have $(i)r \equiv (i)f \equiv (i)f'$ (mod n) for each i and hence, by (C), $s_r \in L_f^{@}$. Thus,

$$(s_r)\bar{\varepsilon}_{[n]} \cdot \rho \in T.$$

Finally,

$$\begin{aligned}(s_r)\bar{\varepsilon}_{[n]} \cdot \rho &= (u_{(1)r}, \ldots, u_{(n)r})\varphi^{@} \cdot \bar{\varepsilon}_{[n]} \cdot \rho \\ &= ((u_{(1)r})\bar{\varepsilon}_{[n]} \cdot \rho, \ldots, (u_{(n)r})\bar{\varepsilon}_{[n]} \cdot \rho)\varepsilon_Q \cdot \delta \\ &= ((1)f \cdot q, \ldots, (n)f \cdot q)\varepsilon_Q \cdot \delta \\ &= (f \cdot q)\varepsilon_Q \cdot \delta \\ &= (f)\varepsilon_{[n]} \cdot qF \cdot \delta.\end{aligned}$$

This concludes the proof. □

3.8. Definition. Let F be a super-finitary varietor. The class of *rational languages* is the least class of languages containing each finite language and closed under union, concatenation and iteration.

Thus, a language $L \subset I^{\#}$ is rational iff there exists a rational expression for L, i.e., an expression using languages $\{w\}$ for $w \in J^{\#}$ and operations

$\cup$ (union),
$\cdot_x$ (x-concatenation in J for some finite set J),

and

$*^x$(x-iteration in J for some finite set J),

finitely many times. For example, the languages $\{v, (x)\eta\}^{*x} \cdot_x \{w\}$ of the preceding theorem are rational.

Main Theorem. For each super-finitary functor F, the following statements are equivalent:

(i) Kleene theorem holds, i.e., an F-language is recognizable iff it is rational;
(ii) F is perfect.

Proof. Since perfectness is necessary by the preceding theorem, it remains to prove that it is sufficient. Thus, let Σ be a type, and for each $\sigma \in \Sigma_n$ let P_σ be a permutation group on $[n]$. We prove that the functor

$$F = \coprod_{\sigma \in \Sigma} H_{|\sigma|, P_\sigma}$$

(see Proposition VII.2.7) satisfies the Kleene theorem. We know that Kleene theorem holds for H_Σ (II.4.11). Thus, it is sufficient to prove that a language

$L \subset I^{\#}$ is rational iff the tree language $L^{@} \subset I^{\#@}$ is rational, and L is recognizable iff $L^{@}$ is recognizable.

I. The minimal presentation

$$\varepsilon \colon H_{\Sigma} \to F$$

of F is given by the canonical maps

$$\varepsilon_X \colon \coprod_{\sigma \in \Sigma_n} X^n \to \coprod_{\sigma \in \Sigma_n} X^n/\sim$$

where

$$(x_1, \ldots, x_n)\sigma \sim (y_1, \ldots, y_n)\sigma$$

holds iff there exists a permutation $p \in P_\sigma$ with

$$y_i = x_{(i)p} \quad \text{for} \quad i = 1, \ldots, n.$$

We can describe the homomorphisms $\bar{\varepsilon}_I$ analogously as follows.

Denote by R_0 the following relation on the set $I^{\#@}$ of all Σ-trees over I: $t\, R_0\, t'$, $(t, t' \in I^{\#@})$, iff t' is obtained from t by a P_σ-permutation of branches of a σ-labelled node ($\sigma \in \Sigma$). More precisely, recall that t and t' are partial maps from m^* to $I \cup \Sigma$ (see II.1.4). Then $t\, R_0\, t'$ iff there exists $a \in m^*$ such that

$$(a)t = \sigma \in \Sigma_n$$

and for some $p \in P_\sigma$, the tree t' is defined in each $b \in m^*$ as follows:

$$(b)t' = \begin{cases} (b)t & \text{if } b \neq ac \quad \text{for each } c \in m^*,\ c \neq \emptyset; \\ (bjc)t & \text{if } b = aic \quad \text{for some } c \in m^*, \text{ and } j = (i)p. \end{cases}$$

This relation R_0 is obviously reflexive and symmetric. Let R be the transitive closure of R_0 [i.e., $R = R_0 \cup (R_0 \circ R_0) \cup (R_0 \circ R_0 \circ R_0) \cup \ldots$]. It is easy to verify that R is a congruence on $I^{\#@}$ and that

$$\bar{\varepsilon}_I \colon I^{\#@} \to I^{\#} = I^{\#@}/R$$

is the canonical map.

II. For arbitrary tree languages $K, L \subset I^{\#@}$, we prove that

$$\langle L \rangle \cdot_x \langle K \rangle = \langle L \cdot_x K \rangle.$$

First, consider a tree

$$t \in \langle L \rangle \cdot_x \langle K \rangle.$$

It is obtained from a tree

$$s \in \langle K \rangle$$

by substitutions of x-labelled leaves by trees in $\langle L \rangle$. For each leaf $d \in (x)s^{-1}$ we have $r_d \in \langle L \rangle$ such that t is obtained from s by substituting r_d for x. We can choose $r'_d \in L$ congruent to r_d. The tree t' obtained from s by substituting r'_d for x in d $[d \in (x)s^{-1}]$ is obviously congruent with t:

$$t \, R \, t'; \; t' \in L \cdot_x \langle K \rangle.$$

Further, we can choose $s' \in K$ congruent to s. Without loss of generality, we assume that

$$s \, R_0 \, s',$$

i.e., there is a node a with $(a)s = \sigma$ and there is $p \in P_\sigma$ such that s' is obtained from s by the p-permutation of the branches of a. Then also $(a)t' = \sigma$.

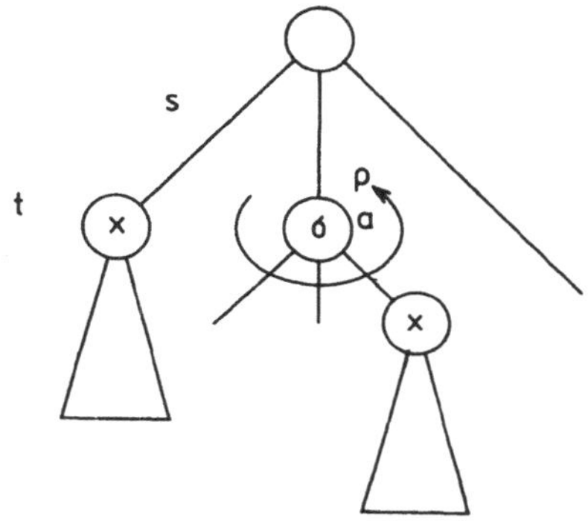

Let t'' be the tree obtained from t' by the p-permutation of the branches of a. Then

$$t \, R \, t' \, R \, t''.$$

Moreover, t'' can also be obtained from s' by substitions of x-labelled leaves by trees in L. In fact, if a leaf $d \in (x)s^{-1}$ has the form

$$d = a \, i \, c \quad (c \in m^*, \, i \in m),$$

then for $j = (i)p$ we have a leaf $a \, j \, c$ of s' which we substitute by r'_d; if d does not have this form, then d is a leaf of s' which we substitute by r'_d. Therefore, $t'' \in L \cdot_x K$, and we have

$$t \in \langle t'' \rangle \subset \langle L \cdot_x K \rangle,$$

so $\langle L \rangle \cdot_x \langle K \rangle \subset \langle L \cdot_x K \rangle$.

To prove the converse implication, consider a tree

$$t \in \langle L \cdot_x K \rangle.$$

We can choose a tree $t' \in L \cdot_x K$ congruent to t. Without loss of generality, we can assume

$$t \, R_0 \, t',$$

i.e., there exists a node $a \in m^*$ with $(a)t = \sigma$ and a permutation $p \in P_\sigma$ such that t' is obtained from t by the p-permutation of the branches of a.

Let

$$s' \in K$$

be a tree such that t' is obtained from s' by substitutions of x-labelled leaves by trees in L. For each $d \in (x)(s')^{-1}$ we denote by $r_d \in L$ the tree used in this substitution.

(i) Let $(a)s' = \sigma$

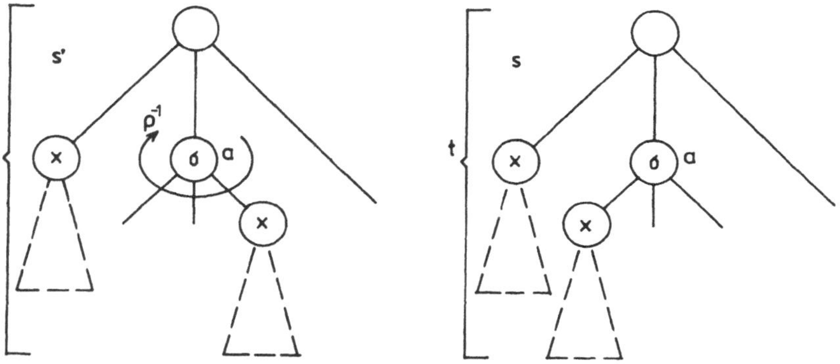

Denote by s the tree obtained from s' by the p^{-1}-permutation of the branches of a. Then

$$s \in \langle K \rangle$$

is a tree from which s' is obtained by the p-permutation of the branches of a. Then t can be obtained from s by substitutions of x-labelled leaves by trees in L: for each $d = a\, i\, c \in (x)(s)^{-1}$ we use the tree r_d for the leaf $a\, j\, c\, [j = (i)p^{-1}]$ of s; for each $d \in (x)(s')^{-1}$ not under a, we use r_d for the leaf d of s. Hence,

$$t \in L \cdot_x \langle K \rangle.$$

(ii) Let $(a)s'$ be undefined or equal to x.

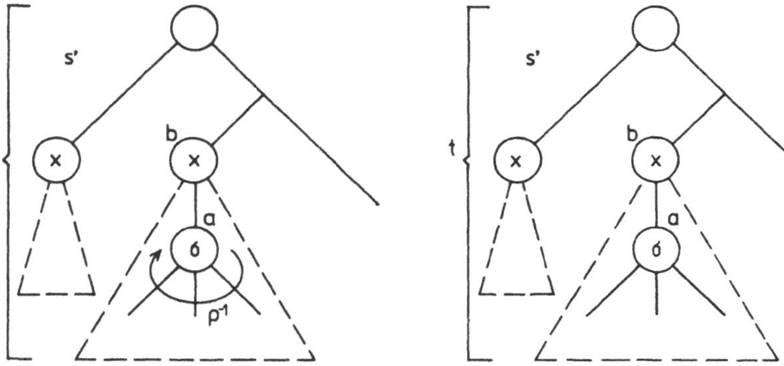

There exists an x-labelled leaf d of s' such that $a = dc$ for some $c \in m^*$. Then clearly $(c)r_d = \sigma$; let $\bar{r}_d$ be the tree obtained from r_d by the p-permutation of the branches of c. Then

$$\bar{r}_d \in \langle L \rangle.$$

The tree t can be obtained from s' by substituting the leaf d by $\bar{r}_d$, and all other x-labelled leaves as before. Therefore,

$$t \in \langle L \rangle \cdot_x K.$$

III. For each tree language $L \subset I^{\#@}$ with $x \in L$ we have

$$\langle L \rangle^{*x} = \langle L^{*x} \rangle.$$

This is a direct conseauence of II.

IV. An F-language $L \subset I^{\#}$ is rational iff the tree language $L^{@} \subset I^{\#@}$ is rational.

First, let us observe that concatenation and iteration are independent of the set of variables: if $K, L \subset I^{\#} \subset J^{\#}$, then for each $x \in I$ the language $K \cdot_x L$ is the same when computed in I or J, and the same holds for L^{*x}. [This follows from the fact that $(I^{\#})\bar{\varepsilon}_J^{-1} = I^{\#@}$.]

For each rational language $L \subset I^{\#}$, the language $L^{@}$ is rational because the operation $(—)^{@}$ preserves

(a) finite languages: for each tree $t \in I^{\#@}$, the class $\langle t \rangle$ is finite and hence, if L is finite, then so is $L^{@}$;

(b) union;

(c) concatenation: by II above and by VII.3.4, we have

$$\begin{aligned}(L \cdot_x K)^{@} &= [(L^{@} \cdot_x K^{@})\bar{\varepsilon}_I]^{@}\\ &= \langle L^{@} \cdot_x K^{@} \rangle\\ &= L^{@} \cdot_x K^{@};\end{aligned}$$

(d) iteration: by III above and by VII.3.5, we have

$$(L^{*x})^{@} = \langle L^{@} \rangle^{*x} = (L^{@})^{*x}.$$

Conversely, for each language $L \subset I^{\#}$ with $L^{@}$ rational, the language

$$L = (L^{@})\bar{\varepsilon}_I$$

is rational because the operation $(—)\bar{\varepsilon}_I$ preserves

(a) finite languages;

(b) union;

(c) concatenation: by II above and by VII.3.4,

$$\begin{aligned}(L \cdot_x K)\bar{\varepsilon}_I &= (\langle L \cdot_x K \rangle)\bar{\varepsilon}_I\\ &= (\langle L \rangle \cdot_x \langle K \rangle)\bar{\varepsilon}_I\\ &= (L)\bar{\varepsilon}_I \cdot_x (K)\bar{\varepsilon}_I;\end{aligned}$$

(d) iteration (analogously by III and VII.3.5).

V. An F-language $L \subset I^{\#}$ is recognizable iff the tree language $L^{@} \subset I^{\#@}$ is recognizable.

(i) First, assume that $L^{@}$ is accepted by an H_{Σ}-acceptor

$$A = (Q, \delta, I, T).$$

We can suppose that A is minimal (II.2.4), i.e., the run map $\rho: I^{\#@} \to Q$ is onto and A has no non-trivial congruence. We prove that

$$(*) \qquad (q)\delta = (p \cdot q)\delta$$

for each $q = (q_1, \ldots, q_n) \in QH_{\sigma}$ (where $\sigma \in \Sigma_n$) and $p \cdot q = (q_{(1)p}, \ldots, q_{(n)p}) \in QH_{\sigma}$, where $p \in P_{\sigma}$.

Since the run map $\rho: I^{\#@} \to Q$ is onto, there exists $t: [n] \to I^{\#@}$ such that $t \in I^{\#@}H_{\sigma}$ fulfils $q = (t)\rho H_{\sigma}$. Then $p \cdot q = (p \cdot t)\rho H_{\sigma}$ and hence both

$$(q)\delta = (t)\rho H_{\sigma} \cdot \delta = (t)\varphi \cdot \rho$$

and

$$(p \cdot q)\delta = (p \cdot t)\rho H_{\sigma} \cdot \delta = (p \cdot t)\varphi \cdot \rho.$$

The following trees

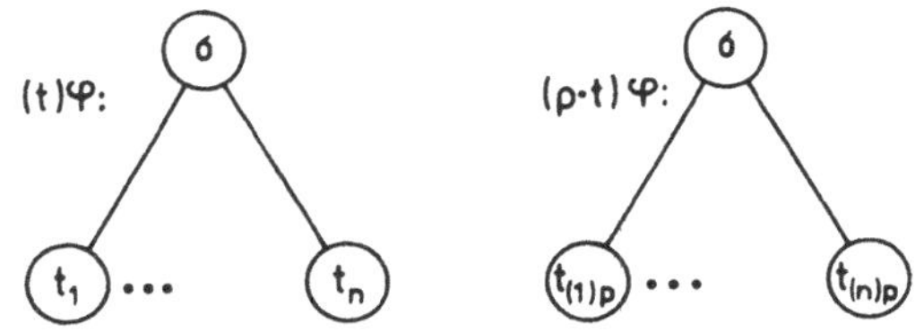

are congruent under R. Hence, it is sufficient to prove that

$$s_1 \, R \, s_2 \quad \text{implies} \quad (s_1)\rho = (s_2)\rho \qquad (s_1, s_2 \in I^{\#@}).$$

Let S be the least equivalence on the set Q such that

$$s_1 \, R \, s_2 \quad \text{implies} \quad (s_1)\rho \, S \, (s_2)\rho.$$

We are to prove that S is trivial. Obviously, S is the transitive hull of $S_0 = \{((s_1)\rho, (s_2)\rho);\ s_1 \, R \, s_2\}$.

(a) S is a congruence on the algebra (Q, δ). To verify this, it is sufficient to prove that for $q, q' \in QH_{\sigma}$ with $q_i \, S_0 \, q_i'$ (for all $i = 1, \ldots, |\sigma|$) we have $(q)\delta \, S_0 \, (q')\delta$. Let $s, s' \in I^{\#@}H_{\sigma}$ be n-tuples of trees with $q = (s)\rho H_{\sigma}$ and $q' = (s')\rho H_{\sigma}$, and with $s_i \, R \, s_i'$ for all i. Then also $(s)\varphi \; R(s')\varphi$, since R is a congruence, and hence,

$$((s)\varphi)\rho \;\; S_0((s')\varphi)\rho,$$

i.e., $(q)\delta \;\; S_0(q')\delta$.

(b) S is a congruence on the automaton A, i.e., if qSq', then $q \in T$ iff $q' \in T$. In fact, let $s\ R\ s'$ be trees with $q = (s)\rho$ and $(q') = (s')\rho$. Then $q \in T$ iff $s \in L^{@}$, and $L^{@}$ is closed under R, thus, $q \in T$ iff $s' \in L^{@}$, i.e., $q' \in T$.

Since A is minimal, the congruence S is trivial, and this proves (∗). It follows that

$$(q)\varepsilon_Q = (q')\varepsilon_Q \quad \text{implies} \quad (q)\delta = (q')\delta \quad (q, q' \in QH_\Sigma).$$

Define

$$\bar{\delta}: QF \to Q$$

by

$$((q)\varepsilon_Q)\bar{\delta} = (q)\delta \quad \text{for each } q \in QH_\Sigma.$$

Then

$$\bar{A} = (Q, \bar{\delta}, T, I)$$

is an F-acceptor, the run map $\bar{\rho}: I^{\#} \to Q$ of which fulfils

$$\rho = \bar{\varepsilon}_I \cdot \bar{\rho}$$

[because $\bar{\varepsilon}_I \cdot \bar{\rho}: (I^{\#@}, \varphi^{@}) \to (Q, \varepsilon_Q \cdot \delta)$ is a homomorphism]. The acceptor $\bar{A}$ recognizes the language

$$L_{\bar{A}} = (T)\bar{\rho}^{-1}$$

which is equal to L because $\bar{\varepsilon}_I$ is onto and

$$L^{@} = (T)\rho^{-1} = (T)(\bar{\varepsilon}_I \cdot \bar{\rho})^{-1} = ((T)\bar{\rho}^{-1})^{@}.$$

Therefore, L is recognizable.

(ii) Let L be accepted by an F-acceptor

$$\bar{A} = (Q, \bar{\delta}, I, T).$$

It is sufficient to prove that $L^{@}$ is accepted by the corresponding Σ-tree acceptor

$$A = (Q, \delta, I, T)$$

with

$$\delta = \varepsilon_Q \cdot \bar{\delta}: QH_\Sigma \to Q.$$

In fact, let $\bar{\rho}: I^{\#} \to Q$ denote the run map of $\bar{A}$. Since both

$$\bar{\varepsilon}_I: (I^{\#@}, \varphi^{@}) \to (I^{\#}, \varepsilon_{I^{\#}} \cdot \varphi)$$

and

$$\bar{\rho}: (I^{\#}, \varepsilon_{I^{\#}} \cdot \varphi) \to (Q, \varepsilon_Q \cdot \delta)$$

are Σ-homomorphisms, also $\bar{\varepsilon}_I \cdot \rho$ is a Σ-homomorphism. For each $x \in I$ we have

$$(x)\bar{\varepsilon}_I \cdot \rho = (x)\rho = x$$

and hence, $\bar{\varepsilon}_I \cdot \rho$ is the run map of A. Therefore, the language accepted by A is

$$(T)(\bar{\varepsilon}_I \cdot \rho)^{-1} = ((T)\rho^{-1}) = L^{®}.$$

This concludes the proof. □

Remark. In the course of the preceding proof we saw that if F is perfect, then the concatenation of tree languages fulfils

$$\langle L \cdot_x K \rangle = \langle L \rangle \cdot_x \langle K \rangle.$$

This property actually characterizes perfect functors: by Proposition VII.3.4, the concatenation of F-languages is clearly associative whenever F has the above property (see Remark II.6.9). We prove now that associativity of concatenation implies that F is perfect.

(We devote the rest of the present section to some of the basic properties of concatenation.)

3.9. Proposition. A super-finitary functor is perfect iff concatenation in each finite set I is associative, i.e.,

$$L \cdot_x (K \cdot_x H) = (L \cdot_x K) \cdot_x H \qquad (\text{in } I)$$

for each $x \in I$ and all $L, K, H \subset I^{\#}$.

Proof. I. Let F be perfect. Concatenation of tree languages is associative by Remark II.4.9, and hence,

$$L^{®} \cdot_x (K^{®} \cdot_x H^{®}) = (L^{®} \cdot_x K^{®}) \cdot_x H^{®}.$$

By part II of the proof of the Main Theorem, it follows that

$$\begin{aligned} [L \cdot_x (K \cdot_x H)]^{®} &= L^{®} \cdot_x (K \cdot_x H)^{®} \\ &= L^{®} \cdot_x (K^{®} \cdot_x H^{®}) \\ &= (L^{®} \cdot_x K^{®}) \cdot_x H^{®} \\ &= (L \cdot_x K)^{®} \cdot_x H^{®} \\ &= [(L \cdot_x K) \cdot_x H]^{®}. \end{aligned}$$

This proves the equation above.

II. Let concatenation be associative. We prove that F is perfect, analogously to the proof of Theorem VII.3.7. Using the notation of that proof, we use the language

$$L_f = \{V_n\}^{*1} \cdot_1 \{w\}.$$

The associativity of concatenation implies

$$\{V_n\}^{*1} \cdot_1 L_f = L_f$$

[see (f) in VII.3.6]. We now generalize (C) of the proof of VII.3.7 as follows:

(C′) For each $\tau \in \Sigma_m$ and all maps $r, q\colon [m] \to \omega$ with

$$(i)q < (i)r \equiv (i)q \ (\mathrm{mod}\ n) \quad \text{for } i = 1, \ldots, m$$

if

$$(u_{(1)q}, \ldots, u_{(m)q})\tau \in L_f^{@},$$

then

$$(u_{(1)r}, \ldots, u_{(m)r})\tau \in L_f^{@}.$$

The proof is completely analogous to (C): since $(i)d = (i)r - (i)q$ is a positive multiple of n, we have $u_{(i)d} \in \langle u_n \rangle^{*1}$. The tree $t = (u_{(1)r}, \ldots, u_{(m)r})\tau$ is obtained from $\bar{t} = (u_{(1)q}, \ldots, u_{(m)q})\tau$ by substituting $u_{(i)d}$ for the 1-labelled leaf in $u_{(i)q}$. Consequently, by (e) in VII.3.6,

$$t \in \langle U_n \rangle^{*1} \cdot_1 \{\bar{t}\} \subset (V_n^{*1})^{@} \cdot_1 L_f^{@} \subset \langle (V_n^{*1})^{@} \cdot_1 L_f^{@} \rangle,$$

and by Proposition VII.3.4,

$$t \in (V_n^{*1} \cdot_1 L_f)^{@} = L_f^{@},$$

which proves (C′).

We are to prove that for $f \in [n]H_\sigma$ and $g \in [n]H_\tau$ with $n = |\sigma| \geq |\tau| \geq 2$ and

$$(f)\varepsilon_{[n]} = (g)\varepsilon_{[n]},$$

we have $\sigma = \tau$ and $g = p \cdot f$ for some $p \in P_\sigma$. We use the fact that

$$(u_{(1)f+n^2}, \ldots, u_{(n)f+n^2})\sigma \in L_f^{@}$$

by (C). We have $\varphi^{@} \cdot \bar{\varepsilon}_{[n]} = \bar{\varepsilon}_{[n]} H_\Sigma \cdot \varepsilon_{[n]^\#} \cdot \varphi$

and hence,

$$\begin{aligned}[(u_{(1)f+n^2}, \ldots, u_{(n)f+n^2})\sigma]\bar{\varepsilon}_{[n]} &= ((u_{(1)f+n^2})\bar{\varepsilon}_{[n]}, \ldots, (u_{(n)f+n^2})\bar{\varepsilon}_{[n]})\varepsilon_{[n]^\#} \cdot \varphi \\ &= (v_{(1)f+n^2}, \ldots, v_{(n)f+n^2})\varepsilon_{[n]^\#} \cdot \varphi \in L_f.\end{aligned}$$

The n-tuple $(v_{(1)f+n^2}, \ldots, v_{(n)f+n^2})$ in $[n]^\# H_\sigma$ can be expressed as $f \cdot h$, where $f\colon [n] \to [n]$ is the given n-tuple and

$$h\colon [n] \to [n]^\#$$

is defined by

$$(i)h = v_{i+n^2} \quad (i = 1, \ldots, n).$$

Thus,

$$(f\cdot h)\varepsilon_{[n]\#}\cdot\varphi = (f)hH_\Sigma\cdot\varepsilon_{[n]\#}\cdot\varphi = (f)\varepsilon_{[n]}\cdot hF\cdot\varphi \in L_f.$$

Since $(g)\varepsilon_{[n]} = (f)\varepsilon_{[n]}$, we get

$$(g)\varepsilon_{[n]}\cdot hF\cdot\varphi \in L_f$$

and, analogously as for f above, this means that

$$[(u_{(1)g+n^2},\ldots,u_{(m)g+n^2})\tau]\bar{\varepsilon}_{[n]} \in L_f.$$

Thus,

$$(u_{(1)q},\ldots,u_{(m)q})\tau \in L_f^{\circledast}$$

for the map $q:[m]\to\omega$ defined by

$$(i)q = (i)g + n^2.$$

Let $r:[m]\to\omega$ be an arbitrary one-to-one map with

$$(i)q < (i)r \equiv (i)q \pmod m \quad \text{for } i = 1,\ldots,m.$$

Then by (C′) above,

$$(u_{(1)r},\ldots,u_{(m)r})\tau \in L_f^{\circledast}.$$

By (B) in VII.3.7, this implies $\sigma = \tau$ and hence,

$$s_r \in L_f^{\circledast}.$$

By (A) in VII.3.7, there exists $p \in P_\sigma$ with $(i)r \equiv (i)p\cdot f \pmod n$. Since

$$(i)r \equiv (i)q \equiv (i)g \pmod n,$$

this implies $(i)g \equiv (i)p\cdot f \pmod n$. The values of g and f lie between 1 and n, therefore, $(i)g = (i)p\cdot f$. This concludes the proof. □

3.10. The last property of concatenation we are going to study is its independence of the concrete set of variables:

Definition. A super-finitary functor F is said to have *absolute concatenation* if for arbitrary languages $K, L \subset I$ and each $x \in I$, the concatenation

$$L\cdot_x K \quad (\text{in } I)$$

is the same language as this concatenation in J for each (finite) set $J \supset I$.

Examples. (i) Concatenation of tree languages is obviously absolute.
(ii) The functor D_2 (VII.2.3.) does not have absolute concatenation. Put

$$I = \{x, z_1, z_2\} \quad \text{and} \quad J = \{y\} \cup I.$$

The concatenation

$$I \cdot_x \{*\}$$

in I is a language in $I^\#$ which clearly does not contain the following tree

t:

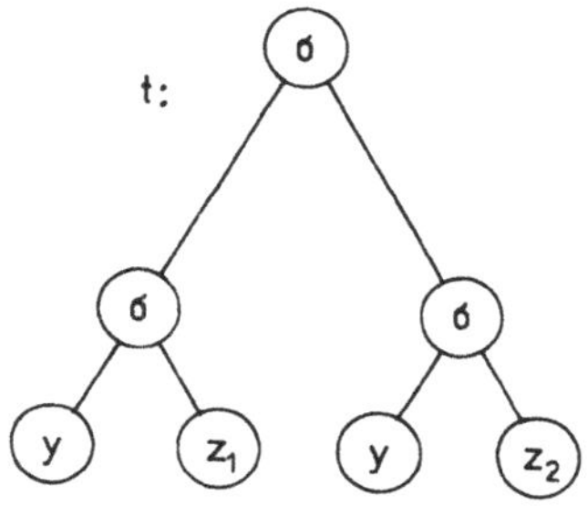

(because $t \notin I^\#$). But the tree

$\bar{t}$:

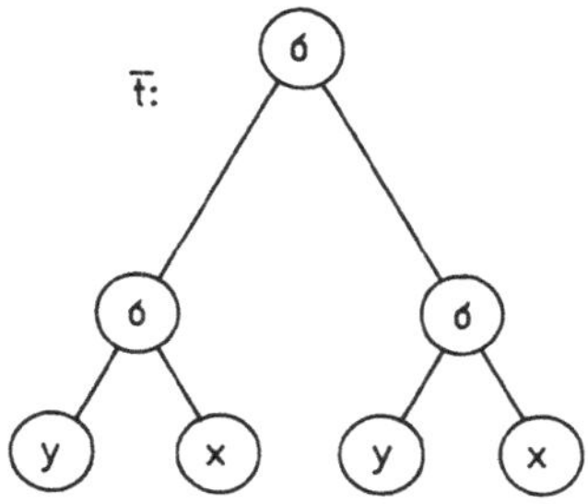

is an element of $\{*\}^{@} = (*)\bar{\varepsilon}_J^{-1}$, and t is obtained from $\bar{t}$ by substituting x by elements of $I = I^{@}$. Hence, $t \in I^{@} \cdot_x \{*\}^{@}$ and this implies that $t = (t)\bar{\varepsilon}_J \in I \cdot_x \{*\}$ (in J).

3.11. Proposition. A super-finitary functor F has absolute concatenation iff F is regular.

Proof. I. Assume that F is a regular functor.

Let $x \in I \subset J$ and K, $L \subset I^\#$ be given. We denote by a^I, $b^I: (I - \{x\} + L)^\# \to I^\#$ the homomorphisms from the definition of $L \cdot_x K$ in I, and analogously a^J, b^J. Denote by

$$v: I \to J$$

and

$$u: I - \{x\} + L \quad \to \quad J - \{x\} + L$$

the inclusion maps. (Then $v^\#$ and $u^\#$ are also the inclusion maps, see Remark

VII.3.3.) Then we prove that

(i) $\qquad a^I \cdot v^* = u^* \cdot a^J$

and

(ii) the following square

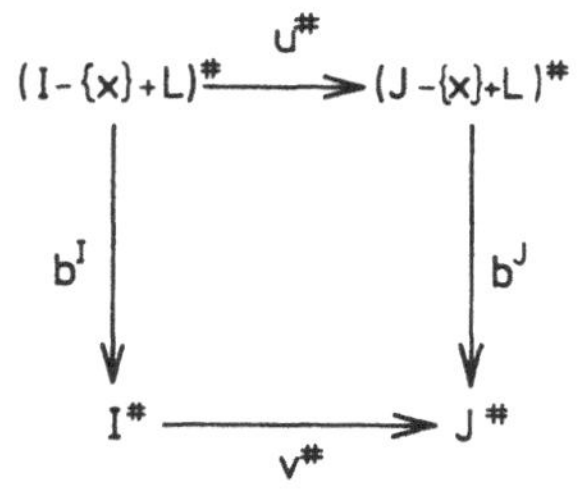

is a pullback. This will conclude the proof: by (i), a^J is an extension of a^I, and by (ii),

$$(I^*)(b^J)^{-1} = (I - \{x\} + L)^*.$$

Therefore,

$$\{(s)a^I;\ a \in (K)b^I)^{-1}\} = \{(s)a^J;\ a \in (K)(b^J)^{-1}\}.$$

(i) Since $a^I \cdot v^*$ and $u^* \cdot a^J$ are homomorphisms, it is sufficient to prove that they are equal on the set

$$\bar{I} = I - \{x\} + L$$

of generators. Since u^* and v^* are the inclusion maps, $a^I \cdot v^*$ is an extension of a_0^I and $u^* \cdot a^J$ is an extension of a_0^J (restricted to I); for each $y \in \bar{I}$ we have $(y)a_0^I = (y)a_0^J$. Thus, $a^I \cdot v^* = u^* \cdot a^J$.

(ii) Put also

$$\bar{J} = J - \{x\} + L.$$

Since F is a finitary varietor, we have

$$I^* = \bigcup_{n < \omega} W_n^I$$

with the inclusion maps $w_n^I : W_n^I \to I^*$; analogously with W_n^J, W_n^I and $W_n^{\bar{J}}$. Define

$$b_n^I : W_n^{\bar{I}} \to W_n^I \quad (n < \omega)$$

by the following induction:

$$b_0^I : \bar{I} \to I$$

is the given map;

$$b^I_{n+1} = b^I_0 + b^I_n F: \bar{I} + W^{\bar{I}}_n F \to I + W^I_n F.$$

Analogously,

$$b^J_n: W^{\bar{J}}_n \to W^J_n.$$

We prove that for each $n < \omega$, the following square

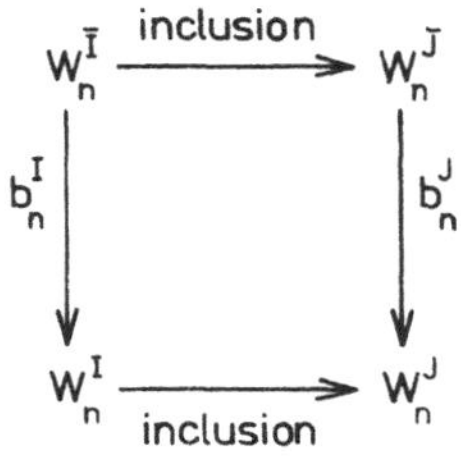

is a pullback. This is clear if $n = 0$. Assuming this holds for n, then it holds for $n + 1$ because (a) F preserves this pullback (in fact, the inclusion maps are monos and hence, the pullback is a preimage, see VII.2.8) and (b) the coproduct square of two pullbacks in **Set** is a pullback.

Pullbacks in **Set** commute with colimits of ω-sequences and hence, the colimit of the sequence of squares above is a pullback. This is the following square

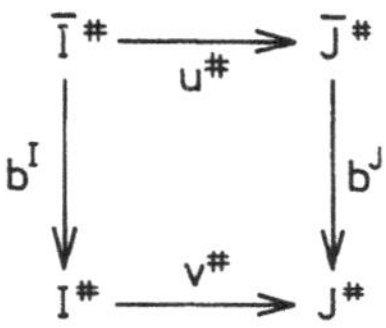

which proves (ii).

II. Assume that concatenation is absolute. To prove that F is regular, let $\varepsilon: H_\Sigma \to F$ be a minimal presentation, and let

$$(f)\varepsilon_X = (g)\varepsilon_X$$

hold for some $\sigma \in \Sigma_n$, $\tau \in \Sigma_m$ and $f \in XH_\sigma$, $g \in XH_\tau$. We prove that for each $r \in X$,

$$r \in ([n])f \quad \text{implies} \quad r \in ([m])g;$$

by symmetry, it follows that $([n])f = ([m])g$. We can suppose that $n \geq 2$ (VII.2.7). Moreover, without loss of generality, we assume that

$$(r)f^{-1} = \{1, 2, \ldots, k\} \subset [n]$$

for some $k \geq 1$.

Put

$$I = \{x, z_1, z_2, \ldots, z_n\}$$

and

$$J = I \cup \{y_2, \ldots, y_n\},$$

where the variables x, z_i and y_j are pairwise distinct. Let $s = (x, x, \ldots, x)\sigma \in I^{\#@}$, and define languages $K, L \subset I^{\#}$ by

$$K = \{(s)\bar{\varepsilon}_I\}$$

and

$$L = I.$$

Since concatenation is absolute, the language

$$L \cdot_x K \quad (\text{in } J)$$

is a subset of $I^{\#}$ (because it is equal to $L \cdot_x K$ in I). We are going to present a tree

$$t \in J^{\#@}$$

such that

$$(t)\bar{\varepsilon}_J \notin I^{\#}$$

and if $r \notin ([m])g$, then $(t)\bar{\varepsilon}_J \in L \cdot_x K$. This will prove that $r \in ([m])g$ [because we know that $(t)\bar{\varepsilon}_J \notin L \cdot_x K$].

III. Let t be the following tree

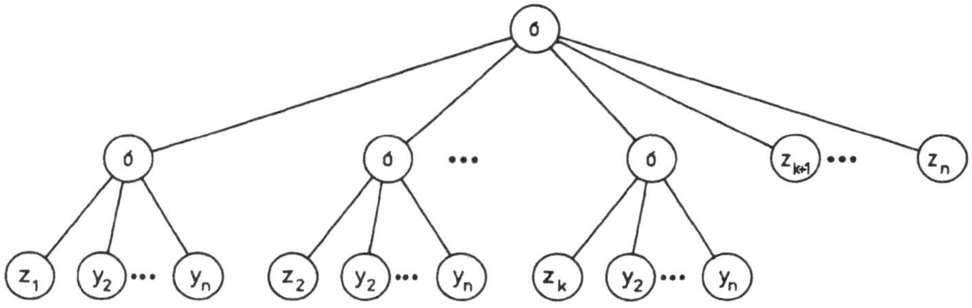

That is,

$$t = (s_1, \ldots, s_k, z_{k+1}, \ldots, z_n)\sigma$$

where

$$s_i = (z_i, y_2, \ldots, y_n)\sigma.$$

By VII.2.5, the class $\langle s_i \rangle$ contains only trees obtained from s_i by a permutation of branches. Each of these trees uses the variable z_i. Consequently, the classes $\langle s_i \rangle$ are pairwise distinct ($i = 1, \ldots, k$) and distinct from $\langle z_j \rangle = \{z_j\}$ ($j = k + 1, \ldots, n$). By Lemma VII.3.6, we conclude that also each tree in the class $\langle t \rangle$ is obtained from t by permutations of branches. Consequently, none of these trees belongs to $I^{\#@}$. It follows that

$$(t)\bar{\varepsilon}_J \notin I^\#$$

because else there would exist $t' \in I^{\#@}$ with $(t')\bar{\varepsilon}_I = (t)\bar{\varepsilon}_J$, and the last clearly implies that $t' \in \langle t \rangle$.

Denote

$$\bar{J} = (J - \{x\}) + L = \{z_1, \ldots, z_n, y_2, \ldots, y_n\} \cup \{x', z_1', \ldots, z_n'\}$$

where primes are used to distinguish the elements of the second summand $L = I$ from those of $J - \{x\}$.

Let

$$a, b\colon \bar{J}^\# \to J^\#$$

be the F-homomorphisms from the definition of

$$L \cdot_x K = I \cdot_x \{(s)\bar{\varepsilon}_I\} \quad (\text{in } J),$$

and let

$$a^@, b^@\colon \bar{J}^{\#@} \to J^{\#@}$$

be the corresponding Σ-homomorphisms with respect to $I \cdot_x \{s\}$. Then

$$a^@ \cdot \bar{\varepsilon}_J = \bar{\varepsilon}_{\bar{J}} \cdot a\colon (\bar{J}^{\#@}, \varphi^@) \to (J^\#, \varepsilon_{J^\#} \cdot \varphi)$$

because these two homomorphisms agree on the set $\bar{J}$; analogously,

$$b^@ \cdot \bar{\varepsilon}_J = \bar{\varepsilon}_{\bar{J}} \cdot b.$$

For each tree in $J^{\#@}$, the map $a^@$ "forgets" the primes and the map $b^@$ changes each z_i' to x. Therefore,

$$t = (t')a^@,$$

where

$$t' \in \bar{J}^{\#@}$$

is the tree obtained from t by adding a prime to each z_i, $i = 1, \ldots, n$. Thus,

$$(t)\bar{\varepsilon}_J = ((t')\bar{\varepsilon}_{\bar{J}})a.$$

To conclude the proof, we show that if $r \notin ([m])g$, then

$$((t')b^{@})\bar{\varepsilon}_J = (s)\bar{\varepsilon}_J;$$

it follows that

$$((t')\bar{\varepsilon}_{\bar{J}})b = ((t')b^{@})\bar{\varepsilon}_J \in K$$

and hence,

$$((t')\bar{\varepsilon}_{\bar{J}})a = (t)\varepsilon_J \in L \cdot_x K.$$

We have

$$((t')\bar{\varepsilon}_{\bar{J}})b = ((t')b^{@})\bar{\varepsilon}_J,$$

and $(t')b^{@}$ is the following tree

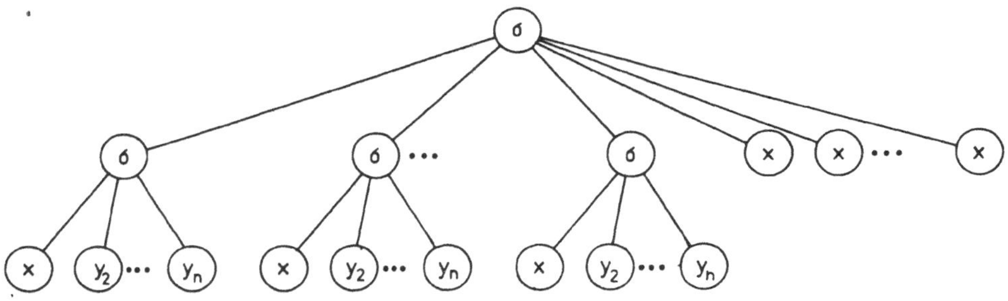

That is, for

$$p = (x, y_2, y_3, \ldots, y_n)\sigma \in J^{\#@}$$

we have

$$(t')b^{@} = (p, p, \ldots, p, x, x, \ldots, x)\sigma,$$

where p is repeated k-times [with $(r)f^{-1} = \{1, 2, \ldots, k\}$]. Define maps

$$c, c': X \to J^{\#}$$

as follows: c is the constant map with value x, and

$$(r)c' = (p)\bar{\varepsilon}_J$$

and on $X - \{r\}$, also c' is the constant map with value x. Then for $f \cdot c$, $f \cdot c'$: $[n] \to J^{\#}$, written as n-tuples, we have

$$(x, x, \ldots, x) = f \cdot c$$

and

$$((p)\bar{\varepsilon}_J, (p)\bar{\varepsilon}_J, \ldots, (p)\bar{\varepsilon}_J, x, x, \ldots, x) = f \cdot c'.$$

Since $r \in ([m])g$, we have

$$g \cdot c = g \cdot c'.$$

Therefore, using the fact that $\bar{\varepsilon}_J$ is a homomorphism (i.e., $\varphi^{@} \cdot \bar{\varepsilon}_J = \bar{\varepsilon}_J H_\Sigma \cdot \varepsilon_{J\#} \cdot \varphi$), we get

$$\begin{aligned}((t')b^{@})\bar{\varepsilon}_J &= (p, p, \ldots, p, x, x, \ldots, x)\varphi^{@} \cdot \bar{\varepsilon}_J \\ &= (p, p, \ldots, p, x, x, \ldots, x)\bar{\varepsilon}_J H_\Sigma \cdot \varepsilon_{J\#} \cdot \varphi \\ &= ((p)\bar{\varepsilon}_J, \ldots, (p)\bar{\varepsilon}_J, x, \ldots, x)\varepsilon_{J\#} \cdot \varphi \\ &= (f \cdot c)\varepsilon_{J\#} \cdot \varphi \\ &= (f)cH_\Sigma \cdot \varepsilon_{J\#} \cdot \varphi \\ &= (f)\varepsilon_X \cdot cF \cdot \varphi \\ &= (g)\varepsilon_X \cdot cF \cdot \varphi,\end{aligned}$$

and also

$$\begin{aligned}(s)\bar{\varepsilon}_I &= (s)\bar{\varepsilon}_J \\ &= (x, x, \ldots, x)\varphi^{@} \cdot \bar{\varepsilon}_J \\ &= (x, x, \ldots, x)\bar{\varepsilon}_J H_\Sigma \cdot \varepsilon_{J\#} \cdot \varphi \\ &= (x, x, \ldots, x)\varepsilon_{J\#} \cdot \varphi \\ &= (f \cdot c')\varepsilon_{J\#} \cdot \varphi \\ &= (g)\varepsilon_X \cdot c'F \cdot \varphi.\end{aligned}$$

Since $g \cdot c = g \cdot c'$, we have

$$(g)\varepsilon_X \cdot cF = (g \cdot c)\varepsilon_{J\#} = (g \cdot c')\varepsilon_{J\#} = (g)\varepsilon_X \cdot c'F$$

and hence,

$$((t')b^{@})\bar{\varepsilon}_J = (s)\bar{\varepsilon}_I.$$

This concludes the proof. □

3.12. Summarization. For each super-finitary set functor F, equivalent are:

(i) Each partially recognizable language is recognizable;
(ii) concatenation is absolute;
(iii) F preserves the composition of partial maps;
(iv) F preserves preimages;
(v) F is regular.

For each super-finitary set functor F equivalent are:

(a) Each nondeterministically recognizable language is recognizable;
(b) Kleene theorem holds;
(c) each language $\{(x)\eta, v\}^{*x} \cdot_x \{w\}$ is recognizable;
(d) concatenation is associative;
(e) F preserves the composition of relations;
(f) F covers pullbacks;
(g) F is perfect.

The proofs have been exhibited above: the equivalence of (i), (iv) and (v) is Theorem VII.2.10, for (ii) see Proposition VII.3.11 and for (iii) Remark V.2.10.

The equivalence of (a), (f) and (g) is Theorem VII.2.12, for (b) see Theorem VII.3.8, for (c) Theorem VII.3.9, for (d) Proposition VII.3.10, and for (e), Theorem V.2.10.

Exercises VII.3

A. Each recognizable language is rational. This statement holds for each superfinitary functor F (whereas the converse implication holds iff F is perfect). Prove this, using the following steps.

(i) Given F-languages K, $L \subset I^{\#}$ and $x \in I$ with $L \subset (I - \{x\})^{\#}$, prove that $L \cdot_x K \subset (I - \{x\})^{\#}$. [Hint: The map $a_0 \colon I - \{x\} + L \to I^{\#}$ factors through $(I - \{x\})^{\#}$.]

(ii) Let L be a language recognized by an F-acceptor $A = (Q, \delta, T, I)$. Prove that for the minimal presentation $\varepsilon \colon H_\Sigma \to F$, the Σ-tree acceptor $A^{@} = (Q, \varepsilon_Q \cdot \delta, T, I)$ recognizes $L^{@}$. Prove that for each $M \subset Q$, if $\rho_M \colon M^{\#} \to Q$ is the run map of $A_M = (Q, \delta, T, M)$, then $\rho_M^{@} = \bar{\varepsilon}_M \cdot \rho_M$ is the run map of $A_M^{@}$. (Hint: VII.1.8.)

(iii) Recall that the proof of the rationality of $L^{@}$ in II.4.11 was performed by finding languages $L_{M,j}^k \subset M^{\#@}$ ($M \subset Q$, $j = 1, \ldots, m$ and $k = 0, \ldots, m$, where $Q = \{q_1, \ldots, q_m\}$) such that

$$L^{@} = \bigcup_{q_j \in T} L_{I,j}^m$$

and (a) $L_{M,j}^0$ is finite, (b) $L_{M,j}^k = L_{M,k}^{k-1} \cdot_{q_k} (L_{M',k}^{k-1})^{*q_k} \cdot_{q_k} L_{M',j}^{k-1}$ for $M' = M \cup \{q_k\}$ and (c) $(L_{M,j}^k)\rho_M^{@} \subset \{q_j\}$.

(iv) Define analogous F-languages $\bar{L}_{M,j}^k \subset M^{\#} : \bar{L}_{M,j}^0 = (L_{M,j}^0)\bar{\varepsilon}_M$ and $\bar{L}_{M,j}^k = \bar{L}_{M,k}^{k-1} \cdot_{q_k} [(\bar{L}_{M',k}^{k-1})^{*q_k} \cdot \bar{L}_{M',j}^{k-1}]$, for the operations $\cdot_{q_k}$ and ^{*q_k} in M'. Verify that each $L_{M,j}^k \subset M^{\#}$ is rational. (Hint: use (i) by induction on k.)

(v) Prove $(L_{M,j}^k)\rho \subset \{q_j\}$. [Hint: use induction on k. For two languages H, $K \subset (M')^{\#}$ with $H \subset M^{\#}$ and $(H)\rho_M = \{q_k\}$ prove that $(H \cdot_{q_k} K)\rho_M = (K)\rho_M$ by verifying that the homomorphisms a, b defining $H \cdot_{q_k} K$ fulfil $a \cdot \rho_M = b \cdot \rho_M$.]

(vi) Prove the rationality of L by verifying that

$$L = \bigcup_{q_j \in T} \bar{L}_{I,j}^m .$$

The inclusion $\supset$ follows from (v); for the inclusion $\subset$, prove that $(L_{M,j}^k)\bar{\varepsilon}_M \subset \bar{L}_{M,j}^k$ by induction on k. [Hint: by Propositions VII.3.4, VII.3.5 if $H \subset \bar{H}^{@}$ and $K \subset \bar{K}^{@}$, then $H \cdot_x K \subset (\bar{H}^{@} \cdot_x \bar{K}^{@})^{@}$ and $H^{*q_k} \subset (\bar{H}^{*q_k})^{@}$.]

B. Defining the iteration. We have defined the iteration by

$$L^{*x} = L \cup (L \cdot_x L) \cup ((L \cdot_x L) \cdot_x L) \cup (((L \cdot_x L) \cdot_x L) \cdot_x L) \ldots.$$

If concatenation is not associative (VII.3.9), then another natural definition would be

$${}^{x*}L = L \cup (L \cdot_x L) \cup (L \cdot_x (L \cdot_x L)) \cup (L \cdot_x (L \cdot_x (L \cdot_x L))) \ldots.$$

(i) Prove that whenever ${}^{x*}L \neq L^{*x}$, then the new definition does not correspond well with the iteration of tree languages, i.e., ${}^{x*}L \neq ({}^{x*}(L^{@}))\bar{\varepsilon}_I$. [Hint: VII.3.5 and $(L^{@})^{*x} = {}^{x*}L^{@}$.]

(ii) The two definitions need not agree even for regular functors: let F be the quotient of H_Σ with $\Sigma = \Sigma_2 = \{\sigma, \tau\}$, given by the equation $(x, x)\sigma = (x, x)\tau$. Define F properly, and verify that F is regular.

Consider the following trees over $I = \{x, y\}$:

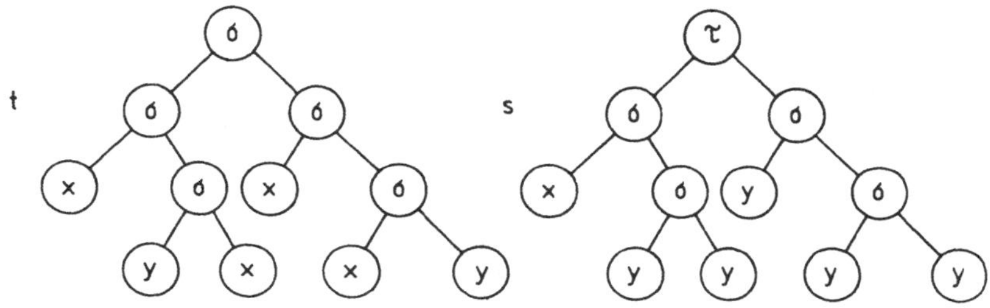

For $L = \{x, y, t\}\bar{\varepsilon}_I$ prove that $(s)\bar{\varepsilon}_I \in {}^{x*}L - L^{*x}$.

Notes to Chapter VII

VII.1

All notions and results of this section appeared in V. Trnková [1980].

VII.2

N-recognizable languages in **Set** were investigated by V. Trnková [1977, 1980]. Theorem VII.2.12 was announced in the former paper, and a proof of a more general result (concerning L-fuzzy F-automata) appeared in V. Trnková [1979a]. P-recognizable languages were studied by V. Trnková [1981], where Theorem VII.2.10 was announced; the proof appears here for the first time. In all those papers, perfect functors were called tree-group functors, and regular functors were called saturated.

VII.3

Concatenation and iteration of F-languages was introduced by V. Trnková and J. Adámek [1979], where the Main Theorem VII.3.8 was announced; the proof appears here for the first time. The discussion of properties of concatenation in VII.3.9–11 is new.

Appendix

Set-Theoretical Conventions

1. We expect the reader to be acquainted with sets and thus, we do not introduce any axiomatic set theory. Nevertheless, we were careful to have all our constructions well-defined within the framework of Bernays-Gödel theory of sets.

Collections which are "too large" to form a set are called *classes.* Thus, classes are more general than sets: each set is a class, but not vice versa. (For example, the class of all sets is not a set.) A class is said to be *large* if it is not a set, and *small* otherwise. That is, small class is a synonym to set.

2. The empty set is denoted by $\emptyset$, the set with elements $x_1, \ldots, x_n$ by $\{x_1, \ldots, x_n\}$. For example, $\{x\}$ is a one-element set. The set of all elements with a property P is denoted by

$$\{x;\, x \text{ has property } P\}.$$

3. By a *pair* (x, y) we always mean an ordered pair. More generally, an n-tuple means an ordered n-tuple; we denote it by $(x_i)_{i<n}$ or (for finite n) by $(x_0, x_1, \ldots, x_{n-1})$.

Each *ordinal* is considered to be the well-ordered set of all smaller ordinals. In particular,

$$1 = \{0\} = \{\emptyset\}$$

and $2 = \{0, 1\} = \{\emptyset,\{\emptyset\}\}$, etc; the set of all natural numbers is the ordinal

$$\omega = \{0, 1, 2, \ldots\} = \{n;\, n < \omega\}.$$

Ordinals are also used to label domains of trees; here they are just labels (but this double use does not lead to any confusion).

The cardinality of a set X is denoted by card X.

4. A *mapping* (or *map*) $f\colon X \to Y$ is a triple consisting of a set X (the *domain*), a set Y (the *codomain*) and a subset $f \subset X \times Y$ such that for each $x \in X$ there is a unique $y \in Y$ with $(x, y) \in f$ [which we write as $y = (x)f$]. For example, given a set X and its subset $Z \subset X$, the *inclusion map*

$$v\colon Z \to X$$

is the map defined by $(z)v = z$ for $z \in Z$. In case $Z = X$, this is the *identity map*

$$\mathrm{id}_X : X \to X \quad \text{or} \quad 1_X : X \to X.$$

In case $Z = \emptyset$, v is called the *empty map.*

We also work with *class maps* allowing the domain and codomain to be classes.

5. *Composition* of maps is writen from the left, i.e., the composition of $f: X \to Y$ and $g: Y \to Z$ yields the map $f \cdot g: X \to Z$ defined by

$$(x)f \cdot g = ((x)f)g.$$

If $f: X \to Y$ is a *bijection* (i.e., both one-to-one and onto), then $f^{-1}: Y \to X$ denotes the *inverse mapping,* defined by $f \cdot f^{-1} = 1_X$ and $f^{-1} \cdot f = 1_Y$.

6. For each equivalence relation $\sim$ on a set X, we denote by

$$X/\sim \; = \{[x];\ x \in X\}$$

the *quotient set* of all equivalence classes

$$[x] = \{y \in X;\ x \sim y\}.$$

The *canonical map* $c: X \to X/\sim$ is defined by $(x)c = [x]$ $(x \in X)$.

7. Each mapping $f: X \to Y$ defines the *kernel equivalence* $\sim$ on X by

$$x_1 \sim x_2 \quad \text{iff} \quad (x_1)f = (x_2)f \quad (x_1, x_2 \in X)$$

and the *image,* i.e., the subset

$$\mathrm{im} f = (X)f = \{(x)f;\ x \in X\}$$

of Y. Then f is composed of the canonical map $c: X \to X/\sim$ the bijection $b: X/\sim \to (X)f$ defined by $([x])b = (x)f$, and the inclusion map $v: (X)f \to Y$.

8. The set of all maps from a set T to a set X is denoted by X^T. If $T = n$ is an ordinal, this is the n-fold cartesian product of X. For example, $X^1 = X$, $X^2 = X \times X$, etc. The set X^0 has just one element, and we use the following convention:

$$X^0 = \{0\} = 1 \quad \text{for each set } X.$$

For each map $f: X \to Y$, we denote by

$$f^{(n)}: X^n \to Y^n$$

the map defined by

$$(x_i)_{i<n} f^{(n)} = ((x_i)f)_{i<n}.$$

9. A *collection* or *family* with the index set, or class, I is a map with domain I (that is, to each $i \in I$ we assign an element x_i from the codomain). We write $\{x_i\}_{i \in I}$ or $\{x_i;\ i \in I\}$. In case I is an ordinal, the collection is an n-tuple, and ordinary parentheses are used.

References

ADÁMEK, J. [1974a]: Free algebras and automata realizations in the language of categories, *Comment. Math. Univ. Carolinae 15*, 589—602.

ADÁMEK, J. [1974b]: Categorical theory of automata and universal algebra (Czech), Doctoral Dissertation, Charles University Prague.

ADÁMEK, J. [1975]: Automata and categories: finiteness contra minimality, *Lect. Notes Comp. Sci. 32*, Springer-Verlag, Berlin—Heidelberg—New York, 160—166.

ADÁMEK, J. [1976a]: Limits and colimits in generalized algebraic categories, *Czech. Math. J. 26*, 55—64.

ADÁMEK, J. [1976b]: Cogeneration of algebras in regular categories, *Bull. Austral. Math. Soc. 15*, 355—370.

ADÁMEK, J. [1977a]: Realization theory for automata in categories, *J. Pure Appl. Algebra 9*, 281—296.

ADÁMEK, J. [1977b]: Colimits of algebras revisited, *Bull. Austral. Math. Soc. 17*, 433—450.

ADÁMEK, J. [1978]: Finitary varietors, Forschungsbericht 74, University Dortmund.

ADÁMEK, J. [1979a]: Categorial realization theory I & II, Algebraische Modelle, Kategorien und Grupoiden, Akademie-Verlag Berlin, 111—136.

ADÁMEK, J. [1979b]: On the cogeneration of algebras, *Math. Nachrichten 88*, 373—384.

ADÁMEK, J. [1982]: Construction of free continuous algebras. *Algebra Universalis 14*, 140—166.

ADÁMEK, J. [1983]: Theory of mathematical structures. D. Reidel Publ. Comp., Dordrecht—Boston—Lancaster, 1983.

ADÁMEK, J., EHRIG, H. and TRNKOVÁ, V. [1980]: On an equivalence of system-theoretical and categorical concepts, *Kybernetika 16*, 399—410.

ADÁMEK, J., and KOUBEK, V. [1972]: Coequalizers in generalized algebraic categories, *Comment. Math. Univ. Carolinae 13*, 311—324.

ADÁMEK, J., and KOUBEK, V. [1977a]: Remarks on fixed points of functors, *Lect. Notes Comp. Sci. 56*, Springer-Verlag, Berlin—Heidelberg—New York, 199—205.

ADÁMEK, J., and KOUBEK, V. [1977b]: Functorial algebras and automata, *Kybernetika 13*, 245—260.

ADÁMEK, J., and KOUBEK, V. [1979]: Least fixed point of a functor, *J. Comp. Syst. Sciences 19*, 163—178.

ADÁMEK, J., and KOUBEK, V. [1980]: Are colimits of algebras simple to construct? *J. Algebra 66*, 226—250.

ADÁMEK, J., KOUBEK, V., and POHLOVÁ, V. [1972]: Colimits in the generalized algebraic categories, *Acta Univ. Carolinae 13*, 311—324.

ADÁMEK, J., and MERZENICH, W. [1984]: Fixed points as equations and solutions, *Canad. J. Math. 36*, 495—519.

ADÁMEK, J., NELSON, E. and REITERMAN J. [1982]: Tree constructions of free continuous algebras, *J. Comp. Syst. Sci. 24*, 114—146.

ADÁMEK, J., and TRNKOVÁ, V. [1977]: Recognizable and Regular Languages in a Category, *Lect. Notes Comp. Sci. 56*, Springer-Verlag, Berlin—Heidelberg—New York, 206—211.

ADÁMEK, J., and TRNKOVÁ, V. [1981]: Varietors and machines in a category, *Alg. Universalis 13*, 89—132.

ANDERSON, B. D. O., ARBIB, M. A. and MANES, E. G. [1976]: Foundations of system theory: finitary and infinitary conditions, *Lect. Notes Economics and Math. Systems 115*, Springer-Verlag, Berlin—Heidelberg—New York.

Arbib, M. A. [1969]: Theories of Abstract Automata, Prentice-Hall, Princeton, New Jersey.

Arbib, M. A. [1977]: Free dynamics and algebraic semantics, *Lect. Notes Comp. Sci. 59*, Springer-Verlag, Berlin—Heidelberg—New York, 212—227.

Arbib, M. A. and Giveon, Y. [1967]: Algebra automata I: parallel programming as a prolegomena to the categorial approach, *Inf. Control 12*, 331—345.

Arbib, M. A. and Manes, E. G. [1974a]: Machines in a category: an expository introduction, *SIAM Review 16*, 163—192.

Arbib, M. A. and Manes, E. G. [1974b]: Foundations of system theory: decomposable systems, *Automatica 10*, 285—302.

Arbib, M. A. and Manes, E. G. [1975a]: Adjoint machines, state-behavior machines and duality, *J. Pure Appl. Algebra 6*, 313—344.

Arbib, M. A. and Manes, E. G. [1975b]: Fuzzy machines in a category, *Bull. Austral. Math. Soc. 13*, 169—210.

Arbib, M. A. and Manes, E. G. [1980]: Partially-additive categories and flow-diagram semantics, *J. Algebra 62*, 203—227.

Barr, M. [1970]: Coequalizers and free triples, *Math. Z. 116*, 307—322.

Barr, M. [1971]: Factorizations, generators and rank, unpublished preprint.

Barr, M. [1974]: Right exact functors, *J. Pure Appl. Algebra 5*, 1—7.

Birkhoff, G. and Lipson, J. D. [1970]: Heterogeneous algebras, *J. Combinatorial Theory 8*, 115—133.

Brainerd, W. S. [1968]: The minimization of tree automata, *Inf. Control 13*, 484—491.

Budach, L. and Hoehnke, H.-J. [1975]: Automaten und Funktoren, Akademie-Verlag, Berlin.

Donner, J. E. [1970]: Tree acceptors and some of their applications, *J. Comp. Syst. Sciences 4*, 406—451.

Ehrig, H., Kiermeier, K.-D., Kreowski, H.-J. and Kühnel, W. [1974]: Universal theory of automata, Teubner, Stuttgart.

Eilenberg, S. [1974]: Automata, langugaes and machines, Vol. A, Academic Press, New York and London.

Eilenberg, S. and Wright, J. B. [1967]: Automata in general algebras, *Inf. Control 11*, 52—70.

Gécseg, F. and Steinby, M. [1984]: Tree automata, Akadémiai Kiadó, Budapest.

Goguen, J. A. [1972]: Minimal realization of machines in closed categories, *Bull. Amer. Math. Soc. 78*, 778—783.

Goguen, J. A. [1973]: Realization is universal, *Math. Syst. Theory 6*, 359—374.

Goguen, J. A., Tchatcher, J. W., Wagner, E. G. and Wright, J. B. [1977]: Initial algebra semantics and continuous algebras, *J. Assoc. Comp. Machinery 24*, 68—95.

Grätzer, G. [1967]: Universal algebra, Van Nonstrand, Princeton, New Jersey.

Grillet, P. A. [1971]: Regular categories, *Lect. Notes Mathematics 236*, Springer-Verlag, Berlin—Heidelberg—New York, 121—222.

Herrlich, H. and Strecker, G. E. [1979]: Category theory, 2nd Ed., Höldermann-Verlag, Berlin.

Isbell, J. R. [1957]: Some remarks concerning categories and subspaces, *Canad. J. Math. 9*, 563—577.

Jarzembski, G. [1982]: Tree ω-complete algebras, *Alg. universalis 14*, 231—234.

Kelly, G. M. [1980]: A unified treatment of transfinite constructions . . ., *Bull. Austral. Math. Soc. 22*, 1—85.

Knaster, B. [1928]: Un théorème sur les fonctions d'ensembles, *Ann. Soc. Polon. Math. 6*, 133—134.

Koubek, V. [1971]: Set functors, *Comment. Math. Univ. Carolinae 12*, 175—195.

Koubek, V. [1973]: Set functors II, contravariant case, *Comment. Math. Univ. Carolinae 14*, 47—59.

KOUBEK, V., and REITERMAN, J. [1973]: Set functors III—monomorphisms, epimorphisms, isomorphisms, *Comment. Math. Univ. Carolinae 14*, 441—455.

KOUBEK, V., and REITERMAN, J. [1975]: Automata and categories, input processes, *Lect. Notes Comp. Sci. 12*, Springer-Verlag, Berlin—Heidelberg—New York, 280—286.

KOUBEK, V., and REITERMAN, J. [1979]: Categorial constructions of free algebras, colimits and completions of partial algebras, *J. Pure Appl. Algebra 14*, 195—231.

KŮRKOVÁ-POHLOVÁ, V. and KOUBEK, V. [1975]: When a generalized algebraic category is monadic, *Comment. Math. Univ. Carolinae 15*, 577—587.

KŮRKOVÁ-POHLOVÁ, V. [1973]: On sums in generalized algebraic categories, *Czech. Math. J. 23*, 235—251.

MAGIDOR, M. and MORAN, G. [1969]: Finite automata over finite trees, *Technical Rep. 30*, Hebrew University, Jerusalem.

MANES, E. G. [1976]: Algebraic Theories, Springer-Verlag, New York—Heidelberg—Berlin.

MEZEI, J. and WRIGHT, J. B. [1967]: Algebraic automata and context-free sets, *Inf. Control 11*, 3—29.

NELSON, E. [1981]: Z-continuous algebras, *Lect. Notes Mathematics 871*, Springer-Verlag, Berlin—Heidelberg—New York, 315—334.

REITERMAN, J. [1971]: An example concerning set functors, *Comment. Math. Univ. Carolinae 12*, 227—233.

REITERMAN, J. [1977a]: A more categorical model of universal algebra, *Lect. Notes Comp. Sci. 56*, Springer-Verlag, Berlin—Heidelberg—New York, 308—313.

REITERMAN, J. [1977b]: A left adjoint construction related to free triples, *J. Pure Appl. Algebra 10*, 51—71.

REITERMAN, J. [1978a]: One more categorical model of universal algebra, *Math. Z. 161*, 137—146.

REITERMAN, J. [1978b]: Large algebraic theories with small algebras, *Forschungsbericht 74*, University Dortmund.

REITERMAN, J. [1979]: A note to *ADJ*'s subset systems, Proc. "Fund. Computation Theory", *Mathem. Forschung 2*, Akademie-Verlag, Berlin, 387—390.

SMYTH, M. B. [1976]: Power domains, *J. Comp. Syst. Sciences 16*, 23—26.

STARKE, P. H. [1972]: Abstract automata, Elsevier, Amsterdam.

TARSKI, A. [1955]: A lattice-theoretical fixpoint theorem and its applications, *Pacific J. Math. 5*, 285—309.

TCHATCHER, J. B. [1973]: Tree automata, an informal survey. Currents in the theory of computation, Prentice-Hall, Princeton, New Jersey, 143—172.

TCHATCHER, J. B. and WRIGHT, J. B. [1968]: Generalized finite automata theory with an application to a decision problem of second-order logic, *Math. Syst. Theory 2*, 57—81.

TRNKOVÁ, V. [1969]: Some properties of set functors, *Comment. Math. Univ. Carolinae 10*, 323—352.

TRNKOVÁ, V. [1971]: On descriptive classification of set functors. I., II. *Comment. Math. Univ. Carolinae 12*, 143—175 and 345—357.

TRNKOVÁ, V. [1974]: On minimal realizations of behavior maps in categorial automata theory, *Comment. Math. Univ. Carolinae 15*, 555—566.

TRNKOVÁ, V. [1975a]: Automata and categories, *Lect. Notes Comp. Sci. 32*, Springer-Verlag, Berlin—Heidelberg—New York, 138—152.

TRNKOVÁ, V. [1975b]: Minimal realizations for finite sets in categorial automata theory, *Comment. Math. Univ. Carolinae 16*, 21—35.

TRNKOVÁ, V. [1977]: Relational automata in a category and theory of languages, *Lect. Notes Comp. Sci. 56*, Springer-Verlag, Berlin—Heidelberg—New York, 340—355.

TRNKOVÁ, V. [1979a]: Behaviour of machines in categories. *Comment. Math. Univ. Carolınae 20*, 267—282.

TRNKOVÁ, V. [1979b]: L-fuzzy functorial automata, *Lect. Notes Comp. Sci. 74*, Springer-Verlag, Berlin—Heidelberg—New York, 463—473.

TRNKOVÁ, V. [1979c]: Machines and their behaviour in a category, Proc. "Fund. Computation Theory", *Mathem. Forschung 2*, Akademie-Verlag, Berlin, 450—461.

TRNKOVÁ, V. [1980]: General theory of relational automata, *Fund. Informaticae 3*, 189—233.

TRNKOVÁ, V. [1981]: Partial and nondeterministic automata in a category, Proc. 7th Nat. School "Mathematical methods in Informatic", Varna 1981 (Centre of Appl. Math., Sofia), 71—92.

TRNKOVÁ, V. [1984]: Kleene type theorems for functorial automata in categories, Proc. 9th Nat. School "Mathematical methods in Informatic", Varna 1983 (Centre of Appl. Math., Sofia), 5—24.

TRNKOVÁ V. and ADÁMEK J. [1977a]: On languages accepted by machines in the category of sets, *Lect. Notes Comp. Sci. 53*, Springer-Verlag, Berlin—Heidelberg—New York, 523—531.

TRNKOVÁ, V., and ADÁMEK, J. [1977b]: Realization is not universal, Vorträge zur Automatentheorie, Weiterbildungszentrum für mathematische Kybernetik und Rechentechnik, Heft 21, Technische Universität, Dresden, 38—55.

TRNKOVÁ, V. and ADÁMEK, J. [1979]: Tree-group automata, Proc. "Fund. Computation Theory", *Mathem. Forschung 2*, Akademie-Verlag, Berlin, 462—468.

TRNKOVÁ, V. and ADÁMEK, J. [1982]: Analyses of languages accepted by varietor machines in a category, Banach Centre Publications 9, Polish Scientific Publishers, Warsaw, 257—272.

TRNKOVÁ, V., ADÁMEK, J., KOUBEK, V. and REITERMAN, J. [1975]: Free algebras, input processes and free monads, *Comment. Math. Univ. Carolinae 16*, 339—351.

TRNKOVÁ, V. and GORALČÍK, P. [1969]: On products in generalized algebraic categories, *Comment. Math. Univ. Carolinae 10*, 49—89.

WAND, M. [1979]: Fixed point constructions in order-enriched categories, *Th. Comp. Science 8*, 13—30.

WYLER, O. [1966]: Operational categories, Proc. Conference on categorical algebra, Springer-Verlag, Berlin—Heidelberg—New York, 295—316.

List of Current Symbols

β	behavior	2, 19, 36, 95, 103
∂	branch	33, 55
δ	next-state map	1
ε	epi-transformation	85
$\mathscr{E}$	class of epis	99
η	injection of generators	93, 238
F	functor	83
f_{in}	from automata morphisms	39
f_{out}	from automata morphisms	4, 39
$[f_1, f_2]$	relation	257
φ	free-algebra operation	93
γ	output map	1, 29
Γ	output object (alphabet)	1, 29, 90
hom	set of morphisms	83
I	initialization object	29, 90
$I^{\#}$	free algebra	93
$\mathscr{K}$	category	82
$\mathscr{K}^{\mathrm{m}}$	class of morphisms	83
$\mathscr{K}^{\mathrm{o}}$	class of objects	83
$\mathscr{K}^{\mathrm{op}}$	dual category	85
λ	initialization morphism	29, 90
$\mathscr{M}$	class of monos	99
π	projection	85, 107
Q	state object (set of states)	1, 29, 90
ρ	run morphism (run map)	2, 36, 95
Σ	type	29, 52
$\Sigma[z]$	module of polynomials	94
T	terminal object	87
U	forgetful functor	126
W_k	steps in the free algebra construction	34, 172
$[\ ]$	equivalence class	469
$\lvert\sigma\rvert$	arity of σ	29, 52
$f^{\#}$	free extension of f	93
$\perp$	initial object	87
$r_{(1)}, r_{(2)}$	projections of a relation	257

Subject Index